CANDID SCIENCE IV

More Conversations with Famous Scientists

CANDID SCIENCE IV

More Conversations with Famous Scientists

István Hargittai

Magdolna Hargittai

Imperial College Press

Published by

Imperial College Press
57 Shelton Street
Covent Garden
London WC2H 9HE

Distributed by

World Scientific Publishing Co. Pte. Ltd.
5 Toh Tuck Link, Singapore 596224
USA office: 27 Warren Street, Suite 401-402, Hackensack, NJ 07601
UK office: 57 Shelton Street, Covent Garden, London WC2H 9HE

Magdolna Hargittai
Eötvös University and Hungarian Academy of Sciences
H-1518 Budapest, Pf. 32, Hungary

István Hargittai
Budapest University of Technology and Economics
Eötvös University and Hungarian Academy of Sciences
H-1521 Budapest, Pf. 91, Hungary

British Library Cataloguing-in-Publication Data
A catalogue record for this book is available from the British Library.

ISBN-13 978-1-86094-414-7
ISBN-10 1-86094-414-0
ISBN-13 978-1-86094-416-1 (pbk)
ISBN-10 1-86094-416-7 (pbk)

Printed in Singapore

Foreword

In this volume, some three dozen physicists offer the reader uniquely personal glimpses into their chosen profession, the Physics they inherited, worked in, and are leaving for posterity. What was Physics like in the 20th century? Scientific progress during these past hundred years has given us an extraordinary legacy — a wide-ranging and comprehensive foundation of understanding, powerful tools and techniques with which to refine that understanding, and practical applications which touch upon every significant aspect of modern life, as we know it.

In its early decades, 20th century Physics centered upon the collective efforts of a modest number of European scientists. By mid-century, however, the balance had shifted to a large, and growing, scientific establishment in North America. Enriched by refugees, and enlarged by government support, scientific research in Physics flourished in scope and diversity — providing foundations for the nuclear and information revolutions, and catalyzing transformative advances in Astronomy, Biology, Chemistry, and much else besides.

Indeed, so much has happened that we may sometimes forget that all these advances took place within a time scale not much longer than a single human lifetime. This connection with what may seem a far-distant past is reflected in the conversations captured herein — several of which combine first-hand accounts of key advances made as senior researchers, with personal recollections of people and events encountered as students and fledgling scientists.

Experimental physicists generally gain their clearest insights into Nature's workings by studying phenomena under extreme conditions: energies so high that symmetries break, or so low that quantum condensations occur; the nearly perfect vacuum that Nature was once said to abhor, or the ultra-crowded insides of a neutron star; the mass components of the entire Universe, or the rest mass of a single neutrino. In individual terms, this might mean anything from writing proposals for time on some enormous machine, to crafting, assembling, and adapting bits and pieces in a single laboratory.

If anything, the pathways to theoretical insights range even further, probing spaces its practitioners can only explain to interested outsiders by use of analogies, if at all. Small wonder then, that success stories in this arena usually include years of formal study at a famous University and apprenticeships with distinguished theoreticians. And yet, some seeming outsiders can gain entry, and even join the front rank of contributors. Consider for example, the London-based military attaché who selected a graduate school because of its proximity to his "day job". Without a subject matter expert to guide him, he persisted on his chosen line of research — independently rediscovering a series of mistaken conclusions made by earlier workers, before making the discoveries which earned him a place of honor in this select group.

In the end, therefore, the manifest diversity of life experience might seem to thwart attempts at all but the most general categorizations. In place of a rigid roadmap, the compilers of this volume have wisely allowed this diversity to manifest itself in an artfully threaded series of anecdotal excursions — much as a reader might get from a personal conversation with the Hargittais themselves. Accordingly, chronology and sub-disciplines play secondary roles herein, as the thread of conversation moves through one topic and then on to the next. Instead, more subtle groupings become evident as one conversation often serves to introduce the next. Note, for example, that Mößbauer follows Bahcall, reflecting their common involvement in cosmic neutrinos. Similarly, we find microwave background radiation grouped with particle physics, reflecting its relation to cosmic nucleo-synthesis.

Appropriately, the book begins and ends with recollections. The opening conversation centers upon Eugene Wigner's account of people and events in earlier years — a clear link to the past. While superficially similar, David Schoenberg's recollections of Landau and Kapitza serve to turn our attention to the future. In a fitting coda to what has gone before, this final conversation serves to reintroduce those of us in the West with our colleagues in the East. Split apart for most of the past century, these two spheres of science once again share in a common community of knowledge — and common hopes for a fruitful future for science and humanity in this new century.

Menlo Park, California, November 2003   Arno Penzias

PREFACE

This is the fourth volume of the *Candid Science* series. It is devoted to famous physicists. Volumes I and III contained interviews with chemists and Volume II with biomedical scientists. Magdolna Hargittai (Magdi, for short) has joined me in authorship for this volume whereas she acted only as Editor for the previous volumes. As it happened, she prepared about half of the interviews in the present volume and I prepared the other half. For a few interviews we acted jointly. Both of us are physical chemists, but she has a special interest in cosmology and in the physics of fundamental particles.

The 36 interviews in this volume present a good, however incomplete, cross section of late 20th century physics. Some of the interviewees were already active and successful physicists in the 1930s. Quite a few participated in the Manhattan Project in various capacities. Whether it was a more prominent role, like Edward Teller or Mark Oliphant, or a less conspicuous one, all who participated in it were impacted by the experience. Some of the interviewees represent a link to such early giants of modern physics as J. J. Thomson, Ernest Rutherford, Niels Bohr, and Enrico Fermi. There is in the volume Laszlo Tisza, whose paper in 1944 was one of the forerunners of the Bose-Einstein Condensation and Wolfgang Ketterle, who was one of its discoverers half a century later. There are Nobel laureates, Wolf laureates, and Templeton laureates in this volume. The number of Nobel laureates is conspicuously high, but the non-laureates are no less great physicists than the laureates.

We would like to make our collection of interviews yet more comprehensive. Thus, we have not stopped making these interviews and further interviews with physicists will appear in *Candid Science V*.

Our interviews are, in most cases, by-products of our scientific and family travels. When we go to conferences or other trips related to our work, as well as when we are visiting our children, we use the opportunity to record interviews. We both use the same approach to these interviews. We contact our interviewees in advance, set up a date, and record the conversation on audiotape. Back at home we prepare the transcripts and send them for correction and change to the interviewees. This procedure is repeated as many times as it takes until the interviewee is happy with the material. The taping is very informal, providing only a framework for the interview. On the other hand, it is important to have this personal encounter, which could not be substituted by exchanging letters. We never engage anybody else in transcribing our tapes. Listening to them brings back memories of the personal meeting, helps us to capture what was said, and sometimes even gives us clues for a few additional questions that may then be posed in correspondence or at a next meeting. This approach gives a rather tight control of the interviewee over what appears as the text of the interview. However, our experience shows that acting in this way, rather than looking to reveal some "dark secrets", has been helpful in getting closer to our interviewees in a human sense than it might be possible by a more aggressive approach. By the same token, during our actual conversations, some of our questions remain unanswered but then we drop the questions as well from the printed version. In this way the reader may sometimes feel that we should have asked a certain question, and, maybe, we did, but there is no trace of it in what is in this book. We are happy with what we have received without any embarrassment.

The interviews are very different in length and depth. Some interviewees were happy to open up more than others; some were willing to talk in greater detail about their science and about their personal lives than others. Again, we were happy to have what we could get. The circumstances of the recordings were also vastly different. Some interviews were made leisurely in a quiet office or home. Others were squeezed into a crowded program of a scientific meeting or, as it happened, into the rich program of the centennial celebrations of the Nobel Prize in Stockholm.

There is one exception among the 36 interviews communicated in this volume and that is the "interview" with Eugene Wigner. It is not a bona fide interview, rather, a summary of a series of conversations I had during Eugene Wigner's visit to the University of Texas at Austin in 1969, while I was a Visiting Research Associate there in the Physics Department. Those conversations stayed deeply ingrained in my mind and I felt there should be an entry from them in this volume augmented by quotations from others.

We note with sadness that three of our interviewees are no longer among us. Eugene Wigner, Mark Oliphant, and Edward Teller have passed away since our interviews with them. I had corresponded with Edward Teller up to a few weeks before his death on September 9, 2003. I am excerpting two letters here as I believe they augment our Teller interview. The original letters were in Hungarian communicated through his assistant, Mrs. Margit Grigory to whom I am grateful for her help in making it possible to keep in touch with the ailing Teller. Here the excerpts from his letters appear in my translation.

From Teller's letter of August 13, 2003:

...

In my life's work, I loved science a thousand times more than its applications. I agreed to the latter because I took the dangers of war very much to my heart. I hope you know that I was always against our becoming the first to deploy the hydrogen bomb. I only wanted to have the possibility of the H-bombs as a deterrent for wars, and this has worked so far.

I am convinced that the business of scientists is exclusively science itself. The application of science is the business of politicians and consequently that of the voters. I had problems with my fellow scientists, especially with those according to whom we shouldn't have worked on anything like the hydrogen bomb.

Incidentally, in this question we bitterly differed with Oppenheimer. Similarly we had different positions with Enrico Fermi but with him our friendship did not suffer from this difference. The same can be said about Leo Szilard, who was the most gifted in treading on other peoples' corn, but he never bored anybody.

However, let's though speak about science. Very few performed true science and I knew two such people, Einstein and Bohr. I would be curious to know, what Einstein thought when he received the Nobel Prize definitely not for relativity. There are things that are unavoidable, like benzene [reference to the first part of the letter] and the hydrogen bomb, although it may be important for the moment who thought of them first. Things like relativity and quantum mechanics far surpass all other intellectual activities.

For me it is important that the same four letters describe the DNA of all living creatures. This may bring us closer to the understanding of what life is.

…

From Teller's letter of August 17, 2003 [focusing on the question about success in science]:

Your question is a difficult one, but my answer is easy. I was not an unsuccessful scientist but my scientific research suffered from my work on weapons. Also, in part this happened when I was in the peak of my energies.

In addition, as witnessed by my productive work, I liked to cooperate with others and in this, our disagreement with Oppenheimer caused a lot of damage.

*****

We are grateful to the Hungarian Academy of Sciences and the Budapest University of Technology and Economics as well as the Hungarian National Scientific Research Foundation for their support of our research activities in structural chemistry. Our scientific research brings us to meetings and laboratory visits whose byproducts are often the interviews presented here. A very fruitful period was the three months we spent at the Cold Spring Harbor Laboratory at James Watson's invitation in 2002, which gave a convenient opportunity for several interviews. We appreciate Jim and Liz's hospitality and personal attention during our stay there. Our family vacations provide additional opportunities to expand our interviews project.

We also appreciate the dedicated efforts of the associates of Imperial College Press and World Scientific Publishing Company in bringing out this volume. Senior Editor Ms. Ying Oi Chiew spared no labor and attention in making this book as nearly perfect as possible. Her friendly cooperation and pleasant care enhanced our pleasure in working on this project.

Both Magdi and I are infinitely grateful to all our interviewees for their patience with us and for their fruitful cooperation. We have learned a lot from them in physics, in science history, and in human conduct. Our highest hope is that our readers will similarly benefit from these interviews.

Budapest, October 2003   István Hargittai

Technical Comment

There is an apparent inconsistency in the way some of the names appear in this volume. However, they were given careful consideration. For example, Peter Kapitsa appears also as Kapitza and his first name as Petr and Pyotr as well. We did not want to arbitrarily change from their original appearances that is the result of difference in transliteration and also in the way he used his name in Cambridge and in Russia. Martinus Veltman's first name also appears as Martin, and Gerardus 't Hooft's first name as Gerard at places. We have consulted about this problem with Drs. Veltman and 't Hooft and we tried our best to follow their preferences. Dr. Tisza's first name appears as László when he is being referred to his time in Hungary. Later, adapting to American usage, his name became Laszlo.

CONTENTS

Foreword	v
Preface	ix
Eugene P. Wigner	2
Steven Weinberg	20
Yuval Ne'eman	32
Jerome I. Friedman	64
Martinus J. G. Veltman	80
Gerard 't Hooft	110
Leon M. Lederman	142
Valentine L. Telegdi	160
Val L. Fitch	192
Maurice Goldhaber	214
John N. Bahcall	232
Rudolf Mößbauer	260
Arno A. Penzias	272
Robert W. Wilson	286

Owen Chamberlain   298

Marcus L. E. Oliphant   304

Norman F. Ramsey   316

David E. Pritchard   344

Wolfgang Ketterle   368

Laszlo Tisza   390

Edward Teller   404

John A. Wheeler   424

Freeman J. Dyson   440

John C. Polkinghorne   478

Benoit B. Mandelbrot   496

Kenneth G. Wilson   524

Mildred S. Dresselhaus   546

Catherine Bréchignac   570

Philip W. Anderson   586

Zhores I. Alferov   602

Daniel C. Tsui   620

Antony Hewish   626

Jocelyn Bell Burnell   638

Joseph H. Taylor   656

Russell A. Hulse   670

David Shoenberg   688

Name Index   699

Cumulative Index of Interviewees   709

By the same authors

Candid Science III: More Conversations with Famous Chemists. Imperial College Press, London, 2003. (IH, edited by MH)

The Road to Stockholm: Nobel Prizes, Science, and Scientists. Oxford University Press, Oxford, 2002 (paperback edition 2003). (IH)

Candid Science II: Conversations with Famous Biomedical Scientists. Imperial College Press, London, 2002. (IH, edited by MH)

Candid Science: Conversations with Famous Chemists. Imperial College Press, London, 2000. (IH, edited by MH)

In Our Own Image: Personal Symmetry in Discovery. Kluwer/Plenum, New York, 2000. (IH & MH)

Upptäck Symmetri! (Discover Symmetry!, in Swedish, with M. Hargittai). Natur och Kultur, Stockholm, 1998. (MH & IH)

Symmetry through the Eyes of a Chemist. Second edition, Plenum, New York, 1995. (IH & MH)

Symmetry: A Unifying Concept. Shelter Publications, Bolinas, CA, 1994. (IH & MH)

The VSEPR Model of Molecular Geometry. Allyn & Bacon, Boston, 1991. (R.J. Gillespie & IH)

The Structure of Volatile Sulphur Compounds. Akadémiai Kiadó, Budapest, 1985. (IH)

The Molecular Geometries of Coordination Compounds in the Vapour Phase. Akadémiai Kiadó, Budapest, 1977. (IH & MH)

Edited books

Strength from Weakness: Structural Consequences of Weak Interactions in Molecules, Supermolecules, and Crystals. Kluwer, Dordrecht, 2002. (A. Domenicano & IH)

Symmetry 2000. Vols. I–II, Portland Press, London, 2002. (IH & T.C. Laurent)

Advances in Molecular Structure Research. Vols. 1–6. JAI Press, Greenwich, CT, 1995–2000. (MH & IH)

Combustion Efficiency and Air Quality. Plenum, New York, 1995. (IH & T. Vidóczy)

Spiral Symmetry. World Scientific, Singapore, 1992. (IH & C.A. Pickover)

Fivefold Symmetry. World Scientific, Singapore, 1992. (IH)

Accurate Molecular Structures. Oxford University Press, Oxford, 1992. (A. Domenicano & IH)

Quasicrystals, Networks, and Molecules of Fivefold Symmetry. VCH, New York, 1990. (IH)

Symmetry 2: Unifying Human Understanding. Pergamon Press, Oxford, 1989. (IH)

Stereochemical Applications of Gas-Phase Electron Diffraction. Vols. A–B, VCH Publishers, New York, 1988. (IH & MH)

Crystal Symmetries, Shubnikov Centennial Papers. Pergamon Press, Oxford, 1988. (IH & B.K. Vainshtein)

Symmetry: Unifying Human Understanding. Pergamon Press, Oxford, 1986. (IH)

Diffraction Studies on Non-Crystalline Substances. Akadémiai Kiadó, Budapest, 1981. (IH & W.J. Orville-Thomas)

Eugene P. Wigner (photograph by and courtesy of Robert Matthews).

1

EUGENE P. WIGNER

Eugene P. Wigner (1902, Budapest — 1995, Princeton) received the Nobel Prize in Physics "for his contributions to the theory of the atomic nucleus and the elementary particles, particularly through the discovery and application of fundamental symmetry principles." He attended one of the famous Budapest high schools before his studies at the Berlin University of Technology, where, eventually he earned his doctorate in chemical engineering. When the Nazis came to power in Germany he left for the United States. He was Thomas D. Jones Professor of Mathematical Physics at Princeton University between 1938 and 1971, when he retired. Wigner worked on the Manhattan Project during World War II. He was a member of the General Advisory Committee to the U.S. Atomic Energy Commission, 1952–1957 and 1959–1964. Among his decorations, he received the U.S. Medal of Merit (1946), the Enrico Fermi Prize of the U.S. Atomic Energy Commission (1958), the Atoms for Peace Award (1960), the Medal of the Franklin Society, the Max Planck Medal of the German Physical Society, the National Medal of Science (1969), and others. He was a member of the National Academy of Sciences of the U.S.A., the American Academy of Arts and Sciences, and other learned societies. We met and had extensive conversations at the University of Texas at Austin in 1969. This narrative (by István Hargittai) is based on these conversations and on other interviews in this volume.

I spent a year at the Physics Department of the University of Texas at Austin in 1969 as a Research Associate. I worked in a research group loosely directed by Harold P. Hanson who was the chair of the Department.

We met in the previous year in Oslo. By the time of Wigner's visit though, Hanson had left for the next position in his distinguished career as a scientist and administrator. When I heard about Wigner's forthcoming visit — his lectures were advertised well ahead of time — I went to see his official hosts. I told them that I had known Wigner if only by correspondence and would like to see him while he was there. They declined, explaining it was very expensive for them to bring Wigner to Texas and they could not waste his time on me. However, they let me leave a note for Wigner in which I said hello and gave him my location in the department.

Wigner must have received my note because he came to see me in my office every morning during his stay in Austin and spent an hour with me from 8 a.m. to 9 a.m., when his official program started. Thus he spent some of his "private" time with me with no loss to his hosts. When he first came to see me having read my note, I might have ascribed his call as a courtesy. But his subsequent visits I took as a genuine expression of interest and magnanimity at the same time. Of course, I also experienced his politeness, but to a lesser degree than others might have. His legendary politeness and modesty on occasions reached such a degree that it irritated people and many saw through it as being a shield preventing others from getting close to him. Freeman Dyson (see elsewhere in this volume) and his wife lived very close to the Wigners in Princeton for many years. According to Dyson, they were friendly, but they never got close. About Wigner's politeness, this was Dyson's observation,

> There is a professor here, whose wife is Japanese. She is a sociologist and writes about Japanese society. One of the papers she has written, and I think it is brilliant, is called "Politeness as a Tool of Repression". It is certainly true for the Japanese society and perhaps a little bit with Wigner, too.

In physics though, his modesty may have helped Wigner to recognize the principal tasks of physics. In his Nobel lecture[1] he stressed the limitations in the ambitions of physics and physicists:

> ... [P]hysics does not endeavor to explain Nature. In fact, the great success of physics is due to a restriction of its objectives: it only endeavors to explain the regularities in the behavior of objects. This renunciation of the broader aim, and the specification of the domain for which an explanation can be sought, now appears to us an obvious necessity. In fact, the specification of the explainable may have been the greatest

discovery of physics so far. It does not seem easy to find its inventor, or to give the exact date of its origin. Kepler still tried to find the exact rules for the magnitude of the planetary orbits, similar to his laws of planetary motion. Newton already realized that physics would deal, for a long time, only with the explanation of those of the regularities discovered by Kepler which we now call Kepler's laws.

The regularities in the phenomena which physical science endeavors to uncover are called the laws of Nature. The name is actually very appropriate. Just as legal laws regulate actions and behavior under certain conditions but do not try to regulate all actions and behavior, the laws of physics also determine the behavior of its object of interest only under certain well-defined conditions but leave much freedom otherwise ...

According to another physics Nobel laureate (1979) Steven Weinberg (see elsewhere in this volume),

Wigner realized, earlier than most physicists, the importance of thinking about symmetries as objects of interest in themselves. In the 1930s, although physicists talked a lot about symmetries, they talked about them in the context of specific theories of nuclear force. Wigner was able to transcend that and he discussed symmetry in a way, which didn't rely on any particular theory of nuclear force. I liked that very much.

Wigner received the inspiration to look for regularities in Nature from his mentor in his doctoral studies in Berlin, Michael Polanyi. We know this from Wigner's two-minute speech at the Stockholm City Hall, following the award ceremony of the Nobel Prize in December 1963.[2] Wigner devoted this two-minute speech to his teachers. He said[3] that

I do wish to mention the inspiration received from Polanyi. He taught me, among other things, that science begins when a body of phenomena is available which shows some coherence and regularities, that science consists in assimilating these regularities and in creating concepts, which permit expressing these regularities in a natural way. He also taught me that it is this method of science rather than the concepts themselves (such as energy), which should be applied to other fields of learning.

Looking for regularities is very much the same as looking for symmetries. In this sense, Wigner's interest in symmetry may have originated from his

At the banquet of the Nobel Prize award ceremonies in Stockholm, 1963: in the middle of the picture, the three physics Nobel laureates, Maria Goeppert-Mayer, J. Hans D. Jensen, and Eugene Wigner (courtesy of Franca Natta Pesenti, Bergamo, Italy).

interactions with Polanyi. There was then a less direct impact by Polanyi in turning Wigner's attention to symmetries. After his doctorate Wigner returned to Hungary and worked in the same leather tannery where his father had a managerial job. Even decades later I sensed pride in Wigner's describing his knowledge of the chemistry processes involved in tannery and the different ways of preparing the leather. Some leather is prepared for the bottom of the shoe, some for the upper part of the shoe, yet some other leather is prepared for travel bags. Wigner had a tremendous knowledge of chemistry, especially materials. This came from his training as a chemical engineer and he made good use of this knowledge later in the Manhattan Project. During the two years he spent at the tannery, he kept subscribing to the German physical journal *Zeitschrift für Physik* and read it, in his words,[4] "in the evenings industriously." Then,[4]

> After two years, to my great surprise, I received a letter of invitation from Professor Becker, the newly appointed professor of theoretical

physics at the Institute of Technology in Berlin, offering me a position of assistant. I suspect that it was my teacher, the person I worked with for my doctoral dissertation, Michael Polanyi, who must have recommended me. He was a famous scientist and excellent man.

...

Before my job started with Professor Becker, I had two months in Berlin and I used that time to learn crystallography, and this was very useful later. In crystallography I learned about group theory, which became the center of my interest for several years. It was a great joy when Schrödinger's paper came out in which he described his equation and quantum mechanics, and which made the application of group theory to quantum mechanics so much easier.

Wigner was not only an excellent pupil; he became an outstanding mentor himself. He served as guide for several major figures in condensed matter physics in the United States. John Bardeen was one of them. Bardeen is the only person to have ever received two Nobel Prizes in physics (1956 and 1972). From Bardeen's description, Wigner emerges as a popular mentor and Bardeen notes that "Wigner was attracting a number of other students and postdoctoral fellows interested in problems of solid state physics."[5] Among them were F. Seitz and C. Herring. Herring was later Philip Anderson's (Nobel Prize in 1977) mentor. However, Wigner may not have been equally good a pedagogue for everybody. Steven Weinberg (see next interview) was a graduate student at Princeton when Wigner was a professor there. This is how he remembers Wigner as his teacher

I took his course in nuclear physics. He was not a very good teacher because he was obsessively worried that there might be someone in the class who wasn't understanding him. So he went very slowly. But he still was very profound. He did me great compliment once, when he had to go out of town, of asking me to take over the class. I didn't learn much from Wigner when I was a graduate student at Princeton. But in the years following, I found that my point of view toward physics was very much in tune with Wigner's, much more so than with other people. Wigner had analyzed, especially in 1939, the nature of the elementary particles and, in particular, the significance of the spin, in a way far superior and much more well-grounded in fundamental principles than what I had earlier learned, say, from the treatment of the spin by Paul Dirac. In that sense I became a disciple of Wigner. In my book on quantum field theory I very much followed Wigner's 1939 paper.

At the time when Wigner received the Nobel Prize, I did not pay much attention to the Nobel Prizes. However, the next year, in the fall of 1964, I read an article by Wigner in the Hungarian literary weekly *Élet és Irodalom* (*Life and Literature*).[6] In the wake of Wigner's Nobel Prize, they translated an earlier Wigner article into Hungarian and communicated excerpts from it.[7] At that time I was a Master's degree student at Moscow State University. The topic of his article was the limits of science and it immensely interested me, so I wrote a response and sent it to the editor of *Élet és Irodalom*. To my surprise, it soon appeared and was quite a sizeable article.[8] I had never written anything before let alone seen my writing and name printed. It was quite a thing for a student. Soon after the appearance of the article, something of even greater significance happened. I received a long letter from Wigner and some reprints of his papers. He agreed with some of what I had written and disagreed with other aspects, all in his most polite way.

In our conversations in Austin, Texas, we talked about many things, but the most remarkable for me was what he taught me about symmetry, and this is what had the longest-ranging impact on me. Wigner had contributed to the application of the symmetry principles in the most fundamental ways and the Nobel Committee stressed this in its one-sentence description of his achievements (*vide supra*). It was fortunate that at the time of our meeting, I was engaged in studying one of the most symmetrical molecules, called adamantane, $C_{10}H_{16}$. The name refers to its high stability. The structure of the molecule resembles the diamond structure. Wigner's interest was in the solid state rather than in molecules, but he found my molecules intriguing. Later Wigner sent me his essay book, *Symmetries and Reflections*,[9] in which he had a diagram of the diamond structure. It dawned on me only later how lucky I was that, if only for a few days, I had Wigner as my mentor in symmetry.

On that occasion we hardly went into specifics, but he had made seminal contributions to the utilization of symmetry in chemistry. Here I mention only a few examples. A fundamental property of the electronic wave function of a molecule is that it can be used as basis for irreducible representations of the point group of a molecule. This property establishes the connection between the symmetry of a molecule and its wave function. The preceding statement follows from Wigner's theorem, which says that all eigenfunctions of a molecular system belong to one of the symmetry species of the group.[10] The first application of symmetry considerations to chemical reactions

John von Neumann (courtesy of Ferenc Szabadváry, Budapest Museum of Technology).

can be attributed to Wigner and Witmer.[11] The Wigner-Witmer rules are concerned with the conservation of spin and orbital angular momentum in the reaction of diatomic molecules. Although symmetry is not explicitly mentioned, it is present implicitly in the principle of conservation of orbital angular momentum. Wigner and Witmer prepared the ground for the breakthrough that followed almost four decades later through the works of Woodward and Hoffmann, Fukui, Bader, Pearson, and others. For two of them (Fukui and Hoffmann) it even resulted in a Nobel Prize in Chemistry in 1981.[12] In modern theory of chemical reactions invoking the symmetry principle, there is a consideration to connect levels of like symmetry without violating the so-called non-crossing rule. This non-crossing rule was introduced by Neumann and Wigner, and independently by Teller.[13] According to this rule, two orbitals of the same symmetry cannot intersect in the correlation diagram. These diagrams provide valuable information about the transition state of the chemical reaction.[14]

Looking back to my conversations with Wigner about symmetry, two things have crystallized for me. Both are fundamental, almost trivial, and they provided a great introduction for me into symmetry considerations. One is that there is the geometrical kind of symmetry and then there is everything else, which includes, among others, the symmetry of molecules.

Eugene Wigner and Edward Teller at the Lawrence Livermore Laboratory during the 1970s (courtesy of György Marx).

The molecules are not rigid bodies and the larger-amplitude is their motion, the more they are away from structures with rigorous symmetry. The adamantane molecule has a rather rigid carbon cage and its T_d symmetry is unambiguous, but there are floppy molecules too that can be described much more loosely with point groups. Wigner then helped me see the yet much broader meaning of symmetry that Hermann Weyl had made broadly familiar through his Princeton lectures and his classic book[15] that symmetry is also harmony and proportion.

The other characteristic of Wigner's approach to symmetry that has remained in my mind as sticking out from many other scientists' interpretations, is that Wigner did not distinguish between chemical symmetry and physical symmetry, and so on. For him the symmetry concept transcended man's subdivision of Nature into subject areas. In various fields and various applications, the models that we utilize may have different emphases on different aspects and may ignore different other aspects, but the symmetry concept itself is universal. Perhaps this universality is what has captivated me most about the symmetry concept and has encouraged me to write broadly about symmetry in various books.

Disregarding the boundaries of various branches of science served Wigner well. It has been noted repeatedly how useful his knowledge of materials proved to be in the Manhattan Project. His studies with Polanyi about the mechanism of chemical reactions and the transition state have also proved useful in other branches of science. John Wheeler (see elsewhere in this volume) provided a nice example to illustrate this point:

> We had to understand this new nuclear phenomenon, fission. It was obvious that the nucleus of such a heavy element as uranium must undergo a considerable deformation before it splits. For that it needs energy. When the uranium is bombarded by neutrons, the neutron can provide this energy; we say that the nucleus is excited. This excitation then could initiate a vibration in the nucleus that could deform it. Our Hungarian friend, Eugene Wigner helped us out. He ate some oysters in downtown Princeton and got sick and was in the hospital on the campus. I went to see him at the hospital to get some help. The questions that Bohr and I were dealing with were like a chemical reaction. Uranium breaking up is like carbon monoxide breaking up into carbon and oxygen. I remembered that he had worked in that field with Michael Polanyi. And he helped us and, eventually, getting also ideas from discussions with other colleagues, such as Placzek and Rosenfeld, Bohr and I saw how fission works. Bohr left Princeton in April of that year and during the following months I wrote the paper and we submitted it to *Physical Review* in June. It came out in the September 1, 1939 issue;[16] by strange coincidence the same day when Germany invaded Poland.

I thoroughly enjoyed my encounter with Wigner and his lectures in Austin were attended to capacity. He gave talks on different topics, his physics and his almost-obsession topic of civil defense. In the latter he mentioned the Budapest subway system as an example of civil defense in Eastern Europe, and he urged the United States to pay more attention to civil defense. Almost 30 years later, Wigner's infatuation with civil defense came up in my conversation with Steven Weinberg (see next interview). Weinberg mentioned that he and Wigner

> ... were quite different about politics. He got very angry with me. He edited a book about civil defense. ... He was a very committed believer in the Cold War, that America should prepare in every possible way for a nuclear confrontation with the Russians. ... The Cold War fed on itself in the sense that both sides were building more and more destructive weapons because the other side was. Wigner saw nothing wrong with that. He just wanted us to build everything we possibly

Princeton Physics Department Faculty, 1962 (photograph by and courtesy of Robert Matthews).

could. He wasn't interested in arms control, and I was. I wrote a rather negative review of his book on civil defense, and Wigner got very angry with me. He attacked me at a cocktail party in Princeton. He was quite hostile to me for a number of years.

Philip Anderson (see elsewhere in this volume) and Wigner never argued about politics but Anderson knew that they were quite different politically. Apparently Wigner became rather isolated in the department at Princeton, at least on one occasion. Anderson remembers when they discussed the David Bohm affair at a party, "Wigner was the only member of the department who was not in favor of keeping Bohm. The department voted for Bohm, but the President of Princeton turned down our recommendation and fired him. But Wigner had not voted for him."

Anderson told me an amusing story from 1954, which was about a trivial personal friction of no particular interest. What I found of interest is how Anderson characterized Wigner: "Wigner was not a Nobel-Prize-winner yet but he was Wigner."

People in Austin in 1969 were very much in awe of him although there is a general opinion that by about the 1950s he had lost his creativity in physics. According to Anderson, he could no longer keep abreast with the latest developments and his influence started waning in the early 1950s. Weinberg told me an awkward story:

> I gave a talk at a symposium in his honor, explaining how much I thought I'd learned from Wigner's approach to elementary particles and why his approach was superior to others. Wigner came up to me after my talk, and he didn't really understand what I'd said. Maybe what I interpreted as wignerism was not entirely wignerism. He may also have been already ill, so I don't know. Wigner is a person I have complicated feelings for.

Anderson's other encounter with Wigner was related to science and, sadly, it confirmed Weinberg's impressions of the late Wigner:

> Then there was one more encounter around 1958 or 59, during one of my talks about the BCS (Bardeen-Cooper-Schrieffer) theory of superconductivity for which I had resolved the problem of gauge invariance. Wigner had written a paper in which he promulgated his super-selection-rules. One of these said that there can be no phase coherence between two states, which have different numbers of particles. A fundamental principle of the BCS theory allows phase coherence between states with different numbers of particles. It explicitly violates Wigner's super-selection-rule. There is though a way you can get around it, but I never believed much in its necessity. I don't think though that I was particularly dismissive of the super-selection-rules in my talk. In fact, I didn't mention them at all. But Wigner saw through me and he was very negative in his polite way. He made it clear that he didn't believe the BCS theory, and to his dying day he never accepted it. This was a deeper problem between us whereas the first two I mentioned were merely amusing incidents. Up until that point Wigner had been very central in theoretical physics but at about that time, in the 50s, he put his foot down and refused to go any further. At that point Wigner, as a person, seemed to become less relevant to physics.

Wigner went to the United States in 1930 for the first time. He had a half-time job in Princeton and a half-time job in Berlin. When Hitler came to power, Wigner decided to leave Germany for good and he helped his parents and sisters to go to the United States from Hungary too. In the mid-1930s, he lost his job at Princeton and spent two years at the University of Wisconsin. This is how he remembered it in 1986[4]:

> In Princeton I never felt at home, but in Wisconsin I felt at home from the second day. After two years, I was asked to return to Princeton. The reason was that the people whom they had wanted to have, and for whom they had fired me, did not get the job in Princeton. Instead, they invited Van Vleck to Princeton but he preferred Harvard and recommended me in his stead. There was another reason for me to leave Wisconsin. I fell in love with a young lady in Madison and we got married, but after eight months of marriage she passed away. That made me very sad and I thought it was better for me to go away from the place where I had lost my better half. Thus I went back to Princeton in 1937.

In the video recording of 1986, Wigner talked to Clarence Larson about his participation in the Manhattan Project and, what I find especially interesting, about his later thoughts concerning the atomic bombs used in Japan in 1945[4]:

> Many scientists came to the United States at that time from Germany, from Europe. Hitler did something very good for science in the United States. Niels Bohr was visiting in Princeton and gave a colloquium. I could not attend because I had to be in a hospital but a friend of mine, Dr. Szilard came to me and told me about it. Bohr talked about Hahn and Strassmann's discovery of nuclear fission. We thought a good deal about it and realized what enormous energies will be available this way. We started to think also of the danger of it. It became evident that those who had said that atomic energy could not be liberated were mistaken.
>
> We soon began to realize also that it was quite possible that a bomb could be made with uranium, based on liberating such energy. We were very much afraid that the Germans would develop it because, after all, they had discovered all that. We were afraid that this would make it possible for Hitler to realize his dream of ruling the whole Earth. For most Americans, Hitler was so far away, in Germany, that they thought that they did not have to worry about

Eugene Wigner and Leo Szilard, winners of The Atoms for Peace Award (1960) (courtesy of György Marx).

that. Szilard and I did not think so. We knew that if we could come from Hungary to the United States, then Hitler would be able to come from Germany to the United States, and the distance was even shorter from Germany.

I persuaded two of my colleagues to work a little on this problem, and one of them, slightly annoyed, told me, "You are pleasantly disagreeable." We also wanted to interest the United States as a whole because we thought that it was the duty of the government to defend the country. We talked with several people, but it was not very effective. Nobody took it very seriously. It was then Szilard's idea to talk to Einstein about it. Perhaps Einstein could persuade the American government that it was an important question. So we went out to Long Island, where Einstein was vacationing. I expected that it would take a long time before we would be able to convince him that this was a serious problem. He did not know about the process, he was not interested in nuclear physics. But it took only 15 minutes and he understood the danger and he dictated a letter in German, which could be taken to the American authorities, even to the President. I took down the letter, brought it back to Princeton, had it translated into English, and had it typed, and I think it was Teller and Szilard who took the letter back to Einstein and he signed it. The letter then was taken to President Roosevelt, who then appointed people to study the question. Eventually, General Groves was put in charge of the program, and it was a great success.

As it turned out, it was not necessary at that time because Hitler, who had been told about the possibility of such a bomb, said that he would win the war long before such a weapon could work. Hitler did, in fact, lose the war before the bomb became effective. At that point we thought that it was not necessary to continue the work on the bomb, but the government was not of that opinion. General Groves also wanted to continue and he said that we could use it against the Japanese and it would shorten the war.

We proposed then to demonstrate the bomb in the presence of some Japanese scientists and military leaders. Groves once again disagreed and said that we should demonstrate it on a city. And that is what happened, but we were against it and were quite unhappy. We thought that many Japanese lives could have been saved if the bomb had been demonstrated on an uninhabited territory. But, apparently, I must admit, and I will admit, we were probably mistaken. Much later, I read in a book that the demonstrations in Hiroshima and Nagasaki may have saved many, many Japanese lives. Since I thought that a demonstration over an uninhabited territory in the presence of Japanese scientists and politicians could have sufficed, I went around and asked my Japanese friends about it. And with one exception, they said, "No, such a demonstration would have had no effect on the Emperor." According to all my Japanese friends, with one exception, "It would not have had the same effect; it was very good that you demonstrated it this way." Maybe that was the way to do it, but I did not think so at that time. Of course, they knew the Japanese politicians, the Japanese Emperor, and the Japanese military leaders much better than we did. But I was very surprised. They thought that many Japanese lives were saved this way even though it led to the extinction of many Japanese lives. Apparently General Groves was right, and the bomb had to be demonstrated the way it was.

In my conversations in 1969 with Wigner, apart from symmetry, we talked about many other things too, for example, poetry. His favorite was the Hungarian poet of the first half of the 19th century, Mihály Vörösmarty, and I introduced him to the poetry of the martyr Miklós Radnóti and later sent him a volume of Radnóti and corresponded with him about it. Wigner was interested in Hungarian politics, much more than I was. I did not think there was any political life in Hungary, but he carefully read even the speech of the defense minister, a truly insignificant figure, and made comments on it. I also noticed Wigner's conservatism. At that time, in 1969, the first year of the Nixon Administration, the notion of

Eugene Wigner and István Hargittai in Austin, Texas, 1969 (by unknown photographer).

Lyndon Johnson's Great Society was still very close and the question of poverty in America came up. Wigner denied that there were poor people in America. As a proof he wanted me to come and see his neighborhood in Princeton! Both Anderson and Weinberg noted Wigner's political conservatism.

Apart from what Wigner taught me about symmetry, what especially stayed in my memory was what he told me about the Hungary of the early 1920s, as he experienced it. Sadly I recognized much of what he said in my own experience in the 1960s. His most important observation was how the Hungarian authorities treated the very Hungarians they were supposed to serve. They were the authority and the rest were the subjects. People stood in their offices, no matter who they were, and whatever they needed they were asking for it like beggars asking for a handout. Wigner said that whenever he felt any nostalgia about Hungary, he only needed to remember how the officials treated the people there, and it sufficed to sober him up.

References and Notes

1. Wigner, E. P. "Events, Laws of Nature, and Invariance Principles." In *Nobel Lectures: Physics 1963–1970.* Elsevier, Amsterdam, 1972, 6–17.

2. There were three physicists laureates in 1963 and they decided among themselves that Wigner should be the one giving the speech. There is one speech only in each category of the Nobel Prize on the occasion of that reception. Wigner had, of course, some preeminence among the three laureates since he received half of the prize while the other two, Maria Goeppert-Mayer and Hans Jensen shared the other half "for their discoveries concerning nuclear shell structure". The composition of the three laureates caused some headache to the people responsible for the protocol of the Nobel ceremonies that year. If there is one physics laureate, his wife goes into the dining room on the arm of the king. Further rules regulate the assignment of the other laureates and their spouses to members of the royal family when going into the dining room, and as regards their seating order. In 1963, according to alphabetical order, Goeppert-Mayer would have been the first among the physicists. However, the protocol people chose Wigner to be the first among the physicists, hence no new precedence had to be established. Wigner's primacy was justified by the above consideration [Hargittai, I., *The Road to Stockholm: Nobel Prizes, Science, and Scientists.* Oxford University Press, 2002, 10].

3. Wigner, E. P. "City Hall Speech — Stockholm, 1963." In Wigner, E. P. *Symmetries and Reflections: Scientific Essays.* Indiana University, Bloomington and London, 1967, 262–263.

4. Hargittai, I. " 'You are Pleasantly Disagreeable': Eugene P. Wigner Remembers." *The Chemical Intelligencer* **1999**, *5*(3), 50–52. This article is based on a video recording on March 4, 1986, by Clarence Larson.

5. Bardeen, J. "Reminiscences of early days in solid state physics." *Proc. R. Soc. Lond. A* **1980**, *371*, 77–83.

6. Wigner, J. "Van-e a tudománynak határa?" *Élet és Irodalom* 1964, VIII, 40. sz., 6–7. old. (1964 október 3). [Here J. stands for Jeno, Wigner's original first name whose English equivalent is Eugene.]

7. Wigner, E. P. "The Limits of Science." *Proceedings of the American Philosophical Society* **1950**, *94*, No. 5 (October).

8. Hargittai, I. "Tudományok határán." *Élet és Irodalom* 1964, VIII, 51. sz., 6. old. (1964 december 19).

9. Wigner, E. P. *Symmetries and Reflections: Scientific Essays.* Indiana University Press, Bloomington, Indiana, 1967.

10. Wigner, E. P. *Group Theory and Its Application to the Quantum Mechanics of Atomic Spectra.* Academic Press, New York, 1959.

11. Wigner, E.; Witmer, E. E. *Z. Phys.* **1928**, *51*, 859.

12. Hargittai, I. *The Road to Stockholm: Nobel Prizes, Science, and Scientists.* Oxford University Press, 2002.

13. Neumann, J. von; Wigner, E. *Phys. Z.* **1929**, *30*, 467; Teller, E. *J. Phys. Chem.* **1937**, *41*, 109.

14. For examples, see Hargittai, I., Hargittai, M., *Symmetry through the Eyes of a Chemist* (Second Edition). Plenum, New York, 1995, 301.

15. Weyl, H. *Symmetry*. Princeton University Press, Princeton, New Jersey, 1952.

16. Bohr, N.; Wheeler, J. A. *Phys. Rev.* **1939**, *56*, 426.

Steven Weinberg, 1998 (photograph by I. Hargittai).

2

STEVEN WEINBERG

Steven Weinberg (b. 1933 in New York City) is Josey Regental Professor of Science in the Department of Physics of the University of Texas at Austin and Morris Loeb Visiting Professor of Physics at Harvard University. He received his undergraduate degree from Cornell in 1954, started his graduate studies in Copenhagen, and received his Ph.D. from Princeton in 1957. He started his university career at Columbia University, spent longer periods of time at Berkeley, MIT, and Harvard before moving to Texas in 1982. Dr. Weinberg received the Nobel Prize in Physics in 1979 together with Sheldon L. Glashow of Harvard University and Abdus Salam of Imperial College, London, "for their contributions to the theory of the unified weak and electromagnetic interaction between elementary particles, including, *inter alia*, the prediction of the weak neutral current." Steven Weinberg has authored over 200 research papers and seven books, including *The First Three Minutes — A Modern View of the Origin of the Universe* (Basic Books, New York, 1977), which has appeared in 22 languages, and *Dreams of a Final Theory* (Pantheon, New York, 1993). He has been awarded many honorary degrees, is member of many learned societies, including the National Academy of Sciences of the U.S.A., and received many awards, including the National Medal of Science. Our conversation was recorded in Professor Weinberg's office at the University of Texas at Austin, on March 3, 1998, and the text was finalized in the spring of 2003.*

---

*István Hargittai conducted the interview.

You just came back from your class. What was your lecture about?

I'm teaching a course on supersymmetry. It's a subject I last taught fifteen years ago here and at Harvard. Supersymmetry is a conjectured symmetry for which there is no direct evidence yet. It would combine particles of different spin in the same symmetry multiplets. This is very different from the kinds of symmetries that we're familiar with, like the symmetries that underlie the theory that unifies the weak and the electromagnetic interactions which only combine particles of the same spin, say, a neutrino and an electron.

There are various reasons to believe that Nature, probably at some scale of energies, exhibits supersymmetry, although it is certainly a broken symmetry which only manifests when you get to sufficiently high energy. There's widespread feeling among particle physicists that the energy at which supersymmetry will manifest is really just around the corner in today's experiments. Most likely, the kind of experiment that will reveal supersymmetry, will use the large hadron collider which is a somewhat smaller counterpart of the supercollider. The supercollider won't be built, but the hadron collider will be built early in the next decade. If supersymmetry is not discovered before then, I would bet that it will be discovered in the hadron collider. The discovery of supersymmetry is imminent and it's a good time to teach the subject.

Would you tell us a little more about the broken symmetries that appear to be broken only under our temperature conditions, and thus are really hidden symmetries?

This is a very common phenomenon in chemistry. The laws that govern atoms are completely symmetrical with respect to direction. There's nothing in Nature that says that one direction in the laboratory, whether it's east and west or up and down, is any different from any other direction. On the other hand, when atoms join to form a molecule, for example, when three oxygen atoms join to form an ozone molecule, that's a triangle that points in a definite direction. It breaks the rotational invariance of the laws of chemical attraction by forming a particular object that has not the full rotational symmetry but a smaller symmetry, just rotations by multiples of sixty degrees. If you had a more complicated molecule, there'd be no symmetry left, yet the underlying laws are perfectly symmetrical. Those molecules only exist below a certain temperature. You can always restore the symmetry by heating them sufficiently so the molecules break up into a gas. If you have a gas of monoatomic oxygen, without worrying about the walls, it is symmetrical; all directions are the same.

Is it perfect disorder?

Yes. Perfect disorder is symmetry. To have order, for example in a crystal, you break the symmetry. You only have symmetry by finite rotations. A crystal of salt is invariant when you change your point of view by rotation of 90 degrees around various axes. It's a cubic crystal. But if you have molten sodium chloride, then there's no preferred direction at all. You've created complete disorder, as far as the directions are concerned. People in condensed matter physics very often use the terms order and disorder rather than broken symmetry and restored symmetry although they are very closely related.

The symmetries that we talk about in elementary particle physics are not broken because of any particular object formed. The physical state that breaks the symmetry is not a molecule or a crystal. It is empty space. The vacuum, although it's perfectly symmetrical with regard to rotations in space, or translations in space, is not invariant with respect to changes in your point of view about which particles are viewed. It's the vacuum that distinguishes the neutrino from the electron, or the weak interactions from the electromagnetic interactions. The reason that photon is massless whereas the other particles on the same symmetry multiplet, the W and the Z particles, are very heavy, is because of the way they propagate through the vacuum.

You stress the importance of symmetry of laws of Nature, but the symmetry of objects is also important, for example, in chemistry and in molecular biology.

Symmetries are, of course, important in lots of areas of science, but they are important as auxiliary facts about the object. It is important that the sugars in living things are right-handed and the amino acids are left-handed, but it's not the most fundamental about them. On the other hand, the symmetries of Nature are the deepest things we know about Nature. It's much easier to learn about the symmetries of a set of laws than about the laws themselves. For example, long before there was any clear understanding of the nuclear forces, it was clear that there was a symmetry that the nuclear forces obeyed that related neutrons and protons and it said that they behaved the same way with regard to the strong forces.

You are saying somewhere that chirality is a property of strong interactions. Why stress the strength of the interactions in this case?

Chiral symmetries is one of the classes of symmetry that used to confuse us very much, because they are not fundamental symmetries underlying the laws of Nature. They're accidents. A particular set of phenomena, like the strong interactions, may for other reasons, be described by equations which are so simple that they naturally, automatically, exhibit a certain group of symmetries, even though those symmetries are not required by some more fundamental principle. At very high temperatures, the strong interactions, that is, the forces that hold together the protons and neutrons in the nucleus and hold together the quarks inside the protons and neutrons, are described by equations that have no mass in them for the quarks. The fundamental equations of the underlying theory of strong interactions have massless quarks. Quarks really do have a mass. The quarks in our models have a mass of about a third of the mass of a proton or neutron. That symmetry is broken. However, on the level of the underlying equations, which describe the quarks as massless, because they are massless, there's an additional degree of symmetry which we call chiral symmetry. The reason it's called chiral symmetry is that it involves transformations of the quarks which are different for quarks that are spinning to the left or to the right, in other words, using the language of optics, there are left-circularly-polarized quarks and right-circularly-polarized quarks. There is a symmetry which says that the laws that govern the strong forces, that bind the quarks together, are invariant if I perform a certain transformation of the wave functions on the left-handed quarks, leaving the right-handed quarks alone; or, on the right-handed quarks, leaving the left-handed quarks alone. The reason why we call it a chiral symmetry is obvious, because chirality refers to handedness. This kind of chiral symmetry is only possible for massless particles because the only reason we can talk, in a sensible way, about a right-handed photon or a left-handed photon, is because the photons never come to rest. If you have a particle at rest, then whether it's spinning to the right or spinning to the left, depends on your point of view. However, if a particle is traveling with the speed of light, and is always moving, then there's an invariant meaning to it spinning to the right or to the left around its direction of motion. It's only the particles that have no mass that can never be brought to rest. A particle that has a mass, can always be brought to rest, just mathematically by moving along with it. You can't move along with a photon.

Steven Weinberg with Martinus Veltman during the 1999 Nobel celebrations (photograph courtesy of M. Veltman).

A symmetry that behaves differently for left- and right-handed particles, will only make sense if the particle is massless. Otherwise you would have to have a paradoxical situation if you moved along with the particle, you would not be able to distinguish between left and right in any invariant way. Therefore the chiral symmetry in the strong interactions only exists in the mathematical sense, on the level of the equations which treat the quarks as massless. In the real world they are not massless, they get a mass from spontaneous symmetry breaking and the chiral symmetry, therefore, is broken. It still has implications, but the fact that a symmetry is broken means that its implications are different from those you'd normally expect.

When you have a crystal, condensed from a liquid, the crystal breaks translational invariance. The crystal is in one location and if you translate the crystal by an infinitesimal amount, you have a different crystal, the atoms are clearly moved, the crystal has a definite location, it's here, not there. That means that translational invariance is a broken symmetry. But it has, nevertheless, an implication, namely, that there is a mode of vibration of the crystal which has zero frequency in the limit of large wavelength. This just amounts to the rigid translation of the crystal. That mode is, of course, the phonon, the sound wave. The frequency of phonons vanishes in the limit as the wavelength goes to infinity. This you can deduce from the fact that there was symmetry to begin with that the crystal has broken. Another example is that the spin waves in magnets arise because the magnet breaks rotational symmetry. The generic word for this kind of phenomenon is the Goldstone boson or Goldstone mode. It is a mode of the oscillation of a system, which has zero wave number limit at zero frequency or vice versa. You always get a Goldstone mode of a system for every broken symmetry. In the case of superconductivity, another broken symmetry, which is actually electromagnetic gauge invariance, the Goldstone mode is the phase of the Landau wave function.

The Goldstone mode for the broken chiral symmetry in the strong interactions is a particle called the pi-meson. It is very light, though not massless, because the chiral symmetry is not exact to begin with. But we learn a lot about the interactions of the pi-mesons by knowing that underlying the equations of strong interactions there's a symmetry which is broken. This is the first example of a broken symmetry in elementary particle physics. It was discovered through studying the properties of the pi-meson. It's a very practical thing. It looks like hand waving and sounds highly speculative, but it's quite practical. We use theorems about broken symmetries to make predictions about the interactions of low-energy pi-mesons which are verified by experiment.

Does it mean that we understand these interactions?

No, actually, it meant that we understood the symmetry without understanding the interactions. We understand the interactions now in the 1990s. When these symmetries were first discovered in the 1960s, we had no idea what the strong interactions were. Nonetheless, by studying the pi-meson, we discovered this broken symmetry without knowing what the underlying equations were. The pi-meson behaved like Goldstone mode associated with broken symmetry. This is the way it often happens. You learn about the symmetries more easily than you learn about the underlying dynamics.

In a way the first example of this was Einstein's special theory of relativity. While it has nothing to do with broken symmetry, it's an example of the power of symmetry. Einstein was working at the time, in the first decade of the century, when the big topic on everyone's mind was the structure of the electron. The electron had been recently discovered and it was generally understood that the electron was a fundamental particle that occurred in all matter. Physicists were trying to explain all sorts of things by making mathematical models of the electron. This was work done by Lorentz, Abraham, Poincaré, and others. They were, in particular, trying to understand how the electron could behave in such a way as to hide the motion of the Earth through "ether". That was the big puzzle at the time: why did the Michelson-Morley experiment not detect the motion of Earth through "ether".

Einstein solved the problem by ignoring it. He said it's premature, we don't understand the structure of the electron; I'm not going to worry about models of the electron, I'm simply going to think about the symmetries that govern any possible theory of the electron. He realized that if you believe in Maxwell's equations, then you have to understand the symmetries that govern space and time. What happens when you change your point of view by moving at a different speed, for instance. Einstein discovered the nature of transformations relating different frames of reference, which is the essence of the special theory of relativity, by just thinking about this kind of symmetry. By symmetry I mean what kinds of changes of points of view leave the laws invariant. People at the time, like Lorentz, were rather put out because, although they were very supportive of Einstein's work, they still felt that somehow or other Einstein had cheated. He had not solved the problem, he had just stated the solution, the solution being that there is such and such symmetry.

Very often you learn about the symmetries much more easily than you learn about the underlying dynamics. This is true also in chemistry. You can understand the distinction between right-handed and left-handed sugars without understanding the details of forces inside the sugar molecule.

Particle physics and astrophysics seem to be converging. There is, however, a huge domain in between.

In a fundamental sense we already understand Nature on the level of the ordinary scale in which we live our everyday life. We have an understanding of the way atoms and molecules behave and the way electric forces and magnetic forces behave. This is sufficient to provide a basis for understanding the behavior of matter at ordinary scales. I'm not saying that we do understand the behavior of matter at ordinary scales; there are all kinds of wonderful things to be discovered in chemistry, in condensed matter physics, and elsewhere. We don't yet understand turbulence and we don't understand intelligence. There are all sorts of things we don't understand. But we have a basis for understanding, which doesn't require much more information. It's not likely that the behavior of chemicals or the behavior of living matter is going to reveal anything to us about the laws of Nature at the most fundamental level.

This makes people angry, when I say it, but by fundamental I don't mean interesting or exciting or valuable, useful, mathematically profound. It just means something that has to do with the rules that govern everything. By this I'm not saying that it's necessarily useful, for instance, for a chemist always to think in terms of solving the Schrödinger equation for a chemical molecule. Chemists have intuitions and expertise which is often much more valuable than just the brute force solution of the Schrödinger equation.

Computational chemistry is making tremendous progress and has become a bona fide tool in chemistry.

That's good, but it depends on the moment of history you're at. Computational chemistry is beginning to take over larger and larger fields. Some biologists like to work on the molecular level, other biologists like to study whole organisms. Both are valuable. I'm not a reductionist who would say that the only kind of interesting biology is molecular biology. And I'm not a reductionist who would say, the only kind of interesting chemistry is solving the Schrödinger equation. But I do think that some things are more fundamental than others.

So you are a tolerant reductionist, but you still manage to upset people.

I know, because I do think that one should have some idea of the order of Nature. And the order of Nature is that things in everyday life, including the chemistry and biology we see around us, are what they are (aside from history, which impacts biology) because of the laws of physics. You may not be interested in that, you may be interested in the cure of cancer or in studying the structure of some very interesting new molecule, like an enormous carbon molecule, but still, the carbon molecule behaves the way it does because of the laws of physics. They provide a complete explanation. It may not be an explanation that is very useful to the working chemist, and it may not be what the working chemist has to concern himself with, but they are still the reason why the buckyball behaves the way it does.

How do you feel about the personal element in science?

I don't think that scientists behave objectively. I don't think they would get very far if they did, because we never know enough so that you can't judge things just on the basis of some mechanical assessment of the experimental evidence. All good scientists have to rely on their intuition, on their taste of what an attractive theory would be, in the historical sense they feel this field is moving. That's very subjective. Then we all argue with each other and that's a social process. We interact socially with other scientists in a complicated way. This is fascinating for a sociologist to study. But I think we converge on an objective truth which is what it is because that's the way the world is. In the end that becomes a stable, permanent part of the body of our knowledge. At any one moment, the work of Kepler, Galileo, Copernicus, and Newton were highly influenced by the times they lived in, by the kind of people they had interactions with, and by their religion. But all converged to a picture of the Solar System we now have, and we have no doubt about it. Many of the sociologists who had studied science as just another social phenomenon, miss this. What they see is a process and the process is like any other social process, but the final result we are driven to is determined by Nature and we can't resist in the end knowing what the answer is.

You write that the task of physics is to provide a simple picture of Nature. This reminds me very much of Eugene Wigner.

I was a graduate student at Princeton when Wigner was a professor there. I took his course in nuclear physics. He was not a very good teacher because he was obsessively worried that there might be someone in the class who wasn't understanding him. So he went very slowly. But he still was very profound. He did me great compliment once, when he had to go out of town, of asking me to take over the class. I didn't learn much from Wigner when I was a graduate student at Princeton. But in the years following, I found that my point of view toward physics was very much in tune with Wigner's, much more so than with other people. Wigner had analyzed, especially in 1939, the nature of the elementary particles and, in particular, the significance of the spin, in a way far superior and much more well-grounded in fundamental principles than what I had earlier learned, say, from the treatment of the spin by Paul Dirac. In that sense I became a disciple of Wigner. In my book on quantum field theory I very much followed Wigner's 1939 paper. Wigner realized, earlier than anyone, the importance of thinking about symmetries as objects of interest in themselves, quite apart from the dynamical theory. In the 1930s although physicists talked a lot about symmetries, they talked about them in the context of specific theories of nuclear force. Wigner was able to transcend that and he discussed symmetry in a way which didn't rely on any particular theory of nuclear force. I liked that very much.

On the other hand, Wigner and I were quite different about politics. He got very angry with me. He edited a book about civil defense. He was a refugee from Communist Hungary.

He was a refugee from the Nazis.

Yes, but Hungary was taken over, after the War, by the Communists. He was a very committed believer in the Cold War, that America should prepare in every possible way for a nuclear confrontation with the Russians.

In hindsight, weren't those worries justified?

To some extent. It's hard to say now. The Cold War fed on itself in the sense that both sides were building more and more destructive weapons because the other side was. Wigner saw nothing wrong with that. He just wanted us to build everything we possibly could. He wasn't interested in arms control, and I was. I wrote a rather negative review of his book on civil defense, and Wigner got very angry with me. He attacked me at a cocktail party in Princeton. He was quite hostile to me for a number of years.

Then I gave a talk at a symposium in his honor, explaining how much I thought I'd learned from Wigner's approach to elementary particles and why his approach was superior to others. Wigner came up to me after my talk, and he didn't really understand what I'd said. Maybe what I interpreted as wignerism was not entirely wignerism. He may also have been already ill, so I don't know. Wigner is a person I have complicated feelings for.

Who was your mentor if you had any?

I didn't have a mentor, at Princeton I had a very good thesis advisor, Sam Treiman.

Who is your hero in physics?

To me the greatest scientist is Newton because he invented the style of modern science. Other people before him had developed the mathematical style of isolating certain features of a problem and working out the problem mathematically, regarding the ability to calculate as the sine qua non. In a sense Galileo started modern science, but Newton had the vision of a mathematical science that would comprehend everything. He made the biggest single step toward that in his theory of motion and gravitation.

Any single overall hero?

The person I admire more than anyone else in history is Shakespeare. Of course, I can't judge other languages, but he has left us with a legacy of plays and poems that is priceless. I know, this isn't a very original judgment.

Why?

He is widely admired by other people.

Do you know anybody in literature whose hero is a physicist?

No, but after all, we all read, but we don't all do theoretical physics.

You write about the challenge of communicating with the "unwilling public".

The best hope of science is to become part of the culture of our times, and yet we are not really, because the public can't read our articles. Even when we try very hard to write in a non-mathematical way, it's only a very limited segment of the public that would consider reading it. I have many dear friends who would never read one of the books that I've written for the general public. This is because anything scientific at all immediately loses their interest. I don't know what to do about it, we just have to try harder.

Yuval Ne'eman, 2000 (photograph by I. Hargittai).

3

YUVAL NE'EMAN

Yuval Ne'eman (b. 1925 in Tel-Aviv, Israel) is Professor of Physics (Emeritus, since 1997) at Tel-Aviv University and Chairman of the Israeli Space Agency. He received his B.Sc. degree in Mechanical Engineering in 1945 and the Diploma of Engineering in 1946, both from the Israel Institute of Technology (Technion), Haifa. In 1962, he received his D.I.C (Physics) from Imperial College of Science and Technology, London, and a Ph.D. (Sciences) from London University. His academic positions included his appointment at the Department of Physics of Tel-Aviv University, which he had in fact founded in 1962. He served Tel-Aviv University in various capacities, including that of President in 1971–1975. He has been very active in committees and other appointments in Israeli academia, as well as on the international academic scene.

Professor Ne'eman is a Member of the Israel National Academy of Sciences and Humanities, a Foreign Associate of the National Academy of Sciences of the U.S.A., a Foreign Honorary Member of the American Academy of Arts and Sciences, an Honorary Life Member of the New York Academy of Science, and a Member of the International Academy of Astronautics. In Israel, he received the Weizmann Prize, the Rothschild Prize, and the Israel Prize; on the international scene he was awarded with the Albert Einstein Medal, the College de France Medal, the Wigner Medal, and the Birla Award (India). He is a member of many learned societies and has received many other distinctions. In addition, Professor Ne'eman has had an exceptional military and political career in Israel.

We recorded our conversations during the *Symmetry 2000* Symposium in Stockholm, September 13–16, 2000.[1] The text was then augmented with references in subsequent correspondence.*

Would you please single out one or two of your scientific achievements?

First chronologically — and in importance too, in my view — I should put the SU(3) Octet Model (1961).[2] This is the classification of the hadrons, namely the strongly-interacting "elementary" particles. As I shall explain, this was a matter of reading a pattern; while my second key contribution[3] provided the first glimpse at the structural explanation of this pattern (1962), namely the existence of a further layer, deeper down in the "onion" of matter (later to be known as "quarks"). Returning to SU(3), this work is often compared to the work of Mendeleev, who classified the chemical elements. Here in Sweden, the one-hundred-crown bill carries the picture of Linnaeus, who classified the plant and animal kingdom. It is the same mode of operation. My tool was group theory. I did this work during the last months of 1960 and it was published in 1961. At the time this model was not taken very seriously, because other models were favored by the physicists. The most popular one grew out of a basic idea of Fermi and Yang in the late forties. With the discovery of the pi-mesons, Fermi and Yang suggested that the proton and neutron and their antiparticles should be taken as the fundamental particles and that the pions are made out of them: the positively-charged pion would then correspond to a proton bound to an antineutron, the negatively-charged pion to a neutron bound to an antiproton, etc. This was then augmented in the so-called "Sakata model" by adding the Λ^0 (carrying one unit of negative "strangeness") and its antiparticle. Sakata was a Japanese physicist at Nagoya University who strongly believed in Marxist dialectical materialism [see my excerpts[4] from his writings and from those of his school]. According to this dogma, there had to be, at the bottom of it all, some little hard basic "bricks" and this was the role they expected the proton, neutron and lambda (with their three antiparticles) to fulfill. Around the same time when I was working on my model (summer 1960), the Sakata team at Nagoya University was visited by Yukawa, who advised them to use group theory. The Nagoya school then identified SU(3) as their classification group, defined by the assignment of the Sakata model [p, n, Λ^0] to be the defining

---

*István Hargittai conducted the interview.

representation. Note that I had heard about the model, but not about their use of group theory. My own approach was phenomenological, without any preconceived notions about which of the particles is more elementary. Returning to my application of Groups, I read about Lie Groups and looked at the rank-2 Lie groups, as I could see that the strong interactions appeared to conserve only two linear (additive) charges, or in other words, all inter actions allowed by Isospin and strangeness conservation indeed occur. There were five such Lie algebras: A(2), the algebra of traceless 3×3 matrices, B(2) and D(2) which correspond to orthogonal transformations in 5 and 4 dimensions respectively, C(2) to symplectic (i.e. orthogonal, but with an antisymmetric metric) transformations in four dimensions, and G(2). When I constructed the root-diagram for G(2), it turned out to be a Star of David. Had I been a believer in miracles, I might have regarded this as some heavenly revelation and G(2) would have been suggested as "it" — except for the fact that the fit with experimentation was rather lacking and the Sakata SU(3) remained as the main competitor to my octet until 1964 and the Ω^- experiment. Meanwhile, I settled on SU(3) but with a different basic identification: the nucleon iso-doublet I assigned to an octet, together with the iso-singlet Λ^0, the iso-triplet $\Sigma^=$, Σ^0, Σ^-, and the iso-doublet $\Xi^0 \Xi^-$. The three Σ were very similar to the Λ and I had decided this meant that they should "dwell" together. The situation with the two Ξ was less clear, as their spin had not yet been measured with due precision, so that one of the first predictions of the model was that the spin $J(\Xi) = 1/2$, which was confirmed within the year. Another conclusion was that the nucleons should be considered as compound structures. I published my first scientific paper along these lines early in 1961. Almost at the same time, the Sakata group published their SU(3) model.

Perhaps I should fill in and explain how I came to be in London at the age of 35 working on this problem in the context of my Ph.D. thesis. My original degrees were from the Technion (the Israel Institute of Technology at Haifa, founded in 1912), in Mechanical and Electrical Engineering, first a Bachelor's degree and after I had submitted a "project" I received a "Diploma", the equivalent of a Master's degree. We were organized in the German academic system, and all my professors had indeed come from Germany. As a matter of fact, they did not speak Hebrew yet and read their lectures from a prepared text with the Hebrew written in Gothic characters. This was happening in 1941–1945 and at the same time I was active in the Haganah (Jewish defense underground militia) and involved in lots of activities. Among other things, I specialized in light-weapons training and as such was busy training instructors or sometimes training particular individuals, such as the parachutists that were dropped in German-occupied Europe in collaboration with the British. For the British, they were doing liaison missions to the various underground organizations; while for the Haganah, they were collecting information on the status of the Jews in each country, on their safety and on Nazi crimes. This is how we started hearing about the Holocaust, very gradually. I trained Hannah Szenes, who was dropped over Hungary, was caught by the Germans and executed, a real heroine. In 1945 I also attended the Haganah Officers' School, which was then commanded by Yadin, the archeologist (Yadin, Dayan, and Alon, the three most brilliant future generals, got their training in Palestine in 1936–1939 under Captain Orde Wingate — the future Allied commander in Ethiopia and in Burma). The officers' course I was attending included two-sided exercises and when my turn came I managed to achieve complete victory (over two forces, each equal to mine — the only such result among 120 trainees), by inventing a new technique of preconceived plans based on an analysis of the terrain. The school umpire attributed the victory to luck. The years 1946–1947 were spent mostly in Haganah operations, in a struggle for Immigration and Settlement (against British decisions).

I had meanwhile decided I would go into science. I had gone to the Technion out of a feeling of family responsibility — my grandfather had founded a Pumps Factory in Jaffa in 1900, my father had taken over in 1939 and some day I would have to. During the Technion years, however, I had reached the decision to go into science — perhaps combining this with my engineering responsibilities. I was not sure whether it should be mathematics or physics, but meanwhile tensions mounted and after working for a few months designing pumps in the family factory, I was fully mobilized in mid-1946. In 1947 the United Nations decided on the partition of Mandatory Palestine into a Jewish and an Arab State; the Jews agreed, but the Arab side refused and attacked the Jewish cities and settlements, in an attempt to conquer the entire territory and nullify the UN's decision. Prospects were very dim. Lord Attlee, the British Prime-Minister in 1947–1949 had written in an article in the (London) Observer in 1958[5] that when his government decided in 1947 to evacuate Palestine, he, Attlee, was doing it under the assumption that the Arabs would win and conquer [and exterminate?] us. He had been presented with this evaluation by General

Montgomery, the Chief of the Imperial General Staff at the time and in 1958 he thus criticizes Montgomery's strategic thinking. I found myself commanding a unit in the field, first in the defense against the Palestinian Arabs, commanded by volunteer Bosnian and other officers who had served on the Russian front in 1943–1945 in the Muslim division recruited for Hitler by the Grand Mufti of Jerusalem all over the Balkans and in some Arab countries (especially Iraq, where in 1941–1942 the dictator Rashid Ali el-Khilani's regime fought on Hitler's side, and Syria, which the Germans took over from Vichy France in 1940–1941). After very difficult times in early 1948, we started having the upper hand — and then in May 1948, six Arab countries invaded Palestine in a last attempt to forestall the creation of the Jewish state. In the first stage of the 1947–1949 War, I fought some tough battles and went through awful experiences, which I should better leave for other times, otherwise we shall never get to my scientific career. On one occasion — countering an attack by a large Arab force on one of our very isolated outposts ("the JNF building at Beit-Dagon") — I did have the satisfaction of full success, resulting from some imaginative tactics on my part — as in the Officers' School exercise. In the second phase of that war, I commanded units on the Egyptian Front and was responsible for some of the more successful strategic or operational war plans, especially our October 1948 counter-offensive. I have a nice testimonial to my role in Ben-Gurion's 1952 diary: the Chief of Staff, recommending me for the position of Defense Planner, refers to my having been "the brains" behind the successes on the Egyptian Front. In the last phase I was deputy commander of a brigade group, when I was transferred to the General Staff in the beginning of 1950. I served within the Operations Section, headed by General Rabin, both as his Deputy and as head of the Operations sub-section. It was then that I organized the Israeli Mobilization System, inspired by the Swiss. In 1951 I asked for leave to study physics in France (I was thinking of working with de Broglie) but I gave in to the entreaties of General Yadin (the archeologist, my commander at the Officers' school, now the Chief of Staff) and his deputy (General Maklev) and instead, I was sent to the French "Ecole d'Etat-Major" — a branch of the "Ecole Superieure de Guerre" in Paris. I returned after a year and was appointed Head of Defense Planning. This position proved to represent a tremendous intellectual challenge, because it covered every aspect, from contingency war plans to designing the country, to selecting the sites for future ports, cities, forests etc. I have told some of this in my evening lecture at the VIIth Marcel Grossmann Conference on General Relativity (Stanford, 1994) and it appears in volume 2 of the proceedings. I was transferred to the Intelligence Branch as Deputy Director General in 1955 and in this capacity in 1956 I dealt with the secret alliance with France and with the collusion with the French and English in the 1956 Suez-Sinai War. After that war there was a relaxation in tension and in the immediateness of the dangers to the country and I decided to go into science. It was my old dream and I felt this might be my last chance, considering my age (I was 32). In addition, I had learned that Nathan Rosen, a former co-worker of Einstein's, had joined the Technion and had created a Department of Physics in it. Now I had taught myself general relativity and was captivated by the aesthetical aspect, due to symmetry. Here was a chance to work specifically in this area! I went down one floor, to see General Moshe Dayan, the Chief of Staff, whose office was just under mine, and told him that I would like to take a two-year leave from the Army and try my hand (or rather my brain) at research in physics at the Haifa Technion. Dayan thought it over and came up the next day with a counter-suggestion. He was then in the process of selecting a Defense Attaché of Israel in London and his idea was for me to do it and combine these tasks with my studies. I accepted — I had heard of Bondi, Gold and Hoyle and their Steady-State Cosmology and that Bondi was teaching in London. The attaché appointment included accreditation to our Embassies in Denmark, Norway, Sweden and Finland, in addition to the position at our Embassy to the U.K. in London. Also, I was Military, Naval and Air Attaché (in all 5 countries). Throughout my 3–4 years in this position, I made about 4–5 trips per year to Scandinavia.

Did it work out, to do your studies parallel to your job?

It went well in the beginning, i.e., in the first semester of 1958. Originally, I had intended to work on general relativity with Bondi at King's College, but it was too far from the Embassy, with the whole of London traffic in between. However, I discovered Imperial College at a five-minute walking distance. The catalogue cited a Professor Brockman in theoretical physics. I went to see him and told him I was interested in Einstein's unified field theory. He reacted "I do not know about the unified field theory, but if you want to learn about field theory, go see Abdus Salam[6] in the old Huxley building", which is how I got to Salam and to group theory.

A younger Yuval Ne'eman, 1972
(courtesy of Y. Ne'eman).

Salam laughed when I showed him my letter of introduction from Moshe Dayan, but he admitted me "on trial" and I became his graduate student. Whenever I could, I attended lectures. For the first time in my life I was the oldest student in my group. Note that I used to be a "wunderkind", having graduated from high school at 15, the youngest ever in Tel-Aviv. I made friends among the other students, Ray Streeter among them, and when I had to miss a lecture, I would copy Streeter's notes — we had a copying machine in my office at the Embassy. Everything worked well in the spring of 1958, but things went wrong in the summer.

In July 1958, the Iraqis assassinated their king and prime minister and the first dictatorship (General Kassem) followed in Iraq. In Egypt, the monarchy had been replaced in 1952 by a dictatorship, first under General Neguib, then under Gamal Abdel Nasser. The latter was conducting a Pan-Arab agitation all over the Arab world, endangering the traditional regimes in the Middle East. The Americans — worried about a possible "domino effect" reaching Saudi Arabia and Kuwait — landed in Lebanon, in an attempt at stabilization. The British wanted to drop two battalions of parachutists in Jordan, to protect King Hussein's regime. To do it, they had to overfly Israel — and I found myself partaking in a negotiation in which we stressed that the new threat was even more a threat to Israel's existence and we requested (as a condition for the overflying) that we be given access to needed weaponry. The result was that we were allowed to purchase two submarines (our first) and fifty Centurion tanks (heavier than anything we had till then). This meant that come 1959, I found I could do no physics at all. I studied how the British trained their submarine crew and I organized an adaptation to our conditions, namely training all levels in parallel, as we had never possessed submarines before. I also had to deal with the fifty Centurion crew. There was no more any time left for physics. I complained — this was not the deal I had agreed to in my "negotiation" with Dayan. In May 1960, my replacement arrived (the new attaché) and I was granted a full year at Imperial College, now free of any other duties, courtesy of the Israeli Ministry of Defense (they would later recover the investment, by sending their bill to the Israeli AEC). As to the contents, Salam suggested a problem in quantum field theory, namely *the acquisition of mass by the vector-mesons in a Yang-Mills gauge theory* — a problem that was solved by Peter Higgs in 1965. I believe I have in fact given it another jump ahead in 1979–1990 with my SU(2/1) "Internal Supersymmetry"[7] based on applying ideas in non-commutative geometry (long before this discipline was invented as such ...). However, by the time I was free to devote myself entirely to physics I had also made my own choice of problem. I knew I wanted to work on the classification (and symmetries) of the "elementary" particles. By Emmy Noether's theorems, symmetry also implies conserved charges, at least some of which would be dynamical and provide some understanding of the forces. With an understanding of the manner in which the electric charge is "situated" within this complex of charges, one might also be able to calculate electro-magnetic mass-differences within a multiplet, magnetic moments, etc. By mid-1960, there were about a hundred "elementary" particles (mostly hadrons, with the new ones coming up mostly as resonance). I studied Salam's lectures on group theory and started applying to the classification issue the groups he had presented in his course. I conceived various such ideas and would go see Salam each time there seemed to be a new way. On each such occasion he would listen to my new idea and react with, "This has been tried by Schwinger in 1957; its weakness was ..." or "This has been done by Tiomno in 1958", etc. Salam was getting a bit impatient and he kept reminding me of the problem he had suggested — rather than wasting my time on the classification. My own reaction was the opposite: if I was thus repeating history, it meant that I was at least neither stupid nor uninformed, since the best people in the trade had followed this path. It was encouraging, at least. When Salam saw that I was indeed determined to pursue this problem, he gave in, after warning me, "You are embarking on a highly speculative project, without the security of the typical thesis topics — and you might fail to find something good within your allotted year! But if you persist, go deeper into group theory, learn more than the little I taught — which is what I know". He gave me some references, wished me good luck and left for the U.S.A. for the summer (of 1960).

When he returned in October, I showed him my SU(3) octet model. He was interested, but also related that at a conference he had attended at Rochester, a Japanese group had reported work on the Sakata model, in which they were using the same SU(3) group, though in a completely different context. He had brought back a preprint, which he showed me. He told me to write up my theory and said he would add an idea he had had about "gauging" their SU(3) and getting an octet of vector-mesons — in this case these would be the same as the ones I had derived in my model as one of my predictions. [Note: all eight mesons were indeed discovered in 1961.] He would write up this bit and add it to my paper as a comment in a joint publication. He later (after I had waited for a month) changed his mind and told me to go ahead on my own and publish my model. One unfortunate result of my inexperience was that although I had worked out a nice notation (for the basis matrices, etc.) I changed to the Nagoya notation (thinking this was the rule, once they had issued their preprint), which was very ugly. Anyhow, I re-edited the October manuscript and submitted it to the journal *Nuclear Physics*. Then, one day, Salam announced excitedly, "Gell-Mann has just issued a preprint with the same model as yours!"

I immediately sent a copy of my manuscript to Gell-Mann. After a few days, my manuscript came back from *Nuclear Physics* with an angry letter, because it was not typed double-spaced, and requiring me to do it again. I retyped it and added that I had learned that Gell-Mann had the same idea. My paper[2] carried a received date in February 1961; Gell-Mann then submitted his to the *Physical Review* in March 1961. In June, Salam went to a meeting in La Jolla and came back with the news that experiments seemed to support the Sakata model, rather than the octet. Gell-Mann was also present at La Jolla and, therefore, in June he withdrew his paper. Later on, in September 1961, he submitted another manuscript, in which he described both the Sakata model and our own octet model, but did not commit himself to either. Gell-Mann's original manuscript was never published in a journal, but only appeared three years later in a book *The Eightfold Way*, a collection of reprints, which we published together, after the model had been proven when I was at Caltech in 1964. This book was republished in December 2000.

When I felt confident about my model, I filled out all the forms for my doctorate. Before that I had only registered at Imperial College, but not at London University for a doctorate, as at the beginning I was not sure I would succeed in my work. Sometime after I had sent in the former, I was notified that the Board of Physics of London University had determined that I could work for my doctorate in electrical engineering but not in physics, considering my background at the Technion. This was already after the publication of my first paper, just as my new results were coming in. Salam called the Registrar of London University and told him about my work. The Registrar invited me for an interview and explained to me that they could not look into the affairs of each of their five thousand students. He invited me to resubmit my application and suggested to indicate "quantum theory" instead of "elementary particles" as my subject. In this case it could go to the Board of Mathematics instead of the Board of Physics,

Yuval Ne'eman (first row, first from the left) in a group of physicists attending the Eighth Nobel Symposium in Göteborg, 1968. Abdus Salam (1926–1996, Nobel Prize in Physics 1979) is second row, third from the left and Murray Gell-Mann (1929, Nobel Prize in Physics 1969) is second row, first from the right (courtesy of Y. Ne'eman).

in order to avoid asking the physicists to go back on their word. Note that the Ph.D. title is in fact "in Science", and does not mention the precise discipline. This is how I ended up with Ph.D. under the aegis of Mathematics from London University.

My first publication thus proved to be important, although at the beginning only a few people believed in it. The Sakata model was very popular. For instance, I discussed my model with Victor Weisskopf, who visited Israel after I had returned home. Weisskopf thought it was nonsense. In his autobiography[8] he mentions our meeting but he does not mention our discussion of SU(3). He only tells a story I told him when we were driving and passing Mount Carmel, the biblical story of Elijah the prophet proving the superiority of his God when the prophets of Baal are destroyed by a sudden fire. I had commented that perhaps Elijah was using a hidden lens to ignite that fire. To Weisskopf, according to his comments, this revealed my bias for science. In any case, he avoids mention of our discussion re SU(3).

My second important contribution was the following. If you compare my first contribution with the works of Linneaus and Mendeleev — let's take the Periodic Table, for instance — it took more than four decades to explain correctly the foundations of the Periodic Table, namely the electronic structure of the atoms of any chemical element. The full explanation and understanding came as late as 1932, after both Pauli's exclusion principle and the discovery of the neutron. As regards my classification of the hadrons, I worried about the explanation from the very beginning; I wanted to understand the underlying principles of my model. It also met strong opposition based on the lack of a structural model. Even Eugene Wigner criticized me for not considering the proton to be a fundamental elementary particle.

I was looking for another level of fundamental particles indeed, and I was working out its mathematical foundations. I arrived at a triplet but I also found that it would have to have *fractional charges*. These particles we now call "quarks". The proton would then already be a 3-quark composite. I developed my structural scheme in a paper. The first issue of *Nuovo Cimento* in 1963 contains a paper[3] by me with H. Goldberg, dated February 1962. It took a long time to appear because at one point the editors lost my manuscript.

The *Nuovo Cimento* paper was submitted from Israel as I had returned home in the summer of 1961 and became the Scientific Director of one of the Atomic Energy Laboratories, which was at the reactor south of Tel-Aviv on a campus close to the Weizmann Institute. I was still on leave from the Army. In 1962, I finally became a civilian.

Haim Goldberg came to work with me. He had a good preparation in group theory from G. Racah. I had come across Racah when I was the defense planner for Israel between 1952 and 1955. As such I represented Defense on the Atomic Energy Commission. In preparation for the development of Israel's nuclear infrastructure, six people had been sent to good schools, e.g. to Zurich to work with Pauli and Weisskopf, and also to other places. They had all completed their Master's degree with Racah in Jerusalem. The only one who was not a former Racah student was Lipkin, who had immigrated from America to work in a kibbutz but then joined the project. Out of the six, five returned to Israel to work on the nuclear project. Originally the Israeli nuclear project was an extra motivation for me to become a physicist, on top of my original interest in the subject. In high school my main interest had been mathematics and my interest shifted to physics only after I finished high school. A series of lectures by Sambursky was an important impact. He was the first physics professor in Israel. He came in 1924 after his doctorate with Theodor Kaluza who is much talked about nowadays because of the Kaluza-Klein approach in string theory or in supergravity.

The other influence which pushed me towards physics was a series of lectures at the Technion by Franz Olendorf who had just come from Germany. His lectures were on wave mechanics and they were beautiful.

When did you become aware of what was going on in Europe during WWII, what is now being referred to as the Holocaust?

During the war I had no idea of what was really going on in the extermination camps. But we were trying to find out and this was also why we prepared and dropped parachutists like Hannah Szenes behind the German lines.

There was the Auschwitz Protocol.

In America they knew but Roosevelt did not let anything like that be published. The BBC too got orders not to mention a word about it. Apparently, they were not very sorry about Jews being exterminated. They did not want it to interfere with their efforts and they did not care. I squarely believe that this was the case and I can illustrate my point with

Attlee's article which I quoted earlier.[5] When I was the Israeli Defense Attaché in England, in 1958, General Montgomery published his memoirs. In his book he criticizes Ernest Bevin, who was the Foreign Minister when the British left Palestine in 1948 — and Israel was fighting the War of Independence. About a week after the British forces' evacuation of Palestine, Montgomery, the Chief of the Imperial General Staff (CIGS) at the time, got a phone call from Bevin asking him to return troops to Palestine because the Jews had just taken Jaffa and that was against the British plans. Montgomery did not comply, explaining that they had left Palestine as planned and they could not reverse such things out of the blue. Some time later, in the London *Observer*, Clement Attlee, the British Prime Minister at the time of the events I just referred to, published a long review of Montgomery's book. The first page is full of praise for Montgomery. On the second page, however, he criticizes Montgomery's strategic abilities, citing as an example that when the British left Palestine, Montgomery had estimated that the Arabs would liquidate the Jews. Attlee says, "and we know what really happened", exposing Montgomery as having given him a wrong estimate. It is thus explicit that the British left Palestine on the assumption that we would be conquered and exterminated by the Arabs — or at least that there was a high probability for such a "happy end." And this is not Germany.

This is why I have no illusions about the motivations of suppressing the information about Auschwitz during WWII. There has been documentation about Roosevelt in America and it is now well known. In Britain it is less clear who gave the orders to the BBC and Foreign Office to suppress all information during the war.

During the war the concentration camps were not bombed, neither were the railway lines. I recorded an interview with Joshua Lederberg, Nobel laureate, a man of broad activities and one who professes his Jewishness conspicuously. When I asked him about this issue of not bombing the concentration camps and railway lines, he defended and explained the non-action of the allies, saying, "It was not an optimum use of U.S. air power."

It was criminal.

Even though the American Jewry did not do much and could not do much then, it appears difficult even more than fifty years later to face history.

It is a shame. We should also be more aware of Jewish resistance as for example, the uprising of the Warsaw Ghetto and the Vilna Ghetto. To me, the preparation for the Manhattan Project was also an act of Jewish resistance. When you read about the "refugees from Central Europe", the year is 1938, and in 1938 only Jews were refugees, and Szilard works his way to the U.S. President through the Jewish lobby. When in August 1939 Einstein signs his famous letter, Europe is still not at war. In Britain, Peierls and Frisch follow the same scenario. All this is part of Jewish history.

The sadder it is when great scientists are afraid of admitting errors of the past.

Dick Feynman was a genius but he fled from his Jewishness.

Let's return to your story.

I started telling you about Haim Goldberg joining me in the Atomic Energy Lab. I had continued to think about the existence of more fundamental structures underlying the hadron list. I found that mathematically we could have something like the Sakata model but the particles would have to have fractional charges. In order to make an octet from triplets you need three triplets. You need three such quanta to make a nucleon. I did not give them names but I explained that they had a baryon charge of 1/3 and how all the hadrons could be made of them. This was in my 1963 paper and it was very mathematically presented, as I had no idea about the forces acting between the triplets, to make baryons. I only saw that structurally it would be a way of starting from a fundamental field. I sent out an Atomic Energy preprint to Gell-Mann and everybody else.

Not giving names to what you discovered may have been a severe omission. How do you view this in hindsight?

In general, I did not give "popular" names anyhow. I called the octet, octet and the symmetry "SU(3)", and I talked about a Sakata-like triplet model.

Could you be sure that what you found to be the fundamental particles would indeed be the last level?

I only said that the present level could be explained by such a level underneath. It would be a fundamental field, yielding a three-quanta realization of the

Murray Gell-Mann (b. 1929, Nobel Prize in Physics 1969), 2001 (photograph by I. Hargittai).

particles. The mathematics is all there, the physics is discussed very briefly because I had no model of what would hold them together and how.

Gell-Mann's story is that about a year later, in the summer of 1963 he gave a talk at Columbia University. Robert Serber was in the audience. Gell-Mann talked about the octet and Serber asked him whether the octet could be made of something more fundamental. Gell-Mann agreed that there must be something like that and then went to work on his model. Lipkin has published a review in *Physics Reports* called "Quarks for Pedestrians",[9] a review of the whole quark model — in which he starts with what we did at the time, Goldberg and I, when very few people believed in the octet. According to Lipkin, had I stressed at that early stage that in the basis of my model there are particles with fractional charges, it would have been considered as a crazy notion.

Did Gell-Mann quote your paper?

He did not. He told me he had read it, but it did not "sink in". Only after Serber's question did he really think about these triplets. The first analysis of the birth of the quark model was done by Lipkin, in that 1974 review. Many years later there is a chapter to this effect in John and Mary Gribbin's biography of Feynman.[10] They study the papers in depth and even end up criticizing the Nobel Foundation for disregarding my contributions.

There was also George Zweig at CERN who rediscovered my idea, although he thought of the quarks as actual particles. The Gribbins say that Gell-Mann chose to go in between the two approaches — or with both: like me, he considers quarks as "a mathematical model" and "a field theory" but like George Zweig, he also allows for the possibility of quarks as observable particles. Luckily, his paper appeared just when the results of the Ω^- experiment were announced — and this attracted physicists. Here I'll have to tell you a story.

In the summer of 1962, about a year afer my return to Israel, I attended the next high energy conference in Geneva, at CERN. I had intended to present experimental evidence for the octet and against the Sakata model, but the organizers of that session considered the octet to be wrong and did not allow me to. However, in the experimental section there was a report from Berkeley by the Goldhaber (Gerson and Sulamith) couple about K-meson scattering on nucleons — they had expected to find resonances but there were none in that channel. They were very disappointed but to me that was an indication about the assignment of the SU(3) representation particles produced in meson-baryon scattering. If you take an octet of baryons and you scatter on them an octet of mesons, they can go into a certain set of representations

$$8 \otimes 8 \to \lfloor 8_F \oplus 10 \oplus 10^8 \rfloor_{as} \oplus [\oplus 8_D \oplus 27]_{sym}.$$

The "oldest" resonance of that type was Fermi's "3-3" (spin 3/2, isospin 3/2) and such an assignment exists both in "10" and in "27". The lack of a positive strangeness (from the K meson) resonance in the Berkeley experiment was now pointing unambiguously to a 10, and thus indicated the existence of the "Ω^-" particle (the one empty box in the 10 at that time) — again a name given by Gell-Mann. When I was 60, there was a *Festschrift* for me[11] and in it Gerson Goldhaber wrote a little article entitled "The Encounter in the Bus", which tells the story of what followed in 1962. I had heard that the Goldhabers had studied in Israel. When I took the bus from the hotel to CERN, I noticed a couple in the bus. There were very few ladies in particle physics so I took the chance, said "Shalom" — and indeed, these were the Goldhabers. I told them that they should not be sad about the missing resonance in their experiment. I gave them in writing the properties of the missing particle. In the Plenary Session the next day, the scattering experiments were being reviewed and I raised my hand. The chair, Professor Bogolyubov looked at me and said, "Professor

Gell-Mann." It turned out that Gell-Mann was sitting right behind me and he also wanted to speak and the chair recognized him. Gell-Mann explained exactly what I had explained to the Goldhabers. The Goldhabers asked if I had told him, but this was not the case; it was just that Gell-Mann knew the theory as I did. That was when I met Gell-Mann, we had lunch, and this is when our association started. Gell-Mann at once suggested that I should come to Caltech and I did, in September 1963, first as a postdoc but within one year they changed my appointment to a Visiting Professor. In 1964 the Ω^- was indeed discovered at Brookhaven, and their official announcement mentioned our predictions.

Can we then summarize that you did the classification of the elementary particles and the prediction of a fundamental particle, Ω^-?

Yes, but also the first glimpse at the structure, quarks. In the concluding part of Luis Alvarez's Nobel lecture in 1968 he talks about us, the theoreticians who identified the pattern, mentions the latter's "beauty" — and assumes we shall soon be there, so that he does not have to explain it, since we can do it best. The next year, in 1969, Gell-Mann indeed got the Nobel Prize (there is no lecture by him in the Annals of the Nobel lectures, although he gave that lecture, because he never submitted his manuscript for publication). However, there is the presentation speech by Ivar Waller and we can read why they did not comply fully with Alvarez's recommendation.

Were you present?

No. He did not invite me although, for example, Alvarez invited his whole group in 1968. This was different, however, and I would not have come, anyhow. Gell-Mann too may have felt bad about me not sharing the prize — I do not know. I stayed with him for two years at Caltech and we were treated there as equally responsible for these discoveries. There was an hour-long BBC film about us and the discovery of Ω^-, with Feynman as the narrator.[12] In 1968 there was a conference on elementary particles organized by the Nobel Foundation. I was there and Gell-Mann and Salam were there. At some point Ivar Waller of the Nobel Committee for Physics came to see me, but he was completely uninterested in what I had done. All he wanted to know was whether Salam had anything to do with it or not.

Then he wrote in his laudatio about what Gell-Mann did and added that this was also done by Ne'eman "somewhat later". This is not true. Even Gell-Mann's biographer George Johnson[13] writes that when Murray asked someone at Caltech about Lie groups (his first step on the way to SU(3)), I had already done my paper. Besides, his paper was finally never published, except in our joint book in 1964.[14]

In 1969, Isidor Rabi — who seemed to have some information about that Nobel affair — told me that the Nobel Prize for Gell-Mann alone was a way out for the Nobel people who were having pressure exerted on them by the British, who were strongly for Salam and either uninterested or even antagonistic in my case. I also got some direct indications of this dislike in 1980, when I was nominated by the U.S.A. and Israel for the vice-presidency of IUPAP and the British vetoed it. I believe this may have changed since.

I would like to mention one more area of my scientific activities. In 1979, I decided to go into politics because I found Begin's territorial surrender policies dangerous for the security of Israel. At the same time I had a new idea about a certain type of symmetry. It is a supersymmetry using a superalgebra. That means that you mix commutators and anticommutators. Ordinary supersymmetry relates bosons to fermions or fermions to bosons. This symmetry SU(2/1) came to me intuitively because of certain work I had done before that. I called it an "internal supersymmetry". It was a completely different thing. There were no transitions between bosons and fermions, instead it goes between left and right chiralities; and yet it works with a superalgebra. About a month after me, David Fairlie in Durham, England had the same idea, though coming from a completely different direction. He published his paper two issues after mine in *Physics Letters.* Both appeared in 1979.[7] I was puzzled by my own work because it was a completely new subject mathematically and yet it gave important predictions. It predicted, for instance, the mass of the Higgs meson, which the Weinberg-Salam model does not predict. People had been looking for the Higgs meson all the way from 1 GeV and it is not there up to 50 or 60 GeV. It has not been found yet.[15] They are waiting for the new accelerator, the LHC, which will be completed in 2005. My model predicts that the mass of the Higgs is twice the mass of the W, which is 85 GeV, so it will be 170 GeV. However, there is a renormalization correction because at high energy the mass is a function of energy. The energy at which this Pythagorean result holds is a higher energy. We calculated the mass of the Higgs at lower energy and it comes out as 130 ± 10 GeV. My prediction is the only prediction in the field. No other theory says anything about the Higgs — except for ordinary supersymmetry, which then requires the existence of lots of new particles.

I worked a lot with a French student I had, Jean Thierry-Mieg and with Shlomo Sternberg, a mathematician at Harvard, on understanding the new concept behind that superalgebra, which produced such a completely new type of symmetry. It was like M. Jourdan, one of Moliere's characters who was "talking prose" all his life without being aware of it. It turned out that I "applied" non-commutative geometry ten years before it was invented. In 1991, two physicists, Coquereaux and Scheck showed that in non-commutative geometry, if you derive the Weinberg-Salam model, you also get my algebra. I had constructed an algebra that is now fully understood in terms of non-commutative geometry. This was one of the two mathematical theories of which my theory of 1979 was a precursor. The other was the so-called "theory of the superconnection" worked out by the mathematician D. Quillen in 1985.

If and when the Higgs will be found and its mass measured, I would now like to advertise my theory and people to know that I had predicted it. I have suggested to my co-inventor, David Fairlie that we should collect our reprints in a volume, for example, or a review in *Physics Reports*. David is still worried about some things and wants to calculate some quantities to make it easier to understand our approach. The name of the supergroup is SU(2/1). It is now all published material, it is all there, and I only want people to know about it.

Compared with Gell-Mann and other physicists, what is the fraction of your career that was devoted to physics?

I only became a physicist at the age of 33. Even then I was also involved in the nuclear project in Israel, as scientific director of a large laboratory. Later I was one of the founders of Tel-Aviv University and I founded its Physics Department. I had a large share in the fight for getting Jewish scientists out of the Soviet Union. I organized a world system for that. For example, Arno Penzias, co-recipient of the Nobel Prize in Physics for 1978, went straight to Moscow from the Nobel ceremonies in Stockholm. He gave the same talk in an apartment seminar in Moscow as he did in Stockholm. These were "Refusenik" Seminars and I was involved in starting and supporting this activity. There were articles in *Izvestiya* considering me as the arch enemy of the Soviet Union. I had called up Penzias one day (he was then at Bell Labs) and told him about our fight, about Levich and others, and he understood the problem at once and joined me in our efforts. There were others, however, who would not touch this problem. My pioneering intervention is even documented in writing.

While I was at Caltech in 1965, a friend, Ivor Robinson of Dallas, Texas showed me the *New York Times* with a letter signed by Landau and Lieberman. Landau was the famous physicist, Nobel laureate and Lieberman, the famous economist in the Soviet Union. The letter is a Letter to the Editor protesting the accusations that the Soviet Union is not tolerant to the Jews. They state that their own careers constitute proof that Jews can make it to the top in the Soviet Union. They suggest that Americans should rather deal with their own racial discrimination. I knew that Landau had had his accident in 1962 and was in bad shape (he was indeed in a "vegetable" state, but I did not know that) at that point. I believed he would not have signed such a letter, except under threats. I drafted an answer and immediately called various people inviting them to co-sign my letter. My Caltech friends did not want to, Rabi did not want to. The non-Jewish Dyson agreed, Chandrasekhar agreed, and Felix Bloch, who was then the President of the American Physical Society, told me, "Make it tougher," and he also signed it.

How do you explain Feynman's and Gell-Mann's reticence?

I have written about Feynman's weaknesses on his Jewish side for the Israeli paper *Ha'aretz* in Hebrew, and it also appeared in the New York *Herald Tribune* in English.[16] Feynman never visited Israel, for instance.

Some prominent Hungarian Jews are famous for hiding their Jewish identity.

But then there are Hungarian Jews like Theodore Herzl, the founder of Zionism and of the State of Israel — although for some time his was also a rather assimilationist Jewish family. Such an attitude of denying one's roots puzzles me especially when a Nobel laureate does it.

At what level of recognition and achievement would such a person feel himself comfortable?

When I was looking for support for Soviet Jewry I learned a lot of lessons in this respect. I had a long conversation with Ernst Chain, the Nobel

Laureate in Physiology or Medicine (1945), who received the prize together with Alexander Fleming and Howard Florey for penicillin. Chain told me that he was a Foreign Member of the Soviet Academy of Sciences, so he refused to help. On the other hand, Melvin Calvin signed the letter and helped in our fight. There were examples of both kinds of behavior. Rabi's justification for not signing the letter was that he was afraid that the Jews in Russia would pay for it. I told him though that I would not do anything like this without first consulting their own representatives, which I did.

So what fraction of your career was in physics?

About a half. I have had three or four parallel careers, but my involvement in physics was continuous from 1960 to the present, whereas other activities came in "phases". Once I started my scientific career I never stopped it. However, the other main "career" was my continuous involvement in the defense of Israel — which was also my motivation in jumping into Israeli politics. The worst year for my science was when I served as Deputy to the Minister of Defense (Simon Peres). That was in 1974–1975. I was President of Tel-Aviv University. I went to some committee meeting and Peres was there. He was Minister of Defense. He sent me a note during the meeting, asking me to become his deputy as Minister of Defense. At that time I was getting more and more involved in university politics and I thought if it should be politics, I would rather be in national politics than in university politics. I resigned from the University and went over to Defense. Once there, I also appointed myself Chief Scientist of Defense when my predecessor in that post left.

Did you have any political affiliation at that time?

No.

Do you still have your party?

No. The party dissolved itself after the election of 1992.

How are you being viewed in Israeli politics?

It is a complicated thing. I believe I am well respected in general. However, not in the extreme left. For instance, I am known to have done a lot for the Israeli nuclear infrastructure project. When the contribution of

Mr. Peres to the nuclear project is mentioned, the left is all praise because "he did it for peace". When I am mentioned in this same context, to the extreme left, I am "Professor Strangelove".

I would like to ask you about Israeli science.

Before I answer, I would like just to mention two other results of mine in physics. I discovered in 1977 the "curved space spinors" (they are infinite-component systems)[17] which were believed not to exist. These are representations of the double-covering of SL(4,R) — which most textbooks used to say does not exist.[18] Another important contribution was my 1965 model for quasars[19] (also suggested by Igor Novikov). It is a precursor of today's "Eternal Inflation" Cosmology (A. Guth, A. Linde).

Now to the role of science in Israel. In 1983, when I was the first Minister of Science I was invited to the exclusive Elsevier "Economist" Club to give a lecture about Israel's science-based industries. This club meets once a month in Haarlem, Holland. Before me the lecturer was

Yuval Ne'eman (right) and Edward Teller during the 13th International Conference on High-Energy Physics held at the Lawrence Radiation Laboratory, 1966 (courtesy of the Lawrence Berkeley National Laboratory Archives).

Helmut Schmidt, the former Chancellor of West Germany and the one after me was Henry Kissinger. We had strong science-based industries in 1983 and since then they have further developed well. They are spread all over the country. They started many years ago next to the Weizmann Institute and next to the Technion. Today there is also a large concentration in the Tel-Aviv area and in other places. They employ people who left the universities, among them scientists who became successful entrepreneurs. As a matter of fact, our key economic strength is in "high tech".

Jewish scientists constitute a large proportion of Nobel laureates, especially in the United States, with Central European background and immigrant parents. Then a few generations down the road, there seems to be no such presence in science. Israel has not produced Nobel laureates so far.

The graduate schools in America in physics used to be full of Jews and now they are full of Asians. I have my explanation about Jews in science. The Jews were out of society until about 1800 and were allowed only to be peddlers. On the other hand, at least all males could read and write. Then came two revolutions, one was the French Revolution, which started Jewish emancipation, and the other was the Industrial Revolution, which created new types of white-collar jobs, such as engineers, accountants, lawyers, and other new professions. European society was generally traditional, with considerable inertia; the sons often following their fathers in their professions. So the new professions were not very popular. On the other hand, the Jews were allowed, just at that time, to move from peddling to other professions. They were also better prepared than many others, with their background in education. They jumped at the new opportunities and by 1900, especially in Central Europe, a large proportion of white-collar professionals were Jewish. That still did not mean science, which was still mostly the privilege of the previous order. Even though the impact of religion was gradually waning, many of the Jewish professors, when they were moving up the academic ladder, had to convert. There are perhaps also other factors. The study of the Talmud prepared the Jewish scientists for abstract thinking. I even have a conjecture about a genetic component. Throughout two thousand years, there was pressure to convert. Those who resisted the pressure were those who were more apt than others to create and live in an abstract world of their own. That may have created some selection. There were more Jews in theoretical areas and less in experimental and engineering fields.

A Dutch author has a book on Jews in science, with a different genetic explanation. According to him, the Christian world for many centuries, between Constantine and the Reformation, and in Catholic communities even longer, bred itself out of talent for abstract thinking. If there was a son who was more theoretically inclined, he was sent to the Church and this was in practice a way of getting the genes out. The Jews had the opposite effect. A rich Jew wanted to marry his daughter to a scholar and they would live at his table and this would breed more scholars.

In present-day Israel the computer science groups in industry are very strong. There we have a large proportion as compared to America and Europe.

Would you say that Israeli science is on a par with their most advanced counterparts in the U.S. and Europe? Israel is small and I am asking about the quality.

Yes, I would say so, although there are weaknesses. Generally, we produce 1% of world science, although we are 0.1% of the world population. The scientific infrastructure is weak. Theoretical physics is very good, there are excellent people in string theory, for instance. Experimental physics is much weaker because not enough is spent on infrastructure. There are highs and lows. I tried to correct it at the time but now it is the opposite.

The Russian immigration helped our physics. There was a decrease in physics graduate students, which was reversed by the immigration. This was so for the other sciences as well. Now they tend to go into business, like everybody else.

A very different question. Do you personally trust any other country besides Israel?

No, if you mean trusting anyone to guarantee our security.

Not even the United States?

For one thing, it would be ineffective. The motivation to liberate Kuwait in 1990 was very strong (oil), and yet it took 6 months of preparations. Had it been Israel, we would have been exterminated in the meantime. Definitely, no. For the fate of the Jews? No. I strongly criticized President Bush's hypocrisy during the Gulf War. I was a minister at that time, member of the Cabinet. Our Prime Minister, Mr. Shamir believed that the safest thing was NOT doing anything rather than doing something. I would come up to him with suggestions for doing something and he would say it was good but not doing it was even better, because not doing it means avoiding mistakes. There was a group of three or four ministers in the cabinet who wanted at least to react to the Iraqi missile showers but the majority was against. Then, one day, after two Saturdays on which we got plenty of missiles on Tel-Aviv, an American general arrived from Schwarzkopf's headquarters. He came for coordination. When he was asked why they were not more successful in destroying the Iraqi missile launchers, he answered, "Sir, I assure you that the Saudis are doing their best". It turned out that for "simplicity's sake", the Americans, who were mainly in the East, south of Kuwait, had charged the Saudis with destroying the Iraqi missile launchers against Israel, which were positioned at the Western end, in the Iraqi Panhandle! You can imagine with what enthusiasm the Saudis were protecting Israel. When this was reported to the Cabinet, I suggested action, including a detailed plan, which was accepted. This time the majority voted with me. Our Minister of Defense, Ahrens then called the U.S. Secretary of Defense Cheney about our intention. Coordination was, of course, important. Cheney informed the President who then called Shamir, gave him hell, and Shamir cancelled the operation. A few weeks later Bush gave a famous speech in which he called Israel "ungrateful" because American soldiers had defended it. But my distrust goes back to John Kennedy, when I was helping PMs Ben-Gurion and Eshkol with their correspondence with the American President about the Nuclear Infrastructure project. We told him that we did not intend to produce weapons, but we did not want to depend on others if someday our enemies would be getting them. Kennedy wrote back that this was precisely what the United States could not allow Israel to do![20]

Now I would like to ask you about your family background.

In the expectation of the Messiah, rabbi Elijah, the "Gaon" (a title given to the "dean" of a rabbinical school, when recognized as an outstanding scholar) of Vilna (around 1790) instructed his students to move to Israel after his death. Our family on my father's side is said to be descended from the Gaon; at least this is the tradition in our family. My mother's father's family derives directly from one of his leading students. They came to present-day Israel in 1807 and settled in Safed (Galilee). They then moved to Jerusalem in 1837 after an earthquake, which destroyed Safed. The first modern press in Jerusalem was founded by one of the ancestors of my mother on her mother's side. My paternal grandfather was one of the 66 founders of Tel-Aviv and of Israeli Industry (he was a self-taught and very talented engineer).

How do you feel, being an atheist, about the political power of the religious right in Israel?

In the middle of the 19th century there was a big massacre of Christians in Lebanon by the Druzes. It was under the Ottoman Empire — and the Western powers intervened to save the Christians. They forced Turkey to give autonomy to each religious community, this is called the "regime of capitulations". When the British took over, these religious communities were autonomous, and the British left this system in place. The Jews were happy with it at the time because it gave us self-administration. When the State of Israel was created, it was a fact of life that all civic matters were in the hands of the community's religious system.

Concerning those small parties — in a democracy, small parties always have greater influence than their proportion because the government mostly depends on them to achieve a majority. So they succeeded in preserving the right of autonomy. It was not created by Israel; it was inherited from previous times.

I, myself, didn't like this as a person. On the other hand, there are people on the left who say that Zionism is a thing of the past and the Jews and the Arabs are in one common state. My position is that Israel is a Jewish state and the only one after two thousand years of persecution. This state should thus care for the Jews — including those outside Israel as well.

In my political career I allied myself with the religious right because the people who live now in Judea and Samaria are presently the true pioneers, doing what the kibbutz movement did in Palestine during the British times. Most of these people belong to religious youth movements. When I created my political movement I spent a whole night with their leaders, I told them exactly where I stood and that I was an atheist. They were pragmatic and interested in practical questions. I told them that I would like them to live their life undisturbed, provided that they did not disturb the life of others either.

How about serving in the Army?

The ultraorthodox refuse to serve in the Army. Immediately after the War of Independence, Ben-Gurion signed a document with the idea that a few brilliant youths would be studying the Torah. Gradually, however, it became a big crowd. It's a shame and I'm waiting for a spiritual leader of the ultraorthodox who will understand that it's a shame not to take part in the defense of Israel. On the other hand, the settlers in Judea and Samaria are mostly of the National Religious Party and all serve in the Army. They make the best units nowadays.

The Rabin assassination was a big shock.

I was very close to Rabin historically. At one time I was his deputy in the Army. It was also when his son was born and he called him Yuval. However, he could be very insensitive politically. The people on the Golan had supported him, but when he started negotiations with Syria, he did very little to explain the situation to them and ask them to be ready to lose their farms for the sake of peace, etc. They were shocked and turned against him. Concerning his assassination, there are always and anywhere some crazy people. However, there are just as crazy people on the left as well. For instance, they demonized Sharon internationally in 1983 after the Sabra- and Shatilah massacres of Palestinian Muslims by Lebanese Christians.

The Israeli security forces made a grave mistake. They could have prevented the assassination. They had an undercover agent who knew everything. He has not been tried yet because the authorities feel very uncomfortable about the whole case. Of course, I condemn the assassination strongly.

How do you feel about the impact on daily life by the religious forces?

Maybe I notice it less than others do because I am conditioned by it, and it has decreased considerably. When I was a child, Friday night was a dead night. Now even cinemas stay open.

Of course, in a modern country you expect complete separation of the state and religion.

There will have to be more of this separation in the future. It is a gradual process and we have to remember, if our forefathers had not been so fanatic about their religion we might not exist today. Thus we must be tolerant. Had I been in Ben-Gurion's place or in that of any other Prime Minister, I would have cultivated a more patriotic religious leadership, more devoted to the existence of Israel, for their own good too. The ultraorthodox are in a terrible economic situation today. They spend their life miserably as Yeshiva students. They have large families and are terribly poor. They should have had a leadership that would open their eyes, tell them to serve in the army and defend Israel. The Bible says nothing about not to fight for Israel. On the contrary.

How do you view the Jews outside Israel who are not religious?

The existence of Israel replaces the need of religion for their identity. Non-religious Jews can identify with Israel from a distance. That is what a large portion of American Jewry does. Otherwise, there is a huge amount of intermarriage, although many of the non-Jewish spouses convert to Judaism.

Your present family?

My wife is a housewife. We have a daughter and a son. Wherever we went, my wife used the opportunity to study. In Princeton, for example, she studied modern Japanese literature. We became both experts because she taught me what she had learned. In Pasadena, she took decorative arts. Our daughter studied art history and later business administration. She is in business now. She is married to a surgeon. They have two daughters; both have served in the army. One is studying law and the other Computer Engineering in Jerusalem. Our son is an electronics engineer, he has been in army research and works now for Motorola. He is also married and they also have two daughters, 12 and 10 years old. The 12-year old is quite brainy, she may perhaps be the next scientist in our family (three of my cousins are physicists, one a mathematician).

Who are your heroes?

When I was young, I was very interested in people like the older Carnot, Lazare. I was torn between science and the army. I tried to join the Haganah when I was 13. I had jumped several classes and I was in the 10th grade at 13. I was very much interested in mathematics and I did some things that I may still write up and publish someday. I finished high school at 15 and joined the Haganah. I took part in my first battle with the British in the streets of Tel-Aviv in 1946. An immigrant ship had just landed and we created a bridgehead to let the people come down and the British wanted to prevent that. Although I was increasingly involved with the military,

I still wanted to do science. My knowledge of history was always very broad. For my Bar Mitzvah, I had asked my parents for the *Encyclopedia Britannica*. This is how I came across Lazare Carnot (1753–1823). Originally, he organized the armies of the French Revolution and gave them their strategic instructions. He was a geometer, became Napoleon's minister of war and of education (he founded Ecole Polytechnique, etc.), but left politics for science in 1807, and did applied mathematics. Then there was a time when Einstein caught my imagination. My heroes have been from different directions.

What will be your legacy?

There will be a part in science and a part in the history of Israel. In 1952–1953 I was preparing Israel for the eventuality of what later became known (when it happened in 1967) as the Six-Day War. There was an article recently in the newspaper *Ha'aretz* with three photographs showing the Prime Minister, the Head of Opposition, and me. The article said that both of them are strong believers in the doctrine, which I established back at that time for the defense of Israel. I wish I could be certain that this is true, even under pressure.

Incidentally, I was appointed head of military planning of Israel by Ben-Gurion although he knew that I had voted Mapam (left of the Labor Party). He also knew that when I was in France, I got disillusioned with the left. My stay in France, at the Staff College, coincided with the Slansky show trial in Prague and that disgusted me. I had read Arthur Koestler's *Darkness at Noon*, and I had almost convinced myself that the aims justified the means. After the Slansky trial, however, I felt otherwise.

What do you do nowadays?

Lately I have been interested in evolutionary epistemology and just published a book on it in Hebrew. I am preparing an expanded English edition. Its title is something like *Order out of Randomness*. It has nothing to do with chaos theory. I start with cosmological evolution, nucleosynthesis, then biological evolution. I classify and characterize the evolutionary process. I distinguish between active and passive evolution: "active" is when you mutate, by a mistake in copying your DNA, and fit better (or worse) into your environment. "Passive" is when it is the environment which mutates and impacts you, like the catastrophe that extinguished the dinosaurs (or gives you an advantage, as it did for the smaller mammals). I have had two main new ideas. One concerns the evolution of society. Ever since *homo sapiens* appeared on the scene, he has not evolved much biologically, but society has evolved. Every evolution must have randomness and selection. Where then is randomness? I claim randomness comes through science. Any really important discovery cannot be foreseen. To move science forward, there must be a lot of dynamics, straightforward research programs. The really important discoveries, however, cannot be put into proposals, they will come by surprise. It also means that a society, which tries to regulate its research people and requires them to keep to planned utilitarian targets is condemned to stagnation. My first point thus concerns the role of science in the evolution of society.[21] The other point is that ideas themselves also develop in an evolutionary process.[22] I give lots of examples for both points. I also discuss other models of epistemology, such as Popper's falsification theory, Thomas Kuhn's paradigms, Lakatos' programs, Feyeraband's "anything goes", and I show how they all correspond to various aspects of my thesis.

References

1. Hargittai, I.; Laurent, T. C. *Symmetry 2000*. Parts 1 and 2, Portland Press, London, 2002.
2. Ne'eman, Y. *Nuclear Physics* **1961**, *26*, 222–229.
3. Goldberg, H.; Ne'eman, Y. *Il Nuovo Cimento* **1963**, *27*, 1.
4. Ne'eman, Y. In *The Interaction between Science and Philosophy*. Elkana, Y. ed., Humanities Press, Atlantic Heights, 1971, 94–105.
5. *The Observer*, 2 November **1958**, 2–3.
6. Abdus Salam (1926–1996) was co-recipient of the Nobel Prize in Physics for 1979, together with Sheldon Glashow and Steven Weinberg "for their contributions to the theory of the unified weak and electromagnetic interaction between elementary particles, including, *inter alia*, the prediction of the weak neutral current".
7. Ne'eman, Y. *Physics Letters* **1979**, *B81*, 190–194; see also Fairlie, D. *ibid*. **1979**, *B82*.
8. Weisskopf, V. *The Joy of Insight*. Basic Books, 1991, 202–203.
9. Lipkin, H. J. *Physics Reports* **1974**, *8C*, 173–278.
10. Gribbin, J. & M. *Richard Feynman, a Life in Science*. Dutton Books, 1997, 192–194.
11. *From SU(3) to Gravity*. Gotsman E.; Tauber, G. eds., Cambridge University Press, 1985, 103–106.
12. *Strangeness Minus Three*. BBC, Third Program film, 1964.
13. Johnson, G. "Strange Beauty," *Murray Gell-Mann and the Revolution in XXth Century Physics*. Knopf, New York, 1999.

14. Gell-Mann, M.; Ne'eman, Y. *The Eightfold Way*. W.A. Benjamin, New York, 1964. Republished in 1999 by Perseus Publications.
15. In September 2000, several groups at CERN found 9 events that could be interpreted as a Higgs meson with a mass of 115 GeV. However, it is only a "2-standard deviation" effect as yet.
16. Ne'eman, Y. *Ha'aretz/Herald Tribune* joint weekend supplement, 8 October 1999, B7.
17. Ne'eman, Y. *Annales de l'Institut Henri Poincare'* **1978**, *28*, 378.
18. Ne'eman, Y.; Sijacki, Dj. *Int. J. Mod. Phys.* **A1987**, 1635–1669.
19. Ne'eman, Y. *Astrophys. J.* **1965**, *141*, 1303–1305.
20. See, Cohen, A. *Israel and the Bomb*. Columbia University Press, New York, 1998.
21. Ne'eman, Y. *Acta Scientifica Venezolana* **1980**, *31*, 1–3.
22. Kantorovich, A.; Ne'eman, Y. *Studies in the History and Philosophy of Science* **1989**, *20*, 505–529.

Jerome I. Friedman, 2002 (photograph by I. Hargittai).

4

JEROME I. FRIEDMAN

Jerome I. Friedman (b. 1930 in Chicago) is Institute Professor of the Massachusetts Institute of Technology (MIT). He was one of the three recipients of the Nobel Prize in Physics for 1990, the other two being Henry W. Kendall, also of MIT, and Richard E. Taylor of Stanford University. The citation said, "for their pioneering investigations concerning deep inelastic scattering of electrons on protons and bound neutrons, which have been of essential importance for the development of the quark model in particle physics". Jerome Friedman received all his degrees in physics at the University of Chicago: A.B., M.S., and Ph.D. in 1950, 1953, and 1956, respectively. He worked in the Enrico Fermi Institute at the University of Chicago and at Stanford University before he joined MIT where he has been since 1960. He has been a member of the American Academy of Arts and Sciences (1980) and the National Academy of Sciences of the U.S.A. (1992). He received the W.K.H. Panofsky Prize of the American Physical Society in 1989 and was President of the Society in 1999. He has been a member of numerous committees and panels. We recorded our conversation on February 5, 2002, in Dr. Friedman's office at the Massachusetts Institute of Technology.*

You entered the University of Chicago at the age of 17. Robert Hutchins was President of the University of Chicago and his name has come up in several of my interviews.

---

*István Hargittai conducted the interview.

Hutchins developed a radically new undergraduate educational program for the University of Chicago. It was a college of liberal arts that I consider extraordinary. I enjoyed it tremendously. The College, as it was called, worked in the following way. One could enter the College after the sophomore year of high school. All entering students took placement examinations. Depending upon how one did on these examinations, one could get a degree immediately or spend up to four years in the College. The College program was based on reading original materials. There were no textbooks interpreting the material we read. We had syllabi to suggest questions and points of view that we could think about as we read the material, but that was all. We would read the material and then have wonderful discussions in class. The instructor would fill in some details about the historical context, lead the discussion, and raise important issues. That was my undergraduate education. The courses I took covered the humanities, social and political science, philosophy and western history. During the year, there were only diagnostic examinations without real grades. At the end of the year we would have a six-hour examination in each subject, which was called the comprehensive, covering the entire year's work. If we were not satisfied with our performance on a comprehensive on a particular course, we could take it the next time it was given, provided we paid $25. The highest grade would remain in your transcript. The comprehensive generally consisted of multiple-choice questions, but sometimes involved writing an essay. The College was a wonderfully intense intellectual environment that I found highly stimulating, and I look back at it as one of the high points of my intellectual development. We were totally immersed in what we were studying. Often when we would go to the local bar to have a beer to relax, we would end up talking about Aristotle, Plato, or whatever we were reading.

If the atmosphere was so important, why did the system allow students to sail through it without experiencing the atmosphere, just by virtue of the placement examinations?

I think that the objectives of this educational program were to provide knowledge in a variety of subjects and have the students develop the ability to analyze material and exercise critical thinking. The placement examinations tested these abilities. I assume that the idea was that passing this exam implied that the student had had an equivalent educational experience in the past.

How long did it take for you?

I was required to take a year and a half of College courses. But I started taking some mathematics courses, which were not part of the College curriculum, while I was still in the College. By the time I received my bachelor's degree, I had spent two years in the College.

Instead of four.

I wasn't an exception, many people finished in two years or less.

When was this?

In 1950, I received an A.B., the Bachelor of Art degree; and I immediately entered the Physics Department. It had an accelerated program, which was very unusual. The department let just about everybody in who applied, but it had a high attrition rate based on a series of exams. I had no credentials whatsoever. In high school, I was an art student and I took very little mathematics. I had one course in physics that was very poorly taught, so I was totally unprepared in physics and mathematics compared to most of the other students — despite the few math courses that I took while I was in the College.

What turned you to physics?

My father who had little formal education was always interested in science and would often discuss scientific matters with me. Despite his influence, I was still going in the direction of wanting to be an artist. As a child, I spent much of my time drawing and painting. In high school, I entered a special art program imbedded in the regular curriculum, in which I spent two to three hours a day doing art work. In fact, I got a scholarship from the Museum School of the Art Institute of Chicago upon graduating from high school. But when I was at the end of my junior year in high school, I visited the Museum of Science and Industry in Chicago and I happened to wander into the bookshop. I picked up a book by Albert Einstein entitled *Relativity* and I was absolutely fascinated by it. After reading it and trying to understand it, I realized that there was so much that fascinated me in the physical world that I didn't understand, and would like to understand, that I had to have more education. I had a difficult decision to make when I finished high school. But against the strong advice of my art teacher, I decided not to accept my art scholarship and instead entered the University of Chicago.

So it was Einstein's Relativity *that turned you to science?*

Absolutely. There is no doubt about it. Of course, Einstein was one of the gods in our house. He was an absolute luminary. I thought that this book might give me some understanding of the mysteries of how meter sticks shrink and clocks slow down when they move fast — things that I had read about in popular articles. I read the book carefully and tried my best to understand these matters; but in the end, I really didn't understand the basic concepts of special relativity. This only made me more curious and more determined to try to understand them. It was clear to me that I would have to study physics to really understand these ideas.

Did your father read it?

My father was an avid reader, and our house was filled with books; but I don't think he ever did. He clearly appreciated Einstein's monumental achievements as they had been portrayed in the press. Relativity had been popularized in many different ways and Einstein was universally admired.

How far did your parents witness your career?

My father died right after I received my Master's degree. My mother lived until I got my Nobel Prize; but unfortunately, by then she was in a nursing home and was somewhat senile. When I was in Stockholm, a friend of ours went to see my mother and brought her a photograph of my receiving the Nobel Prize. I'm not sure if my mother understood its significance, but our friend told me that she smiled when she saw the picture. She died the next night. I came back from Sweden and had to go immediately to the funeral home to make arrangements. So she perhaps knew it, but I'm not sure. My father would have been absolutely elated, had he known. When I consider the kind of environment in which I grew up, I think it would have been totally inconceivable to my father or to anybody else around me that I might be awarded the Nobel Prize. I grew up in a poor neighborhood in the west side of Chicago. The public schools were poor and there were bad influences on the street. My parents had severe financial problems. It was the Depression and often my father couldn't come up with the rent. On one occasion, the landlord wanted to evict us from our apartment, but finally decided to let us stay because if we moved out and another family moved in, they probably wouldn't be able to pay the rent either.

When was it exactly?

In the 1930s and early 1940s. In looking back, I must say that I was very fortunate that my parents always gave me strong encouragement and emotional support. They had little formal education, but for them education was of the greatest importance. They greatly respected scholarship in itself, but they also impressed upon me the idea that there are great opportunities available to those who are well-educated. They always told me to get a good education and do something constructive.

Did you have siblings?

I had an older brother, Harold, who tragically passed away about 30 years ago at a young age.

Did you pay tuition?

No, I had a scholarship. It would not have been possible for me to attend the university without such help. I lived at home for my first two years and my parents supported me during this period — which they happily did despite economic difficulties. They were always very supportive. When I was young and they couldn't really afford it, my mother wanted me to take violin lessons, which I took for a year and a half. Then I decided that I really liked drawing and painting better, so she arranged for painting classes for me every Saturday at the Art Institute of Chicago.

That means she was not very religious.

That's right. She kept a kosher home and observed other aspects of Judaism, but did not strictly observe the Sabbath. Every Saturday, she would take me to the Art Institute of Chicago and would patiently wait for me while I was in my painting class. I was very fortunate in having parents who supported me like that. When I entered graduate school, I got a job at the cyclotron as a cyclotron operator. At that time, my father developed cancer and was unable to work. He became quite ill; and since I was earning a little money during this period, I could help support my parents. This was in the early 1950s and they were very tough times.

How much do your children appreciate your experience of these times?

It's hard for them to understand such conditions. They can't understand what poverty is really all about in a direct sense. They can see it on the streets, of course; but that's different from experiencing it. The only way for them to really understand it is to go through it; and that is something I wouldn't want to happen. I'm happy that our children don't know it. When I was growing up, one could be working very hard and still be very poor. Having been a Depression child, I have always been cautious with money. That is, of course, different from the feeling of not having enough money for necessities. One can be careful about discretionary spending, but the hard thing is not having money to buy the necessary things.

Getting back to your studies, you entered graduate school at the age of 20. Was it hard?

It was a hard program but it was especially difficult for me because I did not have the proper background. As I said, the Department admitted just about everybody who applied, but it had a very high attrition rate. The program was very difficult, and a number of people left because it was just too tough. As I recall, in the first course I took, a year-long course of introductory physics, there were initially 125 students; but only about 35 completed it. After two years, one had to take the qualifying exam, and only about one half of the students passed it. One year later, there was the Basic exam, which qualified the student for Ph.D. research; and again the attrition rate was about one half. It was truly a tough program, but it was certainly worth the great effort it took. And I feel fortunate to have gotten through it. I received an exceptionally good education, and doing research in the Institute of Nuclear Science was an exhilarating experience. The physics department of the University of Chicago was one of the very best in the world at that time. There might have been comparable ones, but there weren't any better. Fermi was the intellectual leader of the department and he had assembled an extraordinary faculty. Many of the students and postdocs were also extraordinary.

I've read that Fermi was a kind person but very tough in choosing his students because he considered being a mentor a big investment of his time and efforts.

That's not my experience, and let me tell you about it. As I said, I had trouble going through that system because I didn't have the right background. I could make no claim to being an outstanding student in the department, because I wasn't. I was just part of the group, but it was the group that survived the series of tough examinations.

Did you meet Fermi?

I took a number of his courses, but that doesn't mean that I really knew him at that time. I knew that he was a truly great physicist and a superb teacher. When I had passed the Basic exam, I decided to ask him whether I could be one of his students. While I wasn't optimistic, I didn't think I could lose anything by asking. After all, it is no affront to be turned down by such a great man. So I went to ask him, and he said yes immediately.

How many students did he have at that time?

Five or six. When he said yes, I couldn't believe it. It was like winning the lottery, and I was absolutely overjoyed.

Who else was there whom you could've chosen?

I could have worked with a number of other physicists. I had had a good deal of contact with Valentine Telegdi, and Leona and John Marshall.

Was Edward Teller there?

He was, and I took a course from him. But he was not a potential research supervisor because I was an experimentalist. And I was an experimentalist because of Fermi. Fermi had told us in a class that in the next 10 to 20 years there would be great developments in experimental physics and he advised us all to become experimentalists. Prior to this, I had the aspiration of becoming a theorist. Despite knowing very little about physics, I had had the *chutzpah* of thinking of Einstein as a role model. However, after Fermi gave that talk, I decided to go into experimental physics. So did most of my classmates. Only one or two people in that class went into theory. This is how I decided on my direction in physics. As I said, the research environment at the University of Chicago was just spectacular. There were wonderful talks, the greatest physicists in the world would come to visit Fermi, and we had the synchrocyclotron, which was the most energetic accelerator in the world at that time. This was the period in which Fermi and his group discovered the so-called three-three resonance, which was the first excited state of the proton.

It must have been a terrible let-down when Fermi died in 1954.

It was a tragedy.

Was it a prolonged illness?

No, it wasn't. He gave a course in the spring of 1954. It was on quantum mechanics and I sat in on the course. When Fermi gave a course, I always sat in on it regardless of whether I had taken it before or not; because I could always learn something from this great man. Fermi was a robust man who appeared to be in excellent health. But he went away that summer to Italy and developed a rapidly growing form of stomach cancer. I saw him when he came back in September. He was about 50 feet away from me in the corridor as he was going into his office. I waved to him and he waved back to me. I looked at him and was terribly startled to see how gaunt he looked. The next day he went to Billings Hospital, where he had exploratory surgery and was found to have inoperable cancer. I never saw him again. He died on November 28. What a loss it was. Besides being a great physicist, he was a very kind and considerate man; and he had enormous patience in explaining physics to his students. He was a wonderful human being.

For my thesis research, Fermi had suggested that I carry out a nuclear emulsion investigation of proton polarization produced by nuclear scattering, an effect that had been observed at cyclotron energies. The objective of this study was to determine whether the polarization resulted from elastic or inelastic scattering. I did not know at the time that Fermi had already theoretically shown that elastic nuclear scattering could produce large polarizations. This calculation was in his famous notebook of problems that he had investigated and solved. The calculation was, as usual, based on a simple model, utilizing a real and an imaginary nuclear potential and a spin-orbit coupling term. This is the same term that he had suggested to Maria Mayer as possibly playing a role in the structure of the nucleus and which was crucial to her development of the Shell Model.

When I had only partially completed scanning my emulsion plates, Segrè visited Fermi and told him that he had observed large polarizations in nuclear elastic scattering in a counter experiment at the Berkeley cyclotron.

According to Segrè, on the morning of his visit, Fermi calculated the polarization expected in this experiment, and his results matched Segrè's measurements beautifully.

I had been scooped and was quite dejected. However, Fermi was very understanding and suggested that I continue my measurements. First, it would be valuable to confirm Segrè's results with another technique; and secondly, I could also determine to what extent inelastic scattering produced polarization.

When Fermi died, I was devastated. What an immense loss it was to all of us. My thesis work was not yet completed, and I clearly didn't want to start all over again on another problem with another professor. John Marshall kindly came to my rescue. He took over my supervision and he ultimately signed my dissertation.

After I received my Ph.D., I continued working in Fermi's nuclear emulsion laboratory, which had been taken over by Val Telegdi, who was an outstanding young faculty member. At about that time there was the so-called tau-theta paradox in which there appeared to be two particles of opposite parity having the same mass. These puzzling results were causing much controversy and speculation in the particle physics community. In a bold paper, Lee and Yang proposed that this paradox was due to the non-conservation of parity in the weak interactions and suggested some experimental tests of this hypothesis.

While most of the community considered the conservation of parity to be sacrosanct, Val was quick to pick up the significance of this paper. He asked me to join him in making a measurement of muon decay in nuclear emulsion to test this radical new idea. Following the suggestion of Lee and Yang, we planned to measure a forward-backward asymmetry in muon decay, which would be sure sign of the violation of parity conservation. In my thesis work, I had developed some expertise on how to make double blind visual measurements of asymmetry. Most of the others in our lab thought that this was a waste of time. I remember giving an Institute seminar on the measurement we were going to make. After the seminar, a distinguished older member of the faculty came up to me and said that I had given a nice talk, but that I should realize that we were not going to find anything.

As it turned out, we were one of the first three groups that demonstrated the non-conservation of parity in the weak interactions. Madame Wu and collaborators were the first to demonstrate this effect in their measurement of the decay of cobalt-60; and Garwin, Lederman, and Weinrich also did so in their measurement of muon decay. These two beautiful counter experiments demonstrated the effect with excellent statistics. While many perceived this as an experimental race, it really wasn't so from our perspective. When we started our experiment in the late summer of 1956, we knew of no other measurements going on at that time to test the non-conservation of parity. Our progress was hampered by the following circumstances. Val had to go to Europe during the early autumn of 1956 on personal matters and remained there for about two months. During this period, I was starting to see a hint of an effect and I wanted to get more scanning help. But the physicist left in charge of the emulsion lab wouldn't give them to me, because the only scanners available were involved in what was thought to be a more promising measurement. Only when Val returned did I get more help. We heard about the other two experiments when our scanning was close to being complete and we were already seeing an effect. All three publications appeared within a short time of one another.

In addition to being a physicist of deep insight and strong opinions, Val has a wonderful sense of humor and an inexhaustible supply of jokes. He was an excellent mentor, and I learned a great deal from him. He also helped me get my first real job, a three-year postdoc position with Robert Hofstadter at Stanford University.

At that time, Hofstadter was conducting his famous electron scattering studies of the proton and other nuclei, for which he was awarded the Nobel Prize. It was there that I learned the techniques of electron scattering and where I met my long time co-worker, Henry Kendall, who tragically died a few years ago. Henry and I established a partnership within Hofstadter's group and worked closely together on a number of different experiments. It was also at Stanford that I met Dick Taylor, who was then finishing his Ph.D. thesis under the supervision of Bob Mosley.

In 1958, Martin Deutsch, Henry's Ph.D. supervisor, visited Stanford and told us about research and employment opportunities at MIT and encouraged us to apply for faculty positions. Of course, MIT was a very attractive place, but what really drew us here was the Cambridge Electron Accelerator, a 6 GeV MIT-Harvard synchrotron that was soon to come into operation on the Harvard campus. We both applied and were exceedingly pleased to have been accepted. I came to MIT in 1960 and he in 1961, establishing our own small high-energy group in the Laboratory for Nuclear Science. I could not have asked for a better scientific partner. From early on, it was quite clear that Henry was an outstanding experimental physicist.

He had a superb command of the technology of the field and an enormous enthusiasm for doing physics. He was an ideal collaborator. One could always work out scientific differences with him, and such differences never became personalized. Though he had strong points of view, he had a great openness for new ideas, irrespective of their source.

The construction of the Stanford Linear Accelerator began in the early 1960s. This new accelerator presented an opportunity to explore electron proton scattering in a new energy domain that provided unprecedented resolution to study the proton. But electron scattering was not regarded by most of the high-energy community as a promising tool to unravel the mysteries of the proton. We were advised by friendly colleagues that pursuing a program at SLAC would not be a good idea. Nevertheless, Henry and I teamed up with Dick and others to design and build an electron scattering facility at SLAC, which was then being constructed. The two huge spectrometers built for this facility were probably the largest measuring instruments ever built for physics research up to that time. We later carried out a series of experiments from 1967 to 1974 that demonstrated that the proton and neutron had point-like spin 1/2 constituents. Using our results and neutrino scattering measurements made at CERN with the Gargamelle bubble chamber, it was clearly demonstrated that these constituents had fractional charges compatible with those of the quark model. These experiments validated the quark model and ushered in the theory of quantum chromodynamics.

These were the most exciting days of my professional life. Before we started, the quark model was in dispute. Quarks had been searched for in accelerator and cosmic ray experiments and in the terrestrial environment, and not found. And there was great skepticism about their fractional charge assignments. After all, there were no other fractionally charged particles found in nature. In addition, bound states of quarks to make up certain known particles violated the Pauli Principle, because this was before the property of color was established for quarks.

We were not thinking about quarks in planning our inelastic scattering measurements. The idea was to explore inelastic electron scattering over a wide range of inelasticity at a much higher energy than had been done previously. We thought that this might shed some new light on the structure of the proton. Some of our collaborators, who worked with us on the earlier elastic scattering measurements, thought that this was such an unpromising program that they dropped out. When we started this program, we quickly noted that something unusual was happening. Our counting rates were a factor of 10 greater than that expected on the basis of rudimentary model based on the picture of the proton prevalent at the time. At first we could attribute this to the crudeness of the model; but as we kept increasing the four-momentum transfer of the scattering, the factor rose to 100, then to 1000 and then to 10,000.

We realized that we were onto something new. I made a plot of the data versus four-momentum transfer that suggested there was point-like structure in the proton. Such a conclusion would have been considered quite bizarre in those days. When I was going off to Vienna in August of 1968 to attend the International Conference on High Energy Physics and present our preliminary results, my colleagues instructed me not to mention anything about the possibility of point-like structure in the proton. Of course, I followed their wishes. However, Wolfgang Panofsky, the Director of SLAC, suggested this possibility in a brief, cryptic comment in his plenary talk. But I don't think it made much of an impression on the audience. Despite the strong doubts of most of the community about the quark model, a continuing series of experimental results made its acceptance inescapable. And by the late 1970s, the quark model became the basis of planning experiments and developing theory and became one of the cornerstones of the Standard Model.

Henry, Dick and I worked well together; and the other physicists working with us, all young and many of them students, were extremely capable and pleasant individuals. There was a wonderful camaraderie among us. We had a highly compatible and well-functioning team. I regard the awards that the three of us received as honoring the achievements of the entire group. And when the three of us were awarded the Nobel Prize, we invited the entire group to accompany us to Stockholm to participate in the celebration. It was a great party.

Panofsky, called Pief by people at the lab, took a great interest in our electron scattering program. He was a collaborator in its planning and construction phases, but had to drop out when we started taking data because of his heavy responsibilities as Director of SLAC. Pief was the best laboratory director I encountered in my career. His leadership paved the way for the many successes of SLAC. I have always been impressed by his wide ranging knowledge and understanding of both the technical and theoretical aspects of physics, and also by his great wisdom in making decisions. He was a strong director but an unusually pleasant and jovial one. I can remember once going to his office to make a request; and despite his turning me down, I walked out with a smile.

Since completing our electron scattering program at SLAC, our MIT group has worked primarily at Fermilab on a series experiments covering a range of topics. Over the past decade, we have worked as collaborators in the CDF group, which has about 500 members and is studying proton-antiproton collisions at an energy of 2 TeV. As members of this collaboration, we participated in the discovery of the top quark. Our MIT group is also presently a part of the CMS collaboration, which is constructing a huge detector to be employed at the Large Hadron Collider. The Large Hadron Collider is a 14 TeV proton-proton being built at CERN and is expected to be commissioned in 2007.

How serious is your painting?

I'm just an amateur painter but I enjoy painting very much.

Have you ever sold a painting?

No.

Would you have liked to sell one?

No. I paint to satisfy myself. Selling is not the objective. Right now I only do watercolors. I also like oils, but oils take much more of a commitment than I am able to make at the present time. Oils take a good deal of

Jerome Friedman and his wife, Tanya, in Stockholm, 2001 (photograph by I. Hargittai).

set-up time; and because of the drying time required in the process of over painting, it generally takes an extended period of time to complete a painting. With watercolors, I can do a painting in one or two hours and the set-up is minimal. But I haven't painted for about a year and a half. When I go on a vacation, I like to find a nice spot and spend a number of days there painting. When I retire, I want to spend much more time painting.

Do you plan to retire?

Yes. But after retirement, my activities may not change all that much because I really enjoy what I am doing. One of the nice things about MIT is that you can retain your office after you retire and continue doing research. I plan to stay here and participate to some extent in the research program of our group, but in a somewhat peripheral way. The main reason for retiring is to make room for a younger and more energetic person in the system. Of course, it also allows one to slow down a bit and to turn one's attention to other activities. I will be able to take time off to pursue other interests such as painting and personal travel.

So you have not retired yet? Do you still give courses?

I have not retired yet, but I presently am not teaching because of my heavy travel schedule. I am a member of many advisory boards, panels and committees, which forces me to travel continuously. Being an Institute Professor, I can take time off from teaching when other responsibilities do not allow me to teach. After teaching at MIT for nearly 40 years, I do miss teaching.

What will be the difference between now and retirement?

The difference will be that I'll be paid by my pension rather than by MIT; and as I said, I will take some days off and cut down on some of my activities so that I can pursue other interests.

What is it that you feel you do too much?

I enjoy what I do, but I probably have consented to being on too many committees and advisory panels.

Will MIT benefit more from your retirement? Will they hire somebody in your stead?

Absolutely. Young people are the life-blood of research. Their energy, passion and creativity are needed to continue teaching and research at MIT. All systems have to be renewed.

Don't you have to apply for grants?

I am currently group leader of one of the high-energy groups at MIT. It is my responsibility to oversee the preparation of the documentation necessary to apply for research funds for our group. There are very talented young assistant professors and students in our group, and I work to help get funding for their research. I see my job as helping them achieve their research objectives. I am not in the main line of research anymore, but function more as an advisor, mentor, and facilitator. When I was young, others did this for me.

Martinus J. G. Veltman, 2001 (photograph by M. Hargittai).

5

MARTINUS J. G. VELTMAN

Martinus Veltman (b. 1931 in Waalwijk, the Netherlands) is Professor Emeritus, John D. MacArthur Professor of Physics at the University of Michigan. He received his Ph.D. from the University of Utrecht in 1963. He was a fellow at CERN and Professor of Physics at the University of Utrecht. He has been at the University of Michigan since 1981 till his retirement in 1997. Now he lives in Bilthoven, the Netherlands. He received the Nobel Prize in Physics in 1999, together with Gerardus 't Hooft, "for elucidating the quantum structure of electroweak interactions in physics". He is a member of the Dutch Academy of Sciences (1981), the National Academy of Sciences of the U.S.A. (2000) and Fellow of the American Physical Society (1984). He has served on policy committees of all major high-energy physics laboratories in the world. We recorded our conversation in Professor Veltman's home in Bilthoven on March 18, 2001.*

What turned you originally to science?

I could not really tell you; it's a complicated affair. Originally I was more interested in technical things, like making radios or making electricity. That is something that's built-in in people; some people like it and some don't.

---

*Magdolna Hargittai conducted the interview.

I liked it from an early age and my uncle gave me a present that worked with batteries and was connected with lamps. I liked it very much and I went on in a technical direction.

Why didn't you become an engineer?

That's funny because that is what I wanted to do, but my parents were not very rich. Everything was a pure accident. First, I wanted to go to a technical school in a nearby town, but for some reason I could not get into that school in that year, so there would've been a one-year delay. Then my high school teacher came along and told my parents that it would be a waste for me not to continue my studies. Then the issue was simply decided by the fact that the University of Utrecht was the nearest place where I could go. I would have preferred to go to Delft, where there is a Technical University. But that was too expensive and so I went to Utrecht, which was doable. I lived at home in Waalwijk with my parents and commuted every day to Utrecht by train. So for a small amount of money I could go there. Thus the only rational reason that I went in this direction was provided by my high school teacher.

Apparently he saw something in you.

He did.

High-school teachers sometimes play an important role in the careers of future scientists. Eugene Wigner mentioned his high school teacher in his Nobel talk.

So did I.

Was it physics that you started to study in Utrecht?

Yes. We had to choose one major subject and two others as well. So I took physics as my main field, and mathematics and chemistry as the two others. Mathematics was important; you can't really do physics without that. As to chemistry, I was a total zero there and remained a total zero, and the way I got through my chemistry exams — I rather not talk about.

How did you wind up with theoretical physics when, based on your original technical interest, experimental physics would have been more logical?

It was a gradual process. At that time, in Holland, I could not make a choice between theory and experiment until I completed my degree that more or less corresponds to the bachelor's degree in the U.S. Generally, it was about three years, but in my case it was five years because I lost two years. Then, I chose experimental physics and I did that for a year. Then I decided to switch to theory because I did not like the type of experiments that I could join in Utrecht. There are also some disadvantages to doing experiments.

What kind of disadvantages?

Experiments require more time to get some results — if you start something you have to keep doing it at least for a year or two before you get results; while with theory if you have a good idea you might get results much sooner and can get on to the next problem. A theorist does his experiments on paper, it is very easy. For an experimentalist, there are so many peripheral conditions — getting material, getting money, etc., etc. Once he has made his choice he is tied to it for a number of years as a rule and I disliked that very strongly. I also discovered that I was not as good at making instruments as I have seen others making them.

As a 4-year-old boy in the backyard of his home (courtesy of M. Veltman).

Why particle physics?

Oh, particles are very interesting fundamental things. Once I got into theory, it was a one-way street. There was no question about it, ever. I started immediately learning quantum mechanics, relativity, and so on. To get access to all these fields was not that easy in Utrecht at that time but I went for it immediately.

Did you have good teachers on these subjects?

No, there were no good teachers there. Later on, yes, but not at that time. The situation was rather bad in Europe, all the good ones had left. Holland is a small country and one or two good people may be enough for a given field, but they were not there at the time. There used to be a few good people, that was the heritage of Hendrik Lorentz, but by then, the only one left over from that heritage was Kramers, a professor in Leiden. Here, in Utrecht there was a professor named George Uhlenbeck, who went to the U.S. already before the war. There was another professor, called Abraham Pais, also in Utrecht. He was here during the war, he "went under," as we say it, because he was a Jew, but he survived somehow. After the war he went to the U.S. to the Institute in Princeton, so why should he have come back to Utrecht? So when I started in fundamental physics, there was just no one around. There were some people who did

Abraham Pais (1918–2000), 1992
(courtesy of M. Veltman).

statistical mechanics, notably van Kampen, pupil of Kramers, but nobody did particle physics here. There were also no experiments done in that direction. It was a very unlucky situation the way it developed in Holland; till about 1960, there was really nothing.

How could you work then in this field? You needed an advisor, for example.

At some point I had to do a thesis. My thesis advisor was Leon van Hove, but he did not know anything about particle physics. So I started to study particle physics under his guidance, but he himself did not really know the subject. Then he switched directions and he became the director of the theory division at CERN and then he started doing particle physics himself. He took me with him to CERN, and that's where I finally got into a particle physics environment. Van Hove didn't really make a good switch because he never achieved anything comparable to the way that he was in his previous field; it is claimed that had he not died he would have gotten the Nobel Prize for neutron scattering. But he died; one of the mistakes you can make if you want the Nobel Prize.

Would you mind explaining to, say, a general scientist what gauge theories are?

It's very hard to say, it needs a longer introduction. But the easiest way of putting it is that in particle physics you look for the forces between particles; there is more than one force. Each of these forces, when looked at by themselves, theoretically, has defects. When you look at a process at very high energy, the forces you get are far too strong, stronger than reasonable. You get probabilities that start exceeding one. But you can actually have these forces be tuned to one another in such a way that the bad things cancel out. Let me give you a (non-realistic) example. In electromagnetism, there is a proton and there is an electron and the electron is attracted by the proton. When they get close enough, this attraction becomes very very strong, unbelievably strong; as strong as you wish. When the proton and the electron come very close, other interactions come into play; like weak interactions. If you want the interaction between the proton and the electron to be finite, you can make the weak interactions to be repulsive so when they get sufficiently close, the forces may start balancing each other. So you can balance forces in order to avoid "bad behavior" at small distances or at high energies, which is the same thing. This balancing of forces, that has to be done in many places — this was just one example — is mainly what you do by using a gauge theory. In a gauge theory all the forces are tuned to each other. They are tuned in such a way that at small distances and at high energies they remain relatively mild and do not explode.

Who gave the name "gauge theory"?

I think it goes back to Hermann Weyl. It was a wrong name for all I know. I think he wanted to reformulate gravity as a gauge theory and the English word "gauge" refers to the meaning of measure. In gravity, that was the way it came about, and he did that by deducing some sort of symmetry; so that name was subsequently used in another context that had nothing to do with gauging; it was just a symmetry idea. So the name was derived in another context by Weyl. I think this is how it happened but I am not a historian of gauge theory.

How did you pick this subject?

When I entered particle physics, at the beginning of the 1960s, there was this domain of weak interactions, for which the theory was in very bad shape. You had infinities all over; you had bad behavior for high energies. Everyone in the field was interested in making a correct theory for weak interactions, not only me. T. D. Lee spent a lot of time on this and so did Richard Feynman, Murray Gell-Mann, but they never got anywhere. This had been a dream for me since about the beginning of the 1960s; to solve this problem. We all wanted it but the idea to get a good theory with gauge theory did not occur. In fact some people tried something this way but not in a serious manner; it was too difficult. Finally, sometime around 1968, I decided to work on this seriously; trying to chase the old dream. I thought of using this particular idea of harmonizing the forces and making them cancel against each other.

Initially a few people realized at this time that the theory was at least an order of magnitude more complicated than the existing theory of electromagnetism. This manifests itself in this way. In electromagnetism we have a photon interacting with an electron and this interaction is very simple, just one term. In weak interactions, as in a gauge theory, or Yang-Mills theory, which is another name for it, you find that there are several vector bosons instead of one photon. When you write down these interactions, you have 6, 7, or even 8 terms in the equation, and the calculation becomes very complicated. It is much more complicated than electromagnetism. Then, you don't have only one particle, but many; at the last count, as I remember it, there were 20 or more particles, and you have to consider all their combinations. You could have an interaction with particle A becoming particle B by emitting particle C, and so on. It's immensely more complicated than electromagnetism. I think that most people, when they realized the complications they got into, just went away from it. So, in 1968, when I looked at it, I decided that it had to be done. I started working out the details. Getting over the threshold was an important part of the process. After you have done it, it becomes relatively easy, but actually doing these calculations was very hard. I think I may have been the first person going over that particular threshold. I don't think that anyone ever had done this calculation but I am not sure, maybe Feynman had done some work on this. But actually doing them systematically and understanding them, I think I was the first and that made a big difference. This happened in April of 1968; I went to Rockefeller University for a month and that's where I did it.

When did your student, Gerard 't Hooft come into the picture?

That was sometime in 1969.

What was his contribution?

When I was busy in the field, I wanted to see if things were all right, so I started computing things. Here I need to be a little technical. In field theory, you calculate things in different steps, in orders of perturbation theory. I had proved that things were all right in the lowest order of perturbation theory, the one-loop approximation. However, I did not succeed in getting to the next order. I realized that there was a way of treating the theory that was better than the way I had learned. This was by using the path-integral method invented by Feynman, but I was not familiar with that. But I realized that this method would be very useful to study this subject. So I started lecturing on this subject in Utrecht. That was the stage when 't Hooft came in. He was my student that time and he prepared the lecture notes and, being younger and being an unspoiled person, he learned all these methods better than I knew them. He is extremely gifted in the way of mathematics. When he started to work on this system, he used those methods I mentioned and that helped him a lot. He became better in the mathematics of the standard model than I was; there is no question about that. He discovered that if he added an extra particle, the Higgs particle, he could do the calculation in two loops. So the way I was looking at the theory, from today's point of view, was incomplete, I did not have this extra particle, the Higgs particle. I could balance all difficulties, except a little bit that I could not and that's what we needed the Higgs particle for, and 't Hooft found out that piece of it. We did not call it the Higgs particle; we just found out that we had to have a particle. It was later that we found that Peter Higgs and two other people, Robert Brout and François Englert, already had it. We should really never say Higgs alone, but that name has stuck in the community.

But everybody always does.

That's wrong. It is my fault, actually. When 't Hooft finished this article, I went back to CERN and I looked up people who knew all about literature. I am not very good in the science literature, but someone mentioned the work of Higgs to me. So even without seeing the work of Higgs I did put a reference to him in 't Hooft's work. I was his thesis advisor and this was his thesis work. I changed some things in his thesis and I told him that he had to put a reference to Higgs. Had I done a proper literature search, I would have found out that there were also Brout and Englert. So I put that reference in there and that is what has survived, that's where Higgs' name became associated with this subject, from 't Hooft's article.

So this is how the particle got its name?

Yes, that's the way things usually go; although Brout and Englert did their work even a little earlier than Higgs. Later on I very much regretted this and ever since I have been working on getting credit to Brout and Englert as well, and gradually it is coming around. People know it now.

This Higgs particle has not been found yet.

No, it has not.

Do you expect it to be found?

I don't know. It has a function in the theory; it has to balance out some remainder. Most of the cancellations I have worked out initially, but there was some remainder that I could not master and then 't Hooft did that last step. But, experimentally, we have not seen that particle yet. This Higgs particle has some funny disadvantages as well, theoretically. It does something funny with the theory of gravitation. For this reason I distrust it, others don't. We'll just see what Nature does; we'll know in another ten years or so.

Leon Lederman called this particle "the God particle".

I don't like that name.

When it's found, will it be the final answer?

Oh, not at all. This is complicated. The theory we have needs it. To make the theory closed, if that particle is found, it may be the last stone in that particular house. But there is a problem. The problem is that while we can now describe exactly all the forces between the particles, and they are all fine, there are, however, still so many things that we do not understand. There are three generations of particles and we don't know why. There is trouble with gravitation and we don't know why. As long as the Higgs is not there, you can have some hope that when you start to investigate it, you will find the key to something else. If they find the Higgs, as the theory naively supposes, in a sense, that closes the door to other possible things that we don't understand. Then the theory is closed and we'll have no idea about all the other problems. All you can do is to do more and more experiments but you don't know where to go.

When you say that if they find the Higgs particle, the door will be closed, you do not mean that particle physics would then be over?

Oh, no. The theory will be closed in a sense that Newton's law is closed. In Newton's law there is nothing, not a hint, about elementary particles and things like that. So if you are investigating Nature, Newton's law won't tell you anything about elementary particles. It's a closed system. It is perfect in itself, but there are many issues it does not address. The same goes with the Standard Model. It's a closed system. It has its pillars and its stands, but there are still so many questions and the Standard Model provides no clue.

Is the Standard Model the right model for the Universe?

It certainly is the right model for a part of it, as we know it. But there is another layer of complications that apparently we don't know about. The Standard Model is a piece of phenomenology, there is too much in it that you put into it, and you would like to know why. For example, the three generations in the Standard Model — we just say we need them, but we don't know why. There are many things in the Standard Model that are there but we don't know why. It is just like a house with five floors; would you not like to know why there are exactly five floors and not six or four? Why is one so small and the other so large; we have understood all the architecture but we don't know why this size.

The Standard Model with all its particles and forces seems to be rather complicated. Many people don't like it and they say that Nature prefers simpler rules, so perhaps there is one, we just don't know it yet.

These arguments are always very dangerous and usually not very useful. What looks complicated today may look simple tomorrow. When Feynman first heard of the theory he did not like it; he said it was too complicated. But for me it seems to be the correct description and I am not sure that Nature makes things simple. Do you think you and I are simple? What's so simple about Nature? I don't know about it. Why should the basic laws be simple? Nature has been nice to us so far. But even so it's not that simple. Before I can explain the present-day theory to anybody, how long should he go to school? Quite some time. It's not so easy. I do not know of any law of Nature that would say that physics equations have to be simple and elegant. It's not an argument, though it's nice if it is simple and elegant. Besides, sometimes things look complicated and then you go one layer further and discover it's easier.

Professor Jarlskog in her introductory speech at the Nobel ceremony said that your guiding star was the concept of symmetry. Was it really?

It was not the guiding star, but I wanted to make a finite theory of weak interactions and started using the Yang-Mills theory, which is connected with symmetry. That symmetry was certainly a key in the whole business.

(a)	(b)

Martin Veltman lecturing (a) at the Fermilab, 1979 (courtesy of M. Veltman) and (b) in Budapest, 2002 (photograph by I. Hargittai).

In the Standard Model we have all these forces having to play together, which, in the end, takes the form of a symmetry. You find that all the forces are related to each other by some kind of symmetry manipulations. The symmetry is there to guarantee a result, so it is really important. It has to be precise, and if you don't make it precise, it does not work.

The symmetry of the Standard Model is broken but not in a way that it is really broken, it only appears to be broken. This is complicated. The symmetry is unbroken but we look at it in a funny way and that makes it appear to be broken. It is as if you see a symmetry but then you move away from it and thus you do not see it symmetrical anymore but it still is — only your point of view has changed. This is how it goes in particle physics. The breaking of symmetry is not in the theory, not in the balancing of forces; it's in the way we look at it, so to say. In Nature, you see things as you would expect to see them; you see so much green and blue and red, etc. But suppose that someone had made Nature already blue. Let's say, the vacuum is blue. Then you would see a lot more blue than green or red. You have something similar in chemistry with the left-handed and right-handed chirality. Even if Nature basically comes completely even-handed, how does it come about that certain things come only as left-handed and others as right-handed? It's because a choice was made somewhere at the beginning. In spontaneous symmetry breaking, we have precisely the same kind of choice. Even if originally it was entirely symmetric, Nature made a choice. The fact that we have only left-handed amino acids does not mean that the laws of physics have a preference for left. You could imagine that there is another planet at the other end of the Universe where everything is right-handed, so the laws of physics are still totally symmetric. It works out. You can have perfectly symmetrical laws of physics; yet, Nature can be polarized towards one particular direction.

I really got to be aware of the role of symmetry in particle physics with the introduction of the SU3 to describe strong interactions. That was discovered by Gell-Mann, that's what is called the "eightfold way." This discovery led to the discovery of quarks by Gell-Mann and Zweig. Gell-Mann and Zweig introduced three quarks and all the particles that were known then were made of these three quarks. That led to a symmetry in those particles. So one sees that there is a symmetrical consequence of a dynamical fact that happens to be three things that enter in equal way. They are almost interchangeable but not quite. So the symmetry arises from the dynamical starting point. The fact that these particles are not really identical, since they have somewhat different masses, breaks the symmetry. Of course, the way it went, first the symmetry was discovered and then the dynamics. It is funny that many people who were working in Europe were happy that there was a symmetry and never worried about where that symmetry came from. So they did not postulate quarks. Well, Gell-Mann and Zweig did postulate them.

Since then we have learned that there are many more quarks, not just three; we know now that there are six quarks, but the mass differences are so big that the symmetry has lost its meaning. In fact, today you have a hard time in trying to explain to a young person that there was a symmetry here at one time. It's not used anymore. So you realize that sometimes the symmetry is just the consequence of a dynamic circumstance and as your knowledge deepens, the symmetry may disappear. Symmetry, to me, is rarely a goal by itself.

It gets a little bit different with gauge theories. Gauge theories require that forces balance each other; they have to be in tune, of equal strength.

One can be obtained by a transformation from the other. This fact guarantees that the coupling between them will be equally strong, and this is a crucial point because they cannot balance out unless they are equally strong. Thus the tunings between these forces are necessary in order to have a good theory. Here we really have the symmetry and we cannot break it because if we did, the forces would not balance any more. As we see the Universe now, that would mean that one of the forces would be a little stronger than the others, and we can't have that. So the symmetry has to be exact. Whatever is broken, it has to be of the type that I described earlier, that is more of a change in perspective than an actual breaking. Thus the symmetry in gauge theories plays a very different role as compared to the SU3 symmetry of Gell-Mann. It has to be there, it has to be strictly obeyed, and you don't look at it with the question of where it comes from, does it have a deeper mechanism, as in case of the quarks, but you just say it has to be there or things would go wrong. Attempts have been made, I myself have tried, to explain why this is so and if this has a deeper meaning, but we could not find any better understanding of this. So the two situations are very different. And the same goes for gravitation, it also has a strong symmetry. The gauge symmetry of gravitation is a very complicated one, you cannot really break it. Again, that symmetry has to be there but you don't see a deep background for it. This is where we stand today. I am not very happy with this, it would be more natural to me if we could understand why Nature chose these symmetries. In the Standard Model we have three symmetries, SU3, SU2 and U1. No one has the foggiest notion why these three, except that we know that we must have the symmetry for the balancing of forces. I think that we can never really be happy until we understand two things: first, why in the Standard Model these particular symmetries play a role and secondly, if there is a deeper layer to understanding the balancing of forces.

Do you think about these things?

Oh, very often, in all sorts of ways. I am not getting anywhere either. There is very likely another layer and that's why we cannot understand these things. We just don't have enough input, there is information lacking. I don't think that it is possible to make that step without experimental input. There will be new more powerful accelerators and they will give these data. But that will take another 10 or 20 years.

Dr. Veltman with Queen Sylvia at the Nobel Banquette in Stockholm, December 1999 (photograph by and courtesy of U-BLAD, University of Utrecht).

We have already talked about the Higgs particle. There are also the W and the Z particles. How much role did you play in introducing them?

The W particle was already suggested sometime back in the fifties by many people. When I went to CERN, in 1963, there was a neutrino experiment whose aim was to discover the W. The Z is another story because that requires neutral currents and there were no experimental hints for that. I think it was Glashow who first talked about the Z in around 1963. The Z is a neutral particle and it is necessary to have it because of symmetry. If you have Ws and if you have symmetry, there must be a Z. Z is the consequence of imposing the symmetry on the weak interactions. When I started writing down the weak interactions, not knowing about Glashow's work I called it a W_0.

Why, do you think, it took more than 20 years for the Nobel Committee to recognize the importance of your work?

Because there was a Nobel Prize given before on this topic to Salam, Glashow, and Weinberg.

That was for the electroweak interaction.

Well, it was, but to a large extent it was the same subject. Sometime after their prize, in the early 1980s, I spoke with the then secretary of the

The 1999 Nobel laureates in Stockholm, from left to right: Gerardus 't Hooft (Physics), Ahmed Zewail (b. 1946, Chemistry), Robert Mundell (b. 1932, Nobel Memorial Prize in Economics), Günter Grass (b. 1927, Literature), Günter Blobel (b. 1936, Physiology or Medicine), and Martinus Veltman (courtesy of M. Veltman). © The Nobel Foundation.

physics Nobel Committee about this and he said, in essence, that the prize has already been given for that subject. That was the thinking in Sweden up till about 1990.

What could be the reason? In 1984, there was the Nobel Prize to Carlo Rubbia and Simon van der Meer, for the experimental detection of the Ws and the Zs.

It's a complicated history. I do know and I don't know. I do know, but I find it difficult to talk about it. It involves all sorts of personalities and the way the physics community perceives the situation. If you look at it now, for the subject of making the Standard Model, from the theoretical point of view, there are five people who got the Prize: Glashow, Weinberg, and Salam in 1979, and 't Hooft and I in 1999. You could say that this is sort of right, which I think it is, although you could dispute a little bit to what extent Salam was part of it, but more or less, it's probably all right. That means that it was clear that more than one prize had to be given. And I think that the Nobel people were not up to giving more than one prize in this direction. When they decided on a subset of these people in 1979, for whatever reasons and ideas as they were perceived in 1979, they had to go over them before they perceived that they would have to give out another prize. Actually, there is some logical reason behind the separation of the subject; namely making the model and having the scheme. To some extent I perceived that quite early, in the end of the 1970s, starting in 1976. I thought that it would be an important question, whether using the mathematical methods we had developed we could actually make physical predictions. I very much devoted my attention towards using the mathematical scheme that we had developed to make physics predictions. There were several such predictions that became true. Perhaps the most spectacular one, in which I was directly involved, was the prediction of the top quark mass. I wrote a paper in the middle of the seventies in which I derived an equation that could be used to predict the mass of the top quark. As the input data became more and more precise, the value of the top quark mass could be predicted more and more precisely. For this we needed the machine that runs at CERN, from about the middle of the 1980s. By 1990 the prediction became very precise and in 1995 the top quark was found in an experiment at Fermilab, at exactly the same mass where it was predicted. It was the first time that they could run a machine at this very high energy needed for the top quark, which is a very heavy particle. So if you think of the Nobel Committee, this was something that the person responsible for high-energy physics in the committee could look at and say; here is this theoretical method, which leads to a prediction for an experiment, which now has been fulfilled. I can very well imagine that being able to make that statement at that point made a difference in convincing the other members of the Nobel Committee that there was something here that amounted to something.

Do you think that those who found the top quark experimentally might get the prize?

I don't think they will. It was a very directed experiment, and they found it exactly where it was predicted. Although, of course, you could argue about that; Rubbia and van der Meer did receive the Prize for the *W* and *Z*, these particles were also very much predicted.

I read somewhere that you expressed some skepticism about the idea of the unified treatment of the electroweak interaction.

It is not so much whether the unification is right or wrong; the question is whether we are dealing with a unification at all. A splendid example of unification is what Maxwell did in writing down Maxwell's law. He truly unified electricity and magnetism. The most splendid example of that is light: an electromagnetic wave, generated by an electron moving up and down. When an electron moves, you have electricity and magnetism. So one implies the other. You can make predictions on the strength of magnetism based on the strength of electricity; these things are very much united. What used to be two sets of laws became one law with only one parameter, the electric charge. That's truly unification. To put it in a very pragmatic way, unification in the first place decreases the number of free parameters that describe the system. The propagation speed of electric fields is the same as the propagation speed of magnetic fields, that is, the speed of light. Unification usually leads to a reduction of the number of parameters; that is one criterion to go by.

In the Standard Model there is no such reduction of parameters related to having at the same time electromagnetism and weak interactions. They always speak about electroweak unification. There is no reduction in the number of parameters associated with that. It's not that what looks like weak interaction in one case, becomes electromagnetic in another case. This is not the case, they are not related with this kind of symmetry. They have separate symmetries. The only unification they have is writing them down on the same sheet of paper. I could take away electromagnetism and the Standard Model would still stand as much as it stands today. I could take weak interactions out and electromagnetism would do fine. If you had true unification one could not survive without the other. You cannot take away magnetism from Maxwell's equations. In the Standard Model you can separate electromagnetic and weak interactions, and that shows that there is no true unification there. The question of unification has gone through a number of falsifications, wrong statements and misguided directions.

How about unification involving gravity?

It's hard to say. The unification idea was started by Einstein. He came with the theory of gravity and with a certain philosophy. The philosophy was that forces of gravitation could be reformulated as a structure of space-time and then there is no gravitation, but instead space-time is curved. Thus the properties of gravitation became the properties of space-time. So his way of saying was that here is the Sun and it curves space-time and the Earth moves through that space-time, making a long spiral. Wonderful idea. The trouble is in the next step when he wants to do electromagnetism. Einstein thought that this idea was so great that he also wanted to express the influence of electricity and magnetism in terms of the structure of space-time. So we would have an electric charge, it would deform space and another particle coming along would go in a funny way because of the distorted space. And this does not work very well because of the following: if we have a proton and we scatter an electron by it, it gets attracted, so it goes down. But if we have a positron, it goes up, which means that two different particles follow a different path in the same space. Apparently different particles see different spaces and that does not make sense. With gravity everyone is doing the same, every particle follows the same space but with electricity that's not the case. These particles have to follow routes that depend on what particles they are. In gravitation the path that a particle follows in a gravitational field is the same and independent of the particle. A satellite going around the Earth or a pebble, they follow the same orbit. With electricity that's not so. Thus it becomes impossible, or at least very complicated, to make electricity and magnetism a property of space-time. But Einstein would not give up. He kept to this idea and spent the rest of his life making space very complicated, making two kinds of it, so to speak, and tried to embed Maxwell's law in it. That was his way of unification. But that's not the same thing as inventing one law for both of them. He had the same philosophy and he wanted the same unification under that; that was Einstein's unification, to write also electricity and magnetism in this very peculiar way of his. That's very different from the kind of unification that we are doing. We have one group and then we want to make a bigger group and encompass everybody in that one group, and everybody should relate to everybody; you may find that Mr. Putin can be obtained by a transformation of Mr. Bush, for example. That's another kind of unification. I do not know why everything should be unified. Theorists have made an attempt in particle physics, called Grand Unified Theory, that was truly a great unification — but it did not work. Personally I don't see why Nature has to do it. It may well be that Nature has three equations, that would then be Nature's choice. When it comes to looking for the Grand Unified Theory or the Grand Theory of Everything, I say that it is just nonsense. That's not a guiding principle of Nature; not visibly anyway. And need not to be.

Einstein introduced his so-called cosmological constant to make his equations right, but later in his life he felt uncomfortable with it and said that it was the greatest blunder of his life. Recently, new astrophysical data showed that there must be something like this cosmological constant; they call it antigravity force. They need this because they found that the Universe is not only expanding but also doing it at an accelerating rate. In order to explain this phenomenon, they found such a "constant" necessary.

I think you have things a little bit upside down, the way astrophysicists do. The history of the cosmological constant is somewhat more difficult. With Einstein's theory, the cosmological constant is nothing else but the curvature of the Universe if there were no matter at all. We have no prejudice on it, but in Einstein's theory, that was a free parameter; he did not know it either. A cosmological constant zero means that in the absence of matter, space is flat. That's the meaning of zero cosmological constant. It's a free parameter in Einstein's theory, and it describes the geometrical properties of space in the absence of matter. Matter changes space, and it curves the Universe.

Now, the situation changed when we got to this Higgs business. I was actually the first to note that in 1973. I found that the Higgs particle and the interaction it has, generates a big cosmological constant. Of course, the Higgs system implies that you have a field everywhere in the Universe and a non-zero energy distribution. So in 1973 I decided that there was a problem with the cosmological constant and the Higgs system generates a cosmological constant. The title of my paper was "The Higgs system and the cosmological constant." What happens is the following: in Einstein's theory it was a free parameter, but we discovered in particle theory that our theory, the Standard Model, influences this parameter. If Einstein or Nature chose to have it zero at the start, particle physics changes that. What made it worse is that particle physics changes it in an enormously drastic way. If space is flat without the Higgs system, with the Higgs system the Universe becomes as big as a football. The Higgs system generates a cosmological constant that is about 50 orders of magnitude larger than what we observe. Now what do the astrophysicists measure? They have been doing very precise measurements. What they found is that space is flat. What do they say then? They say that space cannot be flat because of all the matter in the Universe, which gives a curvature. Therefore, there has to be a cosmological constant to neutralize that. So these people, instead of saying, what in my interpretation would be the logical thing to say, "Hey, look, isn't that remarkable, space is flat! There must be a principle why space is flat," they say that this implies a cosmological constant. I consider this a first class perversion of logic! What you see is that apparently Nature prefers flat space and somehow this business of the cosmological constant and curvature at large distances does not work. Nature makes flat space. The astrophysicists don't think that flat is anything special. Nature makes space flat and it's clear that our theories are wrong there. I think that Einstein's theory is simply wrong at that scale.

I would not say that this was a "blunder" as he put it because in his time this was not that obvious; if he had chosen the cosmological constant to be zero it would've remained zero. It's only with today's knowledge of particle physics that this is no longer the case, it does not remain zero. Quantum mechanics has so far changed the scenario in such a way that even if you start with a zero cosmological constant it doesn't stay zero. Quantum mechanics changes everything; there are radiative corrections to the cosmological constant. If Einstein had believed in quantum mechanics he would have realized that he cannot make the choice that the cosmological constant is zero; we cannot make it zero. You can see that something is very wrong here.

Einstein did not accept quantum theory.

He did not accept quantum theory for his own reasons. I think that he realized that quantum theory was not in agreement with his concept of the gravitational force being a question of the structure of the Universe's space-time. You cannot maintain the idea that forces can be expressed as properties of space-time. He just would not give up his original ideas; he was a stubborn man. He had a beautiful idea and wouldn't let it go. I think that's a large part of the whole affair.

Mass often comes up in your work. There is physical mass; rest mass; it comes out infinite and then "renormalized" — does this mean that mass is not such a fundamental property of matter as we usually think of it?

It's a funny business. Mass is often something that we cannot explain. It is just an input in our theory. The electron mass, for example, or the mass of other particles, they are measured very precisely by the experimentalist and we just use their data as an input in our calculations. It comes into our theories in a very ad hoc way as free parameters over which we have no control. Things have changed slightly with the introduction of the Higgs system because masses are now made in a more indirect way. First, the particles are without mass, then they have the interaction with the Higgs system and that makes them have a mass. So the mass of a particle arises, as we see it, from its interaction with the Higgs field in the vacuum. The particles may also have mass initially. The miraculous thing with the Standard Model is that originally all the particles in the model have zero mass although there is no need for being massless, they could have some mass of their own and some additional mass from the Higgs. So there is something funny with the Higgs system, it seems that all the mass that we know of comes from the Higgs. There is no logical necessity for this, but this is how it works out in the Standard Model. This is one of the discoveries that we have made. And we don't know why, but it gives you the suspicion that in this Higgs system there is probably another layer where the idea of mass gets another interpretation. But we don't know how.

When you get to this other layer, would it give you a new physics? You mentioned before that the Newtonian physics is a closed system and so is the Standard Model. Is this a possibility?

How would I know? It will probably be something that will be very hard for me to accept because I am so much used to the present state of affairs. But I see defects and I see accidents and I think that in a further theory there should not be accidents. Some day we'll understand why we have three families of particles and then we'll probably also know why various particles have different masses. And, maybe then, in that theory, there will be another layer, and we'll get another view, but I do not know.

What is the present state of affairs of physics in the Netherlands?

I can only comment on particle physics, and if we talk about that in Holland, we immediately have to think of CERN with which we are associated. We do all our particle physics at CERN, experimentally at least. Theoretically we are just part of the great international community doing particle physics. So in this respect our particle physics does not differ from that of, for example, France, or Germany; it may be a little different from that in America. The unfortunate thing for particle physics is that experimentation is very difficult and can only be done at very few places and you have to wait very long before there is any new experimental results. Right now we are waiting till 2007 when the new machine will be operational. Nature does not tell us which way to go. Thus theoretical particle physicists are trying to build theories without experimental input and, unfortunately, humankind is too stupid to be able to do that. It's not so much stupidity, it's more the lack of information. You know about Mendeleev's system of the elements, but would you have guessed the existence of protons and neutrons and electrons? Unless people do experiments and discover protons, neutrons and electrons, we cannot understand the Mendeleev system. But it is hopeless to start with the periodic system and expect to dream up all those particles. Today we are in a similar situation. People do all sorts of complicated things, such as strings and supersymmetry, but I think that they are lost enterprises. I don't give them any relevance whatsoever. In my view their success would require that they could explain at least one of those things that we don't understand today. However, they explain nothing. They just add on more complications.

Why are so many people hooked on the superstrings theory, for example?

You tell me; I would not have done it. There are many people in our profession who are happy to run after a leader and most people just run after them. Most people are not original enough to find their own way. Besides, it's very difficult to find something original.

You wrote somewhere that it was a tragedy for physics in the Netherlands when Pais left after World War II. Do you mean that a strong personality can make a big difference for a country?

Absolutely. For example, the influence of Fermi in the United States was unmeasurably big. Fermi did most of the physics that we know today and many of the people you know today were students of Fermi.

You started a school in the Netherlands, but you also left for the United States. Was it also a blow for physics in this country?

Well, I guess it was but they don't know it themselves.

Why did you move to the United States?

Let's just say that I found it very difficult to stay here and remain without an ulcer. Actually, when I think back, it's not so easy. When we went to the University of Michigan in Ann Arbor, we only went for a sabbatical and we absolutely had no intention of staying there. When we were there the chairman of the Department came to my office once a day asking me to stay. I kept saying no to him for about six months. Then we started wavering a bit and all kinds of factors started creeping in, and finally we decided to stay. There were many reasons; one of them probably was my relationship with 't Hooft, which was not the best of all relationships. He was a professor in Utrecht by that time; I worked hard to get him that position relatively early, already in the seventies. I got him to Utrecht as my colleague.

The crucial moment for us in Michigan was finding a house. Anneke told me, if I buy this house, we'd stay. We did not buy that house because it was too expensive, but by that time we had passed the psychological barrier, so we stayed. There were also other elements to it. The climate in this country became rather anti-intellectual — it still is and it became very socialistic, which I don't like at all. Professors became very much the targets of politicians. They said that the professors made too much money and in reality I did not have an easy time financially. I found that climate very disgusting. Life in America wasn't that easy either at the beginning, but after a while it became much easier, financially as well. There were many factors. We never regretted our decision although sometimes we ask ourselves the question whether our kids would've been better off had we stayed in Holland. There is no obvious answer to this question. Our eldest son always liked the U.S. much better and he was very happy that we stayed. Our youngest son was probably too young at that time to understand that decision. Our eldest daughter was already out of the house by that time, and came only later; she never liked it much there. Now she lives in Paris. She is a particle physicist, she wrote a very good thesis in particle theory. She had problems with finding an academic job and get ahead, which, I think, is still rather difficult for women in Europe. In her time it was much more difficult to become professor than it was in my time in the 1960s. She finally decided to do what many particle physicists do; she accepted a job in the financial industry. The calculations that you have to do there are very easy for us. My daughter learned these financial calculations in two weeks. There are a lot of numerical methods there that we also use in particle physics.

Let's go back to your family background now.

The Veltman family in 1991, from left to right: sons, Hugo and Martin; wife, Anneke; Martinus, and daughter Helene (courtesy of M. Veltman).

My father came from a family of elementary school teachers; he and his father, all my uncles and aunts from my father's side were school teachers. They had great respect for science. My mother came from a family with a more practical background; her father owned a café and there were also building contractors in the family. We grew up in this dual philosophy, my mother always being the practical person and my father, the more rigorous one. This different attitude came much more to light during World War II in the way they reacted to German occupation. In my mind science was always of a high status. We were a big family, with six children, originally seven, but one died.

How about your present family?

I've already talked about our eldest daughter. My youngest son is in the movie business and lives in the United States. The other son is a chef, also in the U.S. It is a well-known phenomenon that sons do not want to do what their father did, so they were not interested in science. My daughter was. My youngest son once told me that there was no way that he would do what I do. But now we have started a project together, making a science movie. Now he is reading some particle physics.

Has the Nobel Prize changed your life?

Not much, really. It added some pleasant things. We go to all sorts of places, everyone says all kinds of things about you, which did not use to be the case. But in a real sense it did not penetrate my soul. I have not come around to have the required arrogance.

How do the people in this little town relate to the fact that suddenly they have a Nobel laureate in their midst?

They never expected it. Now, in a sense you walk around and it is as if you were wearing a little laurel wreath over your head.

Have you become a public figure in Holland?

Oh, yes, I think I am a reasonably known person here. Certainly during the first few weeks after the announcement, I got recognized in the street, airport, and other places and it still occasionally happens. But what does that mean if somebody recognizes me? I don't get much of a kick out of that.

Do you have many friends here?

Well, we've been living in this town for quite a while. We lived here before we went to America, in the time when our children were small, so we got to know all the schools and many people. We came back and our life resumed as if we had never been away. We still live near the same house that we lived before. When we went to America I was already 50 years old; at that age one cannot adapt to another country. I never integrated into the society in Ann Arbor. I only knew my colleagues in the laboratory but nobody else. When we left Ann Arbor, I only had to shake hands with the people in the laboratory. We were not part of the social structure over there, but we are here.

What was the greatest challenge in your life?

I have a hard time seeing it in that perspective. The greatest challenge is the things that you want to do. Maybe it was developing the theory of weak interactions — that certainly was the biggest challenge; taking that direction was the crucial step, much of the other stuff was just a logical consequence of this. I knew the topic was important from the moment I started working on it, but the first moment when I saw I was right with my solution — that was a very peculiar thing, a very definite moment of my life.

Did you know that it was of Nobel Prize caliber?

Of course. But there was also an educational element in it that we have to appreciate here because it is important. When I grew up in Holland I did not study with people who were in the front lines of particle physics. Even when I went to CERN, although CERN was the best in Europe, it was still quite secondary to the United States at that time. All the big shots were Americans; Feynman, Gell-Mann, Lee, Yang, they were all Americans, with no Europeans among them. A younger generation was starting to be sure, but we were not aware of that at that time. When you get to that stage and you really want to do an important contribution, you have to be in an environment that puts you in that direction. I was at CERN, which was good, but was not good enough for that — it still is not good enough for that. Then I went to SLAC [the Stanford Linear Accelerator Facility] in America, and from that time I gradually grew up and I went back to CERN. I could not possibly have done the things that I did in 1968 earlier, in 1964, because there was still a considerable piece of education I had to go through. I had to learn what is important, to learn directions, otherwise you just do what everybody else does, a little aspect of a known theory. To start thinking in an original way, to find a new direction, to ask the real question, you have to learn to be fit for. This is something 't Hooft never had to do; of course he would never admit that because he never learned that. I think I started to mature only in around 1966. Then I came to the level when it was possible to choose my direction, understand the complications, know the perspective, everything. I see this very often, when people come from places where they could not learn this; they just don't see the perspectives, you even want to shout at them, but they would not listen. They are manipulating equations and they think that would solve something. For physics, mathematics is important, but the basic ideas come from somewhere else, and once you understand where to go, it becomes a triviality; then you use mathematics to solve the problem — but you can also get another guy to do that part for you. So there is that particular part of learning, which today, in this country, people don't do; they think that doing a lot of complicated mathematics is the right way to do physics. This is the big disadvantage of any place, where there is not a first class person to teach young people the sense of direction, the sense of relative importance, and the sense of relative unimportance of mathematics.

Do you have heroes?

Not really. There are previous physicists whom I admire. I certainly admire Einstein, the more so as I get older because I am more appreciative now of the ideas he had versus the mathematics he did, which was mighty complicated. But the wonderfulness of his paper in which he introduced the photon in 1905, in which there is essentially no mathematics — I appreciate that more than almost anything else he has done. That is one aspect of it. The other guy I appreciate is Feynman. I appreciate them enormously but I do not put them on a pedestal or put their pictures on my wall.

You told me that you were very good friends with Telegdi. Could you tell me something about him?

I met him first in 1964. Of course, he is an experimentalist and I am a theorist. My friendship with him has something to do with the way I think a physicist differs from a mathematician. When I went to the United States for the first time, in 1963, I got to know a number of people who had this feeling for physics and Telegdi was one of them. From that time I became friends with these people and met with them occasionally. When I was at CERN in 1996, I met Telegdi on a daily basis and we discussed things on a daily basis. To describe him is very hard for me because he is an experimentalist, so even though I know more or less what he has done, it is difficult for me to appreciate it fully.

What do you do these days? Are you retired?

Yes, I am retired. To a large extent I devote myself to the popularization of particle physics. I just finished a popular book and I am giving lectures. I am also being used by people for publicity purposes. For example, in Hamburg they want to make a new machine, called Tesla. They had a kick-off day, and I was invited to give a talk there. So my name serves mostly PR purposes, but that is fine.

Concerning the future of particle physics, what should be pursued in the future, the ever-larger accelerators or the underground detectors?

Well, the underground detectors and anything else beyond the accelerators have rarely added much to our knowledge. Of course, the proton decay experiments in the mines in Ohio did show that the proton did not decay so that is an input, but it is only a negative result, and not a very spectacular one. So by and large we don't gain much from these things, so I have the tendency to be much more interested in accelerator experiments. This has a very basic scientific reason. Accelerator experiments are experiments where we have full control of all circumstances of our experiment. We control the beam, the energy, the detector, and we can change everything and see the effect of the change. With the underground detectors we have no control of the sources where the particles come from. In the neutrino underground experiments, there is no control of the Sun where the neutrinos come from. So I prefer the accelerators. But of course, if you get clear results from whatever experiment, that is very welcome.

But accelerators are extremely expensive.

I think that is all right. We have a very rich time in human existence and can easily afford those things.

For an outsider, it seems that astrophysics and particle physics are converging.

They are not converging. The astrophysicists use a lot of our knowledge but we learn very little from them. Astrophysics is much less hard science than particle physics. They speak about the Big Bang but much of it is basically speculation. Very little of it ever leads to better understanding of particle physics. It's a one-way street. Then, on top of it, astrophysics is not a very strong science.

There was, for example, the double pulsar experiment, which was a nice "laboratory" check for general relativity.

That check of general relativity in astrophysics is rather superficial. I don't remember sitting up when I heard the news of the double pulsar. It was a rather limited contribution that did not add to our understanding of gravitation.

Wasn't it a much better "laboratory" system than what you can get on Earth?

It was, to be sure, but that's not the same question as saying, to which level are you testing the theory of general relativity. Thus, did we learn anything new from that "experiment"? Not anywhere near where we particle physicists need new knowledge. Even if this experiment was a peculiar one that could not have been made on Earth, from the particle physicist's point of view what it tested was quite on the surface of the theory.

Do you have hobbies?

I like to do electronics, but it has become a difficult hobby nowadays when everything is made of chips. I made an intercom for this house and that was a hobby. I always liked to make computer programs.

Gerard 't Hooft, 2001 (photograph by M. Hargittai).

6

GERARD 'T HOOFT

Gerardus 't Hooft (b. 1946 in Den Helder, The Netherlands) is Professor of Physics at the Institute for Theoretical Physics of Utrecht University. He obtained his Ph.D. from the University of Utrecht in 1972 and has stayed there ever since. He received the Nobel Prize in Physics in 1999, together with Martinus Veltman "for elucidating the quantum structure of electroweak interactions in physics". He has received numerous honors, among them the Wolf Prize of the State of Israel (1982), the Franklin Medal (1995), the Commander in the Order of the Dutch Lion (a national distinction) (1999) and the Officier de la Légion d'Honneur of France (2001). He is a member of the Royal Dutch Academy of Sciences (1982) and foreign associate of the Belgian Academy of Sciences (1981), the National Academy of Sciences of the U.S.A. (1984), and the French Academy of Sciences (1995). An asteroid was named after him, "9491 Thooft", in 2000. We recorded our conversation in Professor 't Hooft's office at the University of Utrecht on March 18, 2001.*

What turned you originally to science?

From a very early age on, as long as I can remember, I was interested in science and in Nature in general. Socially I may have been a little bit backward, but I was always very much intrigued by the material world around me, more than people usually are.

---

*Magdolna Hargittai conducted the interview.

How did you pick physics among the sciences?

Well, when I was very young I wanted to become an inventor, I wanted to invent new miraculous machines. But I realized that that would require understanding the laws of Nature and so the laws of Nature were to interest me more and more. This meant especially the fundamental laws of Nature, the laws that move particles, the Universe, the forces.

What made you choose theoretical physics rather than experimental?

I considered physics to be a challenge of the mind, and my idol was Albert Einstein who by pure thought could understand so much around us. This was something that I also wanted to do. To me theoretical physics is a more pure science than experimental physics; in experiment, you also have to deal with management, machines, and lots of people. You can't do an experiment all by yourself. With theory you can devise all of it by yourself. In some sense I am an individualist, I like to work all by myself. This doesn't mean that I don't like experimental science. It's fun to see how marvellous, big, or accurate experiments have been done in the past. I admire and envy people who can do that.

Would you care to tell me something about the work that eventually led you to the Nobel Prize?

As you can imagine, that is a long story and it happened quite some time ago. It started when I was an undergraduate student and was given an assignment on elementary particles, the tiniest building blocks of matter, by my then teacher, Martinus Veltman. Later, when I became a Ph.D. student, I again had Veltman as my advisor, and he gave me several possible topics to work on. From among these topics there was one that interested me most, and that was exactly what he was working on himself. It was the construction of a theory for vector particles. These are particles that are controlled by fundamental fields, which are not Dirac fields or scalar fields, like the ones for the electron or spinless particles, but vector fields, like that of the photon. He had already given me the famous paper by Yang and Mills to read, their fundamental paper in which they show how you can generalize the theory of photons, which are particles with spin 1. But not all particles with spin 1 are described properly by the Yang-Mills theory. Veltman was working on the theory, trying to get it under control, so to speak. He then explained to me what the problem was, and he assured me that it was very hard. The difficulty was that if one tried to calculate the effects these particles had on other particles, you would nearly always get "infinity" as an answer. He had already been working on it for ten years and he did not expect that I could do anything to shed further light on this subject but if I were interested, I could give it a try; there were still many questions to answer. I saw his problem but I also had a fresh mind and I had ideas about how to solve it. He did not pursue these, but I thought that I could indeed make progress.

That is how our discussions started. He had a quite different way of dealing with problems of this sort than I had. This problem was very technical, so he devised a computer program — you have to remember that this was before 1970 and around that time computers were huge and slow big machines compared to what they are today. It was an outright heroic effort of his to try to devise an algebraic computer program to enable him to do the necessary calculations — but that was what he was after. This took him a lot of time at first. Let me cut this long story short. At the very end the computer program worked perfectly. It appeared to tell him that there is no way that a modified Yang-Mills theory can describe vector particles that, unlike a photon, have mass. He had hoped that the infinities would cancel out somehow, but this did not seem to happen. This was mystifying, because it seemed that you need such particles to understand a special kind of force among the subatomic particles, called the "weak force".

He was about to give up when I gave him one more thing to try; which is exactly the equations that I had come up with in the very beginning. I said I was sure that what was missing was one extra particle, a particle without spin, described by a scalar field, and interacting in a way he could not have guessed by looking at his computer output alone. He did not really believe me, I think, but it so happened that his computer program was now ready to deal with the situation, and he certainly believed his computer, which was confirming what I had said. That is how an intense collaboration started, and we realized that a breakthrough was made.

The breakthrough was that my simple arguments worked. The weak interactions among elementary particles can indeed be attributed to the action of a vector particle, now called the weak intermediate vector boson. The weak interaction takes place when this particle is exchanged. That vector particle, which had to be treated relativistically, had given all sorts of problems in the older theories that appeared to be insurmountable. Now we understood that this extra spinless particle is needed to fix the discrepancies, to remove all infinities. We call such a theory "renormalizable".

The extra, spinless particle is what is now known as the Higgs particle. Indeed, it had been described earlier, not only by Peter Higgs, in Scotland, but also by the Belgian François Englert and the American Robert Brout. They, however, had not attempted to handle the infinity question.

So now we understood that we had a theory, which would do the entire job. It became clear very soon that we could make vector theories for the weak interactions, which were renormalizable, and which agree very well with detailed observations. Before that the only renormalizable theory that was applicable in Nature was pure elecromagnetism. Electrons and photons are controlled by a renormalizable theory, called quantum electrodynamics. Now, for the first time, we had a renormalizable theory for the weak force, which arises from the exchange of the Yang-Mills particles.

Later it became clear to me why the competition, so to speak, was not ready for these theories. That was because there were a number of misconceptions in the world, which I did not know about — I just did not share these points of view. There was a generally-felt mistrust about quantum field theory. Could quantized fundamental fields control the forces of Nature? An insight that seemed to be impeccable was that quantum field theories should not allow interactions to become very strong. If interactions became strong these field theories would carry their own seeds of destruction in them, in a sense that the interactions would grow infinitely strong at smaller distances and the theory would become basically self-contradictory.

Furthermore, the vector theories were very complicated and it was not understood how to renormalize those properly. Altogether, people thought that field theories do not work. What is typical in such circumstances is that then people try to find arguments why theories would *never* work. Long papers were written about why quantum field theories would always fail. This view had become practically settled; everybody had his own idea why quantum field theories would not work. Veltman was one of the few people who was still pursuing quantum field theory, and I had made my own observations regarding quantum field theory. In fact, now I know that one of the investigations I did at that time, in 1970, was what later became known as *asymptotic freedom*. In 1973, it was rediscovered independently by others in the United States. It so turned out that this was the essential new ingredient without which we would not be able to understand how strong quantum field theory would work.

Asymptotic freedom was not so important for the weak force because the weak force is sufficiently weak to be valid in a large domain of scales.

The quantum effects that we talk about now would be small anyway for this system of forces. So this objection against quantum field theory would not have been valid anyway for the weak force. But people had other objections as well. Quite generally, they didn't suspect that the whole idea of renormalization would be so fundamental. It was generally thought that renormalization was an ugly feature of the existing theories. Renormalization means that the theory tends to generate infinite expressions, which are totally meaningless and the theory has to be repaired in order to allow these expressions to make sense. Although in the case of electrodynamics it was understood how to repair the theory and how to make it totally consistent. Notions such as renormalizability were regarded as being suspect: you start with a theory, which is bad and then you "repair" it! That cannot be the way forces of Nature work! So, quantum field theories were being ignored by a majority of physicists.

I wonder about this mistrust toward quantum field theory. It was as early as the late 1940s that Freeman Dyson used quantum field theory to unify the theories in quantum electrodynamics by Tomonaga, Schwinger, and Feynman.

As you say, it worked for electrodynamics, but it did because using expansion techniques, very accurate results could be obtained. However, for stronger interactions the same expansions did not work and it was not understood how to deal with the problem beyond what had worked for electrodynamics. Many people had tried and failed and the general feeling was that this was not the way to go. For strongly interacting particles, the general feeling was that all the strongly interacting particles are basically equal; the proton, the neutron, the pion, the rho-meson, they were all considered to be equally fundamental and it was unlikely that some would be more fundamental than others. In quantum field theory you consider a few particles to be fundamental and all others to be bound states. So in quantum electrodynamics, the electron and the photon were the fundamental particles, but the hydrogen atom was a bound state. Here you can easily make the distinctions. However, in the case of strongly interacting particles, all of them are equally fundamental and you can't single out any of the particles for which you write down a field, while all the others have to be regarded as composite particles, whose fields can be found by combinations of the elementary ones. That was considered to be ugly, therefore unlikely to be correct. We now know that in some sense, these people were right. The fields we should use are not the fields of protons, pions and rho-mesons, but the fields of quarks and gluons.

Then came another problem when Murray Gell-Mann, some time around 1964, came up with the quark theory, which proved that it is impossible to isolate those quarks. This enhanced the mistrust, which was now compounded with the notion of some non-existing particles. Now we know that the quarks are there and now we know why they could not have been isolated. By now, there is no mistrust anymore and this is primarily because everything agrees so beautifully with observation. A long chain of insights was needed to understand how quantum field theory would work at all. But in the beginning, up to the 1970s, there was a general feeling that things should be done differently.

Did you know it right away that your work was of Nobel caliber?

We realized that this was a breakthrough in the weak interactions; we also saw our colleagues making a 180-degree switch in their research due to our work. It created a large industry, and in a few years it led to the Standard Model of fundamental particles. With this theory it became possible for the first time to write down theories for the elementary particles, which worked with an accuracy of many decimal places. When I grew up as a student we had just sort of accepted that the strong interactions would be a black box: particles go in, other particles emerge, and we had no idea what goes on inside. We had only very approximate, very ugly theories that gave you only some general feeling about the processes that can occur, but nothing of numerical accuracy. So this was a big change and I think at that time it was immediately clear that this was a very important new development.

Then, in 1979, Glashow, Salam and Weinberg received the Nobel Prize for unifying the electromagnetic and weak interactions. In what way was their work different from yours?

They had done their work earlier and all three of them in their own way had written down some ingredients of the model that would work. There are various aspects of it. Abdus Salam had a very good broad vision about the way things would probably work. He had his somewhat Eastern approach — he was from Pakistan, which, when he was born, was just a part of India. He had the feeling that science should be elegant, it should be holistic, it should be universal, we should have fundamental equations, not just approximations, and so on. In view of this he had concluded that there had to be something like the Yang-Mills theory to describe the weak interactions. Then there was Sheldon Glashow, who was much more pragmatic, an American, who also knew the experimental data. He realized what the most likely symmetry group should be for the weak interactions. He wrote down a dynamical model, which is nearly but not quite right. This was a sufficiently practical model to do some calculation with and see what comes out. Steven Weinberg wrote down more precisely the leptonic sector of the Standard Model, including, in particular, the Higgs particle. But none of the three knew how to renormalize the theory. So they had the general theory but not the renormalization prescriptions. When our work came out, it was realized that, indeed, what Weinberg, Glashow and Salam had produced was going to be the basis of the real theory. Then, at that time, the decision was made in Stockholm. This theory is called the Glashow-Weinberg-Salam theory. Charm was discovered at that time and Glashow had been a very strong advocate of charm, for which he certainly deserves credit. He had very strongly emphasized that the strongly interacting particles can only be incorporated in the Standard Model if you have another quark. He had done this work with other people, John Iliopoulos and Luciano Maiani. He did not really get the prize for charm, he got it for the model he had written down earlier. But it was clear that these laureates had not understood how to renormalize the theory; they had written down the most elementary equations for the theory. When it comes to the question to whom to give a prize, a natural thing to ask is: who thought of the idea first? And, of course, we came later with our prescription about how to renormalize the theory. An important consideration had been that the earlier experimental verifications were not so much the details of the higher order calculations — which required our analysis — but the more basic aspects of the theory, which had already been laid down by these gentlemen.

Do you think that your work helped them to get the Prize?

Veltman was very angry; he felt very strongly about this issue. I thought that, well, if this is the way they decided, then so be it. We are biased, but in Stockholm, they're supposed to have an overview of everything that's going on and if this is their decision, I'll accept that.

Did it occur to you that, by their getting the Prize, you perhaps might have lost it for good?

Many of our friends said at that time that our turn will come, but of course, we wouldn't have thought that we'd have to wait another twenty years for our turn to come. So by that time I already thought that, well, Stockholm apparently has other priorities. Why should they give a prize for something that happened thirty years ago? So I did not expect anymore that they would think of us.

What do you think it was that made them think of you again? Was it the top quark in 1995?

That is how many people like to interpret the decision. What we have here is what we often see in science. First comes the general idea, which is a model for the forces; and you see that the experimental observations confirm the general idea. This was the case when neutral currents were discovered, which is a special kind of weak interactions, predicted by the models of Weinberg, Salam and Glashow. They particularly predicted neutral current events, which are events where neutrinos collide against other particles

During the opening of the Institute for Theoretical Physics at the University of Utrecht in 1991. From left to right: Victor Emery (guest), Bernard de Wit, Leonie Silkens (secretary), Nico van Kampen, Hans van Himbergen, Gerard 't Hooft, Ben Nijboer, Henk van Beijeren, Mathieu Ernst, Theo Ruijgrok (courtesy of G. 't Hooft).

without changing into, for example, an electron or a muon; they just scatter elastically. That was a very important prediction. It was confirmed a few years after their prediction, in about 1973 or 1974. There was a theory and there was experimental evidence explained by that theory, but this prediction did not require long calculations. These experimental confirmations related only to the most basic structure of the theory, which did not require renormalization to work. However, we do have some reservations against this argument, which *is* the official argument. Our objection is that everybody in the field understood that these models were so successful *because* they were renormalizable.

Later, in the early1980s, came the observation of the W and Z particles by Rubbia and van der Meer, and they got the Nobel Prize for this in 1984, but that was an experimental Prize. It was a very difficult thing to do, to detect these very energetic particles. This was, of course, much after the time when we had predicted these particles; in fact it was due to our work that everybody started to look for these particles. Our theory predicted exactly the masses of these particles and they were found exactly where they were predicted.

Then, you are right, in 1995, the top quark was discovered; and to predict the top quark we not only needed the Standard Model but, this time, we also needed to do many higher order corrections and for those calculations, renormalizability was absolutely essential. If you couldn't renormalize these expressions, you could've never done these computations. The Standard Model computations worked, up to many decimal places. This time, many experiments, of all sorts, at lower energy had been checked against the theory. These were precision experiments and precision calculations. Since the top quark itself was heavier than the energies that could be reached by the machines at that time, one could not observe the top quark directly. However, one could include the indirect contribution of these particles to the amplitudes of the lower-energy events, which *could* be measured. Thus measuring these events with great precision, one could infer some information about the top quark and one could predict its mass. The first prediction was already good and later the predictions became more and more accurate, and eventually the top quark was discovered precisely as predicted by the theory. From this analysis it became clear that the theory works quantitatively, to a great precision. That's when many people said that we should be given the Prize for our theory because we were the first to describe it quantitatively. My only reservation against this argument is that the realization that renormalizabilty was essential for these theories occurred very much earlier; they could have given this Prize 20 years before. Another point is the success of the theory for the strong force; the strong force was another example of a Yang-Mills theory, which also had to be renormalized. The strong force could not be understood without renormalization.

I read the presentation speech at the Nobel Award Ceremonies by Cecilia Jarlskog and she also mentioned the W, the Z, and the top quark.

Yes, but I think that it also sounds like an excuse to me why the Prize came so late, but, well, I am happy we got it at the end.

Is the Standard Model the right model for the Universe?

It is right in a way. In science, quite generally, once a certain insight has been obtained, and that insight is right, it won't change anymore. However, what is very likely going to happen is that people will find improved ways of saying things. Newton's laws for gravity are correct. They describe very accurately how the planetary system works. However, Einstein gave further refinements. Quantum mechanics was also not included by Newton in his laws of the planets. Now we know, if things become very tiny, you have to make quantum mechanical corrections in your calculations. There is no existing branch of science that is absolutely valid. There are always domains for which corrections are needed. The Standard Model is just like that. It is a branch of science, which is not absolutely correct. But it has a large domain of validity to a certain number of decimal places. If you want to do calculations with 20 decimal places of accuracy, I am sure there will be many effects that the Standard Model cannot account for. Or if you go to energies much, much greater than what can be reached by today's accelerators, there will be effects that the Standard Model cannot account for. That is just like in any branch of science. The Standard Model is not an absolute truth, but it controls very accurately a large domain of physics, which is the particles we talked about today; the W and Z particles, the strong, the weak, and the electromagnetic force. In the domain where people are doing measurements today; that's where the Standard Model is very accurately valid.

There is always a lot of talk about symmetry and also about symmetry breaking. Which is more important?

Both. During most of the history of particle physics, advances came because people understood symmetry patterns and then, subsequently, the breaking of symmetry patterns. We see a symmetry and we see a violation of this symmetry. By identifying those, both the symmetry and the way it is broken, we have a powerful device to describe what's going on. In the case of the strong interactions, in particular, the symmetry structures are very well understood. Also, the ways symmetries are broken, which, again, hinted very strongly towards the quark theory because with that theory these symmetries are very natural to explain. At first, we couldn't quite understand why they were broken exactly in the way observed — that could only be understood by understanding the details of Quantum Chromodynamics. Generally, by studying symmetries and the ways they are broken, we can figure out how the dynamics work.

Both symmetry and symmetry breaking are examples of patterns that you see in Nature. It goes back to Eugene Wigner, for instance, to understand that you can break the symmetry in many different ways. If you study the way Nature breaks a symmetry, you can learn a lot. For instance, in the weak force, people introduced a rather spurious "particle", called the *spurion*, which was just an imaginary particle. Its use was to describe symmetry breaking as if it was due to interactions with this spurion. It enabled us to understand the patterns that we see. So people tried to derive the properties of the spurion. That is just a way of speech but it is very useful to talk about things this way. You could say, look, the symmetry property is carried away by this one particle. The weak force breaks the symmetries of the strong force. But by looking at how the symmetries are broken you can find the symmetry properties of the particles such as the W and the Z boson and the way in which they are coupled to the electrons and to the strongly interacting particles. If you understand how the symmetry is broken you can understand how the particles are coupled to the weak force carriers and this way you can make a dynamical theory for the weak force. That's the way you get started on these matters. And the weak force is just an example. There are many other symmetries in particles and equally as many symmetries that are broken. For instance, you can combine spin with internal symmetries, but the field theories in the earlier days told you that you couldn't do that exactly, you can only do it approximately for the strong force, and only if you ignore relativity. So people realized that here we have a symmetry that will be broken by relativistic effects. There are many examples of such games that you can play and quite generally this gives you a lot of insight.

Is there such a thing as a Grand Unifying Principle or the Theory of Everything?

As soon as the Standard Model was more or less clear on paper — its equations, its peculiarities — it was clear to everybody who studied it that there was a grander pattern in it, it's not just random. There appear to be quarks and leptons, and they show much more of a pattern than what we'd expect if the Creator of the Universe just took a bag of rubbish and put it all together saying that this is the Model. It is not that. There is an enormous amount of systematics in it. It was also clear to everybody that there must be further explanations why the model is built this way, there must be an explanation of the structure, there must be a further unifying principle. It's much like what the chemists must have discovered when Mendeleev came along with the Periodic Table of the elements. It was quite clear that there was a pattern in this; it was not just arbitrary. But the chemists of that time could not understand what made the system work, they only recognized the pattern. Here we are now, we see the Standard Model and we see that there is a system in it. It's obvious to anybody who looks at it that there must be a more unified theory that explains this particular pattern that we see in the Standard Model. Yes, we all expect that there will be a more unified description and I have sufficient trust in human ingenuity that people will figure out how it works. It will be figured out what the deeper reasons are for this particular pattern that we see in the Standard Model. But we haven't figured it out yet — although there are many ideas and suspicions.

Will it be a simple explanation or a complicated one?

I think it will be a surprising explanation. It happens all the time that Nature outsmarts people. I expect that it will be a beautiful smart argument or reason and I'm also sure that once people put all this together many of us will say, we should have figured this out long ago — it is so simple! But we just did not think of it. Now it looks very complicated and mystifying to us.

Do you find Nature simple and elegant or complicated?

In a way, both. We had this marvelous experience with the strong force. The strong force before the 1970s was thought to be an extremely complicated theory, with very messy equations that we could not possibly understand — although perhaps there were some people who had some foresight and thought that it could be something simple. Now we know that it is very simple; we know exactly the equations, which are very simple, they fit on one line. Such a simple equation can describe tremendously complicated consequences for the strong force. Generally speaking, I think that we will see the same thing in the future; that we'll have a basically very simple set of equations. But the consequences of these equations will be very complicated.

There is the Higgs particle and there is the graviton, the force carrier of the gravitational force; neither of them has been found yet. Is it only the question of the size of the accelerators when they will be found?

For the Higgs particle, yes. In fact, most of us are convinced that the observation of the Higgs particle is just around the corner. In fact, you may have heard the rumor that at CERN they were just about to make the discovery but unfortunately the machine had to be shut down. There is going to be a more powerful machine there. We just keep our fingers crossed that they were probably right and the mass of the Higgs is around 125 or so GeV. If not, it might then be a little bit heavier but even then it will be detected fairly soon, say, within about five to ten years.

In the theory, there is one big question: whether there is only one Higgs particle; many theorists would like to see at least two of them. If there are two, there will also be charged components, not only neutral ones. If this is so, the situation will be a little bit more complicated than in the earliest version of the Standard Model with just one Higgs particle. It is so surprising that the easiest version of the model still appears to work quite well. We often say that if Nature is economical, then there is just one Higgs particle, and we don't need any more. But as soon as we try to understand why the Standard Model is the way it is, and whether there is a deeper theory there or not, most of those deeper theories will tell us that there has to be more than just one of those particles. If economy would be the only lead, we would say that there is only one Higgs. But if we say no, try to be more realistic, we'll find that there is more than one Higgs. My worry is that if more than one Higgs particle is found, there will be different groups of people standing up saying: we predicted this! But the fact is that many different theories predict that there is more than one Higgs particle. In fact it would be very mysterious if there were just one Higgs particle because it would mean that Nature is more economical than any of our theories that we try to use to explain what we see.

Is the graviton a different story?

The graviton is quite a different story because the gravitational force is so tremendously weak. So the graviton will not be discovered in any laboratory in the foreseeable future. Unless a miracle happens of such a dimension, it is unlikely. One of such miracles has been proposed not so long ago, which was the existence of extra dimensions at a large scale. Not extra dimensions at tremendously high energies where nobody can do an experiment, but extra dimensions at relatively low energies where experiments can be done. If that is true, then people will possibly, conceivably, detect features, which have to do with gravity at the laboratory scale, and I don't mean high-energy accelerators but, perhaps, even tabletop experiments. The tabletop experiments that have been done so far, in particular the ones in California, show that the most daring theories of extra dimensions are probably not true. They predicted deviations from the gravitational force already at the millimeter scale. If that were true, that would be very exciting, but it's probably a too optimistic idea. So it is conceivable that maybe gravitational effects will show up in very high-energy accelerators, but it is very unlikely. It's much more likely that we'll have to deal with gravity using a purely theoretical approach, which, history has shown, is extremely treacherous. We theoreticians tend to think that we understand exactly how to proceed, but often we don't understand it at all and we can always be surprised by Nature. In fact I would find it much more interesting if Nature was entirely different at high energies than all of us are thinking right now; because then there would be interesting work to do with new challenges. But the standard view is that quantum field theories are valid in very, very large domains of physics, which means from the present energy scale where accelerators are working — which now is close to the TeV domain — to tens of orders of magnitudes beyond that. The gravitational effects come only after that at the Planck scale, which is at tremendously high energy and tremendously small distances, and these will never be reached by ordinary accelerators. If this is true then it is only with our minds that we can try to penetrate those domains of physics. It's a very challenging idea and I don't know whether the human mind is powerful enough to penetrate those domains of physics — but if you look at what history has shown us so far, you never know.

We are talking about things that have not been figured out yet. What do you think of the possible unification of gravitation and quantum mechanics?

This is a very important subject and it is here where I have some views that deviate from what most of our colleagues are saying. It's not only that I really believe that my approach is correct, but my belief is also that there are so many physicists among us that we should all try something different. If we all bark against the same tree, we would not make any progress. We should all try to do things that we believe in most, particularly if they differ from what other people are doing. My belief about the roles that quantum mechanics and gravity are playing in that very high-energy domain is that, very likely, we'll have to change our views about what quantum mechanics really is. Perhaps even change the views about what general relativity really is. It is here where I have theories which look very much like what people used to call the *hidden variable theories*. There is a theory underlying quantum mechanics and this underlying theory is not quantum mechanical. It's basically deterministic. The thing that we call quantum mechanics today may be nothing but a formalism enabling us to deal with statistics. I think now that there must be some fundamental theory of Nature that we don't know about at all yet, where quantum mechanics does not enter any of the equations. The theory is totally deterministic, causal, coherent and consistent — having nothing to do with quantum mechanics. But the theory is also very subtle and very special and it is likely to be extremely complicated to work out the equations. So complicated that at some point basically the solution of the equations become chaotic. Then we are forced to make some approximations. I think that any approximation of the conventional type is insufficient to understand what's going on so we need a very powerful kind of new mathematics to deal with the sophisticated properties of these equations. Now I've turned to the belief that the very special mathematics to deal with the statistics of these equations is what we call quantum mechanics today.

The remarkable thing about physics is, of course, that physicists are always confronted with the solutions of the equations that they are trying to understand, not the equations themselves. When you study atoms or electromagnetic phenomena, you see solutions to the equations, you don't see the equations. So physics is about uncovering the equations of which we're seeing the solutions. Mathematics is the other way. Mathematicians write out the equations and they want to know what the solutions are. In physics, we see the solutions and we want to know what the equations are. Quantum mechanics, in my mind, is a beautiful solution to a very deep problem: how to write down *approximate* solutions to equations, equations of which we still know very little. These approximate solutions have some statistical elements in them, in the sense that they show some stochastic behavior that we cannot keep under control. It's our job to figure out what the original equations are.

Until the present time physicists have not been able to write down the underlying equations in any other language than what is now called quantum mechanics. So quantum mechanics is generally considered to be a fundamental doctrine in Nature. This is where I disagree. I suspect that, what is called quantum mechanics today is the regular part of the solutions to equations that are partly stochastic, partly regular. But I realize very well the danger of my attitude. It could be totally wrong, of course; it could be totally correct; but there is also an option in the middle, which is that although the attitude I am taking is probably formally correct it could be that I am one thousand years too early with it. It could be that first we have to go another route, take quantum mechanics as it is in order to make further progress and then only a hundred or thousand years later when all the bits and pieces fall into place, people might hit upon theories where my attitude will turn out to be correct — it was however premature. I see an example of such a situation, for instance, in the physics of the old Greeks.

The old Greeks had ideas about matter being made of atoms — they had an atomic theory, but they didn't understand anything about dynamics, they didn't have Newton's laws or anything like that. So although they were formally correct — matter is made out of atoms — their arguments were totally wrong; these were just based on numerology, the philosophy that nothing can be totally continuous, as Pythagoras thought. What they should have done is not worry about atoms but worry about dynamics first, figure out what the equations of falling objects are, figure out the laws of chemistry and electricity and magnetism, and then ask questions about atoms. So maybe I am formally correct in a similar way, but hopelessly premature. Maybe the theories I am thinking about cannot be developed further because there is so much in between that has to be discovered first. Maybe all the way from here, from the Standard Model, to the Planck scale, there is an enormously big domain of physics that first has to be understood before we should even worry about what quantum mechanics really is. But I am continuing anyway, just because I think that maybe a healthier attitude towards the question what quantum mechanics really is could help us in finding new alleys towards quantizing gravity. I have ideas how to do this and that's why I am doing it. But whether it'll lead me to any success, I don't know.

You also mentioned general relativity, and that you would look at it differently.

General relativity has it that the coordinate frame that we use in describing space and time is only locally defined. If you try to understand the gravitational force you must realize it as a symmetry relation between the description of space-time with a gravitational field in it and accelerated space-time. That symmetry is absolute and most people conclude from this that there is absolutely no reason to introduce a universally flat coordinate frame. If such a preferred coordinate frame were to exist, then that would remove the foundations of general relativity, and we no longer understand the gravitational force if we take a fixed coordinate frame. General relativity is based on the idea that there exists no such thing as a flat coordinate frame. That theory is extremely successful, it is as successful as quantum mechanics is in some sense. Why should we possibly abandon that idea?

One reason to be a little bit worried about general relativity is the cosmological constant. It is a constant added to Einstein's equations and it appears to be either absent or extremely tiny. Nowadays, astronomers say that there is a cosmological constant in the equations because this fits the present observations better than if you did not have it. But it is extremely small. This is again a problem with general relativity: why is this constant so small? One possible reason could be that there is a symmetry. Flat space would have a symmetry that is absent in curved space-time. If Nature would somehow respect this symmetry, flat space would be singled out, and this way we could understand why there is no cosmological constant. But the preference for flat space appears to imply some preference for flat coordinate frames, which would be a deviation from general relativity. Maybe we should reconsider the possibility of globally flat coordinates.

General relativity may be seen as a special case of our gauge theories. The gauge degree of freedom is normally taken to be invisible. However, this could be a question of *information getting lost*. My theories now have, as an essential ingredient, an element I refer to as "information loss". There is information that we cannot see in Nature because this information is not preserved by the equations. The information about the flat coordinates is not preserved at the local scale. But it is there in principle. This could be a sort of intermediate approach where there could be some special role for flat coordinates.

Recently some astronomers at Princeton told me that their recent observations indicate that the Universe is flat and that it expands with an accelerating rate. This could only mean that there is an antigravity force, which more or less is what Einstein called the cosmological constant.

Yes. It is exactly the cosmological constant. The nice thing about these observations is that there is no theory for such a cosmological constant at present. It's a great mystery.

Does this mean that Einstein's theory was not good? He himself eventually threw out the cosmological constant saying that it was his greatest blunder.

If he said so, Einstein overestimated his own importance. I feel sorry for him, but it was not his blunder. He was absolutely correct in introducing the cosmological constant. If he had failed to do that, somebody else would have had to put it in. The fact that we are dealing with is not that Einstein had introduced the cosmological constant, the fact is that the thing is there. The possibility to put it into the equations is there, regardless of whether or not Einstein had said so. The problem is there, and it is our job to understand what causes this peculiar situation in Nature. Einstein could not have known why there would be no cosmological constant, so it would have been wrong if he had not introduced it, even if it later would turn out to be zero.

I read somewhere that you are interested in black holes.

Black holes originally were primarily interesting from an astronomical point of view. It was understood that if we have a large amount of matter, a little more — about 4–5 times — than the Sun, then we can have gravitational implosion and then we can just integrate the equations and find that the only stable solutions are black holes. For that to happen, gigantic amounts of matter are needed. However, just considering the laws of physics, there is no fundamental law about the mass of a black hole. We can solve the equations no matter how small the black hole is. If we leave out quantum mechanics for a moment, a black hole can be as small or as large as anything, its mass is a free parameter. Einstein's equations allow solutions for a black hole of any size. For a large black hole of astronomical size we can easily imagine initial configurations of matter. Such a black hole would just spontaneously be formed by implosion. In principle we can also imagine initial configurations for a form of matter of much smaller dimensions, to make very tiny black holes. In principle we can imagine black holes to be as small as the Planck length and as light as the Planck mass; which would be just a couple of micrograms and it would be 10^{-33} cm in size. That would be the tiniest black hole for which we would not really need quantum corrections. If we want a black hole smaller than that we need to reformulate entirely the laws of quantum mechanics. Then there is an interesting domain of physics where black holes are just large enough so that the classic equations of Einstein more or less apply to them but they are small enough to see that there are quantum corrections. The question then is, what would be the fundamental laws for such things?

Here one finds very surprising features; in fact, conflicting features. The apparent conflict between quantum mechanics and general relativity is most severe for black holes. The way I see it, the gravitational force is an unstable force, it tends to amplify itself also because the gravitational field energy is basically negative. Gravity is the only field that has a negative field energy and that is begging for difficulties. We don't have any other situation in Nature where energy can be negative. This negative gravitational field energy causes the equations of gravity to be unstable. We see that as the deeper origin of gravitational collapse. The question that has been asked since the 1980s is: how do we deal with this situation in particle theory, in quantum theory?

Giving the Dies Lecture at the University of Utrecht in 1987 (photograph courtesy of G. 't Hooft).

Try to imagine the strongest possible gravitational force. That is exactly where there is a black hole. If we understand the black hole, we'll also understand the strongest possible gravitational fields, and that would be the core of our problem. If we understand that, we'll understand everything about the gravitational force. Then we should be able to get a more complete theoretical insight concerning the dynamics of gravity, including quantum mechanics. The idea is that if we manage to understand the quantum mechanics of black holes, then we may have all what we need to construct a complete theory for the gravitational force.

At first sight it seemed to be not only not very difficult but also a beautiful problem. Around 1975, Stephen Hawking had realized that black holes emit particles; this is due to a fundamental aspect of quantum field theory in a strong gravitational field. This particle emission happens spontaneously. Black holes emit particles, and not only could Hawking compute the particle emission rates of black holes, but they appeared to be controlled by absolutely beautiful equations. It was natural to assume that there is some fundamental truth in there. All we have to do is to understand the dynamics of this emission; if you understand that, then we have a fundamental theory for the strongest gravitational fields. Then all that's left to do is to fill up the middle region, handle also the moderately strong gravitational fields, and we have the whole theory.

This turned out not to be so easy at all. Hawking's argument why black holes emit particles was a beautiful one, but somewhat indirect. It did not really explain the details of the mechanics and in particular not if we would demand that a black hole *as such* obey the laws of quantum mechanics. The rather contradictory situation that we had landed into was that although Hawking could derive that black holes emit particles, the same argument would also tell us that a black hole all by itself is denying the laws of quantum mechanics; it is not obeying the laws of quantum mechanics because we could not identify the quantum states where the black hole can be in. This question was not properly addressed by Hawking's theory; the theory was basically contradictory. This has everything to do with the fundamental instability of the gravitational force. Since the gravitational force carries negative energies, we could put an infinite amount of negative energy in a black hole and the situation becomes hopelessly unstable. Quantum mechanics requires energy to be bounded from below, and if that condition is not obeyed, we have a much more fundamental difficulty in understanding what goes on than one would think at first sight. Black holes at first sight appear to allow an enormous amount of negative energy inside them. On the other hand, a black hole itself has a positive mass and *in total* does not have negative energies, so it should appear in the laws of quantum mechanics without negative energies. This was a contradiction. In other words, there was a very basic difficulty in this theory of particle emission by the black holes. Once we saw that difficulty we thought that we'd just solve that difficulty and we'll be done. But it turned out that there still does not exist a very satisfactory theory about the black holes emitting particles. Now, of course, we arrive at the topic of the string theories, which do claim to have the answers here — but String Theory does not have the complete answers. Surely, there are ideas, and indeed some vague indirect answers as to what the quantum states of a black hole could respond to have been suggested. So there is a very deep problem here and I still think that if we really understand the detailed dynamics of particle emission by black holes, we'll be close to a more complete understanding of general relativity's connections to quantum mechanics.

Will there ever be an experimental observation of these emitted particles?

Very probably not. There is one kind of experiment that people tried to do, which is with very high accelerations. A gravitational field is basically an acceleration field. If we accelerate something, we should see that the object was affected by apparent radiation coming out of its own horizon. Such experiments could be carried out but I would object that this isn't general relativity; we would just be studying the basic physics of accelerating objects, but we would be saying nothing at all about gravitational fields caused by massive objects.

I was thinking about astrophysical observations.

No, unfortunately, that is not possible either. This is because the black holes that emit sizable amounts of radiation are extremely small ones, of millimeter size or even smaller. The wavelength of the radiation is the same size as the black hole itself. From a black hole of 1 mm, the emitted radiation will be of millimeter wavelength with a temperature of a milliKelvin or so. That radiation would be very difficult to detect. We need black holes of a micron and then it would emit light but even so, think of something of a micron size emitting light — it would be a very, very tiny thing; it would be just a little speck that we could see with the naked eye in a microscope. Of course, this black hole, although of a size of just a micron, would still have a tremendous amount of mass. But the astronomical black holes are of the order of kilometers, so their radiation would be the inverse of a kilometer in wavelength, which is thermal radiation with a Wien wavelength of a kilometer that would correspond to a ridiculously low temperature, of the order of picoKelvins. Those intensities would be tremendously weak, so much lower than the microwave backgrould radiation of the Universe that there is no chance that this radiation would ever be detected.

There is the remote possibility that even tinier black holes might exist somewhere in the Universe, in which case we could try to observe their decay, but I don't see how one could ever measure both the mass and the decay intensity, so as to check the validity of Hawking's derivation. And finally there is talk of creating black holes in a particle accelerator. These would be a different kind of black holes than I have been talking about so far: higher-dimensional black holes, not the ones I am interested in.

So for all practical purposes a black hole is still a black hole.

Absolutely. The only "observed" black holes are the astronomically large ones, which present no formal difficulty for us.

I would like to ask you about your interest in holography.

It is a phenomenon in black holes, which reminded me of what you do when you make a hologram. We are discussing the information contents of a black hole. The black hole has a surface called the horizon; and when we ask questions about the quantum states of a black hole, we find that the quantum states should be described as if the horizon of a black hole was covered with information. Imagine information coming from the two-dimensional surface of a chip; there is a memory grid on a chip. A black hole's horizon behaves like the grid on the chip, on which you can put bits and bytes, except that these bits and bytes are very dense, about one for every square Planck length. This information is spread over the surface. This is a mind-boggling aspect of black holes, because the same horizon, the same surface is just a figment of our imagination. Remember that I can make a coordinate transformation and say that an object falling into a black hole does not know where the horizon is, so it does not know where the information is either, so this information could be somewhere else. It now appears that, according to the theory of general relativity, the information should be spread over the surface and not the volume. This is the strangest thing about the whole situation; one would have thought that in a decent theory of Nature, information is spread all over the volume of the Universe, like in our brains where information is stored, this is a volume effect and not a surface effect. The quantum theory of black hole suggests that information is spread over its surface and not over its volume.

This is a situation comparable with what we do when making a holographic picture. We take a three dimensional scenery; a flower, a human face, or anything else, and make a hologram. Its "information" is stored on a two-dimensional photographic plate. But when we shine a laser onto it, a three-dimensional image of the original object becomes visible again. The information on the black hole is just like a hologram of the three-dimensional world around it. Somehow the information density of the black hole horizon is sufficiently great so as to be able to account for the entire three-dimensional Universe around it. Basically there is a mapping from three dimensions to two dimensions and back, which is miraculous because at first it seems to be impossible but we must realize that the Planck length is so tiny, that the amount of information on the surface can be enormous, one piece of information per square Planck length, which is more than what we'll ever need. We should be able to handle such a situation in a good model. That's what I call the holographic principle. Immediately after this, I also explained that this principle is going to be at odds with locality. You can't map a three-dimensional Universe onto a two-dimensional Universe without giving up some versions of locality. Locality is something different from what we thought of before. If you want a local theory then you encounter a contradiction. I've tried to emphasize this contradictory situation.

However, my colleagues in string theory reacted in an unexpected way. They managed to write down theories, which appeared to show the same features; that is, that three-dimensional information seems to be packed onto a two-dimensional surface of a black hole. People argued that string theory reproduced this feature, so there is nothing to worry about. But my reaction to that argument is that if their theory reproduces this that means that their theory is not local. How to restore locality in this theory is a question that has not been properly answered. The holographic principle is an indication that locality is not what we think it is. In a complete understanding of the laws of Nature we should also address this issue and this has not yet been done. This is still a complete mystery. This has been one of my motivations to investigate the question of the foundations of quantum mechanics, finding that the mismatch between information on the surface and information in the volume must have something to do with information loss. The original equations of the Universe have information distributed over a volume but after a certain amount of time most of the details about the configurations inside the volume are lost. What survived is just the information on the surface but not what's inside the volume of the Universe. This would be an explanation how holography would work for black holes, but, unfortunately, this argument is so indirect that it is not good enough to make a decent model. I now want to see and understand how to make a model, in which these vague and ill-understood features can be made more palatable.

Is it these topics that you are mostly involved with nowadays?

Yes, mostly it is these topics that I want to understand. How can one make deterministic models that show stochastic solutions looking like quantum mechanics? I have many examples of crude models. As yet, they do not look like the real world and there are very basic problems with these examples. Therefore, I am not happy with my examples. But the examples do appear to show to me that at least the general idea has a chance of being correct. I try to find better examples, better models, more realistic ones, ones that give us some indication as to how Nature could be realizing the situation that I want to describe. Maybe there are models that show an apparent curvature of coordinates so that we can mimic the gravitational force. In particular, there should be models explaining the holographic principle, and from there it will be still a long way to understand how to deal with black holes and general relativity. One of the problems is that my ideas are too vague; they have to be made more concrete. We'll have to try to devise a more rigorous mathematics behind it. It seems that there should exist a very powerful rigid mathematical scheme to discuss the situation, but we do not have it right now. Hopefully I can make more progress there.

A totally new mathematics?

Yes, there has to be a better mathematical scheme to put all this in a proper framework. After all, we have Newtonian mechanics, classical mechanics. It is built upon a very fundamental principle, the so-called action-reaction principle. The action principle led to what is called the action function in classical dynamics, which led to a very powerful mathematical scheme for the general discussion of mechanical problems. That scheme could then be incorporated in quantum mechanics, so we have the Hamilton formalism for quantum mechanics. We have all these extremely powerful mathematical principles that give us the general features of classical mechanics in a concise form, helping us to understand statistical mechanics. Classical mechanics could be brought into such a powerful mathematical scheme that people like Boltzmann could use statistical mechanics to explain such notions as temperature, entropy, and so on. All of that works because we understand how to deal mathematically with classical mechanics and then the same mathematics could help us to formulate quantum mechanics, and understand thermal properties of quantum liquids, etc. But this kind of mathematics is lacking for the systems that I want to assess; where there is no action principle.

Do you have students?

Yes. Most of them are not directly working on these issues, because I have difficulties in allowing students to work on such vague concepts; I rather work on them myself. They might not be able to put any of that in a thesis and their job is to produce a thesis for which they need to have more concrete subjects.

Has the Nobel Prize changed your life?

Yes, it has. It's by far the most significant distinction one can get in the field of science. One change is the impact it made on the public and family members, but this perhaps is not so surprising. What surprised me more is that it had the same effect on fellow scientists, including those who knew me well before. The fact that I got the Nobel Prize should not make any difference to them. But it does. There is more respect — the Prize underscored what we've done, it now seems more important to them. I've seen such a thing happen before, with fellow scientists. I've seen how Weinberg and Salam and Glashow gained in reputation just because they received the Prize for their model, while we knew very well what they had done and what they have not done. I see now the same thing happen to myself, what we've done has been given extra weight.

You don't mind that.

No, I don't. It's just interesting.

Is there much demand on you for public appearances or committees?

Yes. The other question is, of course, how often do you agree with such requests. I could just say no. In fact, I realize that often I should say no because many of such invitations are really not in my field of science; often they have more to do with publicity, or glamour, and they do not have anything to do with real research. So why should I accept them? I do accept some of these things because I do feel the need to be an ambassador of science. I like to represent my field of science towards the public just to tell them what it is like to do elementary particle physics, what it is like to try to understand some of the laws of Nature. There is enormous misunderstanding about what science is, what physics can explain and what it does not explain about the world around us. I think I have some duties here. Because of the extra weight of the Nobel Prize, people believe more what I say than they did before. Of course, I can't meet with all the requests because I should also try to continue my research.

Are you involved in teaching?

Yes, I do teach, although the teaching load is very light here; I only teach one course. I like to teach.

What is the general level of research in physics in the Netherlands? Can you do big science in a small country?

Absolutely. Even more so in modern times than what used to be the case earlier. Two things happened: one is that science, and especially experimental science is becoming more and more voluminous, and the other is that theoretical science becomes much more complex. As far as experimental physics goes, we have large machines in international laboratories, such as CERN, which has many member states. Germany also has big laboratories, with international groups working there. Even if you live in a small country, you can work in these places. As for theoretical physics, although theories become more and more complicated, there is the Internet, and there is no reason why a theoretical scientist in Holland should be less informed than one at Harvard or anywhere else in the United States. Traditionally, being in a small country never had any negative effects on our research. Well, our share in science is modest, of course, but still for a country of barely 16 million people, we have had quite a scientific impact.

You have a very elaborate website.[1] How do you find the time to do that?

It is a relaxation for me every now and then to update my website.

You seem to have some very pessimistic views about the future. I also saw the drawings you have on the web.

That's not pessimistic. I was just amusing myself. I like to make these little drawings. Once I had a whole book with such drawings. Then a journalist came to our house to interview me — that was long before the Prize — and my wife showed him that book. Then the journalist asked if they could use them in their special New Year's edition with people's views about the future. So these drawings filled almost the whole newspaper. But these drawings are not very serious. They are just based on the philosophy that humans often copy animals in many things they can do, such as swimming or flying. Then we do things that animals can't do. I concluded that that's only because our culture is evolving faster than natural evolution.

Do you believe in having human colonies in the future outside the Earth?

Yes, I like to dream about that. In principle there is nothing standing in our way. Particularly the Moon itself is within reach. From a purely technical point of view, there is nothing against colonizing the Moon. The only real reason why we do not do it yet is that even on Earth itself there are so many places that could be colonized but have not been colonized yet. The deserts are still empty; in Canada, there are enormous territories that are empty. There is no reason why those places are uninhabited other than purely economic ones. It is not yet worthwhile to live there because life is harsh and you cannot get a good job — it is just uneconomical. To inhabit the Moon will be much more difficult than to live in any desert or tundra on Earth.

Do you believe in extraterrestrial intelligence?

No. There are two questions to be asked. One is whether there exists extraterrestrial intelligence, and the other whether it is possible to make any contact with them or whether there has already been some contact with them. That second question, I think, is to be answered negatively; they are probably just too far away. There might be a one-way contact in the sense that one day we'll find some signals. It's not totally inconceivable — maybe we can even send signals, but we won't be able to communicate because they will be too far away. As to the question of their existence, I think that very often people are just far too optimistic about the possibility of intelligent life arising anywhere. I find it extremely improbable, though not impossible, that there is life anywhere inside the Solar system other than on Earth. Some people think that there might be life on Jupiter's moons, or on Mars — I find that very improbable. We have seen on Earth that although the conditions here were extremely hospitable and in spite of all the good chances, life developed extremely slowly. Most of the other planets in the Solar system are much less favorable for life. Yes, there are probably planets orbiting distant stars that are favorable for life but their number would be very difficult to estimate. People appear not to understand the implications of large numbers. This question is determined by many large numbers that might cancel each other out. There is a large number of planets in the Universe, and you can estimate this number. There are large numbers describing how probable or improbable it is for life to emerge. Those large numbers are expected to cancel each other out, but there is no reason for them to cancel out completely. It is possible that there are billions of planets with intelligent life like us or maybe we are a one out of a billion chance for the entire Universe, it could be either way. I am a bit pessimistic and I would guess that this Earth is a very, very lonely place. Maybe there are entire galaxies without any sign of life whatsoever.

Are you religious?

No.

We haven't yet talked about your family background.

My mother comes from a family of teachers and scientists. My grandmother studied biology and then married her professor of zoology, who was very devoted to science. My grandmother's brother was a physicist, one of the last physicists who were both experimentalist and theoretician. His name was Frits Zernike and he received the Nobel Prize in 1953 for the phase-contrast microscope, more generally for the phase contrast principle, allowing one to make visible phase differences in light in microscopic samples. Of course, he was the pride of the family. My mother had another sister and two brothers; the oldest brother was also a theoretical physicist, he was my prime example as a scientist and I wanted to be like him.

My father was a naval engineer and he's worked all his life in the vicinity of ships, sometimes in oil refineries. My father wanted to arouse my interest in technical sciences, wanted me to know how a car works, or a ship.

I was not really interested in these things. I said that cars were already invented by someone else. I was much more interested in more basic theories of Nature. But then, eventually, he gave me a set to make radios and that I found very interesting. I learned much of my physics while I was in primary and secondary school from my uncle.

Please, tell us about your present family.

I married a fellow student at the time — she is a medical doctor, specialized in anesthesia and for nearly twenty years she worked as an anesthesiologist. We have two daughters. My wife continued to work for sometime after they were born but then she quit. She followed a couple of different courses, and then she became a doctor for an occupational health company. In 1994, a new law was introduced in Holland that required occupational health to be monitored. She continued this until a couple of years ago when she decided to quit and now she can accompany me more often

During the Nobel award ceremony with his family in Stockholm, 1999. From left to right: daughter Ellen, G. 't Hooft, wife Betteke, daughter Saskia (courtesy of G. 't Hooft).

on my travels. Maybe in the distant future, she will again find some work to do but for the time being she is quite busy at home.

What was the greatest challenge in your life?

I always wanted to be a theoretical physicist and understand the laws of Nature, to understand how she works.

Do you have heroes?

Not directly; all people have their strengths and weaknesses. Of course, for a long time Einstein was a great example for me. He did the kind of physics I was interested in. Then my uncle was in a way a hero; he has been an example of a very solid and reliable scientist. I learned from him that you should only make statements if you are sure of what you are going to say. Rembrandt used to be a hero; musicians, like Horowitz, for example, too. I have hobbies — you've already seen one, drawing. I used to have a very good art teacher. There is also music; when I was about ten years old, I wanted to have piano lessons and I still play the piano every now and then. My other hobby is collecting shells on the beach.

Do you ever wonder about their chirality?

Actually, I do. That has always amazed me because the large majority of shells is right-handed, but there is a small minority of left-handed shells. For most of the species, it is easy to decide which is right-handed or which is left-handed. It is clear that chirality change can be brought about by a mutation; it produces exactly the mirror image of the more common species. It also happens that within one species there are both left-handed and right-handed forms but the abnormal forms are very difficult to find.

Even if you have both right- and left-handed shells that is just morphology. If you go deeper ...

The biologists say that all the internal organs are mirror reflected, however, of course, at the chemical level they are not. Once my father took me to a factory where chalk is made from shells. There are ships that go out to the sea, dig and collect shells from the bottom. The factory has huge piles of shells, they are like mountains. Of course, it is somewhat disappointing that most of them are broken, but every now and then you can find an unbroken specimen. Of course, most of them are fossils. They are at least thousands of years old. Among these fossil shells there were one or two species that also exist today, but the distinction was that they were rotated the wrong way. So either their chirality went back to normal during the ages, or these were just a bunch of shells that were fairly rare and got extinct since then.

Is there anything else I did not ask but you would like to convey?

I have written a book, a popular book, on elementary particles.[2] There are also other books I have written. Oh, I did not say anything about my daughters. One of them was born in the U.S. and now she works at the American Embassy in the Netherlands, she has a degree in political science. She has worked with the American Embassy in connection with the Lockerby tragedy and she was assisting the families of the victims, who attended the trial in Zeist. The youngest daughter is studying to be a veterinarian; she does her studies in Belgium. In Holland we have the awful system of limited admission, the admission being based on a lottery system. I am much opposed to this system. People, who are most devoted to study should get in. But the philosophy behind the system is that they do not want fierce competition. Then some weight factors were introduced that were even more unfavorable, and unfair, to my daughter, and so she went to Belgium to study. There the first years are very tough and it requires a lot of stamina not to stop. At least there she could show her determination to become a veterinarian.

References and Notes

1. http://www.phys.uu.nl/~thooft
2. 't Hooft, G. *In Search of the Ultimate Building Blocks.* Cambridge University Press, 1996.

Leon M. Lederman, 1997 (photograph by M. Hargittai).

7

LEON M. LEDERMAN

L eon M. Lederman (b. 1922 in New York City), is Professor Emeritus of Physics at Columbia University, Director Emeritus of Fermilab, and currently Pritzker Professor of Physics at the Illinois Institute of Technology. He received the Nobel Prize in Physics in 1988 jointly with Melvin Schwartz and Jack Steinberger, "for the neutrino beam method and the demonstration of the doublet structure of the leptons through the discovery of the muon neutrino." He graduated from City College of New York, majoring in chemistry in 1943. Three years of service followed in the United States Army. Between 1946 and 1951 he was in the Graduate School of Physics at Columbia University. He stayed on at Columbia University where he became full professor in 1958. Between 1979 and 1989 he was Director of the Fermi National Accelerator Laboratory. Leon Lederman is a member of the National Academy of Sciences of the U.S.A. and was awarded the National Medal of Science (1965), the Wolf Prize in Physics (1982, with Martin Perl), and the Fermi Prize (1993), among many other distinctions. Our conversation was recorded in his office at the Fermilab in Batavia, Illinois, on May 16, 1997.*

*István Hargittai and Magdolna Hargittai conducted the interview. This interview was originally published in *The Chemical Intelligencer* 1998, 4(4), 20–29 © 1998, Springer-Verlag, New York, Inc.

Your B.Sc. degree was in chemistry. Then you changed fields.

Last week I attended a meeting in New York for the 150th anniversary of City College, which is where I graduated from. City College has had 12 Nobel laureates, of which eight are alive and five showed up for the celebration. All five Nobel laureates majored in chemistry, and all five Nobel laureates changed to other fields after the Bachelor's degree. The others were Julius Axelrod [Nobel Prize in Physiology or Medicine, 1970], Herbert Hauptman [Nobel Prize in Chemistry, 1985], Jerome Karle [Nobel Prize in Chemistry, 1985], and Arthur Kornberg [Nobel Prize in Physiology or Medicine, 1959]. The Chemistry Alumni Association invited us to give talks about the same question you asked me.

I came into college interested in chemistry. My interest was fanned by very good teaching in high school and by reading, as a child, various books and having a little chemistry laboratory in my own house. One of the books was *Crucibles* by Jaffe and another was *Microbe Hunters* by de Kruif. All of us read the same books.

I started City College as a chemistry major and I finished as a chemistry major. By the time I finished, my interest was much more in physics than in chemistry. I took a lot of physics courses as a background to chemistry. They were easier; they didn't deal with complex objects like benzene rings. Organic chemistry was very discouraging to me. Our fellow students who were physics majors were a more interesting, more engaging group of people than the students who were chemistry majors.

Then I had three years in the Army to think it over. When I got back from Europe after World War II, I immediately registered for graduate school at Columbia in physics. My mentor there was I. I. Rabi, the Nobel laureate physicist. Teaching was very important for him.

What was your family background?

Both my mother and my father had never gone beyond grade school. They emigrated from Eastern Europe to the U.S., both from the Ukraine. The five Nobel laureates around the table at the City College reunion had very similar backgrounds. Today, the superstars of the future may come from a small town south of Tokyo. You need certain elements such as strong family tradition, devotion to education, some healthy amount of insecurity to realize that you need to have hard work, and some drive for success. These are common elements of Asian families. By the third or fourth generation, they are no longer so different from all the other American families. It's lucky that there is always a new influx, and the next may come from Zaire or Brazil.

It is an interesting question why the relatively large Jewish share of excelling in education is diminishing. I think it was the combination of a tradition, which made learning a very valuable thing and the insecurity of people who were persecuted and who escaped and strived to be more successful through their children. In addition there may be some genetic disposition to learning, which may have come out of the disputations of interpreting the Bible. The kind of thinking that went on in Jewish schools was probably conducive to theoretical physics. But we didn't excel only in physics; City College also excelled in basketball and there were many Jewish players.

It was crucial to us that City College was free. My father was the proprietor of a laundry store. We lived comfortably but it would never have paid a tuition. City College had high standards but I had a good enough high school record to get in. It was during the post-Depression period so we had superior faculty because getting a teaching job was not easy to do. Although City College was just a teaching college, many of the faculty were doing research at Columbia or New York University.

At Columbia, Professor Wu was one of your colleagues.

When I was a student, she was already a professor and I took classes from her. She was a very careful scientist, very thorough, she was a leader, she had many students, she had ideas and instincts for solid measurements. One of my good friends was her graduate student and was doing something with pipettes and got some rather radioactive material on his lip, and she quickly took some Scotch tape and absorbed it on the adhesive layer instantly, to prevent any damage to the student.

Didn't she deserve the Nobel Prize for her experiment proving the violation of parity?

It was a complex thing. You can think of dozens of experiments that are deserving of the Nobel Prize but that won't get it. The Nobel Committee gave the Prize to the theorists who had suggested the experiments. That certainly was correct. Almost simultaneously with Wu's experiment was my own experiment and the Telegdi experiment in Chicago. They appeared within weeks of one another. Since T. D. Lee was a professor at Columbia, we were all under his influence.

What was unique about Wu's experiment was that she saw a way to do it. Lee and Yang published their paper in 1956. I spent the summer in Brookhaven; there was a lot of discussion of that paper. Lee and Yang suggested many specific experiments, and she saw a way of doing one. It required her to convince the people at the National Bureau of Standards to use their already existing cryogenic techniques so as to achieve the low temperatures at which the cobalt could be polarized. There were backgrounds, and she was continually worried whether she was observing a real effect. At Christmas 1956, or the New Year's party, she tells about her results to her colleagues and for the first time she says the intrinsic effect may be large. Until that time, the notion of parity violation in the decay of particles was that it may be a one percent effect. As we looked at the possibility of doing this with pions and muons, you would have to look at two successive effects: parity violation in the pion decay and parity violation in the muon decay. That would give rise to a one percent squared, 10^{-4}, beyond any possibilities of detecting it. That's when I got the idea that we could try to do this using an experiment that was already on the floor taking data. This was the famous weekend experiment, in January 1957, just the three of us. The experiment ran Saturday and Monday, after which we had a huge statistical significance. We could have published it immediately but we consulted with Wu and she still needed a week to confirm the data she had. We waited a week very nervously because the news was traveling all over the world. Finally, she agreed that she had the correct data and we published together. The Chicago data came a couple of weeks later. The Nobel Committee, faced with that, would have had to either give the Prize to five or six people or decide that once the theorists had had the idea, the experiment was trivial.

Was it?

No. Her experiment was clever and difficult. Ours was trivial. You could almost hear parity violation by clicking the switch, changing the magnetic field. The machine was off Sunday for repairs. So we spent Friday desperately modifying the experiment of this poor helpless graduate student whose apparatus was being rearranged. We got some data. By the time the machine was turned off midnight Saturday the data didn't look convincing at all. We went down to the accelerator and found that some coil we had wound on a plastic form had melted the plastic and the whole apparatus had collapsed. We spent Sunday making a more elegant winding to make a magnetic field within a block in which particles would stop. Monday, there was trouble with the machine. It finally came on late Monday night and by 6 a.m. Tuesday morning we had a 20 standard deviation effect. We had done 10 different experiments, with positive mesons, negative mesons, and so on. With five or six hours of data taking, we had established that parity violation was huge and it occurred both in the decay of pions and the decay of muons. That doesn't happen very often.

How did Dr. Wu react to the rapid success of your pion-muon experiment?

What I admire about her is that even though our experiment was extremely clear, she insisted on finishing her experiment carefully, and not being pushed into publication before she was ready. That took a lot of determination and scientific character.

Do you think the experimental proof was decisive in the recognition of Lee and Yang's theoretical contribution?

Eventually it would have been proved, if not by us then by other people. Their work was certainly worth the Prize. They asked *the Question*. How do we know that parity is conserved? Parity was such a useful thing, it was used by chemists, by many molecular scientists. It always worked. The key to Lee and Yang, the breakthrough was that they could consider that there are different forces and that different forces could have different symmetries. That was a tremendous insight.

Leon Lederman with an Einstein mascot (courtesy of L. Lederman).

For thousands of years there has been a quest for the ultimate atom. There are grand unifying theories, supersymmetry, the Higgs bosom. Would you care to comment on the perspectives of this quest?

My colleague Steven Weinberg wrote a book, *Dreams of a Final Theory*. Most of the theorists who think about these things are open-minded. Weinberg says it's not inconceivable that some day we'll find the Higgs particle and understand its law. By that time, some of the very mathematical problems with superstrings will be solved, and we'll have a theory which won't give you all details but which will tell you, yes, the laws of physics say there had to be a big bang. The consequences of the cooling of the universe was a Standard Model. It may come out of these supersymmetric string theories. If the string theories can produce the Standard Model as we have it now with the Higgs, with gravity nicely connected in there, then, more or less, particle physics has finished its job. It does not mean the end of physics. The complexity probably will never finish. Right now, physics has a lot of hyphens — there is astro-physics, geo-physics, bio-physics, and many other subjects in which physics is mixed with other disciplines. It may be that by that time we'll have to reorganize the way we are looking at the discipline of physics. On the other hand, I love the Newtonian metaphor:

> It may be that we are really like children,
> finding shiny pebbles on the shore of a vast ocean
> that we know nothing about.

We're just beginning to understand. However, I share Weinberg's optimism that we've come such a long way, that quarks are in fact points and, therefore, the simplicity comes from symmetry.

There are parallels between modern physics and the ancient natural philosophy, for example, between today's matter/antimatter and Anaximander's apeiron and the opposites separating out. Even superstrings may be related to Kepler's Harmonices Mundi.

They had the right instincts, it is very exciting. Even the notion that everything is made of water, it is not a bad guess. Hydrogen is a fundamental object. They looked very carefully. When you heat anything, water comes out. The Greeks were unique. They made an incredible contribution to Western civilization, i.e. science. Democritus was an amazing guy from the fragments you read about in, for example, Aristotle. The destruction of the Great

Library of Alexandria by fire was a tragedy. All 60 of Democritus' books were lost.

The idea of symmetry is a very Greek idea. Weyl's *Symmetry* is a wonderful little book that tells you how important symmetry is in all human endeavor. The symmetry in physics is crucial. It carries out that Greek idea that the world is simple and by simple they mean beholden to symmetry. However, it is the broken symmetry which makes the world more interesting.

Astrophysics and particle physics seem to be converging.

This is one of the most exciting things. It came about because ultimately the data that came out of the telescopes and space observatories are convincing. In the beginning, there was a small hot singularity, or almost singularity. The temperature was so high that matter was decomposed into its most fundamental components, presumably the quarks and the leptons, subject to the forces. Suddenly, the astronomers, astrophysicists, and cosmologists had to understand what the fundamental forces and the fundamental particles were. They had to come here to learn what we are learning from our re-creation of the temperatures and the collisions that only existed in the very young Universe. When the two particles are colliding you make temperatures that don't exist anywhere in the present universe but which existed shortly after the Big Bang. They also told us about that big accelerator in the sky which has no budgetary constraints. We are learning about that, and that has been useful for particle physics. When I understood this connection, we immediately invited astrophysicists here to work together with us. The mass of the neutrino is a good example of such an interaction. If the neutrino is massive, we understand that the Universe will not be open but might be closed. Imagine that someone in the middle of the winter, in the middle of the night, somewhere in Fermilab discovers that the neutrino has mass, and he knows the future of the Universe!

You say in your book, The God Particle, *"Thousands of years hence, archeologists and anthropologists may judge our culture by our accelerators."*

If you look at the standard photograph of Stonehenge as seen from above, obviously it is an accelerator. But it is a major catastrophe that the superconducting supercollider was stopped. The government made lots of mistakes, the scientific community made lots of mistakes, and we also had bad luck. The Cold War was over and people said, we are not going to spend money on things which don't advance victory in the Cold War. For scientists of my generation, there was a startling discovery that while we were studying quarks, we were not trying to find out how the universe works. We were winning the Cold War. It was a total shock because we naively thought, now that the Cold War is over, huge resources will become available to do useful things, like particle physics, education, and so on. In fact, we came to an era in which budget deficit reductions have become the ruling mood. This was bad luck. A week before they voted to kill the supercollider, they voted on the space station and it was successful by one vote. Therefore, Congress was very angry that it failed to kill something. All of a sudden, here came the supercollider with management problems. Nobody understood it anyway. Once upon a time, the Texas Congressional Delegation was very powerful; it was then (October 1993) no longer very powerful. All the reasons accumulated for Congress to say, this is a good opportunity where we can show we are tough enough to save money. Yet another failure was that it was not planned to be an international project. The one being constructed at CERN will be an international project.

You left Columbia in 1979.

I was very happy at Columbia. Being a professor at a major university is the best métier invented in Western civilization. In 1979 I was asked to be Director of Fermilab. The first director, Robert Wilson, who resigned, made a specially appealing proposal to me, almost emotional. I had been involved in managing and doing research at Fermilab from the beginning, from 1973. He thought I was unique to carry on his ideas. He also pointed out that I had never had an administrative job and it was time for me to be of service to my community. Nasty arguments, I found. It was also a good time for me to change. I spent 10 years as director and enjoyed them. We got the top quark. Fermilab has also become a center of education.

You have been interested in symmetry. It seems to me that there is a tendency to overemphasize broken symmetry. I read recently that even an ideal crystal is an example of broken symmetry.

A crystal would not be broken symmetry. Parity is a good example of broken symmetry. You see it. There is a mirror image, and the mirror image does not correspond to Nature, that symmetry is not respected by Nature. The particles and anti-particles — that is broken symmetry. We thought for a while that their product, i.e. particles and, in the mirror, anti-particles, would be a new,

Magdolna Hargittai interviewing Leon Lederman at Fermilab (photograph by I. Hargittai).

more powerful symmetry which would connect electric charge and space. The product means, reflect everything in the mirror but change particles to anti-particles at the same time; CP it's called. That was found not to be a valid symmetry back in 1965. The broken CP symmetry has a very deep cosmological significance. If the world was symmetric and you had the Big Bang, you created a huge number of quarks and an equal number of anti-quarks. Then initially there would be an equilibrium since the Universe was very dense, quarks and anti-quarks would annihilate and make radiation, but the radiation could quickly materialize into quarks and anti-quarks. You would have this dynamic set of reactions. But the Universe is expanding; it was not an equilibrium situation. The quarks and anti-quarks can always make radiation, but the radiation would get too cold to make quarks and anti-quarks, so you would have only radiation. Eventually, all the quarks would find anti-quarks and annihilate, and the Universe would be total radiation. This would make it very difficult to make planets and galaxies, and physicists and even chemists. So the fact that there is not a perfect symmetry between matter and antimatter in the CP business is the reason why we exist. That is, there is a small asymmetry in the process by which you make quarks and anti-quarks. That small asymmetry means there is a surplus of quarks over anti-quarks. This explains that we can't find antimatter anywhere in the Universe.

The other puzzle was that there are many more photons than there are quarks — quarks meaning protons and neutrons, and so on — like a hundred million times more photons. That was a big puzzle. In the Big Bang when you had an essentially unlimited energy, you should have made equal numbers of everything. The question is, where did that asymmetry come from? Now we know. Most of the quarks ate up all the anti-quarks and made radiation from that. The little left over made the planets and the galaxies. That's why broken symmetry is important — because it enabled us to exist.

There are more abstract symmetries now which don't have anything to do with space, they have to do with other kinds of coordinates. Those broken symmetries are of great interest now in helping us understand the basic laws of Nature. There are four forces: electromagnetic, gravity, strong nuclear, and weak nuclear. There is a deep symmetry, which enables us to understand the electromagnetic force, the weak force, and the strong force. Gravity is still a mystery. In order to understand the possibilities of distinguishing these forces, we look at the symmetries. This particular symmetry comes from quantum field theory, and its breaking makes the electromagnetic force different from the weak force. In other words, we would like to see a unified picture under a perfect symmetry. The picture now is, if we look at the Standard Model, we have six quarks, six anti-quarks, six leptons and six anti-leptons, and four forces covered by twelve particles; it's a mess! It's not a beautiful picture. How do we simplify that picture? There is an idea, which says that there will be a higher symmetry in which things are incredibly simpler. That symmetry is broken. The mechanism for breaking that symmetry is where the Higgs field enters. The Higgs field breaks symmetry. Because of the presence of the Higgs field, the weak force and the electromagnetic force become very different. If we had a way of magically destroying the Higgs field then the electromagnetic force and the weak force would become identical. That's the symmetry called electroweak. That's why breaking the symmetry creates the complex world. If there were no symmetry breaking, the world presumably would never evolve; it would have maintained a completely clear simplicity with maybe one particle which carries its own force, and that's all. Perhaps it's too boring. So it's breaking the symmetry, which creates the interesting complexity of the world in which we live.

Just as Pierre Curie said, "Dissymmetry creates the phenomenon." He didn't elaborate very much on that.

He had the right feeling. There is a connection.

Is there any evidence for the Higgs field?

No. There is only the fact that if you postulate the Higgs field, you then predict the masses of the *W*'s and *Z*'s, and that came out right back in the early 1980s.

Last year we visited the GALLEX experiment, but there are other experiments for detecting solar neutrinos. Why do we need more than one such experiment?

The solar spectrum is complicated. It has lots of different reactions. Each reaction produces neutrinos of a different energy. To find those energies, you need detectors of different materials. GALLEX is gallium. The Canadians use deuterium, heavy water. The solar neutrino flux is still a big puzzle. The original solar neutrino experiment used a cleaning fluid, carbon tetrachloride. The neutrino reacts with chlorine, producing argon, which is then detected by chemical means. The experiment established an anomaly. There was a big conflict between the people who understood the nuclear physics, the nuclear fusion of the Sun. From the known energy flux on the Sun, you could calculate the neutrino flux, and it was off by a factor of 3. So there was a problem with either the understanding of the nuclear physics of the Sun or the neutrinos. The consensus now is that the problem is with the properties of the neutrinos. That gave rise to the possibility that the neutrinos were doing something in the Sun that we did not have any evidence for on Earth. This is the so-called neutrino oscillations.

There are different neutrinos.

Three. My experiment in 1961 established two different kinds of neutrinos. There is also a third kind, which has never been seen directly, but the indirect evidence for it is very strong.

Would you explain your two-neutrino experiment to us?

In physics, either something is compulsory or it is not compulsory, it is forbidden. This is called a totalitarian law. So things are either forbidden or compulsory. The reaction was the decay of a muon into an electron and a photon. We studied it at the Columbia accelerator in great detail.

Leon Lederman lecturing at the centennial meeting commemorating the discovery of the electron in Cambridge, U.K., 1997 (photograph by I. Hargittai).

Many of my students got Ph.D.s looking for a reaction in which a muon goes to an electron plus a photon. Either it is compulsory or it is forbidden, and we did not see it. So we say it is forbidden. We could calculate how often it should go, knowing everything we know about these particles, and that was one time every 10,000 decays. However, we did not see one even in 100 million decays. This was a crisis, and physicists love crises. When it is resolved, you will learn something. How could you forbid this reaction? We know the muon goes to an electron plus a neutrino plus an anti-neutrino. We know that matter and antimatter can annihilate. In some way the neutrino and anti-neutrino with some probability can annihilate and make a photon, conserving energy and all the other things you have to conserve. But something in that sequence was forbidden. One possibility is when you make the neutrino and the anti-neutrino, they cannot annihilate because they are different kinds of neutrinos. We had discussions going on at Columbia in late 1959 and early 1960 and designed an experiment to see whether the neutrino was as simple as we thought it was or there were two different kinds of neutrinos. The experiment was done at Brookhaven. We set up a very intense beam of debris coming from a target with everything that the machine could make: pions and lambdas and protons and anti-protons and everything that could be made. The uniqueness of this experiment was building a wall about 40 feet thick made of iron from some old battleships and absorbing all the radiation from the machine, except neutrinos. You cannot absorb neutrinos. So on the other side of the wall we had only neutrinos. We built a massive detector, some 10 tons, of aluminum, into which we could look and watch these neutrinos reacting. The neutrinos we obtained were made in the decay of heavier particles, pions, which would decay into muons and neutrinos. The neutrinos are leptons so they go with electrons and muons. Electrons, neutrinos and muons are three kind of leptons, known in 1960. Whenever you had a radioactive decay, you would produce a muon and a neutrino, or an electron and a neutrino. When a neutrino would react, it would produce muons or electrons. We said, if there is only one kind of neutrino, then we will see a reaction: the neutrino would hit a nucleus and out would come an electron, or in a different case, out would come a muon. If we are patient, we will see equal numbers of muons and electrons. But if there were two kinds of neutrinos, the kind we were making would only produce muons. So we ran the machine for eight months; neutrino reactions are very rare, and in the eight months we had about 50 neutrino collisions. That is with a hundred million neutrinos coming every second at our detector, for eight months, we had fifty collisions. It took a lot of patience. In the 50 collisions, we only saw muons. Never an electron. We clearly then were dealing with a neutrino that was different from the neutrino that Pauli had talked about. He predicted only one kind of neutrino in reactions in which electrons were involved. In the beta decay — and beta is electron — the electron would come out and there was missing energy and Pauli said, aha, that's a neutrino, we don't see it but it's there.

He did not live to learn about the second neutrino.

No, he had already died. But he was terribly shocked by the violation of parity.

You did this experiment with Melvin Schwartz and Jack Steinberger. What happened to them?

They got the Nobel Prize. Before he received the Prize, Schwartz went from Columbia to Stanford and at some point he went into business. He resigned his professorship and became an electronics manufacturer. Somewhere in the middle of that industrial career, he won the Nobel Prize. In the business community they don't know what the Nobel Prize is. So he went back to Columbia University as a professor and now he is a Nobel laureate. In business he was just a businessman.

Steinberger, even at that time, had already moved to Switzerland, at CERN, and he is still in physics, using the machine at CERN.

You have two wonderful books with co-authors.

The first book was on particles and cosmology, so I wrote it together with a cosmologist. The second book was not my idea. It came to me from Teresi, who is a professional writer. He said, you ought to write a book. I said, I am too busy. He said, no problem, I will write it, you just make sure the physics is right. We had a long interview for three or four hours, but he did not understand physics. But he was a good writer, so we made a collaboration. I would write a chapter and he would make it better. He is a very funny guy and we had a similar sense of humor. Every chapter went back and forth between us several times, eight times, nine times, and eventually we wrote the book. It was by correspondence.

One of your favorite topics seems to be public understanding of science. But does the public understand science? In any case, physics seems to be much more successful in this regards than chemistry.

Chemists have a burden, the word "chemical". It is chemical contamination, poison. The chemical industry puts a lot of money into the American Chemical Society to change that perception. The chemists are very active. They spend 5 or 10 million dollars a year, given by industry, to change the perception of chemistry. My theory is, and this is why I became President of the American Association for the Advancement of Science, that the way to teach the public is not to do it by discipline but to talk about science. What you want is for the people to have an understanding that science is a way of thinking. If you have this way of thinking, you will do better in your personal life, make better decisions about your community, and form better opinions about the two major problems facing us, environmental decay and population explosion. The use of technology is growing exponentially but its understanding decreases exponentially. That is a prescription for disaster. We have to start improvements from the kindergarten, not for future chemists or physicists but for future citizens and voters.

I read once a short story by Aldous Huxley, *Young Archimedes*. I'll never forget that story. It has dominated my life. An English mathematics professor is vacationing in some remote part of Italy. On a walk he sees a farm child near the river with a chopstick making triangles. He looks at the triangles and he realizes this child is about to prove Pythagoras' theorem, $a^2 = b^2 + c^2$. He starts to talk with the child. The child is shy at first. Eventually, over the long summer, he teaches the child formal mathematics, geometry, algebra, trigonometry, calculus. The child is a genius. There is nothing too fast for him. At the end of the summer, the professor goes to the farmer and says, I would like to take your child to England, to educate him. He will have the best food, the best schools, he will come home four or five times a year for holidays; I think he will be a great man. But the farmer says, I need him to help me in the farm work. The end of the story is that the English professor and his wife go away and the child waves.

How many Amazon kids, how many African kids, how many children in some remote village in China or India get lost and could be a Newton or an Einstein?

*****

Art at Fermilab

What is the significance of the sculptures around the Fermilab?

Robert Wilson was an accomplished sculptor. He was interested in creating these sculptures. He had to join the welders' union to be able to participate in welding the tower in the pond in front of our main building. It is modeled after "Cleopatra's Needle", which was, as the legend goes, moved from Egypt to Rome. When it was being erected in Rome, they were controlling it with ropes and the ropes were tight and hot and the sculpture was in danger of falling over. So the commander of all of this said, throw water on the ropes. It was an emergency solution to the problem. They threw water on the ropes, the ropes shrunk, and the sculpture stood up. Wilson liked that story and felt that as a director of the laboratory, he often had to throw water on the ropes. He used to come in at 7 a.m. and do welding until 8:30, when he would become Director. He also designed

View of Fermilab in Batavia, near Chicago (courtesy of Fermilab Visual Media Services).

The "Tower" in the pond in front of the main building of the Fermilab. The "Tower" was modeled after "Cleopatra's Needle". The main building was modeled after a cathedral in Southern France (photograph by I. and M. Hargittai).

Möbius Strip behind the main building of the Fermilab (photograph by I. and M. Hargittai).

the sculpture called Broken Symmetry at the entrance to the Fermilab. He also did the Möbius strip behind the building, on the top of the auditorium.

What is the meaning of the shape of the main building?

There is a cathedral in southern France, Bouvier, after which this building was modeled.

Valentine L. Telegdi, 2002 (photograph by M. Hargittai).

8

VALENTINE L. TELEGDI

Valentine L. Telegdi (b. 1922 in Budapest, Hungary) is Emeritus Professor of Physics at the Federal Institute of Technology (ETH) in Zurich. He received his Ph.D. at the ETH in 1950. He was at the University of Chicago between 1951 and 1976, and at the ETH from 1976 till 1990. Since retiring, he shares his time between CERN in Geneva and Caltech in Pasadena. He has honorary doctorates from the University of Louvain, Belgium, the Eötvös University in Budapest, Hungary, and the University of Chicago. He is a member of the National Academy of Sciences of the U.S.A. (1968), the American Academy of Arts and Science, the Academia Europaea, and foreign member of the French, Hungarian, and Russian Academies of Sciences, the Royal Swedish Academy of Sciences, and other learned societies. He received the Wolf Prize in Physics (1991) and the J. E. Lilienfeld Prize of the American Physical Society (1995). We recorded our conversation in our home in Budapest, on November 14, 2002.*

Please, tell me something about your family background.

I was born in Budapest in 1922. Both my parents came from other parts of the country; my father from Pécs and my mother from Békéscsaba. My father went to a commercial high school and at that time he, just as many other youths, was very much attracted by French culture, the so-called "Nyugat" [West] movement and as soon as he had gotten his

*Magdolna Hargittai conducted the interview.

bachelor's degree he went to Paris; this was in 1914. A few months later, together with many other Hungarians, he was detained as an enemy alien. Two years of this internment he spent on an island in Normandy. There is a Hungarian novel, called *Fekete kolostor* [*Black Monastery*], written by a man who was interned there just as my father was. They kept my father interned for five years, and he came back in 1919. He arrived in Hungary under very turbulent conditions and worked for a while as an interpreter. He was extremely gifted in languages, so eventually he decided that he would leave the country — then there was a great surprise for him in store. Before World War I, one did not need a passport to travel in Europe, now you needed one and also a visa, which was a new invention. So wherever he decided to go he needed a visa, which was practically impossible to get. Only one nation seemed to have been so backward as not to have even heard about the existence of a visa and this was Bulgaria, so he went to work there. He worked for a transportation company. He came back in 1921, picked up a wife and went back to work in Bulgaria. I've spent the first few years of my life in Bulgaria. Then he went to work in Romania, in the delta of the Danube, for a shipping company. About a year later the family moved back to Budapest and that's when I went to elementary school there. We stayed for two years and that was the only continuous period of time that I ever spent in my native country, Hungary. Then my father got a new job in Vienna and so we moved there. A few years after that, my parents went to live in Milan but they left me in Vienna, where I attended school. I was extremely fond of that school and have very pleasant memories.

In 1938 Austria was annexed by Germany. During the Easter vacation I went to Milan to join my parents and never returned to Austria. I did not go to a real Italian school but learned Italian, which I speak to this day. About a year later, my father sent me to Belgium to continue my studies.

Why Belgium?

My father initially wanted me to go to England. He received a catalog of various boarding schools there. There were very strange remarks in these catalogs; don't forget, we are talking about 1938–1939. Remarks like: "First year students are not permitted to have their own motorcars." Well, my father never had a motorcar, so he did not worry too much about me having one. But he decided that these schools would be the wrong thing for me. He came back from England to Milan through Belgium, where he met a businessman who told him how good the schools in Belgium were and also much cheaper than in Britain. So he thought: why not? There was then a funny scene. He took me to the Belgian consul in Milan to get a Belgian visa. Of course, he displayed his beautiful French and then the consul said: one of the conditions of getting a student visa is to know one of the two national languages. Your son does not speak either. Then my father told him rather arrogantly: Monsieur le Consul, why would I send my son to Belgium if he knew French? So he sent me there. I had a hard beginning, listening to the courses, and taking down the notes phonetically because I did not know how to spell. But I learned it in three months.

In Belgium I studied chemical engineering. When I was there, it was the second time it happened that the Germans occupied the country I lived in; I was 18 at that time, so I decided that the safest thing to do would be to return to Italy. At that time Italy was not yet at war; it was in June, 1940. I managed to get evacuated with a group of Italians; Italians did not look for such fine details as to whether I was a citizen or not, so we travelled through Nazi Germany to Italy in a sealed railway car. I arrived in Italy on the 10th of June 1940, the day Italy entered the war, with a Hungarian passport. Had I arrived two days later, I could not have entered Italy. After a while my father went to Switzerland and my mother and I stayed in Milan. I could not enter any school there.

Why?

Well, because there was no direct correspondence between the kind of degrees I had and so forth; well, a good number of reasons, anyway. So I lived there with my mother and thanks to my knowledge of languages I did a lot of translations. Only technical translations, texts and patents, nothing literary. I worked for a patent attorney for three years and these were three very important years in my life. Life was very hard because Milan was bombed practically every day but I had a job and I could not leave my mother alone. I was making good money, at age 20 I already had a secretary and life was pleasant in Italy for a young bachelor. Finally, the same thing happened as it had in Austria and Belgium, namely northern Italy was also occupied by the Germans and there came a lot of complications. Anyway, in October of 1943 my mother and I managed to enter Switzerland illegally with the help of paid smugglers. There, I was interned as were all the other refugees. But the Swiss are clever, they were not going to feed you if there was somebody else who would do that, namely my father, who lived in Lausanne. So they let us out after about 3 months and we went to stay with my father. In 1944 I entered the Engineering School of the University of Lausanne. I did not have all the necessary entrance papers, but being a refugee, they were very lenient and they overlooked this. In fact, I got a scholarship from an international fund and I got my diploma in 1946 with a little master's thesis, which already had to do with physics, with radioactivity. Then I decided that I should learn more physics and more mathematics. I thought that even if I wanted to do chemistry, I needed some more basic knowledge of these two subjects. I applied to the Department of Physics of the Federal Institute of Technology (ETH) in Zurich, which was a very famous institute, and I got a reply from the head of the institute, Prof. Scherrer, that they had no room for me whatsoever.

Then I went to see a professor of theoretical physics whom I knew and told him, "Look, I want to go to ETH but Professor Scherrer sent me such a negative letter, what should I do?" He said, it was very simple, "I'll talk with him." So soon, I received a second letter from Scherrer in which he did not refer in any way to his first one but said that they expected me at the earliest possible date in his institute. So in October of 1946 I left Lausanne and went to the physics department in Zurich. At the beginning I was assigned to do chemistry for the physicists; they had radioactive substances and I had to perform certain separations. I must add that the physicists at ETH considered chemists a lower form of mankind. Anyhow, they discovered that I was doing something that no other chemists they had ever seen could do; namely, there were sheets with problems to hand out to the students and every week I solved all of them. This was something they had never seen before and never expected from a chemist. After that the head of the Institute, the same Professor Scherrer, decided that I could become a teaching assistant in physics. I also started seriously working for my Ph.D. By 1950 it was finished. I got married that year and was planning to go to the United States.

How did you meet your wife?

Well, the answer is very brutal; most people are shocked when I tell them. I picked her up on the street in Zurich. Not perhaps under particularly shocking circumstances but anyway. I was standing in front of a bookstore

Valentine Telegdi and his wife, Lidia, in 1991. Courtesy of V. Telegdi.

looking at the books in the window but in actual fact I was looking for a shop selling zippers because my briefcase had to be repaired. Then I saw a startlingly beautiful young woman, a typical Italian movie actress. So I asked her in Swiss German if she knew where such and such a zipper shop could be and she answered me: "I would prefer to speak French." When I heard her accent in French I told her, "Perhaps you would really prefer to speak Italian?" — and she said yes. We just exchanged a few words, and I decided that one has to "beat the iron while it is hot", so I invited her for lunch. I paid about a week's salary of mine on that lunch but even now I think that it was an excellent investment. We have been happy ever since.

We wanted to go to the States and it took a long time to get the visas. It was the McCarthy era and first I was refused the visa but then under pressure from the University of Chicago they granted it. They then refused to give it to my wife for the reason that she was "fascist trained". I said to the consul, look, every single one of 20 million Italians is fascist trained, that was the only school they had! Anyway, eventually that worked out.

How did I get my job in America? Most of my fellow graduate students at ETH also went to America; there were no jobs for physicists in Switzerland at all. At that time a very well-known physicist of Austrian origin came as a guest professor to ETH, Victor Weisskopf. He was a professor at MIT. He liked me and asked me what I wanted to do after I finished my degree. I told him that I wanted to go to America. "Why?" he asked, "Switzerland is such a wonderful place!" I told him, "Just look around, this is a small country; they do not even have enough jobs for the Swiss, let alone for refugees like me. The Swiss have only one desire, to see me go." They were not very friendly. In 1956 they were very decent to the Hungarians but neither before nor after that. Basically it is a country where foreigners, except hotel guests, are not welcome. So Weisskopf promised me that when he got back to America, he would try to get a job for me. A few months after he had returned he wrote me a letter and said, "I am sorry, I am unable to keep my word because next fall there is no junior faculty position opening at MIT." Then he continued, "Under the circumstances I decided to do the best that I could do for you and I have recommended you to Fermi." So I got accepted in Chicago, which at that time was the absolute top place in physics in the whole world. I've always called it the Mecca of physics. It was a totally unbelievable place.

What was your thesis work in Zurich?

It was on nuclear physics, which was a hot topic at that time. I worked on a certain photonuclear reaction, on a reaction induced by photons. I don't think that my thesis was so great but there was something special about it: half my thesis was experimental and half of it was theoretical. I wanted to do my own theory for my own experiments. That brought me great difficulties. When my thesis was ready, I had it typed and bound and presented it to Scherrer. He told me, "You know, Telegdi, we should have Pauli as your second examiner."

Well, there are two things I must tell you. Wolfgang Pauli was one of the greatest physicists who ever lived. He was professor of theoretical physics at that time at ETH, and he was extremely critical of everything and everybody. People were terribly afraid of him, not as a human being but as an authority, which he was. I don't mean just the physicists next door, but even people like Niels Bohr. So the fact that Scherrer wanted to have Pauli as my second examiner was not a source of great joy to me. Fortunately, everything went very well; there were no problems with Pauli during my doctoral exam; in fact after the exam he personally invited me for a glass of wine. He told me about his own doctoral exam with Sommerfeld, how it went, and I had good relations with Pauli ever after, till his death.

What did you start working on in Chicago?

At that time they had two accelerators in Chicago. They had the cyclotron, which had been completed just a few months before I arrived, and that was really Fermi's machine; it was *the* avant-garde thing. Then they had another accelerator that they had bought a few years before from industry. It is very rare that you buy a whole accelerator from industry but there was one, called the betatron and it was just standing there once the cyclotron started to operate. Everybody went to work on the cyclotron that time, so I decided to work on the betatron. Of course, the betatron makes photons, and reactions generated by photons were my specialty, so for a little while, for at least a year, I continued to work on what I did in Zurich. People were very happy that at least somebody was using this older machine. However, soon I switched.

The main purpose of building the cyclotron was to produce a type of particle called pion. These pions generally came from cosmic rays and this cyclotron was, after Berkeley, the second machine to produce dense artificial beams of pions. These positive and negative pions were of great interest to Fermi and his collaborators and they concentrated all their energies on studying them.

Unidentified person, Valentine Telegdi, Wolfgang Pauli, Philippe Choquard and Aloisio Janner (from left to right) during the Padua-Venezia Conference in 1957 (courtesy of Valentine Telegdi).

Beams of pions also came with other particles, called muons, which were considered just dirt, simply background. A muon is a kind of heavy electron. Pions were the useful particles and muons were the background; in fact the muons were the decay products of the pions. So after a while I decided that I would like to work on those background particles that nobody seemed to have any interest in. That's what I did for the next 15 years; my name is completely connected with this particle called muon. In fact there was a time when some American physicists referred to me as "Mr. Muon".

I think that we are getting close in time to the discovery of parity violation. Lee and Yang came up with the idea that parity might be violated in the weak force. They also suggested experiments, for example, the one that Madame Wu did with beta decay...

You know, nothing upsets me more than when you or anybody else refers to Madame Wu in this way because I can tell you the whole story.

That is exactly what I wanted to ask you to do!

OK, you want to know the story? Of course, the whole story could take hours, but I'll tell you a short version. First of all, let me tell you that the experiment that we did was trivially simple. It was embarrassingly simple and anybody could have done it. However, what was not simple was the idea itself, the decision to do it. I was advised by the most senior people in my department, the most senior experimentalist and the most senior theorist, that doing it would be a waste of time. Lee and Yang themselves were not so sure; parity violation was just one of the possibilities they explored. Lee and Yang did a wonderful thing; they said, we all believe in parity conservation, we all believe in mirror invariance, but it really has never been tested, so let's indicate a few crucial experiments that could test it. One of the experiments was the symmetry of electrons to be emitted from an aligned polarized nucleus, to which you and most of mankind refer to as Wu's experiment. Well, that is very romantic but it is false.

You see, in order to do this experiment, you have to align nuclei, which in 1956 was an art, a technique known only to a handful of people in the whole world. In fact, hardly to anybody in the United States. The two people at the Bureau of Standards, one of whom was Ambler, had been imported to the Bureau of Standards from Oxford, because they had this monopoly in their hands. So Madame Wu had to look for somebody who knew how to align nuclei. People like T. D. Lee give out ideas but they do not worry about such small details as how to align a nucleus; that is not their province. They soon found out that the Bureau of Standards had two Englishmen who knew how to do that. So she proposed to these people that they do this experiment together. The heavy part, the significant part, the difficult part was done by these people at the Bureau and not by her. Her specialty was radioactivity, she knew how to count the beta rays that would come out; but about the alignment technique that was the crucial part of the experiment, she knew strictly zero. So to give full credit to her is a crime!

Why did Lee suggest it to her?

He suggested it to the whole world; it was in their paper. But of course, they were both at Columbia. Also, she was a beta-decay expert. He also told her what nucleus would be particularly suitable, as not all nuclei can be aligned, so it was a well-chosen radioactivity. But anyhow, she was hardly ever in Washington; she would come every now and then for a few days but to call it the "Wu experiment" is criminal. At least give all the members of the group equal credit and do that in alphabetical order.

All three of them were on the paper, weren't they?

Yes, but she did the horrible thing; she put her name first, although it starts with a W.[1] The others were English gentlemen and did not have the courage to object. Actually I must say that Ambler made a big career out of this; he became the director of the Bureau of Standards. There was another experiment Lee and Yang suggested. That is that when a pion decays into a muon, the muon comes out polarized, which means it comes out spinning in the direction of flight. So the suggestion was; if you could demonstrate this, then you would have proved parity violation. That is the experiment that Jerry Friedman and I did together. We did it with so called nuclear emulsions — these are thick photographic emulsions that can reveal the path of a particle. It is like footprints in the sand. You let the pions get into the emulsion, they come to rest and then decay into a muon yielding another track, about half a millimeter long. Then in turn the muon decays emitting an electron. We studied the symmetry of this decay. If the muons are polarized, the electrons come out in an asymmetric way. We observed that. Unfortunately, some time was wasted and lost. It was just at the time when my father died and I had to come to Europe and I left Friedman alone. This created a certain amount of problems. A similar experiment was done at Columbia University with electronic techniques, by Garwin and Lederman.[2] Because of my traveling we got to the journal with our paper a day or two later than they did.[3] So there was a question of priority. But the important thing was that most people thought that parity violation was completely crazy, something that would never happen and the second point was that many people thought that it would be a small, tiny effect, therefore the experiment would be a difficult and long one. Well, my intimate conviction was that if parity violation was present the way Lee and Yang thought it might, it would not be a small effect but a large one, so we could undertake the experiment and we just ought to look into the microscope for these tracks; so I concluded that it could be done in three months. So when people told me that I would waste my time, I told them, I could afford to waste three months of my life on this. The real decision was to do it and not how to do it.

Lederman said their experiment was completed during a weekend.

That's correct.

Was that basically the same as yours?

Yes. But that was done by electronic means. It turned out that there was a piece of equipment that one of Lederman's students had built and it could be used for it right away; they had to build nothing. All they had to do was to make a coil. But all the key ideas in that brilliant experiment of Lederman were not due to him, they were due to Richard Garwin, who also happens to be of Hungarian extraction.

Apparently they started their experiment much later than you did yours.

Much later.

Have they heard about yours in the meanwhile?

I have no idea. I only know that for a long time Lee tried to persuade Lederman to do this experiment and Lederman never knew how to do it. Then Dick Garwin heard about this during a lunch conversation and told them how to do it and they did it in one night. But the key thing which we do not read very much about in the literature, is that when Garwin and Lederman did this experiment, they already knew a preliminary result of the cobalt experiment in Washington. They knew that parity was violated in cobalt decay — which I did not know. It is also said that they wanted to get into print before Madame Wu but they were stopped from doing that.

By whom?

By their colleagues. That really would have been too much. I can tell you things about Lederman. For example, he tried to defame our experiment. He tried to tell people that we never did our experiment, and that I invented our results, that I faked them!

Why would he do that?

So that we could not get into print.

How did he learn about it before it was published?

I called Garwin on the phone and told him. I said to him, "I know that you have done this work, we have done it too and what is your result?" Then Garwin said, "What is your result, you tell me first!" Then I gave him our result and he said, fine.

So there is a certain animosity between you and Lederman, as I see.

Of course.

Have you discussed this with Leon Lederman ever?

I don't think so. There is also a book he wrote, *The God Particle*. After that book came out I sent him a horrible letter. In that I wrote, among other things: "This book contains a degree of misrepresentation that one would not expect even from an Iraqi diplomat." He never answered. I do not even exist in that book. In January of 1957 I had reached the point that I wanted to leave physics forever. I thought that if all you get for something good you've done is such a degree of mistreatment, then what's the point. Anyhow, I obviously did not leave physics, because some other people encouraged me to stay. Also, some time later, I even collaborated with Lederman's co-author, Dick Garwin, on a very important experiment at CERN. There has never been bad blood between us.

Have you ever talked with him about these parity experiments?

Of course.

What did he say?

Basically that he had nothing to do with what Lederman did. You know, the intellectual ratio between Lederman and Garwin is roughly the same as between me and Enrico Fermi. Worlds apart! Richard Garwin is one of the most brilliant people I have met in the last fifty years.

The Nobel Prize for Lee and Yang was one of the fastest Nobel Prizes ever; within a year after their publication.

Yes, that happened because they had proposed something negative. Instead of saying how things are, they said how things are not. They simply said that people believe in parity conservation, but they have no reason to do so because it had never been tested. None of the previous experiments had any bearing on this question. So they gave suggestions for experiments to test it.

By the way, they are not perfect either. They forgot one of the simplest experiments one could do. That is the polarization of electrons emitted in beta decay. That they overlooked although it was such an easy one that people in Albania could have done it.

It is still interesting that they got the Nobel Prize so soon. Nominations for the Nobel Prize have to be submitted by the end of January of the same year. Were the experimental proofs known by that time?

Oh, yes. The New York meeting, where all these results were presented was at the beginning of January of 1957.

Why do you think the experimental people were not included in the prize?

I don't think that anybody among the experimentalists deserves the Nobel Prize very much in this case. If an experimentalist performs an experiment with known techniques and on top of it that experiment has been clearly suggested by the theorists, where is the merit? This is true for me, too. It could very easily have been the case, with slightly different circumstances, that we would have gotten the result first. If we could have done our emulsion work a little quicker, if we had done a few more scanners, etc., even then it would not have contributed to our cleverness, or glory. None of the experimentalists deserved the Prize in this case.

Anyway, the Nobel Prize often prevented people for the rest of their life from working. I think it is a mixed blessing. Once you receive the Nobel Prize, you are forced to participate in activities for which you have no competence whatsoever. Enrico Fermi was not a very humorous man but once he made a humorous comment. When somebody asked him what Nobel Prize-winners have in common, his answer was, "Not much, not even intelligence."

I must say that the discovery of parity violation had one important consequence. It created an enormous number of new experiments that you could not have done before. There was there the possibility of all kinds of measurements that people never had thought were of interest. In the space of one and a half years, these new experiments taught us more about the nature of beta decay than the forty years before. We found a real law of beta decay, the so-called "V-A theory". I contributed to that. I organized an experiment at Argonne National Laboratory, which was the kind of experiment that Wu and her colleagues did, but not with cobalt but with the free neutron, which is much more important. There we measured the symmetry up and down for the electron and also for the neutrino. That experiment did an enormous amount to reveal the real nature of beta decay. So I got into the field of weak interactions. That neutron experiment was much more important in my mind than the parity experiment. I did not put my name first on the paper. The people who knew how to polarize neutrons were the main authors of the paper.

Is this experiment you just mentioned the most important you have ever done in the field of weak interactions?

Probably. It revealed the structure of the coupling of beta decay, that people call the V-A interaction. You see, all β^- decay is fundamentally due to the neutrons except that the neutrons are bound. Why look at the complicated situations when you can have it in free form? Again, for the reason that not everybody knows, that is how to polarize neutrons. That experiment has been repeated 3 or 4 times but nothing new came out of it. I liked that experiment very much.

Which other experiments of yours would you like to mention?

In 1959 I took a year's leave of absence from Chicago. Dick Garwin and I who had been competitors in the U.S.A. went together to CERN in Switzerland. There we did a very beautiful experiment. We were surrounded with young people but Garwin and I were the leaders of the experiment, called the g-2 experiment. It measured the magnetic properties of the muon with a high accuracy. We really established at that time that the muon is a heavy electron and that was very important to recognize. It was a very elegant experiment. It was very funny, you know. When the experiment was finished, I said to Garwin, "You know, Dick, I was afraid of working with you because I found that if you were with an experiment there would be nothing left for me to do." He said, "Well, any time you want a job at IBM you can have it."[a]

Other important contributions from my group in Chicago were the various properties of muon that we investigated between 1956 and 1972 non-stop. All kinds of properties, decay properties, magnetic properties, weak properties, you name it and we studied them; there is no point going into the details.

Parity violation is all about symmetry. What do you think of the importance of the symmetry concept?

Modern physics is completely dominated by symmetry considerations but I think that symmetry considerations, even if rarely so, could be misleading. For instance, it is the strict formal application of symmetry conservation that had prevented people from discovering parity violation before. Somehow people thought that mirror invariance was a principle, but it was not a principle, it was just an experimental fact. When people would write down an equation the first thing they would check was whether it was symmetric under inversion and, if it was not, they would throw it out of the window. I think that it had become a habit of thinking that restricted, in that particular case, what people would do.

When parity violation was discovered I was interviewed over the phone by a man from *TIME* magazine. He asked if this was important. "Well," I told him, "I'll give you an example. We've all been taught in school that two parallel lines meet only at infinity. Now if somebody suddenly discovered that they actually meet much earlier, would not that be an important discovery?" Then he understood it. There was supposed to be

---

[a]By that time he was already at IBM.

Russian academician, Alexander Skrinsky, presents Valentine Telegdi with the medal of the members of the Russian Academy of Sciences at CERN in 2000 (CERN Photo, courtesy of V. Telegdi).

a large article in *TIME* magazine about our later, very important neutron experiment. But the day they had this article more or less ready, the sputnik went up, so we never made *TIME* magazine.

The sputnik was probably a great shock for American science, was it not?

Oh, yes. But not only for science, for the general public, too. In fact more for the public than for science. It also had a positive effect in helping science, but not as deep an effect as some people think. People hoped that it would do a lot for improving American education. Not so much American research but American education. People realized that a certain type of schooling was much better in Russia than in the United States and they wanted to do something about that.

The support of science in the U.S. had much to do with the war and Los Alamos. It was for the first time that the military had understood how valuable science could be to the country. That is the origin of the large scale of support for science. The sum that scientists spent during the war years was still very small compared to military expenditure. There was a certain feeling of gratitude towards the scientists. Unfortunately, all that is gone now, all those people who were grateful are by now dead. So we have to start all over again. In America many people confuse science with technical progress. They think that science will produce faster cars and better toothbrushes, but that has very little to do with science.

This also may have to do with the fact that science education in the United States is not at a very high level.

Not only science education, all high-school education! There are many people in America who are trying to cure that; Leon Lederman is one of them. But this is completely wrong. He knows of the school to which he and many other Nobel Prize-winners went, called the Bronx School of Science. He thinks that he could create such a school in Chicago for the black children; he wants to save the black children through science. I say that what the black children are lacking is not good schools but good parents. He is confusing the black ghetto of Chicago with the Jewish ghetto of New York. The Jewish ghetto of New York consisted of uneducated people who wanted their children to get educated. That is not true of the black ghetto.

How can that be changed?

Not by teaching them physics! I mean the children come home, nobody looks after them, the girls become mothers when they are 15 years old …

Maybe they ought to try to change the attitude of the parents.

But there are no parents! Or maybe even worse, there may be three children from different fathers, who never show up.

So what's the way out?

I don't know, that is not my line of business. But I assure you that the idea of copying the Bronx School of Science is not going to change this. In my opinion that is pointless.

What do you think of the relationship between experimental and theoretical particle physics in general?

One cannot give an absolute answer, it depends on where and when. There are certain places where there is a great deal of interaction between theory and experiment. This is true to many parts of the United States. In other parts there is very little communication. There are theoreticians who cannot do useful work unless they have a puzzling experimental result; there are people whose mind is stimulated by some questions that are produced by experiment. There are, of course, people who invent questions that can then be tested by experiment; that is another thing. European physicists tend to be much more mathematical than the Anglo-Saxons, the Anglo-Saxons are much more pragmatic. It is interesting from the sociological point of view why parity violation was discovered experimentally in the United States and not in Europe. It could have been done in Europe. Imagine these Englishmen, who went from Oxford to the Bureau of Standards. There are two labs in Europe where the grandfathers of nuclear alignment were: Leiden in Holland and Oxford in Britain. But both failed to do the experiment. I don't know why. In Oxford, I think, one particular reason was that the local theoreticians were absolutely not favorable to this parity violation idea, they did not think much of it, they thought this was too wild. That I know. In Holland they had other problems. At least, there was a very powerful lab in Holland and I think they did not make a big effort, only a very small one.

What is your relationship with the theorists?

I have a very good relationship with a certain group of theorists. But then, again, there is a problem. There are very few theorists who understand an experiment. Some of the greatest theoreticians do and some of the greatest ones do not. I'll name two. Feynman understood experiments perfectly. He would look at an experimental paper and he would discover sources of errors for the experiment that the experimentalists had not thought about. He really understood it. On the other hand, a man like Pauli would not spend one minute on an experimental paper, he just did not understand these things. There are many theorists who decide that they are not even going to try. Just like many experimentalists refuse to learn certain kinds of mathematics; they would say why should I learn this, it is useless. I am very fond of theory, I've written several theoretical papers myself. My mathematics is not very good but I can think of certain problems in theoretical physics, so my papers in theoretical physics I've always written with somebody. Generally the idea was mine, but the technical execution I could not do by myself, I did not have the mathematical power for that.

When we are talking about the future of particle physics, what do you see there, larger and larger accelerators or more underground laboratories?

I think there is room for both. Some of the most exciting things are what a Japanese and an American just got the Nobel Prize for and those were done with underground experiments. But they can also run out of things to do. The trouble with experiments of that kind is that you have no control of your sources. If you have a hundred neutrinos per second you can do anything you want and there is nothing else to do but make larger and larger detectors because the incoming flux is completely fixed. But we have certainly learned much more about the Universe in the last 20 years than we did about particles. Now you can see the sky with X-rays, which is a completely new thing, and then the orbiting telescopes show things that you could never imagine to see. I think that they are making tremendous progress, maybe they are today like particle physics was in the sixties. But they also tend to be collective enterprises; people who want to do space physics sometimes have to wait three or four years before they can get their instruments into space.

I see a lot of problems in the way experimental particle physics is being done today. There is a style that may not attract young people. If I were a young man today probably I would not go into particle physics. I don't like to work in a group of 500 or 1000 people. I am more of the nature of an artisan than anything else. I like my independence; if I had the possibility to chose between two experiments, I would choose the one in which I have more independence and not the one that is "more important". I never had a boss in my life, I was very fortunate, and I could always do what I wanted to do.

So you are saying that today there is not much room for individual contribution?

Well, much less than there used to be. Of course, this is understandable with the big accelerators and the space experiments. There are wonderful experiments, for example, which people do nowadays with cold atoms and which were not possible earlier. I did not say that if I were young I would not go into physics at all. I do not know of any subject that would attract me as much as physics; but I cannot see myself in such a large group. Nowadays nobody knows how to do the experiment, everybody has to rely on somebody else, on great experts in this or that. But again, building an accelerator is like building the Brooklyn bridge; it is a tremendous work.

I think that there never will be built another circular accelerator of any importance, unless somebody makes a great invention. I think that the rule that the next one will be the same as this one but ten times larger shall be abandoned. Of course, you can abandon that rule only if someone invents a totally new principle. There, I have great faith in mankind. I have found that every now and then somebody comes up with something totally new and then that advances the field in an unexpected way. I have enormous faith in human ingenuity. Now, it's possible that if I were a young man I would not do any particle physics because of these large groups. But it is possible that I would work with two or three people and invent some new kind of instrument, one which I would be happy to give to others to make new discoveries with. But I would not abandon my desire to work with a small group. I think that the most important thing in doing research is to be original. If you have a smaller group, a smaller time-scale, it is easier to be original.

What do you think of the future of physics in general?

I think that during the next ten years the fantastic developments on the border between physics and astronomy will continue. They will bring us more in the next ten years than we have learned in a century. Maybe there will be completely new technologies; so many people try now to make computing devices which involve only individual atoms. I think that there is a chance that they could succeed. It would be an unbelievable revolution. That research has billions of dollars behind it; more than the particles.

Do you think about possible new experiments nowadays?

Today I would not have the energy nor the money to put together a little group.

Is it possible that in order to explain all the new findings in cosmology, for example the anti-gravity force and others, there will have to be a totally new physics?

Maybe, but maybe not. It is not clear that any of these require new physics. Maybe one just needs to complete the physics that we already have but not change its foundations. Today general relativity is not an academic subject, it is practically engineering. You know this GPS system one has, the so-called General Positioning System, the map in your car helping you to orient yourself? This instrument receives waves from satellites. If they had not taken general relativity into account none of them would work. So it has become really engineering.

Why do you think Einstein did not like quantum mechanics?

Because he was putting philosophical considerations ahead of physical considerations. He included a certain type of philosophy, which extended to Nature. He had no real prejudice about physical laws; in fact he could do the wonderful things that he had done because he could think completely freshly about any subject. But he had deep philosophical principles. He deeply believed in determinism. He did not say that quantum mechanics was wrong, he just said it was not satisfactory; that he could not accept a theory that makes only probabilistic predictions. This is really an expression not of his thinking of physics but about his own philosophy. Why not? There are many puzzles at the foundations of quantum mechanics that people keep discussing and there are many people still discussing these difficulties. There is a thing which I call "Alka Seltzer physics". What is "Alka Seltzer physics"? A man tells me, "When I think about such and such a proposition in quantum mechanics, I feel sort of ill in my stomach." Well, take Alka Seltzer. I cannot judge physics on the basis of anybody else's emotions. It is very individualistic what bothers one's emotions and what does not. It is not when we say that something contradicts an experiment. Determinism is a way of looking at the world. I am not thinking about these things nowadays. There was a time when I used to discuss them with Gell-Mann at great lengths but that was more than 10 years ago.

Fermi had a very simple point of view: quantum mechanics is correct because it works. Here is a machinery that gives results and agrees with Nature, so he never discussed these problems, they never bothered him at all. He was completely pragmatic and he was a very great man. He also had a very charming wife, did you read her book, *Atoms in the Family*? It is a wonderful book, read it.

You received the Wolf Prize together with Maurice Goldhaber. Is there any reason in the citation why you received it together?

They said that we had done similar work, similar in style if not in practical details. I can tell you a little story: when I got the Wolf Prize, and I was to split it with Maurice Goldhaber, many people asked me whether it annoyed me to share it. I said, quite the opposite. When I was a young student in Zurich, one of my great models in life was Maurice Goldhaber. He was the kind of physicist I always wanted to be, so to stand up there together with him was one of the greatest honors. The money is not so important — I have a good retirement — I did not need it.

Why would people think that you would be annoyed?

Because many people would be annoyed to get only half of the money. Then there was a very amusing thing. When we received the prize, we were given two minutes to speak and to thank. There was a little discussion between me and Maurice as to who should give that talk. I said to him, you are the older, you should talk but I will write the text. So that's what we did; of course, he approved the text beforehand.

In retrospect, which of your work do you remember with the greatest pleasure?

It is as if you asked, which is the most beautiful woman that you ever saw? Well, I would say that it certainly is not parity violation. Because, I love experiments, where there is some cleverness, where it was not obvious how to do it. For that experiment, if we had not done it, somebody else would have done it. This is less so for the Garwin-Lederman experiment that was very clever.

Well, I think that there are four papers, which I like. Certainly, the paper on neutron decay had the biggest impact in terms of physics results. Then there is a paper that did not have a big impact, it's a paper on the helicity of the muon neutrino. It is the same experiment that Goldhaber and company had done for the electron but we did it for the muon neutrino. That paper was done in a collaboration with one of my Hungarian friends, Laci Grenács. Further, there is our work, both theoretical and experimental, on the spin dependence of muon capture. The fourth is a theoretical paper, about what governs the motion of spin in an electromagnetic field, which is called BMT, after its three authors, Bargmann, Michel, and myself. It is used all over in many experiments; it is a theoretical paper that had an enormous usefulness. It has gotten into the textbooks, too. I am glad that I also have a theoretical paper that had an impact.

Do you have heroes?

You know that I have declared publicly that the true religion of scientists is Shintoism, because Shintoism has two important principles. One is the worship of ancestors; here, I mean Newton, Einstein, and some others. The second principle of Shintoism is the admiration of Nature. These two principles make Shintoism a very appropriate religion for a scientist.

While we are at it, what is your relationship with religion?

Zero. But not quite. I am in no way anti-religious in the sense that Steven Weinberg is. He has published one or two violent articles against religion. I know him very well, he is a brilliant man, and he writes well. I wrote him that I agree with his articles, except for one small point; everywhere where he puts religion, he should insert one word, "organized". I am opposed to organized religion; I have absolutely no objection against people having some kind of personal belief. That is a very personal thing and I don't think that science has to say anything against religion. Just as religion has no reason whatsoever to say anything against science. I think that the religious thought is a strictly personal matter. I spoke to you about the BMT paper; M stands for Michel, who was a French physicist friend of mine, and an orthodox Catholic. Very rare among physicists, but he was not clerical, just religious. It has been a great experience for me to know him.

Were your parents religious?

No, absolutely not. My father thought that it was evil to let your children go to religion class until at some age, say 14 or 15 when they could choose for themselves. Of course, if you have been educated in this way, by the time you are 14 of 15 you are lost.

What was their religion?

Well, I suppose that my mother was clearly Jewish and my father only remotely.

When you mentioned that you had to move so much during your childhood, was it mainly because of this?

Of course.

Do you feel Jewish?

Rarely.

Your wife is probably Catholic.

Yes.

Do you have children?

No.

    Coming back to your previous question, of course, if I identified strongly with the Jews I could not identify as strongly with the Hungarians.

How about the many Hungarian Jews?

Well, they seem to be a little mixed up these days.

What was the greatest challenge in your life?

I would say not to be thrown out of the Institute at the University of Chicago after having been there for a year or two. Most people who came there as young persons never expected to stay longer than a year or two, it was like being a postdoc nowadays. So the greatest challenge was to be permitted to stay there longer.

You have met many of the great men of physics during your life. Would you mind telling me something about them?

Yes, I've been very, very lucky. First of all, I want to tell you that I am one of the very few people of my generation who have known both Pauli and Fermi. What luck! I knew Fermi quite well, and my wife knew Mrs. Fermi very well and kept in touch with her after Enrico's death. I joined the Institute in Chicago practically at the same time when Murray Gell-Mann did. He is one of the greatest physicists in the world — or at least he was. I also became very early good friends with Dick Feynman.

    Fermi was a man who was 100% devoted to physics. There was very little else for him in life, although I think he was a good husband, perhaps even a good father. But he was not interested in human relations at all. He was very reserved; he lacked many of the qualities that characterize Italians; he was not musical, he was not interested in food, he was not interested in being well dressed; but he was a genius. He acted as a very, very modest man, very understated, but that does not mean that he did not know who he was. I don't think that he was such a wonderful human being except for the fact that he was very simple and unpretentious and very available, but the essential contact with him had to remain professional. You could discuss with him any, I mean really any, topic in physics, whether it was geomagnetism, or particle optics, he was interested and he had a good knowledge of every type of physics. At age 50 the man was still extremely ambitious and working like one who still had to find a job. But I found Pauli to be more interesting as a human being.

I would say that the most interesting physicist I met in my whole life was none of those whom we mentioned here. Unfortunately, I only met him for a few hours, but I will never forget that meeting. That was Lev Landau of Russia. He was really a phenomenon, I was extremely impressed by him. I don't remember what exactly we talked about but it left me with a great impression. He was a very argumentative person as opposed to Fermi. You could not involve Fermi in an argument in a subject that did not interest him. Szilard was very similar in this; you could never talk to him about a subject of your choice; it was impossible. You could either talk with him about what interested him or nothing, "Nem érdekel!" (I am not interested, in Hungarian) — he would say.

My wife also met most of these people; she still thinks that Szilard was the most interesting among them. She worked for quite some time as a secretary to him. Indeed, Szilard was very impressive; not only because he had such an original mind, but also because his interest went far beyond physics. It involved biology, statesmanship, etc. But he was very self-centered; and basically, I think, he did not understand people at all. He was super-ultra logical; he did very few experiments himself but he loved to generate ideas for experiments. He had many good ideas and he would run around the country telling people what they should be doing. But he never understood, in my opinion, that people prefer to work on their own stupid ideas rather than his brilliant ones. I think that he would have had a much bigger impact on science if he had just sat down and done these experiments himself. It must have been very strange when he collaborated with Fermi because they were the exact opposites of each other. I don't know how it worked but the result speaks for itself; after all Fermi and Szilard share the patent on the uranium reactor. (Let me tell you, Fermi was not the kind of guy who would share credit with anybody if he did not have to.) Fermi would have been absolutely fair; he would never have done what Madame Wu did, put his name first; but he was very ambitious. Interestingly, when I was in Chicago, both Fermi and Szilard were there, but I have never seen them together, not for a minute.

What kind of a person was Szilard socially?

He could be very charming. He was always a man who lived by himself in a hotel. You know that he had for many years a lady friend and then finally he married her. But after they got married, they went to live in two different cities. So I used to say: often a man who has a girlfriend for many years is under terrible pressure to marry her and nobody knows what to do about it except Szilard. In my opinion Szilard solved this problem. He probably offered her either to see him but not to marry or to marry him and not to see him, and this must have been what she chose.

What is your opinion about the large number of excellent scientists of Hungarian extraction at that time?

It had to do with a certain Budapest milieu and the very good high schools. But I also would like to point out that all these people got their university education outside Hungary. I always say that the *numerus clausus* seems to have worked in their favor because that's why they got a better education. I wrote about this somewhere; I like paradoxes.

During the war they all worked together, after the war they went different ways; Szilard was involved with the peace movement while Teller with the thermonuclear weapons.

In a very simple way you could say that both of them, to some extent, had the same idea; how to save the world, and they reached opposite conclusions. Many people ask me about Edward Teller and I have an analogy, which is wrong perhaps. I think that Edward Teller suffers from a disease that has not been uncommon among Jewish people in the past two or three hundred years; he thinks that he is the messiah. That he, Edward Teller, personally has to reveal how to save the world. He does all kinds of things that only a person would do who is deeply convinced that he knows what's good for the rest of mankind. Szilard was never as messianic as Teller.

Whose side would you take?

I believe that in peacetime I would take Szilard's side. When I went to the United States in 1951, I knew about secret work. Of course, for the first 5 years there was no danger that I would be involved with secret work because I was not a citizen. But after that I could have been. However, I decided then that as long as there was peace, I would not work for the U.S. government on a secret project because I would have no say about what's done with my work. Now, in wartime anything you do can be considered as defense. You could kill one or two million people and you could always write if off as defense. There are people who tell you that dropping the bombs on Japan saved a lot of lives. Well, I've seen only the ones that have been lost. One can theorize that lives were saved. But I have seen how many people came back from Los Alamos with a certain amount of crisis of conscience. So I decided not to work for the government in time of peace. I may be wrong; nobody is perfect.

What is your opinion of Teller's testimony at the congressional hearings on Oppenheimer?

Well, maybe he thought that Oppenheimer was a hindrance to his way of saving the world. Oppenheimer was also an extraordinarily arrogant person — maybe that irritated Teller too. Fermi did not care too much for Oppenheimer either. Another problem may have been that Teller used a very unfortunate way of phrasing at the hearings. It was all innuendo. The only man who gave a good testimony was Bethe. He said, "I have been asked to do this work, and after a lot of soul-searching I decided that I would do it. Some other people reached the opposite conclusion. That does not mean that I am the good guy and they are the bad guys." He is an honest man. When Fermi was testifying, the man asked him whether he wanted to hear about the consequences of perjury. Fermi said, "I don't think that I shall ever have a need for that."

What is your opinion about the 2002 Nobel Prizes in physics?

They were good choices. Although I must say that of the three winners I am very familiar with the work of two and hardly at all with that of the guy who got the Prize for X-ray astronomy. But I know the Japanese very well; in fact he invited me to accompany him to the Nobel ceremonies. Davis is unfortunately already 87; a little too old. I think that Davis and Koshiba make a wonderful pair because without the work of Koshiba and his group, people would not have accepted Davis' result.

Why not?

He found a deficiency in the solar neutrino flux and many people were not prepared to believe it on the basis of one experiment, even if that experiment took thirty years. You know the thing I regret is that they let Pontecorvo die — he was the author of every significant idea in neutrino physics. He died around 1990. By that time Davis' experiment was already done but not the Japanese's. Pontecorvo was a wonderful physicist. That he could for so long believe in the Soviet system is a mystery to me. I knew him quite well. In fact one of the things I like best is a letter I got from him congratulating me on the Wolf Prize.

Was he an Italian?

Yes, he was born Italian but then around 1947 decided to move to the Soviet Union and lived there ever after. He was a very rabid communist. For a long time they would not let him out, of course.

What do you think of women in science?

Why don't you ask me what I think about German people in science?

I could ask that, too, but now I am asking about women.

To me that question does not exist. However, I made a very nasty remark at one time.

That's why I wanted to ask you about it.

Once I went to a feminist meeting and I wanted to get the attention of the people so I got up and said, "I think that the role of Madame

With Magdolna Hargittai during the interview at the Hargittais' home, 2002 (photograph by I. Hargittai).

Curie has been greatly exaggerated. If I had been married to Pierre Curie, I would have been Madame Curie, too." There arose, of course, a big noise in the room. The whole discussion was about role models; boys have so many role models but girls have only Madame Curie. Then I explained my theory: if you go to the movies and see Greer Garson as Madame Curie, maybe you would like to be a scientist for a whole week, if you were a girl. But if you had a high school teacher, perhaps a woman, who was good, perhaps you would like to become a scientist for life. It would be very important to have excellent women science teachers. Many of us men were launched into our careers, and this includes Feynman, by a brilliant physics teacher, and this would work for girls as well.

This, of course, is very important. But it does not mean that a successful woman scientist as a role model would not be important to have.

That just makes the statement more precise but nothing will change my point of view on the impact on a boy, or on a girl, of a good science teacher. When I was a little boy in Vienna, my physics teacher was wonderful. When he found out that I was interested in atoms and was reading popular books about them, he said, "You know, you are wasting your time. You should not read these books. There is only one that is good about atoms and that is Sommerfeld's *Atomic Structure and Spectral Lines*." That was, of course, a real professional book. I was about 15 years old. So I went to the library and took out the book. I started to read it but then I went back to my teacher and told him that I did not understand the book because there were all kinds of snake-like figures in it. These were, of course, the integral signs. "Ah," he said, "I forgot that you do not know calculus. Well, in that case you'll have to learn it." So I did it by myself and then I got something out of Sommerfeld. That physics teacher later became a very strong Nazi and he lost his job after the war because of that. I went to see him in 1964 and it was a very interesting meeting. He explained to me how he had lost his job and I explained to him how I was fleeing from the Germans; it was a very interesting occasion. I think I was probably the only student who ever corresponded with him after leaving Austria.

Was it due to this teacher's influence that you became a physicist?

Only partly. Later I also had other influences. But already at age 14, it was very clear to me that it was some sort of science that I wanted to do; there was no hesitation in my mind. My parents much preferred that I go in the direction of chemistry because there it would be easier to find a job.

Coming back to Madame Curie, do you think that she was not a great scientist?

She was a great scientist. She did not have a great mind but she was a great scientist. Her husband was a great scientist and also had a great mind. But she was a great scientist by what she had done; you do not have to be a great mind to do great science. Many of the Nobel Prize-winners are not great minds. But Pierre Curie would have been one of the greatest scientists of his time even without radioactivity, just on his own. He had also this very unique combination of theory and experiment, which was very rare and has remained extremely rare.

Why did you leave the University of Chicago?

Well, it would take hours to explain. But I had decided that I should end my life in Europe. In the United States there is one thing you are never forgiven for and that is your age. People there are absolutely obsessed by youth. When you are young and talented, you will succeed in the United States; but if you are an older person, especially if you don't have any children; your position is difficult. I also felt that the University of Chicago was decaying. I made many attempts to stop that decay but when people did not listen to me, I said, well, I have become a prophet in the desert, so I should be leaving. I also realized that the time was near when nobody would make me an offer anymore, so I decided to go back to Europe. I also did not like the way research was funded in America. I also had some problems with my retirement. So I came back to Europe. I had two groups, one in Zurich and another one in Geneva. I also liked the idea of starting from zero. Back in Chicago I started to feel that I was becoming part of the furniture; people were just thinking that Telegdi was always here. Here, in Europe, I started different experiments, some even in atomic physics, which is not so much my field. But my wife was very unhappy to leave the United States. She threatened me with a divorce. She did not like Switzerland; she remembered how badly we had been treated before we left.

Is she happy now?

No. It is an armistice. But we go to America every year for three months, to Caltech; that is as much America as I need.

What do you spend your time with nowadays?

I never really had any hobbies. Physics has been my profession and also my hobby. Sometimes I listen to music. Now I will read Mr. Kertesz's book, which I hope is not boring. Most of the books I start to read I put down after 30 pages because I really don't care; most of the books are about people whom I consider irrelevant. I have retired from work 10 years ago but still go to CERN every day; I go to seminars, I read the journals on the screen now. There is one thing I miss, that is teaching. I've always loved to teach and I think I was good at it. I miss the teaching more than miss the research. I also like the history of physics very much, so I keep myself busy with that.

Reading or writing?

I have written articles on the history of physics. This is, of course, an old-age activity for a physicist, but I can say, as my excuse, that I also did it when I was young. For example, I like family trees. I like to put them together. For example, who was Wigner's teacher, who was the teacher's teacher, and so forth. I am really interested in these intellectual connections, who comes from where, etc., etc. At conferences, on Wigner, you hear so many false statements. Some guy declares that Maria Goeppert Mayer was a student of Teller. No, she was not; it makes my blood boil. People do understand that physicists are physicists, and that they work hard and they try to be self-critical. But then when the same person tries to do history he loses all his credible faculties. It is very hard work to do the history of physics if you want to be serious, just like anything else. But if you don't do it seriously just forget about it.

Thinking back on my life, I have always thought that the real goal of experimental physics is to get the best and most precise results with the simplest and most economical means. I don't like to shoot with cannons at sparrows; some people consider this a form of elegance. I did achieve that. When I was 40 years old, I told my wife what the goal of my life had been: it's always been to be respected by those people whom I respect. I have achieved that goal so I am very happy. Most people think that they have not gotten enough recognition from the world. I feel that the world has given me much more than I ever thought I should get. So

I do not feel neglected by my colleagues and my peers. Many people at my age are very bitter; I feel no bitterness, no reason for it.

References

1. Wu, C.S.; Ambler, E.; Hayward, R.W.; Hoppes, D.D.; Hudson, R.P. *Phys. Rev.* **1957**, *105*, 1413.
2. Garwin, R.L.; Lederman, L.; Weinrich, M. *Phys. Rev.* **1957**, *105*, 1415.
3. Friedman, J.I.; Telegdi, V.L. *Phys. Rev.* **1957**, *105*, 1681.

Val L. Fitch, 2002 (photograph by M. Hargittai).

9

VAL L. FITCH

Val L. Fitch (b. 1923, 42.98 degrees north latitude, 101.71 degrees west longitude, Nebraska, U.S.A.) is Emeritus Professor of Physics at Princeton University. He received his Ph.D. at Columbia University and has been at Princeton University since 1954. He received the Nobel Prize in Physics in 1980, together with James Cronin, "for the discovery of violations of fundamental symmetry principles in the decay of neutral K-mesons". He is a member of the National Academy of Sciences of the U.S.A. (1966), the American Academy of Arts and Sciences (1966), and the American Philosophical Society. He received the E. O. Lawrence Award (1969), the Research Corporation Award (1967) and the John Price Witherill Medal of the Franklin Institute (1967, both together with J. Cronin) and the President's Medal for Science in 1993. We recorded our conversation in his office in the Department of Physics at Princeton University on October 30, 2002.*

Perhaps we can start with your family background.

I was born on a cattle ranch in northwestern Nebraska in an area called the sandhills, which accounts for giving my birthplace above in navigational units. This is about 5 miles from a small town (population about 300), Merriman, Nebraska. I had an older brother by 10 years and a sister by 6 years — they have both recently died. They went to a one-room school, getting on their ponies and riding to the school about 1.5 miles up the

---

*Magdolna Hargittai conducted the interview.

valley from our ranch home. When my brother entered high school, I was 4 years old at the time. We moved to the nearby town of Gordon, Nebraska so I never had the experience of attending the one-room school, as they had. I received all my elementary education in Gordon, Nebraska, through high school.

After that I went to a local college, Chadron State College, which is about 50 miles west of Gordon. There I took science and mathematics courses until I was drafted into the U.S. army; I had had about 3 years of college at that time. This was in 1943. In the army I had the usual basic training and then I was sent to an army specialized training program (ASTP). With nearly everyone a young person in the military, I suppose they were trying to guarantee the continuation of a certain amount of

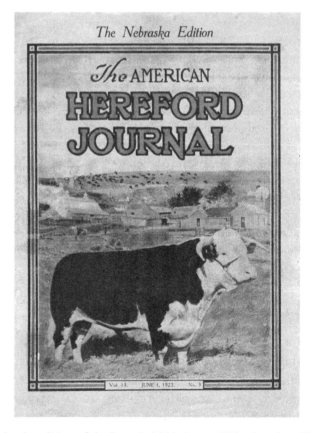

Cover of the Nebraska edition of the June 1, 1922, issue of *The American Hereford Journal*. It shows the ranch of Fitch's parents a year before he was born. His older brother and sister are playing in the hay (courtesy of V. Fitch).

technical expertise in the country. I was sent to Carnegie Tech, now Carnegie Melon University, in Pittsburgh. I was there for 7 or 8 months. In 1944, with the invasion of Europe imminent, nearly all of the army people from Carnegie Melon were sent to join the 95th infantry division. However, I was an exception and was sent to work at Los Alamos. For me that was a most fortuitous army assignment. I was assigned to work for a member of the British mission, Ernest Titterton (after the war he emigrated to Australia, and later became Sir Ernest). He was involved with many aspects of the development of the implosion gadget. We were a small group; Titterton had two civilian technicians and two so-called SEDs working for him. The SEDs, or special engineering detachments, were the army personnel who were assigned to work on the Manhattan Project. By the end of the war about half of the technical staff at Los Alamos were SEDs. With Titterton we did many interesting things. Our most important assignment was making the fast timing measurements of detonation phenomena. This activity lead to our participation in the test of the implosion gadget (the bomb was always referred to as the gadget) at Alamogordo in the summer of 1945. At the time of the test Titterton and I, along with a dozen others, were at the main control bunker.

We had two jobs at the time of the test. One was to measure the degree of simultaneity when the detonators went off around the gadget. The other job was to send out the signals to various experimental groups in the last milliseconds before we sent out the pulse to detonate the thing. These were our two main responsibilities and that's how I happened to be at the main control bunker during the test. Of course, everything operated automatically, so a minute or so before the scheduled detonation Titterton suggested that since there was nothing else for me to do, I might as well go out of the bunker and witness the test first hand. So I took the piece of dark glass that somebody had handed to me; a piece of dark glass that is used in arc welding helmets, it is really dark and very good for watching solar eclipses. I took that glass and went out along side the bunker; there were three or four other people there, Kistiakowsky was one of them, I remember. We stretched out on the ground, the dark glass cupped over our eyes, and watched the explosion. Of course, it was an indescribable experience. We were 10 thousand yards (somewhat less than 6 miles) south of the tower on which the gadget was exploded. I remember there was a military policeman at the door of the bunker, I suppose to control access to it. Nobody had warned him of what to expect and I remember his face just being absolutely white after witnessing the test. My only comment to him at the time was that the war would soon be over — and of course, it was.

Recently I went to a memorial service for Boyce McDaniel at Cornell University. He was also involved in the test, and at the service I was asked to describe McDaniel's contribution to the project at Los Alamos. At the time of the test he had the job of monitoring the neutron activity in the gadget. This was done by running a manganese wire down a thin hypodermic needle-type tube into the center, into the core of the gadget, next to the plutonium. Every four hours he had to extract the wire and replace it with a fresh one. The neutron activity in the gadget would induce radioactivity in the wire. This was monitored with a Geiger counter, which provided a measure of how many neutrons were bouncing around in the plutonium core. It was all to make sure that nothing was getting out of control. The gadget was placed on top of a one hundred foot tower. In order to change the wires and to reach the gadget he had to climb to the top of the tower on an open steel ladder; and this every four hours. I had been up that ladder a number of times because we were making our own measurements on the gadget, and I knew what it was to climb it. The

Young Val Fitch riding a horse (courtesy of V. Fitch).

last time McDaniel went up was at 2 a.m. in the morning on the day of the test. At that time there was a thunder and lightning storm over the desert and there he was, all alone, climbing that ladder and tinkering with the bomb in the middle of that storm. One can only imagine what went through his mind at that time.

After the explosion, I turned off the apparatus in the control bunker and with Titterton drove back to the base camp, which was about four miles further south. When we arrived back at the camp, there was McDaniel and Rabi getting a bottle of scotch out of the trunk of the car. We joined the little group. McDaniel, Rabi, Titterton and I all had a good swig of the whisky and then all of us went to bed — we had been up all night. My comment at Cornell was, "Imagine what a bonding experience it was, after just witnessing the successful testing of that thing we all had been working on so hard."

After the war I still had some undergraduate work to do, so I went to McGill for a year. Why McGill? Titterton had come to Los Alamos by way of McGill and he had some good things to say about the school; that's why I went there. After that year I went to Columbia for graduate school.

Just staying with your Los Alamos days for a moment. How do you feel about the bombs dropped on Japan?

I know only too well all the arguments, pros and cons, about that. I was in the Army. I considered myself lucky not to be out there being shot at. I had many friends who were preparing for the invasion of Japan. Because of the use of the bomb, as I told the military policeman, the war was soon over. I think that the use of it must have prevented the death of a very large number of not only Japanese but also U.S. military. I don't have any reservations about whether that was a correct decision or not; I simply think that it was under the circumstances.

At Los Alamos you must have met many important people. Is there anything interesting to tell about them, about their working style or anything else?

Certainly, I met many physicists there. Titterton always had the pleasant habit of introducing me to various people, after all I was a GI, most of the time in a fatigue uniform, and to a civilian we all looked the same. The guesthouse of Los Alamos — it was actually a guest room — was immediately adjacent to the house that Titterton lived in. So whenever a guest would come, Chadwick or Bohr, he had the happy habit of introducing me and I appreciated that. At the same time I met many people on the ski-slope of Los Alamos; I learned to ski there and that was a very good occasion to meet. I remember meeting Fermi there. In fact, I met more famous people on the ski-slope than at work.

What is your opinion of Edward Teller and his activities?

I am not very sympathetic. I thought that he did not want to cooperate with Oppenheimer at Los Alamos, so Oppenheimer let him do his own thing and he chose to think about fusion weapons, in those days called the "super". He would not work as part of a team, trying to do the principal job at Los Alamos. Also, after the war was over, we were all still there in the Army; the SEDs were still there in the fall of 1945. The academics decided to start a small university program at Los Alamos, and various people agreed to give courses. I signed up for a course under Teller. Almost immediately I took a dislike to his lecturing style; which was always directed toward a particular person in the audience, the son of a well-known physicist. He ignored all the others. But after a couple of lectures, he never appeared again; he always had some subordinate give the lecture. This was not true of the other lecturers. I also took a course with Rossi, the Italian physicist, and he was always there and never assigned that job to somebody else. So very early I developed a not-very-positive attitude towards Teller; rather negative, as a matter of fact. Then, of course, I was in a position where I always admired Oppenheimer greatly. I thought that he had done such a super job at Los Alamos under very difficult circumstances. Then, of course, after I came to Princeton, I got to know him better as an individual as well. I have great difficulty in trying to rationalize Teller's behavior at the time of the Oppenheimer hearings in the early 1950s.

You said that you were working with the British colony at Los Alamos. Did you meet Mark Oliphant there?

No, I did not. But Titterton had been a student of Oliphant and had an enormous respect for him. Oliphant was also one of the leaders of the movement that involved most of us who had worked on the bomb. After the war we wanted to stop the proliferation of nuclear weapons and keep the genie in the bottle.

When did your interest in science start? It must have been before Los Alamos; you mentioned that you took science courses already in college.

I've been reflecting on this; it's an obvious question to which I don't have an answer. You can imagine that in a small town in northwestern Nebraska there was no external force that would especially encourage me to do science. I just had an intrinsic interest. I had a small laboratory in the basement, I had radios, I had a chemistry set, and an enormous curiosity. Science was something that I found extraordinarily interesting. One of the difficulties I had when I was growing up was that there was no one who could answer the many questions I had. I had to search for answers in the small local library. But that is a rather interesting situation because, even today when faced with a question, my first instinct is to go to the literature and search for the answer instead of asking somebody. It was something I was forced to do when I was very young and that habit continues.

You wrote somewhere that your thesis work at Columbia on the η-mesic atoms was a pioneering work. Would you care to say something about that?

The history of how I happened to get involved with that is somewhat interesting. My thesis advisor was Jim Rainwater. Incidentally, one of the reasons I went to Columbia was that the Physics Division leader at Los Alamos was Jerry Kellogg, who'd worked with Rabi in the molecular beams laboratory at Columbia before he went to Los Alamos. He was the one who encouraged me to go to graduate school there. He volunteered to write a letter of recommendation to Rabi for me, so when I first went there I went to the Chairman's office, and the chairman happened to be Rabi. He immediately called Rainwater and sent me upstairs to Pupin laboratory to work with Rainwater. Rainwater was to receive that Swedish prize in 1975.

It was in 1949, as I recall, when Rainwater called me into his office, which he shared with Aage Bohr. He introduced me to Bohr and said that Bohr had just received the preprint of a highly interesting paper by John Wheeler. It looked like there were all kinds of possibilities there for a thesis. So I took the preprint and that's how my work on η-mesic atoms started. John Wheeler had written this paper in which he had pointed out that the energy of the radiation emitted as a negative muon comes to rest in matter would be a sensitive function of the size of the nucleus. The radiation is emitted as the muon cascades down through the various Bohr orbits ending in the lowest $1s$ state. The radii of the muonic atoms are 212 times smaller than the corresponding electronic orbits; the 212 being just the ratio of the mass of the muon to the electron. In heavy elements, such as lead, the orbit of the muon in the $1s$ state is so small that it spends 50% of the time inside the nucleus. Wheeler pointed out that if the lead nucleus was a point charge, the $2p$ to $1s$ energy difference would be over 20 MeV whereas with an extended nucleus the energy would be 4.5 MeV.

We were fortunate enough that the Columbia cyclotron was just then coming into operation to provide the muons and also detectors sensitive to photon energies of a few MeV were becoming available. So we set about designing an experiment and it was very successful. We observed this curious radiation for the first time and discovered that the nucleus was very much smaller than had been shown before. It created quite a stir at the time and for me it was just a marvelous Ph.D. thesis. To see something new that has never been seen before and then to have an impact as it did was just a wonderful experience for me.

You also wrote somewhere that you almost missed it.

Oh, yes, that's another story. Wheeler had calculated that the $2p$–$1s$ transition in lead should be something like 4.5 MeV and so we were looking around in that region. In fact, it turned out to be 6 MeV because the nucleus was so much smaller. Our initial scans of the energy region extended up to about 5.5 MeV and we were getting a little frustrated by not seeing anything. It was Rainwater who suggested that, well, let's look over a wider energy region. And he turned out to be right! So when you are looking for a new thing, you should not restrict yourself.

Considering your Nobel Prize-winning discovery, were you actually looking for CP violation and if so why? When, after Lee and Yang's suggestion, symmetry was found to be violated in parity, it was a very comforting idea that CP was still symmetrical. Why were you looking for violation in that?

Before parity was found to be violated, parity invariance was a sacred invariance principle. It was just because of the tau-theta puzzle that people, in particular Lee and Yang, started to think about this. The idea that invariance principles were always good things to test was actually stimulated by the proof of parity violation. We learned that we should not take anything as a given. Physics, after all, is an experimental science and one should not accept anything

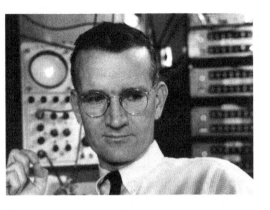

With the apparatus at Brookhaven, 1965 (Photograph by Bruce H. Frisch, with his kind permission; courtesy of V. Fitch).

unless it is backed up by an experiment. So the work was done in that context. In our proposal to Brookhaven to do the experiment, which was only two pages, there were a number of things that we had listed, one of them was testing CP violation, which we felt we could do much better than had been done before. The other things we proposed to test was, for example, coherent regeneration, especially in the case of hydrogen because there had been some reports from a group at Yale that funny things were happening in hydrogen.

You mentioned that there had been other previous attempts by other people to test CP violation.

Yes. Physicists are always interested in setting new limits. The decay of the long-lived neutral meson to two pions was long recognized as a signature of CP violation. At the time we did the experiment there were two results that I remember, but perhaps there were more. One was by the original Lederman group, which had observed some 300 decays with none of them fitting the two-pion mode. There was also a Russian group that had done a similar thing. I think they had seen 700 decays and no evidence of π^+ and π^-. These are the two results that I remember now. But, again, among the cognoscenti there were people who were aware of the fact that this was the signature of CP violation. We certainly felt that this was something that was very interesting to look at.

How long did it take for you to find CP violation?

Things went along much faster in those days than they do now. We proposed the experiment early in 1963, and because of a fortuitous set of circumstances, we were able to start the experiment in June 1963. We took data in June and July, then analyzed them. The data was all on photographic films. It took a while to go through all the rolls of films we had. It was in December of 1963 that the first evidence that there was something funny going on in the part of the data that came from a big helium bag which was our vacuum tank. That signature was obvious. Then we spent the next six months trying to explain our results some other way, but we never could, so then we published it.

Did you know already then that this is a Nobel-caliber discovery?

Oh, yes. We were sufficiently sensitive to this; that we did everything we possibly could to find an alternate explanation. I think that's the best way to put it.

Why do you think it took the Nobel Committee about 16 years to recognize this discovery?

Well, they are, in general, rather conservative. Of course, this is not a question for me to answer. The fact that CP violation implies time reversal violation, which is something that is difficult for anyone to accept, that certainly must have affected the Nobel Committee. That they were so slow in recognizing it, it must have been that this is a discovery that has such conceptually profound implications. I think that's what made them very hesitant; they wanted to be really convinced that it was correct.

This is especially interesting because the Prize for parity violation to Lee and Yang was one of the fastest prizes ever.

That's true. But time reversal violation transcends many ideas.

You used the expression "profound implications".

One of the reasons is that the idea of time symmetry is so built into one's psyche, I suppose. For example, in chemistry you learn that in a chemical reaction A + B gives C + D and then you learn that when everything is properly taken into account, C + D can, in the same way, go back to A + B. This sort of reciprocity in all chemical reactions is taken for granted. This is taught in all beginning chemistry courses. To have something suddenly show that it's not always true is a little hard because it strikes such a deep feeling against what one has always been taught to believe. Whereas

Five Nobel laureates in 1984, sitting, from left to right, Chen Ning Yang (b. 1922, Nobel Prize in Physics 1957) and Isidor I. Rabi (1898–1988, Nobel Prize in Physics 1944), standing: Val Fitch, James W. Cronin (b. 1931, Nobel Prize in Physics 1980), and Samuel C. C. Ting (b. 1936, Nobel Prize in Physics 1976). Courtesy of V. Fitch.

parity itself is a little mysterious in its own right, I guess it did not affect me so much when it was found to be violated. Eugene Wigner was highly disturbed. I think until the end of his life he was just terribly disturbed by the notion of parity violation; he just could not understand how that sort of thing could happen. He always thought that there must be some other explanation for parity violation — and he probably felt the same way about CP violation too.

Pauli was also much disturbed; he reportedly said that "I don't believe that God is a left hander".

That's right. And it all started with the tau-theta puzzle. One of my colleagues, Sam Truman and I at one time had organized a meeting here to discuss the problems caused by the tau-theta puzzle. We had the great experts all there; Wigner, Bargmann, Wightman and a few other people from the department. The purpose was to explain what was going on and whether they had any thoughts on the subject. Wigner was absolutely convinced that something like parity cannot be violated and there had to be something else; so he was not at all trying to consider any possibility like parity violation. It took a couple of young Chinese to worry about that.

Perhaps this is because Wigner was so convinced about the importance of symmetry. But symmetry is still important in much of the world.

Oh, of course. And so are violations of symmetry.

Do you see a connection between symmetry violation in particle physics and in biology? Chirality is a symmetry violation in living organisms.

People have tried to tie one with the other but to my knowledge that has never been done successfully. But many people tried. It's an intriguing idea and I have the feeling that somehow at some level one must be able to make the correlation — but not at our present level of understanding.

What is your idea about this biological asymmetry?

Ultimately there has to be something driving it at the fundamental level. Whether it is the weak interactions or something else, I don't know. As I said no one has been able to successfully make that correlation with the weak interactions. But I should think that there is something that's driving it; like the weak interactions. Maybe someday we'll be able to make that correlation.

As I understand, your discovery made it possible for the first time that an absolute distinction between matter and antimatter could be done.

Oh, yeah.

I can imagine a scenario that this could be important when an alien comes to Earth and then we can explain what we are made of. Otherwise what's the importance of this?

It was Sakharov who first pointed this out. He wrote a famous paper in 1967 and he spelled out the three ingredients necessary to explain the absence of antimatter in the Universe. One of them was CP violation, the other the non-conservation of baryons, which still has not been observed. The proton has not been observed to decay; everyone expects that it will be at some time. The current limit on the proton lifetime is something like 1033 years; depending on the decay mode you're talking about. That experiment, of course, is one of the main reasons why "K" and super-K was built in Japan. There are other experiments, the IMB experiment in the salt-mines in Ohio. Those experiments were all stimulated by attempts to see proton decay, but so far it has not been observed. I suspect it will be some day. Certainly everyone expects it. The third ingredient was the requirement that the universe be in a non-equilibrium state. With big bang dynamics, this is easily achieved.

Can CP violation shed some light on the origin of the Universe?

Well, we're getting back to Sakharov, because he was the first one to point out that CP violation is required to generate this matter-antimatter asymmetry in the Universe. One can turn the argument around saying that the first evidence for CP violation was this matter-antimatter asymmetry in the Universe but we did not recognize it. If one makes the simplest assumptions within the framework of the Standard Model then the CP violation observed in the weak interactions is not nearly large enough to account for the matter-antimatter asymmetry in the Universe. That's a bother. At the same time one has to know more about the Standard Model before one can have any great confidence in these estimates based on the simplest possible models that we now deal with. Do they have any validity or not? We don't know how many Higgs particles there are, for example, and so on and so forth. It's all a wonderful and interesting question and a very difficult one to resolve.

Is the Standard Model the right model for the Universe?

No one has found any departure from it so far, including these new experiments on CP violation involving B mesons that have been conducted at KEK in Japan and at SLAC. It appears that the CP violation that is being observed in the B-meson system is just what one expects on the basis of the Standard Model. That's been a disappointment. One was hoping that there will be a totally new regime of Nature to look at but, at least at the present time, it isn't showing us anything new.

CPT invariance still holds. Is anyone looking for a violation in that?

There have been a number of experiments very specifically looking for that. Actually one of the best systems to look for that kind of thing is in the K-meson system. It has been explored exhaustedly. Through exhaustive analysis of the decay, one can answer the question, is it CP or CPT that is being violated? The answer is that it is CP and time reversal. CPT is a very solid symmetry principle at the present time. One is talking about only 1 part in 1019 or 1020, it is just a very solid symmetry.

People are hoping that the Higgs particle will soon be found. When will it be, and will it give an answer to such questions as, for example, where mass comes from?

Well, of course, that is the reason for the Higgs-fields to exist, to have some mechanism for particles to have mass. First one has to find the Higgs and then we can really address these questions in detail. Does the Higgs exist, how many Higgs particles are there, etc. etc.? The fact that there is a mechanism for producing mass, I think, is extremely interesting, but precisely how this mechanism manifests itself depends on all the details of the Higgs, and we don't know that.

Do you expect it to be found?

If I were a younger person I would be looking for it!

Isn't it just the question of the energy you can reach with the accelerators?

Before LEP was turned off, there was a lot of excitement because it appeared that there was a suggestion, at the very maximum energy LEP could reach, of a signal characteristic of the Higgs. The evidence is not compelling.

What is the theoretical explanation for CP violation?

That is a very deep question. CP violation now has a home, in the so-called CKM Matrix, which is part of the Standard Model. But no one really understands it yet. What is the fundamental cause. Maybe it's tied to the fact that honeysuckle grows up the tree as it does. It certainly exists and it has a natural home in the weak interactions, but as far as a fundamental understanding, it still has to be found, we don't know it yet.

I would like to go back for a minute to that original parity violation paper by Lee and Yang. Interestingly, in that case only the theoreticians got the Nobel Prize who predicted it, but not those who proved it by experiment. Why did that happen?

Let us see, there were four people, I think, who did the Co-60 experiment and they all contributed to it in a major way. Madame Wu is often given the credit but I think that the most dispassionate view would be to recognize that those other guys were very important and it would not have happened without them. And, of course, once that experiment showed the effect, then it was natural to look for it elsewhere. In particular, parity violation became so obvious in the muon decay, which Lederman and Garwin showed. Spectacular results, but they did not do it until after they knew the results of the Wu experiment.

There was also the experiment by Friedman and Telegdi and there seems to be some controversy between the stories of Telegdi and Lederman.

I know. One of the difficulties is that the results of Telegdi and Friedman, which were obtained in photographic emulsions, standing alone, were not very strong in my opinion; statistical certainty was not there. They needed more data. I don't think, as a stand-alone experiment, it demonstrated parity violation. Eventually it did but not at the time. But I know that story too well and I am not sure, what the story really is. But I heard and read all variations. I don't believe Telegdi ever did another emulsion experiment.

Do you consider yourself an experimental physicist?

Absolutely.

Isidor Rabi and Val Fitch on the occasion of Fitch receiving the Columbia Distinguished Alumni award, 1985 (courtesy of V. Fitch).

Yet, you still think about these questions.

Well, of course. They are so intriguing, so intriguing, that you can't avoid thinking of them.

What is your current work?

Writing biographical memoirs and such. I've reached that age. I am not doing any experiments, my last graduate student left about six years ago.

I've read somewhere that you were working on the question of the "hexaquark".

That was my last great experimental interest. We did not find it, otherwise you would have read about it in the newspaper. If it exists it's a new form of matter, stable, with six quarks involved instead of just the three that we know of now. I was intrigued by the whole idea of a new form of matter, and depending on the mass of the hexaquark, if the mass should turn out to be close to that of the deuteron, for example, it would live long enough that it would have cosmological implications. So the whole idea was intriguing and we pursued it for a number of years but never did find it. The hexaquark has all the quantum numbers of two lambda particles, so there are two up-quarks, two down-quarks and two strange quarks in this particular *H*-particle, *H* for hexaquark. If it exists, one should be able to produce it in machines like the one in Brookhaven, that's where we looked for it. We looked at it by trying to dissociate it into two lambdas when passing through matter. We did the experiment, we did see two lambdas, but unfortunately they did not show the characteristic mass. We were happy to see two lambdas but it did not show the characteristics of a single particle decaying.

Are other people still looking for it?

No, that experiment is over. The graduate student involved is a very good guy. Fortunately my last graduate student was one of the best, and he has gone off to do other things. You probably heard about the SNOW experiment in the mine in Canada. The student, Josh Klein, is the head of the data-analysis group there. But I think it is still worth looking for the hexaquark; probably at present we do not have enough sensitivity to find it. All the theoretical calculations suggested that it should exist. If one has any confidence in QCD, then we should still look for it and one just has to be clever enough to find it. If I were somewhat younger, I would still be after it.

In what way would it be a new state of matter?

In the sense that one is so used to neutrons and protons and three quarks, so if we have something that is composed of six quarks and is stable, I would call it a new state of matter.

The Nobel Prize-winners last year [2001], for the Bose-Einstein condensate, also characterized that as a new state of matter.

Certainly, that was a very clever experiment. Probably one can characterize it that way; it's OK, I wouldn't take anything away from them.

What do you think of this year's [2002] Nobel Prize in physics for the neutrino counting experiments?

Ray Davis should have been given the prize years ago. Now, as you know, he is not really alive.

I tried to interview him some time before his Nobel Prize, but I was too late.

This is such a tragedy. When given the Prize, one should at least be able to enjoy it for a while. It happened to Fred Reines, who got the Prize for doing the original detection of the neutrino. He was in that same state as Davis is today. I think that it's true, they don't give the Prize posthumously, but how do you define it when one's brain is gone.

I was interested to see that they did not give it for neutrino oscillations. I would guess that there is a Prize for that in the future also because I think that is a profound development. Then, of course, Giacconi getting it for essentially opening up X-ray astronomy is important. Giacconi was here at Princeton back in the 1950s for a couple of years as a postdoc before he went off to Rossi's group at MIT.

Do you think that John Bahcall should have received the Prize together with Ray Davis?

Well, that's the reason I said something about neutrino oscillations. It's very difficult when one is restricted to three and I think that it's a good idea that they do restrict it to three because otherwise it gets diluted so much. But they may have had something in mind when they did not give it for neutrino oscillations, so John should be patient.

Bahcall did much together with Davis.

But he did not have anything to do with Davis' experiment, which I find such a remarkable experiment. It is true, John was calculating what the flux should have been, and that certainly helped encourage Davis to make the measurements. Davis probably needed some support. But I know from Maurice Goldhaber that he had a lot of support in Brookhaven. I was such a profound admirer of that experiment, pulling 3 or 4 atoms out of those tons and tons of material.

I have visited the GALLEX experiment at Gran Sasso in Italy and I was very impressed.

Yes, but Davis did this experiment for the first time and that is the hardest thing. Experiments are much easier when someone has led the way.

Talking about neutrinos. The identity of dark matter is still not known because even the neutrinos cannot account for all of it. What do you think of this question?

Do you believe in axions?

What is that?

There is the possibility of axions. There is the possibility of super symmetric particles. My theoretical friends don't see anything wrong with the argument that suggests that there should be a super symmetric Universe. We have dark matter and dark energy. These are exciting times in cosmology, trying to answer all these questions and the experiments are getting better and better.

Which of your work are you most proud of?

The η-mesic atom work; I felt very good about that. Of course, CP was just a wonderful thing to be involved with.

Who are your heroes?

Val Fitch at his desk in his office in Princeton, 2002, with a poster honoring Andrei Sakharov (1921–1989, Russian physicist, Nobel Peace Prize 1975). Photograph by M. Hargittai.

Fermi is a good example. Just a superb physicist in all respects; he did experimental as well as theoretical work. He probably is my number one hero, even if I did not know him well personally.

Did you have mentors who meant a lot to you?

Mentor is a word I did not know until about 20 years ago. I suppose it was Rainwater because I did my Ph.D. with him. He was a very good physicist but not a person you could get to know well. I don't think I ever knew him really. I don't think I ever asked him for a letter of recommendation. In fact, I have been in a happy position of never having had to prepare a curriculum vitae; it was just an older time and somehow I got my degree from Columbia, I had a chance to stay at Columbia, then people at Princeton offered me a job; I did not apply for it. I came here and it's been so comfortable, it's been really a wonderful time. I think that by mentor in the modern context you mean someone who is sponsoring you, helping you get ahead, shaping your career and beyond, finding jobs for you and such.

What was the greatest challenge in your life?

Oh, heavens. I don't know. I have had some. My first wife died 30 years ago and we had two sons who were still growing up at that time and I suppose that's a personal challenge. My oldest son has subsequently died so there have been these difficulties that I learned to cope with.

What does your other son do?

He is a freelance writer and lives in the San Francisco Bay area. He seems to make a living at that. He writes mostly for medical publications.

Are you married now?

Yes, I happily remarried, but you never forget the previous happy life either.

Has the Nobel Prize changed your life?

Well, of course, you hate to admit it but you can't avoid it. I try not to broadcast the fact, as you know. But I did not stop research after I received the Prize. I retired in 1993, I thought it was time to make way for the younger people. I have basically given up experimental work.

Do you have hobbies?

No. I don't collect stamps. I have many interests but I don't know if you can call them hobbies.

What do you do when you are not thinking of physics?

We have a very large collection of classical music at home, I do enjoy Beethoven, or Schubert, for example. But I don't know. What do I do? Now I have problems with Eudora working, and I am failing the University's security tests. I am usually pretty good with this stuff. Experimental physics has seized on every computer development as fast as it appeared. I started back in the 1950s writing programs in machine language for the IBM 650, so it's just been part of my soul. The software has become so complicated, I never think in terms of doing anything with the source code anymore.

Do you find Princeton an intellectually challenging place to live?

I think it's a wonderful place to live. I have lots of friends. It's been too comfortable perhaps. The University and the Institute for Advanced Study have provided a tremendously rich environment. I normally have lunch with a group of mathematicians but often people from history or comparative literature join us so it is an eclectic bunch of people and we appear to enjoy each other's company.

Is there anything else I should have asked you and did not? Anything else you would like to convey?

I rather expected you would ask, "if I were young today would I go into physics?" I find this question hard to answer because if I were young today I would be a different person from the one I was 60 years ago. The environment of today is distinctly different from what it was when I was young. I would be making decisions today based on points of views quite different from my perspective 60 years ago.

Maurice Goldhaber, 2001 (photograph by I. Hargittai).

10

MAURICE GOLDHABER

Maurice Goldhaber (b. 1911 in Lemberg, then Austria-Hungary, now the Ukraine) is Director Emeritus of the Brookhaven National Laboratory in Upton, New York. He studied physics at the University of Berlin. Because of the growing anti-Semitism, he left Germany in 1933 and became a research student at the Cavendish Laboratory headed by Ernest Rutherford in Cambridge, England. He did his doctoral work under James Chadwick and continued working at Cavendish after he had completed his Ph.D. thesis in 1936. In 1938, Dr. Goldhaber accepted an Assistant Professorship at the University of Illinois at Urbana and moved to the United States. In 1939, Maurice Goldhaber married Gertrude Scharff, a fellow physicist and fellow refugee. She had a distinguished carrier in physics and died in 1998. They have two sons, Alfred and Michael. In 1950, Maurice Goldhaber moved to the Brookhaven National Laboratory. At one time he was Chairman of the Physics Department and between 1961 and 1973 he was the Director of the Laboratory. Following his directorship he was appointed Distinguished Physicist, while continuing his research. He is best known for his contributions to nuclear physics and to the physics of fundamental particles. Dr. Goldhaber has received many awards and distinctions of which only a few will be mentioned here. He is a member of the National Academy of Sciences of the U.S.A., a member of the American Philosophical Society, and a fellow of the American Academy of Arts and Sciences. He received the National Medal of Science (1983), the Wolf Prize in Physics (1991, Israel), and the Enrico Fermi Award (1998).

We visited Dr. Goldhaber on November 8, 2001, and recorded a conversation with him in his office at Brookhaven National Laboratory, and returned for a brief visit to finalize some points in March 2002.*

I would like to ask you about your origins.

I was born in Lemberg, which at that time was in Austria-Hungary. Then it was successively in Poland, the Soviet Union, and now in the Ukraine. My parents spoke Polish, Yiddish and German. At the end, my father spoke 14 languages. When I was 3, we left for Egypt, where my father became a travel agent. When my parents did not want us children to understand them, they spoke Italian. So when Fermi's papers appeared written in Italian, I could read them. During World War I the British interned my father as an enemy civilian. My mother and us children were sent to Austria by boat. My maternal grandparents lived then not far from Vienna in what later became Czechoslovakia. There I started school. The instruction was in German. My father returned after the war and we moved to Chemnitz in Germany. I graduated from "high school", the Realgymnasium, in 1930. Then I went to the University of Berlin to study physics. In the colloquium there were Max Planck, Albert Einstein, Max von Laue, Walther Nernst, Erwin Schrödinger, Otto Hahn, and Lise Meitner in the first row. There were also often-famous visitors. Among the younger staff and the students there were also many future famous scientists, including Leo Szilard. It was very stimulating. In 1933, everything changed with the Nazis coming to power. My father and the rest of the family left for Egypt, but I stayed in Berlin. I didn't want to leave until I was accepted somewhere else. Ernest Rutherford accepted me as a research student in the Cavendish Laboratory in Cambridge, so in June 1933 I arrived in England. First I stayed in London and went to Cambridge in August. I was the first refugee in Rutherford's laboratory. Later on others came. Rutherford was active in the movement of saving refugee scientists. Szilard was instrumental in starting the Academic Assistance Council. Back in Berlin, Szilard was one of the few older people who would talk to the young students. We met in London occasionally and discussed world affairs and nuclear physics. In time we developed a "mutual admiration society". At one time when we worked on a nuclear problem while Szilard was in Oxford, a British physicist, passing by asked, "Who

---

*István Hargittai and Magdolna Hargittai conducted the interview.

of you is the brighter?" I was struck dumb, but Szilard deflected the question and pointing to me, said, "He has more imagination." Thinking of this remark I realized that "my strength is my weakness, my weakness is my strength". When I read the Szilard biography, *Genius in the Shadows*, I was surprised to learn that he had converted to Calvinism at age 21. I knew him well enough to know that it was not an "internal conversion".

In Cambridge, I met James Chadwick who looked after the students especially when Rutherford was away, as it happened on my first visit. Chadwick was a man of few words. I mentioned to him that I would like to join the inexpensive Fitzwilliam House, but he advised me to join one of the colleges instead. David Shoenberg, who was already a research student and who later became a well-known solid-state physicist, suggested to try Trinity, St. John's, and Magdalene; but there was no space available at Trinity and St. John's would only let me know much later. When I tried Magdalene College, the Senior Tutor, V. S. Vernon-Jones, looked at me, realizing that I was a refugee and said, "I suppose we ought to have one." He also offered me financial support. I stayed at Magdalene through my Ph.D. studies, and then for an additional two years as a Junior Fellow (Charles Kingsley Bye-Fellow).

In principle, Rutherford was in charge of all the students, but he would distribute them according to their interest. In 1933, I still wanted to be a theorist. He suggested that I work with Ralph Fowler, his son-in-low, who was the theoretician at the Cavendish Laboratory. Fowler let you alone if you knew what you wanted to do. His former students included Dirac and Chandrasekhar. With Dirac, he gave him a preprint from Heisenberg and that was enough.

Back in Berlin, in March 1933 the Nazi students were often rioting and one day the University closed down, but the library was still open. I read there, in a popular magazine, that an American scientist, G. N. Lewis, had produced 1 cm^3 of heavy water. This excited me because I knew that isotopes are not easy to separate. I jotted down a few ideas and one of them was to look for a nuclear photoeffect, to disintegrate the deuteron into a proton and neutron. In Cambridge, I wrote a theoretical paper where I needed to know the masses of the light elements. Neville Mott told me that Chadwick knew everything about masses. As I discussed the matter with Chadwick, I noted that he had a mass for the neutron, which was lighter than a hydrogen atom. Rutherford had predicted a kind of neutron as early as in 1922 and predicted that there would be a strong binding of an electron into a proton. He didn't call it neutron; he called it "a neutral atom". The word neutron was used by W. D. Harkins, an American chemist. When Chadwick discovered the neutron in 1932, he measured its mass crudely, finding it was lighter than the hydrogen atom, and Rutherford was pleased. Later, there were two measurements of the neutron mass published. One was by Ernest Lawrence at Berkeley and another by the Joliots in Paris. Lawrence had a much lighter mass and the Joliots had a much heavier mass. Chadwick was keen to know the correct mass when I proposed the photodisintegration experiment that would give the mass very accurately. The gamma energy was known and so was the mass of the deuteron and the mass of the proton. Thus, measuring the energy of the proton from the photodissociation of the deuteron, and knowing that the neutron has about the same energy, the binding energy can be obtained and hence the neutron mass. At the first discussion, however, Chadwick did not commit himself. About six weeks later, he saw me in the corridor and told me that the experiment I had suggested worked on the previous night. At that point he suggested that I work with him. I went to see Fowler. It may have been that Rutherford had already learned about a transfer to Chadwick, and the transition happened very smoothly. In the Cavendish hardly anything happened without Rutherford knowing about it. This was in the spring of 1934. Thus more than a year passed from my original idea, but things were not moving as fast at that time as today. This is the story of how I became an experimenter.

We did the photodisintegration of the deuteron and determined that the mass of the neutron was heavier than that of a hydrogen atom. This showed that the neutron was a new particle (in modern language) and not a compound, it was not a neutral atom. It took the theorists a while to digest it. Heisenberg still wrote papers in which he treated the neutron as if it was a proton plus electron. It was a puzzle why the neutron should be heavier than the neutron with the proton having the extra Coulomb interaction. This puzzle has been around for 65 years, but I think I may have solved it this year. I've just completed a paper ("A closer look at elementary fermions") containing the proposed solution and sent it yesterday to the *Proceedings of the National Academy of Sciences of the U.S.A.*[1]

Where did you publish the original finding of the neutron mass?

In *Nature*. When we were writing it up, I wrote a draft containing an interesting speculation. Chadwick had another student, D. E. Lea, who had published a paper about shooting fast neutrons into paraffin and looking for gammas. He found a very large cross section. Knowing the cross section for the deuteron disintegration, I could work out the inverse cross section, when the neutron is captured by a proton and the gamma is emitted. I came to the conclusion that Lea's cross section was nearly a thousand times too large. So I suggested that maybe the neutrons are first slowed down and then captured. I wrote to Bethe and Peierls about it and Peierls answered that it would not be so because the photoeffect goes to zero at zero energies. This was July 1934. In October 1934, Fermi came out with the discovery of the slow neutrons. Rutherford rushed to me and told me that Fermi should've given me credit, but I didn't think so. I didn't think that Fermi would've known about my suggestion. Then something interesting happened. Well after the war Chandrasekhar asked Fermi whether the subconscience has ever played a role in his discoveries. Yes, Fermi answered, in my greatest discovery, that of the slow neutrons. It happened like this, he said. We had some anomalies. Some effects were different on different tables and we couldn't understand them. I asked the shop, Fermi continued, to make me a piece of lead just right. I think what he had in mind was to place a piece of lead between the source and the detector. This by itself was unusual for Fermi because usually he would just find or cut a piece he needed. A technician brought the piece of lead to Fermi. At that point, Fermi continued, I changed my mind and asked for a piece of paraffin instead of a piece of lead. This was Fermi's example of how his subconscience may have worked.

Did he know about your estimations?

He had only read our paper and congratulated me on the photodisintegration experiment in the summer of 1934 at a meeting. In my paper I originally inserted my suggestion about the slow neutrons and Chadwick accepted it. He rewrote it in better English and better handwriting. In those days journals accepted handwriting. A few days later, however, Chadwick told me that we should not speculate. Rutherford did not like speculations. He speculated only in talks. His famous neutron speculations had occurred in his Bakerian lectures in 1920, not in his papers. Chadwick may have mentioned it to Rutherford and Rutherford may have talked him out of it. What we left in was then a typical Cavendish understatement. We talked about the discrepancy between Lea and our work and instead of giving an explanation we said that it needs more work. And we should have followed this up with new experiments, but we didn't. I think that when Fermi read our paper he must have had a similar idea. Although he didn't act on it, I believe that they may have remained in the back of his mind, and must have come to him when he saw the piece of lead and he changed it subconsciously to a piece of paraffin.

When I read Chandrasekhar's story, I wrote a little note on what had happened from my perspective. I was not looking for a posteriori recognition, but I thought the story was interesting historically. I sent a copy to Mrs. Fermi, to Amaldi, and to Segrè. By then Fermi had died. Mrs. Fermi wrote me that it was a pity I never talked with Enrico about it. Amaldi was so protective of Fermi that he lost his logical mind. He wrote back that Fermi was a very systematic man and that's why he started with paraffin! Finally, Segrè wrote that they did not read my paper! But Fermi had congratulated me on the paper. There was one more person to whom I sent a copy of my note, Bjerge, a Danish physicist, who was at the Cavendish when all this happened. He remembered that I suggested an experiment to him to check for slow neutrons, but when he came back from a vacation Fermi had already discovered them.

Moving to other topics, I would like to ask you about Edward Teller.

I know him well, we published a paper together. He was rather obsessed with the hydrogen bomb, which Ronald Reagan called the peace-keeper because it was so terrible that nobody would've dared to use it. Teller was very afraid of the communist danger and so was Eugene Wigner. Wigner wanted everybody to build a shelter, which was childish. People turned against Teller because of the Oppenheimer business.

Where would you position yourself politically between Teller and Szilard?

I'm nearer to Szilard. I knew Szilard very well. Szilard had a general idea of a neutron-induced chain reaction very early, but the specific nuclei he considered would not have worked. He was an idealist who tried to help stop the use of the bomb.

Szilard was not part of the Establishment but you were.

It came later when I became director of Brookhaven. During the war they did not use my knowledge of neutrons. Szilard wrote to one of his friends that had I been on the atomic bomb project, it would have been finished a year earlier. I am a little more modest; I would say, only half a year. It would've made a difference, because Roosevelt was still alive.

Why didn't they use you?

There were various excuses, but it came out indirectly. One man who was in charge early on was a theoretical physicist, Breit. I saw him in 1940 during a meeting in Washington. He asked me what ideas I had and I told him some ideas. He literally threw up his arms and said "But Fermi and Szilard had the same ideas." My feeling was that they thought that they had already enough of these "important" foreigners. We worked at the University of Illinois at the time; the "Manhattan District" followed my neutron work carefully, and they asked for my permission to copy my neutron counters. They were then used at the first pile in Chicago. Breit acted as a censor. When I wanted to publish a paper, they would send it to him from *Physical Review*. Then the editor would write me a polite letter whether we would be willing to leave the paper unpublished during the war, and, of course, we did. We had about half a dozen papers that were published after the war. However, our data were used in the project although there was never an acknowledgement for it. In 1943, Breit wrote to me saying that it's a pity they couldn't use us because my wife's parents were still in Germany. But there were others in a similar situation who worked on the bomb. This was just an excuse. I regret very much that I was not aggressive enough to barter my expertise for getting Trude's parents a visa to the U.S.A. The Nazis had already killed my wife's parents in 1941 and we know the exact day when they did it, because the Germans kept good records!

How did you get to the University of Illinois?

Rutherford died in 1937. In 1938, I visited the United States and met the head of the Physics Department of the University of Illinois at a Washington physics meeting, where he offered me a job. I visited Urbana-Champain and initially I accepted the offer for one year. Back in Cambridge I went to see Appleton, the acting director of the Cavendish, who was much relieved by my news because they didn't have a job for me. I stayed in Illinois for 11 years and then I came to Brookhaven. I've never applied for a job.

Your wife was also a famous physicist.

She graduated from the University of Munich and had a small job at Imperial College London. I went back from Illinois to London to marry her in 1939 and then we moved to Illinois. Illinois had a strong nepotism rule

Gertrude and Maurice Goldhaber (Brookhaven National Laboratory, Photography Division, courtesy of M. Goldhaber).

and my wife could not have a job there. She supervised some of my students informally. After the war she became an Assistant Research Professor using grant money from the Navy. The nepotism rule at the Brookhaven National Laboratory was much milder; it was only that husband and wife could not work at the same department. First we came here on a temporary basis. Then, in 1950, we came for good. When they dropped even this weak nepotism rule, we became the first couple at the same department. Now there are many. First we both worked in nuclear physics, but eventually I moved to the physics of elementary particles.

One of your students at Illinois was ...

Rosalyn Sussman and so was her future husband, Aaron Yalow, and she became famous as Rosalyn Yalow. She was a very determined student. Once I characterized her to the other professors as rather aggressive, but they took it as a compliment. When she got the Nobel Prize, she introduced me to a man from Stockholm who had something to do with it. He told me that they had a hard time giving her the prize because they don't like to give it to a survivor. He meant that Yalow and Solomon Berson had developed jointly the technique of radioimmunoassays of peptide hormones, but Berson died in 1972.

I would like to ask you about your experiment in which you and your wife showed that the beta particles were electrons.

It was a tabletop experiment. People were interested whether the beta particles were identical with atomic electrons and they measured e/m more and more accurately. However, there would always remain some small undetected difference. We thought of a different experiment. The Pauli exclusion principle states that an identical particle would be kept out when the shell is full. If it is not identical, it can still fall into the shell. So the beta rays, if they are not identical with the electrons, can fall into the K-shell and give X-rays. This was a simple experiment and we showed that more than 97 percent of early on β-rays gave no X-rays when stopped in lead. Then someone repeated the experiment with a better source and achieved 99 percent, and there was no reason to try to get 100 percent.

You have been in physics for seven decades. Would you care to make some observations?

A lot of progress comes when somebody makes an extrapolation or a generalization. De Broglie, for instance, had an important generalization. To predict a new particle is an extrapolation because you see a scheme and you notice that it ought to be augmented. This is how Glashow became famous. There is then the use of analogy. These three approaches guide the human mind. This Laboratory made important contributions in the 1960s and some of the university physicists were awarded the Nobel Prize for work done at our accelerator. The new discoveries in the 1970s and especially later needed large teams, which somewhat reduces their popular appeal.

Your heroes?

Einstein and Rutherford and from an earlier time, J. J. Thomson, who is sometimes underestimated. He nearly started modern quantum mechanics by suggesting an important experiment to one of his students, to find out how many photons are needed to get interference.

I would like to ask you about your discovery that the neutrino has negative helicity.

That's an amusing story. The question was which beta-decay interaction is the correct description out of two possibilities? Columbia University sent us two students, Ruby and Rustad to do an experiment with helium-6, which is best made in the reactor; from lithium-6 with fast neutrons you get helium-6 plus a proton. Helium-6 lives for only 1 second. Columbia professors supervised the students long distance and they concluded that the interaction was what is called tensor and scalar. I was not involved. Then came the Christmas meeting of the American Physical Society in 1957 at Stanford University. I was asked to give an invited paper on beta-decay, which was not my field. I overcame my original reluctance, accepted the invitation, and decided to read the papers from Columbia on their Brookhaven experiments, and also another study from Illinois by a man called Allen. His conclusion was different. He came out in favor of vector and axial vector. I stayed home one Friday morning to read the Ruby and Rustad paper and the Allen paper. I was halfway through the Columbia paper and found it so dull that I told myself that there must be a better way of doing this. Within 20 minutes I thought of the neutrino helicity experiment. People often ask me how did I come to this idea and my answer is, it is very simple, you work 20 years on isomers and then think for 20 minutes. When I thought of the experiment I first wanted to rush to the lab to do it, but decided instead to wait until Monday and see whether it still looks good. On Monday I told my associates, using Rutherfordian language, "Boys, drop everything." The same afternoon we started the experiment and 10 days later it was finished. It came out in favor of vector and axial vector, which was opposite to what was widely expected. At that time people thought that neutrinos were right-handed, which could have meant that the interaction was tensor and scalar. Tensor and scalar go with right-handed neutrinos and vector and axial vector go with left-handed neutrinos. We decided that it was left-handed. On the plane to Stanford I finished reading those papers that I had started reading before our experiments. It is interesting psychologically that since I now knew that the tensor and scalar conclusion was wrong, it was easy to see the mistakes. Ruby and Rustad had carried out two experiments, confirming each other. The first experiment, however, had poor statistics and the second poor geometry. Our results were accepted quickly; it is now part of the folklore and we rarely get credit for them.

Your experiment came soon after Lee and Yang's discovery of parity violation.

This discovery made it possible to set up experiments like ours. Lee and Yang, Landau, Salam, and others believed in the so-called two-component theory of neutrinos: that they have zero mass with an intrinsic helicity and that they are born either left- or right-handed. They preferred right-handed which agreed with the Ruby and Rustad conclusion. Of course, we should have been more cautious and not call it the helicity of neutrinos but the helicity of neutrinos emitted in beta-decay. Now that we know that they have mass, we know that they have no intrinsic helicity. It is connected with their interaction.

I would like to ask you about your wife.

She was part of the experiment in which we showed that the electrons are identical with beta-rays. She continued in nuclear physics and she did systematics. She built a model, which serves as a good guide for theory. This is a model of the first excited states of even-even nuclei. The heights of the columns are proportional to energy. This illustrates the need both for the shell model and the collective model. At a closed shell the energy is very high and in between there are some very low energies. My wife discovered the first very low energies at a time when people thought that even-even nuclei could not have low energies. Such discoveries change concepts, but are usually quoted only in tables.

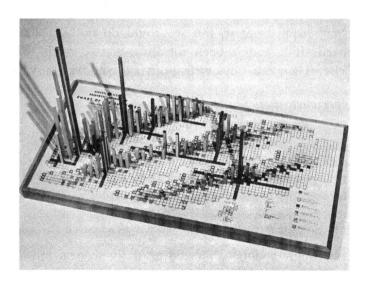

Maurice Goldhaber explains: this is a lego model of the first excited spin-two states of even-even nuclei built by my wife. This model illustrates the influence of magic numbers predicted by the shell model. The regions in between the magic numbers show very low energy states, ascribed to collective nuclear motion. This model was helpful in the development of the presently accepted ideas about nuclear structure (courtesy of M. Goldhaber).

Do you know what made her originally interested in science?

Her father was interested in chemistry as a young man. His father died when he was very young; he had to give up the idea of studying science and had to enter the family business. At first he wanted his daughter to become a lawyer, but she became interested in science and he supported her in this. She started her studies in Munich. In Germany, at that time, you could go each semester to a different university. At one point she came to Berlin and that is where we met. After further stops she returned to Munich to do her Ph.D. work. She studied the magnetic behavior of materials. She was still in Munich when the Nazis came to power in 1933. First she was not taken to be Jewish, but when everybody had to fill out the questionnaires about parents and grandparents, they were very surprised. Her professor let her finish her Ph.D. When she came to England, she started working with G. P. Thomson, the son of the famous J. J. Thomson. G. P. Thomson shared the Nobel Prize in Physics for electron diffraction. When we moved to the United States after we married, she switched to nuclear physics.

It is well known that the nepotism rule prevented her from having a university position at Illinois. How did she take it?

She wasn't very happy with it, but we could do nothing about it. The war had started. If it hadn't been for the war, we might have moved somewhere else. Then our boys were born and she was busy with them.

Did you experience anti-Semitism?

In Illinois, it was very strong. It was not against foreign Jews so much. They treated me well, but they did not treat the American Jews equally well in some departments. Then, during the war, anti-Semitism slowly diminished.

How did she feel about women in science?

She felt very strongly about it. She did a lot for women in science, gave talks, organized parties. I still get things addressed to her about women in science.

With two children, how did she manage?

It wasn't easy, but we usually had some help at home. The children may have felt a little neglected.

What kind of a person was Gertrude Goldhaber?

She was much liked here. There is now a prize named after her, which is given out annually to women scientists who are either students at Stony Brook or have done their work at Brookhaven. There are other prizes named for both of us. There is one at Harvard University called the Maurice and Gertrude Goldhaber Prize and there is another one at Boston University called the Gertrude and Maurice Goldhaber Prize. Then there is the Gertrude and Maurice Fellowship here, which was started when my ninetieth year was celebrated here.

What do your sons do?

The older one is a professor of physics in Stony Brook. The younger one had also studied physics, but he currently writes about social issues. I have a grandson who is a physicist in solid state research and he just received a big prize.

You have characterized your joint work with your wife as too intense.

This was after many years of being involved day and night. From 1939, we worked together for about 15 to 20 years. Then I deliberately moved into fundamental particles, in which I had become interested, but I always kept an interest in her work.

Do you think that men and women do science in different ways?

There may be some difference. Those women who are deeply involved in science pursue it even more intensely than men, as far as their time permits it. Meitner, Wu, Yalow, and my wife were all very committed.

Do you think that it is more difficult to be a successful scientist for a woman than for a man?

It used to be so, but things are slowly becoming even. We are not there yet, but getting close. Some promotions have been hard to get for women, but it too is getting easier.

Science has changed during your lifetime.

When I became a physicist, science was a calling; later, for many, it just became a profession. I have occasionally joked that many now do science as if it were a luge race, where skill, speed and persistence is all that matters, since the path is already laid out. Others do *Karaoke* (sing along) science, and some decide to measure the height of a "Mt. Everest" more precisely, or climb a theoretically-predicted new "Mt. Everest"!

What was the greatest challenge in your life?

I was lucky to leave Germany in time, I always got job offers at the right time, so there was no hard challenge there.

Did you and your wife have a different style in research?

She was very meticulous in working on experiments and thinking about the results. I was more jumping around, although I also pursued some things to the end. To some extent we complemented each other.

What is your opinion about the Nobel Prize?

It is somewhat out of date. People are not isolated enough to judge their achievements separately.

Soon after you had joined Brookhaven National Laboratory, you became its director.

My older son wrote me an 8-page letter why I shouldn't accept it. He was right because it interrupted my scientific research. However, I am one of those few who returned to research after having served as director. I considered it a challenge to be director. I could influence the great things that happened here. Some people flatter me that the decade of my directorship was the golden decade of the Laboratory. CERN was our big competitor at the time, but we were more successful then. CERN concentrates its efforts mainly on high-energy physics whereas we have more diverse areas of research.

You were born in Lemberg, then came all the hurdles and prosecutions, then one day the President of the United States awarded you with the National Medal of Science. What did you feel on that occasion?

I did not feel anything special. In this country this is not so unusual.

One of the Goldhabers' sons handing over the Diploma of the Gertrude Goldhaber Prize in the presence of Gertrude and Maurice Goldhaber (Brookhaven National Laboratory, Photography Division, courtesy of M. Goldhaber).

Have you ever been asked to submit nominations for the Nobel Prize?

In 1947 I was asked for the first time. I first thought to nominate Goudsmit and Uhlenbeck, but I happened to run into Teller and he said what they did was trivial. Then I suggested Born, and Teller said trivial, so I decided to nominate someone whose work I had followed closely and that was Leo Szilard. I nominated him for the physics prize but I should have written that he could get the physics, chemistry, medicine, or the peace prize! I prepared a very careful recommendation. Later they asked me again many times but since they did not follow up my careful recommendation, I resorted to writing a few lines only. These brief nominations proved successful quite a few times. They were usually obvious proposals and I did not have to make detailed justifications. Szilard was not obvious to most, but he was to me. When Szilard had bladder cancer he said to me that he did not understand why he was not a member of the National Academy of Sciences. At that point I immediately proposed him and he was at the top of the list in that year. When he read up on the literature on his type of cancer he found that there were two ways to treat it. Either he could have an operation or he could have radiation treatment, and in each case the outcome was rather uncertain. The operation was not always successful and the radiation usually made people very sick. I suggested a new method to put a radioactive source inside the tumor. It was soft radiation so the rest of the body would not suffer. Szilard liked the idea. He was at Sloan-Kettering and I went there for consultation. There was Szilard, his cancer doctor, some radiologists and I. First Szilard agreed with my suggestion, but his private doctor told him that it would be very painful to put the source in. In the course of the discussion Szilard had the brilliant idea of rather than receiving a large dose of radiation, to get little doses but many, many times. This cured his cancer and they do it like this now, giving small doses many times, which, of course, necessitates many visits. When Szilard died, his widow wrote to me that he died of a heart attack and the autopsy showed that his cancer was completely cured, and she thanked me for my indirect contribution. Today, using a source of soft radiation is an accepted treatment in some cases, e.g. for prostate cancer, and this is how Mayor Rudy Guiliani is being treated.

How do you feel about Heisenberg?

Heisenberg has this myth that he spread that he didn't want to build the bomb instead of saying, "I didn't know how to build the bomb." I always said that the Germans would not succeed in building the bomb. What happened was that I had a student in Illinois who measured the aluminum cross section for slow neutrons. At that time everybody thought, including me, that the way to use uranium was to put it in an aluminum can, something which would not absorb slow neutrons much. So my student measured its cross section better and at about the same time two Germans also measured the aluminum cross section and published it in *Naturwissenschaften*. They thanked Heisenberg in the paper, which was already a nice give-away at that time. They were off by a factor of 2. This is why I thought that they would never build a bomb. They did not have enough experts in neutron physics.

How did you feel about the bomb?

It was used too late. The Japanese were already losing the war. The damage from the fire bombing of Tokyo was heavier than in Hiroshima. The atomic bomb gave them a convenient excuse to give up, but they might have given up because of the raids. In this country they have always said that the atomic bomb saved all the boys who would've gone to Japan to conquer it, but there may not have been a need to go in. I think that the atomic bombs speeded up the process by a short time only. I know that I am nearly alone with this opinion.[a]

At 90, you appear fit in every respect. How do you do it?

I swim regularly. I used to play tennis. I exercise a little at home while watching the news on TV. Once I answered a question like yours saying that I have no time to age. I also did a little climbing or rather, just hiking in the mountains. My wife was a much stronger climber, but she took me along. Earlier this year I had a bad car accident and I had to learn to walk again, but now I am cured. My sons made me promise that I would not drive anymore.

What do you read?

Unfortunately, I have little time now to read books. I read *Nature*, *Science* and *Physical Review Letters*, and there is much more I should read. For my latest paper I had to read a lot of references. I read *The New York Times*, which is a killer. I try to read a little less of it. I watch the news and I am reasonably up-to-date on what's going on in the world.

How do you find it?

The news are depressing, but I am not depressed.

References

1. Goldhaber, M. *Proc. Nat. Acad. Sci. U.S.A.* **2002**, *99*, 33.

---

[a]On March 14, 2002, *The New York Times* published on p. A4 a report from Tokyo by Howard W. French, entitled "100,000 People Perished, but Who Remembers?" This includes the statement: "For Japanese leaders, remembering the firebombing victims could mean explaining ... the prolongation of the war for many months after its outcome was clear ..." [added by Dr. Goldhaber in March 2002].

John N. Bahcall, 2002 (photograph by M. Hargittai).

11

JOHN N. BAHCALL

John N. Bahcall (b. 1934 in Shreveport, Louisiana) is Richard Black Professor of Natural Science at the Institute for Advanced Study in Princeton. He received an A.B. in physics from the University of California, Berkeley (1956), an M.S. in physics from the University of Chicago (1957), and his Ph.D. from Harvard University (1961). He was on the faculty of the California Institute of Technology till 1970 and he has been at the Institute for Advanced Study since 1971. He is a member of the National Academy of Sciences of the U.S.A. (1976), the American Academy of Arts and Sciences (1976), the Academia Europaea (1993), and the American Philosophical Society (2001). He has received numerous awards, among them the NASA Distinguished Public Service Medal (1992), the Heineman Prize (1994), the Hans Bethe Prize (1996), the National Medal of Science of the U.S.A. (1998), the Russell Prize (1999), the Gold Medal of the Royal Astronomical Society (2003), the Dan David Prize in Cosmology and Astronomy (2003), the Benjamin Franklin Medal in Physics (2003), and the Presidential Enrico Fermi Award (2003). He was a member of the Hubble Space Telescope Working Group for more than 20 years. He was president of the American Astronomical Society (1990–1992). His home page contains relevant material, including popular articles: http://www.sns.ias.edu/~jnb. We recorded our conversation in his office at the Institute for Advanced Study in Princeton on October 23, 2002.*

---

*Magdolna Hargittai conducted the interview.

What turned you originally to science?

My path was unconventional. I did not show an early interest in mathematics or science. I was not a particularly interested student. My family and my family's friends were not intellectuals. They concentrated on making a living, a full time occupation.

In our high school, athletics were highly valued. I was excused every day at noon, instead of 3 p.m., to practice tennis. I did not take any science course in high school; my only technical training was a first-year course in algebra. There was, for me at least, no particular academic guidance. I only discovered that I had an academically-related talent in my senior year, when I joined the high school debate team. I rapidly became good in debating and my colleague, Max Nathan, and I won the national high school debate tournament (held in Boston in the summer of 1952). We beat a couple of students from some prestigious New York private school in the finals. This was the first and only time I know of when representatives from Louisiana went to the national debate tournament.

I went to Louisiana State University for my first year of college and studied primarily philosophy. I initially thought that I might become a reform rabbi, but I pretty quickly realized my interests were more academic than pastoral. Growing up in Shreveport, Louisiana, our reform rabbi and his wife, David and Leona Lefkowitz, were both inspirational and supportive. Looking back on that time, I think I must have believed that intellectual activities were practiced primarily by rabbis and their families. I was advised that the best preparation for a rabbinical career was to study philosophy, so that's what I studied. I was a straight A student in my first year of college, but I still did not study any science or mathematics.

During the summer after my first year of college, I attended summer courses at the University of California at Berkeley and I loved it. My mother elicited the financial help of a cousin, Clifford Strauss, who paid for my tuition. Because of Clifford, I was able to stay at Berkeley and finish there. I studied philosophy and made rapid progress. I began to think that maybe I could be a philosophy professor some day, but I ran into a severe problem. At U.C. Berkeley, there was a requirement that you had to take a college science course of some kind in order to graduate. I didn't have the academic pre-requisites necessary to take any of the college science courses, because I did not take any science courses in high school. So my advisor told me that I had to go to high school in the evening and take a high school science course.

I had been reading Bertrand Russell and Wittgenstein. Both of them described their great admiration for the achievements of physicists and the future of physics. I even imagined that I perceived some sense of regret on their part because they themselves did not become scientists. I remember statements to the effect that contemporary physics offered a great opportunity for an individual to make significant intellectual advances.

I really wanted to learn physics but I did not want to go back to high school. At Berkeley, there were three or four different courses in physics. One was for non-scientists, one for engineers and medical students, and one for those who wanted to become professional scientists. The person who taught the physics-for-physicists course was Burton Moyer (I believe). I went to him and said that I would like to take his course, but that I did not have any of the pre-requisites. I told him I was fascinated by the way physics was affecting philosophy. He told me, you are crazy if you want to take this course; it is for real science students. However, he said he would let me enroll in the course provided I dropped out as soon as I realized the course was too difficult for me. I did find that first physics course enormously difficult; it was the most difficult thing I had ever done in my life. I got a C in the course, but I fell in love with the subject. I loved the fact that you could use physics to understand phenomena in the world of experience, like why the sky was blue. I loved the fact that after a while everyone agreed what was the right answer to a question in physics. Because the course was so challenging for me, I loved it even more.

Incidentally, I don't ever remember thinking about a job in physics. To a first approximation, there were no academic jobs in physics when I decided to go to graduate school. Those were pre-Sputnik days. I just wanted to have fun learning physics. However you look at it, my path to science was neither direct nor conventional. It was not well planned. I definitely would not recommend my trajectory as a way to become a scientist.

Then what turned you to astronomy?

I think it was again a combination of Bertrand Russell's influence and chance. I remember an article in one of his books of essays in which he wrote about two things that he thought were the most important in educating human beings. One area was to learn about the majesty of what the human mind was capable of, which was exemplified by what have been accomplished in atomic and subatomic physics. The other was to understand the insignificance of human beings by recognizing their place in the larger scale of the Universe that was revealed by astronomy. This statement made a huge impression on me. I wanted to somehow be associated with one of these two great enterprises.

After completing my undergraduate degree at the University of California at Berkeley, I got a Master's degree at the University of Chicago and a Ph.D. at Harvard, all in physics. I was fully supported financially by university grants; otherwise, I could not have attended graduate school. By accident, I did a thesis in atomic theory. David Layzer gave me a summer job, finding a way to calculate the energy levels of highly-ionized atoms that had recently been measured by Edlen. He then went off to England to work with Bondi on cosmology. I solved the atomic physics problem over the summer. When David came back to the States at the end of the summer, I showed him that my solution was in agreement with the abundant spectroscopic data. He agreed and suggested that I write it up as a thesis.

Then I went in the fall of 1960 to the University of Indiana, where I wanted to learn weak interaction theory. I listened to a course on weak interactions by Emil Konopinski, a great physicist and a great pioneer in the subject of beta-decay. In order to teach myself the theory, I made up problems for myself that I solved. My first published paper, in 1960 or 1961, was on different ways to determine the mass of the muon's neutrino. Then, I calculated the rate of electron capture from continuum orbits, which is different from the usual (in the laboratory) electron capture rate from bound atomic states. Konopinski discussed bound electron capture in his course. I also calculated the effect of the Pauli exclusion principle on beta-decay rates and the probability of beta-decay into bound, not continuum, orbits. These were all variations on the usual themes considered by physicists and they were useful exercises to check that I understood weak interaction theory. I had lunch one day with a friend, Marshall Wrubel, who was an astronomer. Marshall asked me how my work was going and I told him what I was doing. I told him that I was disappointed to find that, when I put numbers in the equations I had derived recently, it did not look like any of the processes I had considered (continuum electron capture, bound state beta-decay) could ever be measured. Marshall suggested that I look at a famous paper by Burbidge, Burbidge, Fowler, and Hoyle, universally known as B^2FH, on the formation of the elements by nuclear processes in stars. B^2FH was the equivalent of the bible for nuclear and stellar astrophysicists. He suggested that maybe there would be applications to what happens in stars.

This lunchtime suggestion was a turning point in my career. There was a table at the back of the B²FH paper prepared by Willy Fowler, which listed the characteristic properties of nuclei that were involved with the formation of heavy elements. The beta-decay rates were particularly important because they were the slowest processes in the buildup of heavy elements and thus set the time scale for the slow transformation from light to heavy elements. Willy had assumed that the beta-decay rates in the stars were the same as in the laboratory. It was obvious to me that this was not the case. I recognized that ions would be stripped of their electrons at the high temperatures in the interiors of stars, so for example, they would not capture electrons from the bound atomic orbits as they do on Earth. They would capture electrons from continuum orbits. There were also some other differences, like the effect of the Pauli principle. I pointed out in a short paper sent to the *Physical Review* that, based on my calculations, the rates of beta-decay processes would be different in stars than the ones that currently were being used by astrophysicists and physicists. Probably, I did not write the paper too tactfully, at least that is what Konopinski suggested to me. Anyway, I never got a referee's report for this paper, just a formal acceptance letter after some time.

I did get a handwritten letter from Willy Fowler, which turned out to be very characteristic of him. Willy wrote that he had seen my paper — which meant that he had been the referee, because they were no preprints in those days — and he would like to invite me to come to Caltech as a senior research associate to work with himself, Fred Hoyle, Dick Feynman and Murray Gell-Mann on problems in physics and astrophysics. When I first showed up at Feynman's office in CalTech, he threw me out saying he had never heard of me. Willy had neglected to tell Feyman or Gell-Mann that he had used their names in inviting me. About the same time he wrote to me, Willy wrote to Ray Davis at Brookhaven, who was thinking about whether it was possible to detect solar neutrinos. Willy wrote Ray saying that there is a guy, John Bahcall, at Indiana University who knows about weak interactions in stars and suggested Ray get in touch with me about solar neutrinos.

Ray wrote me and asked if I could calculate the rate at which ⁷Be captures an electron in the solar interior, thereby producing neutrinos in the Sun. He wondered if those neutrinos would be detectable. I remember thinking about the question for a while, because there was quite a bit of nuclear physics that I had to learn in order to do the calculation. Finally, I realized that this was a unique opportunity to study the interior of stars with neutrinos and therefore decided that I'd like to spend a few months working on that possibility. When I did the calculation, I realized, but only after I wrote up my results and sent them off for publication, that this was really only the beginning. What I had done was to calculate the rate for neutrino emission from electron capture by ^{7}Be as a function of the stellar temperature, density, and chemical composition. This was probably what Ray had in mind. But I did not have a solar model to put those nuclear reactions into in order to predict what Ray should really measure.

So I decided that I would go to Caltech and try to utilize the existing stellar evolution computer programs and expertise in Willy Fowler's group. I wanted to use those programs to calculate the neutrino fluxes. That's what I did in 1962. I added the nuclear physics required to calculate the neutrino fluxes predicted by the solar model. Willy again played a very important role. I had difficulty in getting the experts to agree to run their programs with my nuclear physics. They were not interested in the Sun. They were interested in the frontier astronomical problems involving the evolution of giant stars and the explosions of supernovae. So, at one point, Willy had to use his authority to get my programs run the first time.

I also continued studying the effects of atomic electrons on beta-decay rates; processes that I called "overlap and exchange effects". (These effects arise because of the Pauli exclusion principle and because the initial and final state Hamiltonians are different.) There were a lot of laboratory data with which to compare my results; my calculations were successful in explaining measurements that were previously not understood. I got a lot of recognition, especially from experimentalists, for this work. But, on one occasion, Willy Fowler came into my office, asked what I was doing, and listened rather impatiently while I told him with great enthusiasm about my successful calculations. Willy rotated his head back and forth like he always did when he wanted to make some pronouncement that he thought was important. He said something like: "This stuff is wonderful, but you really need to do something in astrophysics if you want to have an influence on a wider scale. What you have done is an interesting intellectual problem but it's not going to change significantly how people think about the big questions."

I deeply resented Willy's advice at the time; it made me mad. But, when I cooled down and thought about it, I realized he was right. I regret that I never told him how crucial this conversation was to me. Anyway, after our conversation, I started to look around for other problems in astrophysics. There was a lot of work going on at CalTech with the newly discovered quasars, so I made, together with Ben Zion Kozlovsky, the first models in 1964 that explained quasar emission line spectra using photo-ionization calculations. Then in 1965, Ed Salpeter and I suggested that there would be multiple absorption line systems observed in the spectra of quasars and later I developed an empirical method of analyzing data to reveal those systems. I also did a variety of problems in atomic physics applications to astronomy. X-ray astronomy was new, so in the mid-1960s Dick Wolf, my first graduate student, and I calculated the cooling rates for neutron stars by neutrino emission caused by nucleon-nucleon collisions and by pion-like decays. Rates very similar to the ones we derived then are still being used today to discuss the temperatures of neutron stars observed by the Chandra satellite. And, of course, I continued to work on solar neutrino problems.

The bottom line answer to your question is that I got into astronomy by accident, found interesting problems by good luck and by being in a place where new discoveries were being enthusiastically discussed, and had excellent mentors to give me useful advice along the way.

Solar neutrinos came into the limelight earlier this month, when the Nobel Prize in Physics for 2002 was announced. You just told me how you started to get involved with them. Would you care to tell us your further involvement with this topic?

My involvement has lasted more than 40 years. I am still enjoying doing new problems in this subject. I told you how it started. We calculated the neutrino fluxes. But, to know whether Ray Davis could make a measurement or not, I also had to make myself an expert on calculating the cross sections for the capture of neutrinos in chlorine; which was the detector that Ray Davis wanted to use. I took the neutrino fluxes that we calculated from the solar model and I calculated what the rate of capture of neutrinos would be in a chlorine detector. When I first did this calculation in late 1962, the rate was much below what Ray thought that he could ever detect. That was very discouraging to both of us. It looked like the experiment could not be done.

For some reason, I did not publish my calculated cross sections. I remember, I thought at the time about whether I should include the neutrino cross sections in the paper that we did publish on the neutrino fluxes. But I didn't. I can't say precisely why. I don't remember. But, I continued to worry about the cross sections.

I guess about a year or so later, in August of 1963, I visited the Bohr Institute in Coppenhagen where Tommy Lauritsen, a senior nuclear physicist from Willy Fowler's laboratory, was spending a sabbatical year. I spent a week there and gave a talk about solar neutrinos. I described the neutrino fluxes and also the calculations that I had done but had not published on the rate for neutrino capture by the ^{37}Cl nucleus. Ben Mottelson, who together with Aage Bohr formed the best theoretical nuclear physics team in the world, asked a crucial question during my talk. He asked whether there could be a significant contribution from a transition to an isotopic analogue state that must exist, analogous to the ground state of chlorine, as an excited state of argon. I am sure that I must have answered something like it probably would not matter much because only 10^{-4} of the flux had enough energy to reach excited states in argon. That was my initial reaction, but I did not really understand the question. Probably, I had never heard before of analogue states. I talked to Mottelson after my talk and I decided to try to understand enough to evaluate the effect quantitatively. I studied some books and papers that were available at the Bohr Institute; I learned how to estimate the energy at which the analogue state would lie and I calculated the cross section approximately using some analytic approximations. I estimated that the transition to the analogue state, and other excited states in argon, would increase the expected neutrino capture rate by a factor of about 20. That was enough to make the chlorine experiment look feasible.

When I got back to Caltech, I wrote to Davis about this and we both got very excited. I refined my calculations and we gave two related talks, one after the other, at an astrophysics conference in the fall of 1963 at the Goddard Institute in New York. We wrote up a short joint paper on this subject that was published in the proceedings of the Goddard conference, but only came out two or three years later. We concluded that the enhanced rate that I calculated could be detected in an experiment that Ray thought was feasible.

After that, Willy encouraged us to write up our results fully as a joint paper for a refereed journal. We started to do this, but eventually we separated the results into two papers because the description was too long for one paper in the *Physical Review Letters*. The papers appeared back-to-back, theory and experiment, sometime in early 1964. That's how we got started. Incidentally, I never happened to be again at the same place at the same time as Ben Mottelson. I wrote him a couple of times updating him about the cross section calculations. I don't think I ever got a response. I have no idea whether he remembers asking the crucial question that led to the analogue state calculation.

The experiment got funded; I could tell you stories about how it happened. Ray and I collaborated on that also. In 1968, Ray obtained his first results, which were significantly less than what I predicted. This became known as the "solar neutrino problem". For the next 20 years, Ray and I tried to persuade people to do other experiments. I continuously refined my calculations, trying to find errors and estimating more and more accurately the uncertainties. In the refinement stage, I primarily worked with Roger Ulrich, for about 20 years. Roger had initially been a postdoc with me at CalTech, but he soon moved to nearby UCLA as a faculty member. Roger and I learned an enormous amount working together and I enjoyed that experience enormously. The first exploratory solar model calculations were done largely by Dick Sears with me supplying the nuclear physics. Much later on I worked with Marc Pinsonneault to include element diffusion in the solar model calculations.

During the time that Roger and I were refining our calculations, Ray made many tests of his experiment to see if it was possible that he was not measuring some of the argon atoms produced by neutrinos. He convinced everyone that looked carefully at what he was doing that he was not missing argon atoms.

Sometime around 1988, the first of the Japanese-American results with a large water detector also found fewer neutrinos than my calculations indicated. Matoshi Koshiba originally proposed this detector and the experiment was led by Yoji Totsuka, with important contributions from Gene Beier and Al Mann at the University of Pennsylvania, and many talented collaborators in Japan and the U.S. These results confirmed that there was really a "solar neutrino problem", that there were fewer neutrinos than predicted.

Incidentally, the Japanese-American detector was not built to see solar neutrinos. It was called Kamiokande, where the "*nde*" stands for nucleon decay experiment. The detector was sensitive only to high-energy events, such as would accompany nucleon decay. When I first heard about it, I did not think that the conversion to detect low energy events could possibly be successful. But, it was. Koshiba wrote me a letter which I still have saying that he made the conversion in order to resolve the discrepancy between Ray's results and my calculations. He didn't do that single-handedly, but he certainly strengthened greatly the case for a solar neutrino problem.

Subsequently, solar neutrino experiments with gallium, involving large international collaborations of physicists and chemists, were performed in Italy and in Russia and reached a similar conclusion. These gallium experiments were beautifully done and they detected neutrinos from different reactions than those observed in the chlorine and the Kamiokande experiments. The gallium experiments were primarily sensitive to low-energy neutrinos whose flux I could calculate more accurately than the neutrinos that are observed in the chlorine and Kamiokande experiments. There was really a solar neutrino problem. Incidentally, there was an incredible collection of experimental talent that did the gallium experiments, led by the spokesmen, Vladimir Gavrin and George Zatsepin in the Soviet Union and Till Kirsten in Germany, with many expert collaborators.

By the early 1990s, it was certain that there were fewer neutrinos reaching us than predicted by our solar model calculations. But, I think that most physicists not in the field were betting that my solar model calculations were at fault.

According to the Standard Model of electroweak interactions, the neutrino is supposed to have a zero mass. Now all these experiments indicate that the neutrino has a small mass. What about the Standard Model then? Is it a wrong model or does this only mean that we got to a deeper level?

It's not a wrong model. It's an extraordinarily successful model. It predicts precisely the results of many sophisticated and probing experiments. It describes very accurately many phenomena and unifies electricity, magnetism, and the weak interactions. So it is a great theory, even if an incomplete theory. There is not a natural way in the Standard Model to give the neutrino a mass but one can, without going far beyond the model, incorporate a finite mass neutrino and that's adequate for all the phenomena we know about so far.

Does the fact that the neutrino seems to have a mass predict that there will be a new physics?

It's hard to know what the role of neutrino mass will be in a future, more complete theory until we have a more complete theory.

What do you need for getting to this new generation of physical theory? Just thinking or more experimental data?

I don't know. My colleagues here at the Institute for Advanced Study mostly concentrate on getting there by thinking. They are mostly string theorists; they are not experimentalists or phenomenologists. If you look back at the history of physics, very often breakthroughs have been achieved by experiments revealing things that were unexpected and that lead to new theoretical developments.

What will be the next step in neutrino astronomy that you foresee?

We want to study both very low and very high energies. The very low energies are characteristic of solar neutrinos. According to the standard solar model, more than 99.99% of the flux of neutrinos that is expected to come from the Sun is below 5 MeV in energy. So far we only have direct measurements of solar neutrinos with energies about 5 MeV. We must test at low energies our theories of neutrino physics and of stellar evolution. The astrophysics predictions are most precise for low energy neutrinos. At low energies, we expect to see dramatic and characteristic effects of new neutrino physics, not all of which may have been anticipated. The frontier for solar neutrino astronomy lies at energies less than 1 MeV.

For extragalactic and galactic neutrino astronomy, you really want to go to energies above 100 TeV (10^{14} eV) and look for sources using very large underground detectors; under ice in Antarctica, under water in the Mediterranean, and in Russia in Lake Baikal. One plausible possibility is that high-energy neutrinos may be seen from the Gamma Ray Bursts that come to us from some of the most distant observed regions of the Universe. I am sure that with proper instrumentation we will observe neutrinos produced by distant cosmic rays interacting with the cosmic background radiation.

The next step in neutrino astronomy will be to increase the average distance from which the observed neutrinos reach us by fifteen orders of magnitude, from 10^{13} cm (the Sun) to 10^{28} cm (Gamma Ray Bursts or unknown distant sources).

What do you expect to learn from these very high-energy experiments?

We want to know what else there is in the Universe that we can't see in ordinary light, i.e. with photons. If we observe distant astronomical sources, like Gamma Ray Bursts, with neutrinos, we may learn entirely new things about the physics of neutrinos. This is possible because of the long propagation times, 10^{10} years, for cosmic neutrino sources instead of 10 minutes for solar neutrinos. There is more time for something exotic to occur. There is a large arena of unexplored territory that can be studied with high-energy neutrinos. There may well be surprises, unanticipated results, as there were with solar neutrino studies.

John Bahcall standing with Ray Davis at the tank of the Homestake Mine in the mid 1960s (courtesy of J. Bahcall).

There are different neutrino experiments: one in the Homestead gold mine in the U.S. that uses chlorine in the form of perchloroethylene, there are experiments in Russia and Italy that use gallium as a detector, there are experiments that use pure water in Japan and even an experiment that uses heavy water, deuterium, in Canada. As I understand, they all measure a somewhat different energy range. Why is that so?

The chlorine and the gallium experiments are both radiochemical experiments. An electron neutrino is captured by either a chlorine or a gallium atom and transforms that atom into either a radioactive argon atom, for a chlorine detector, or a radioactive germanium atom, for a gallium detector. That can only happen above a certain energy threshold and the thresholds are rather low. But the radiochemical detectors do not record the energy of the captured neutrino; any energy above the threshold, which in the case of chlorine is 0.8 MeV and in the case of gallium is 0.2 MeV, can cause the same reaction. You measure how many radioactive atoms are produced, but you don't know what energy neutrinos produced the signal.

In addition to having different energy thresholds, the chlorine and gallium detectors have very different responses, or sensitivity, as a function of energy. Chlorine has a great sensitivity to high-energy neutrinos, due to a so-called super-allowed transition, which I first recognized in 1963.

How does it work?

In 1963–1964, I predicted that there would be an isotope, not previously discovered, ^{37}Ca, which would not decay rapidly by ordinary nuclear processes but rather would decay to ^{37}K by relatively slow beta-decay processes. Using the same ideas that led to the prediction of an enhanced sensitivity of chlorine to high-energy (^{8}B) neutrinos, I calculated the lifetime of ^{37}Ca, which was potentially measurable in the laboratory. If it were found in agreement with my predictions, then the laboratory measurement would confirm my calculation of the neutrino sensitivity of the chlorine (Davis) detector. The phone call that I got telling me that ^{37}Ca had been discovered and that its properties were in agreement with my calculations was the most exciting event in my scientific career.

Anyway, with regard to your previous question, the water detectors in Kamiokande use a different kind of reaction; they use the scattering of neutrinos by electrons, which was also something that I studied in 1964. I calculated the angular dependence of the scattered electrons that would result from neutrinos from the Sun hitting a target that contained electrons in the water. I showed that the recoil electrons from neutrino-electron scattering are very forward peaked, lying preferentially in the direction of the Sun-Earth axis. That directionality is used today to detect solar neutrinos in the Japanese experiments. Electrons scattered by solar neutrinos move away from the Sun, while background events are essentially isotropic.

These neutrino electron scattering events are sensitive not just to electron type neutrinos as is the case for the radiochemical chlorine and gallium experiments, but they are also sensitive somewhat to muon and tau type neutrinos. So they measure not only electron neutrinos but, to a lesser extent, also the other type of neutrinos.

A much larger version of the Kamiokande detector, called Super-Kamiokande, began operating around 1995 or 1996. This detector was used to study solar neutrinos with great precision; the work of a very large and talented team of physicists was led by Yochiro Suzuki and Yoji Totsuka.

Finally, the Sudbury Neutrino Observatory, universally called SNO, in Sudbury, Canada, uses 1000 tons of heavy water, in which deuterium replaces ordinary hydrogen, as a unique detector. This experiment has two great advantages. SNO can measure separately just neutrinos of the electron type and determine their approximate energies (above about 5 MeV). SNO can also use their heavy water to make a separate measurement of the total flux of neutrinos of all types. Together, these two measurements are a supremely powerful test of whether something new happens to neutrinos on their way to the Earth from the interior of the Sun. This experiment is a Canadian-US-British collaboration, with Art McDonald as the spokesperson for a superb team of physicists and chemists.

The first direct proof of new physics with solar neutrinos was obtained by comparing the results of the SNO measurement of just electron type neutrinos with the Super-Kamiokande measurement with measured electron type plus other neutrinos. The difference between the two measurements indicated that about two-thirds of the electron type neutrinos that originate in the center of the Sun change to other types before they reach detectors here on Earth.

What makes these different types of neutrinos transform into each other; that is, what makes them oscillate?

Oscillation is a quantum mechanical phenomenon. It is caused by the fact that neutrinos of different types can have different masses and the neutrinos we normally see in the lab, like electron type neutrinos or muon type neutrinos, are really linear combinations of neutrinos with different, definite masses. One factor causing oscillations to occur is that neutrinos of different masses travel at different speeds so the phases of the terms in the linear combinations can change. Another factor causing oscillations is that neutrinos of different types interact differently with the electrons in the Sun and in the Earth; this is what is known as the Mikeyev-Smirnov-Wolfenstein effect.

Please, tell me something about Raymond Davis. I understand that you cooperated with him for over 40 years.

Ray is an extraordinary human being. He is helpful to everyone; he is respectful and pleasant to everyone; he treats the janitor who works in his building with the same courtesy, respect and kindness as he does the most famous professor. Ray is modest; he has no personal ambition. His goal in doing science is to understand the way the Universe operates. He is very close to all the members of his family.

Ray always looked for ways to help other people do their science. He would often make measurements that would be useful to others in their experiments. He is absolutely honest and extremely stubborn. If he hadn't been so unambitious and so stubborn, he would never have devoted himself to such an out-of-the-mainstream experiment.

In the early days, the 1960s and the 1970s and into the early 1980s, all of the people who were seriously committed to solar neutrinos could, and frequently did, ride in the front seat of Ray's car. Only Ray and I

John Bahcall and Ray Davis on the occasion of Davis' Tinsley Prize and Bahcall's Heinemann Prize in 1995 (photograph by and courtesy of Jacqueline Miton).

were committed in the sixties and seventies. I rode with Ray because I had confidence that Ray could do a reliable experiment, one that would justify my spending so much time on making precise theoretical calculations. Only Ray was willing to take the risk and devote himself to such an exotic experiment. And, probably, only Ray had the set of talents and the character to make it work.

Ray did an experiment which most people thought was impossible when it was proposed. He treated every criticism of the experiment with seriousness and he made measurements to demonstrate that each criticism was not valid. He is not a person who is particularly quick. He instinctively distrusts theory, but is curious about ideas. We gave hundreds of talks together in the first thirty years of the subject. We were a good complement to each other. He addressed the experimental side and I covered the theoretical side. I was privileged to work with Ray and to learn from him about science and about how to be a decent human being. I admire him immensely. He is a role model for me.

I would like to talk a little about this year's Nobel Prize.

All three people, Davis, Giacconi and Koshiba, richly deserved the Prize. Both Ray Davis and Toshi Koshiba made extraordinary contributions to science. They've written not just new sentences in the history of science but whole new chapters. Their contributions are made all the more extraordinary by the fact that they are such extraordinary human beings, who are enormously appreciated and admired by their colleagues.

Many particle physicists do not know the revolutionary role Riccardo Giacconi played in the field of X-ray astronomy. I worked in X-ray astronomy for many years and I know both Riccardo and his contributions very well. Riccardo energized and inspired two decades of X-ray astronomy, which has led to a multitude of discoveries. X-ray astronomers throughout the world greeted his receipt of the Nobel Prize with jubilation.

You said that Riccardo Giacconi energized discoveries. What was his direct involvement?

Yes. He was at the center of the group that was making the discoveries. When there is a group of people collaborating on a particular project, they each have different roles. Riccardo led collaborations building telescopes and detectors and collaborations making discoveries. In fact, for about two decades Riccardo led the entire field of X-ray astronomy and much of astronomy in general. I called him at home on the day the Nobel Prize was announced, congratulating him and telling him that I shared the widespread view that he richly deserved the award. I have great admiration and affection for Riccardo.

Was it logical to combine these two fields, neutrino astronomy and X-ray astronomy?

In both fields, the prizes were awarded for opening a new window to the Universe.

I talked to people before the Prize was announced this year and you were mentioned as possible candidate. How do you feel about it?

I am pleased to be mentioned in this distinguished context.

What do you think, had you written up your results back in 1963 in a joint paper together as originally planned and not in two back-to-back papers, would the outcome be different?

I don't think so. The Nobel Committee quoted both Ray's paper and mine from 1964 in their technical notes.

Let's go back to science. The solar neutrinos are supposed to account for part of the dark matter in our Universe. Any ideas about what the rest of it is?

If I had any idea about that, I would immediately excuse myself and rush away to write a paper on the subject!

You have also been involved with different models of the Galaxy. What are these?

I served for more than 20 years as the member of the scientific guiding committee that was responsible for the Hubble Space Telescope. To help us think about how to make the inevitable tradeoffs between different instruments or developments, I began calculating, together with a young colleague from our Institute, Ray Soneira, what we expected the telescope to see with the HST cameras.

At one point, the science committee considered how the telescope could be pointed accurately as required. I began looking into the question of how well known was the density of bright stars on the sky. The bright stars could be used as guide points to fix the telescope's direction. Ray Soneira and I found that there were errors in the standard published tables of these data. That was in the late seventies. We concluded that the star density was much less than what had previously been published. There was a lot of resistance to our results at first because they required an expensive change in the guide-star equipment and because we were just amateurs

John Bahcall receiving the National Medal of Science from President Clinton in April 1999 (courtesy of J. Bahcall).

with no previous record in the subject. But, eventually our pessimistic calculations were accepted and the guide-star system was redesigned to reach the level of sensitivity that we specified. In fact, the Hubble Space Telescope requirements document stated, among many other things, that the guide-star system had to be sensitive to the lower star density that was specified in our paper.

As a natural generalization of the answer to the practical question of what was the bright star distribution on the sky, we made a model of the star distribution of the entire Galaxy at different brightnesses, colors, distances, and stellar types; we used that model to compare with lots of data. The basic premise of our model was that our own Galaxy was like other galaxies that we could see which had disk and spheroidal components. We took the initial model parameters from whatever measurements were available. We refined our model iteratively by comparison with the observed distribution of the stars in the Galaxy.

I had enormous fun with the Galaxy model project because there was a continuous interaction between the modeling and the data. We could summarize huge amounts of data that had been previously accumulated with just a few meaningful parameters. The model is still useful today, almost a quarter century later.

What I love most to do in science is to explain quantitatively things that are measured or observed.

Other topics that you've been involved with?

In 1965, Ed Salpeter and I suggested that we could use absorption lines in the spectra of quasars, which had recently been discovered to be the most distant known objects in the Universe, to learn about the gaseous material along the line of sight between the distant quasars and us. The basic idea was to use the quasars as a sort of flashlight to illuminate the medium between the quasars and us. We predicted that gas clouds or clumps along the line of sight would produce discrete absorption lines. We predicted which lines would be strongest.

Within about a year, the absorption lines were discovered and ever since quasar absorption lines have been observed in abundance. Quasar absorption lines are now a standard cosmological tool. But, in the early years, most astronomers, including the astronomical experts, believed that the lines were produced by material associated with the quasars, i.e. they thought that the lines were not produced in the way Ed and I had suggested, by the cosmologically distributed material between the quasar and us. So, I invented a technique of analyzing simulated absorption line spectra that looked like the real spectra to prove statistically that the real spectra had clumps of gas at many different redshifts along the line of sight. I am not sure, but this could have been the first application of Monte Carlo simulations to cosmology or astrophysics. This was a multi-year project because the opposing viewpoint was held by people who were experts in astronomical spectra.

I also got involved, I think in 1964, with making the first models of quasar emission line regions that were ionized by the strong light emitted by the quasar itself. Ben Zion Kozlovsky and I did this work together. All of the quasar work was a lot of fun because there was a lot of interaction between new data and the modeling.

You mentioned earlier that when you first went to CalTech, both Gell-Mann and Feynman were there. Did you know them?

Yes. They were the great scientists to whom all of us looked up. We knew that we were privileged to be working in their vicinity. In retrospect, just thinking about what they understood and created, they seem like even greater giants than they did in person.

I can tell you a personal story about Feynman from 1968. Before Ray Davis' first result was published, Ray came to CalTech for a week. We were again writing papers to appear back-to-back in *Physical Review Letters*. Willy Fowler arranged for a small, private presentation by Ray and myself. He invited Dick Feynman, Murray Gell-Mann, Bob Leighton, Maarten Schmidt, and maybe one or two nuclear physicists. It was a small group in a small classroom in Bridge Hall.

That was a very tense time for me. I was a young assistant professor, without tenure. My best-known work was predicting the rate of neutrino capture in Ray's tank and Ray was getting an answer different from what I calculated. So I presented my theoretical discussion of what was expected and Ray described what he'd measured and why it was clearly less than my calculated rates. There were questions during the talks and a follow-up discussion; then the meeting broke up with no particular conclusion.

I was enormously depressed. I was young and ambitious. I had made a striking prediction that was not confirmed. Dick Feynman was not a person who wasted his time. According to some of his biographers, he was also not characteristically altruistic. But on this occasion, he saw that I was depressed and he did something about it. He said to me, "Let's go for a walk." We did that. We walked for more than an hour; I still remember where we walked. He mostly talked to me about things that were not very substantive.

After walking for a while, Feynman told me that he could see that I was upset. He said I should not feel bad because no one had found an error in my calculations. He said he did not know what the explanation would be for the discrepancy between Ray's measurements and my calculations, but it could be important. He tried to cheer me up by saying that I had not performed badly.

That walk, and that talk, his kindness on that difficult occasion, meant an awful lot to me.

Talking about mentors and advisors, whom would you like to mention?

A lot of people have taught me science and helped me along in my career. Emil Konopinski, at Indiana University, introduced me to weak interaction theory. Willy Fowler was a very strong mentor at CalTech, as I already mentioned. Dick Feynman was always helpful when I had a particular scientific question. Like everyone else at CalTech, I went to Feynman to get his insight and advice whenever I thought I had a new idea that might interest him. Murray Gell-Mann was always extraordinarily generous to me at CalTech. He was the one who arranged for my coming here, to the Institute for Advanced Study. I regret very much that I have seen very little of him since I left CalTech. He was a great inspiration. Conversations with Murray were very different from conversations with Feynman. Feynman would listen and work out for himself what you told him. Murray would tell me what I should be working on, what problems I should solve.

When I came here, there were several people who were very helpful to me. Marshall Rosenbluth who had the office next to mine for almost a decade, is widely regarded as the world's leading plasma physicist. But, he is also a great human being. Marshall was the person I most talked to about technical things in astrophysics. Martin Schwarzschild was a great astrophysicist, a stellar evolution theorist. He didn't do things in the detailed style that was required for the solar neutrino problem. But, he was always enormously supportive and insightful. He was the one I could talk to about "big-picture" aspects of the solar modeling. Lyman Spitzer, with whom I worked for about a quarter of a century on the Hubble Space Telescope, was a close scientific friend and a mentor. Even today, when I have a tough decision to make, I ask myself: how would Lyman have addressed this problem? Very often, that question makes obvious the answer to my problem.

John Bahcall with Willy Fowler at Fowler's 80th birthday celebration at CalTech (photograph by and courtesy of Stan Woosley).

Lyman and I worked together to sell the Hubble Space Telescope to the Congress; we lobbied every relevant congressman and senator together. We went together on innumerable trips to NASA meetings.

Lyman Spitzer, Martin Schwarzschild, and later Jerry Ostriker, Scott Tremaine, and I worked actively together to maintain a supportive and collaborative scientific environment here in Princeton. Astrophysics at the University and the Institute are very closely linked to each other.

In astronomy, my principal scientific mentors were Ed Salpeter (Cornell University), Peter Goldreich (CalTech), Jerry Ostriker (Princeton University), Jim Peebles (Princeton University), Martin Rees (Cambridge University), and Scott Tremaine (Princeton University). But, I learned an enormous amount from all of the young people who were postdocs at the Institute for Advanced Study.

You chaired a committee about the future of astronomy and astrophysics. What was the outcome of that?

We set priorities for the next decade, from 1990 to 2000, in astronomy and astrophysics. We recommended a prioritized list of 20 projects; the top 18 were funded. We didn't recommend a lot of other good projects; I think we may have turned down as many as 10 ideas for each project that we put on our list of priorities. Congress and the federal agencies were all appreciative that we ranked things according to priorities. They understood that we had made tough choices and that they would get good return for their money. Our top priority was the space infrared telescope facility, which is to observe in the infrared region from space in the same way the Hubble telescope observes in the ultraviolet and visual. The infrared telescope is going to be launched within months.

Does your group need more funding?

I was asked the same question once before, in a public forum, by the president of the World Bank, who is also the Chair of the Board of Directors of the Institute for Advanced Study, Jim Wolfenson. He was obviously very supportive when he asked: "Do you and your group need more funding?" I thought about it and I understood that this might be an opportunity for fund raising, but I answered him honestly. We need additional brainpower more than we need more funding. We can use more funding, but the real bottleneck is our limited theoretical understanding.

How about experimental research?

I think that the projects I most believe in, and to which I am willing to devote myself to help make sure that they will happen, are going to get funded by the normal processes. That was not true in the late 1970s and early 1980s, when Ray and I were trying to get a gallium experiment funded in this country. That experiment did not get done in the U.S., but gallium experiments were eventually performed in Italy and in Russia, with international participation.

It is not generally known, but I was a principal investigator on a proposal to do a gallium experiment in this country. The proposal was essentially Ray's proposal, and included both techniques that were used later in Russia and Italy. The reason why I was the principal investigator was that it was a proposal to the National Science Foundation. Since Ray was at a Department of Energy institution, he could not apply to the NSF.

If you really told me now that I had a budget with which I could do anything I thought was important, I would establish the National Underground Science and Engineering Laboratory. I think that for the world and for the United States that's a great opportunity. We need a large, flexible, deep, and dedicated national and international facility that can do all the scientific and engineering experiments that require a deep underground facility. The scientific program includes biology, the discovery and study of species which are unique and exist without sunlight, which may even be the most primitive forms of life; we don't know, they have very different energy cycles. The program includes studying the transport of water over large distances and the geophysics of rocks under high pressure. It includes experiments to study dark matter, perhaps discovering dark matter experimentally, to study double beta-decay, to help determine the characteristics of neutrinos, to do solar neutrino astronomy at low energies, to do many different kinds of science in a unique environment.

The National Underground Science and Engineering Laboratory is the highest priority future project with which I am currently involved. I think it is going to happen; I don't know how soon, but it will.

You run an astrophysics group at the Institute for Advanced Study. Can you tell us something about that?

When I first came to the Institute, there was no astrophysics program, just particle physics and plasma physics. But, over the years, we have built a postdoctoral program in astrophysics that is generally regarded as one of the best in the world. We have been really lucky; many outstanding young people have chosen to work here. We try to make the environment supportive and stimulating. Since we don't have students, we can focus entirely on the postdocs. My primary job as an IAS professor is, in my view, to be helpful and supportive to the postdocs.

Running the astrophysics group has been enormous fun. I get great pleasure from being helpful to the young people and watching what marvelous things they come up with. Moreover, the postdocs are extremely stimulating for me; I learn an enormous amount from talking to them. They have become an extended family for Neta and myself.

I would like to ask you about your family background.

I grew up in Shreveport, Louisiana, and lived there until I went away to college. My mother got a bachelor's degree in music at the University of Illinois and later a Master's degree in social work. She was the only person in our close family with a college degree. My dad was born in Appleton, Wisconsin a few months after his parents came to the U.S. from Russia. He grew up in Maywood, Illinois, near Chicago. He also went to the University of Illinois for a year or two, where he met my mother. My dad worked as a traveling salesman for a wholesale produce company, which sold fresh fruits and vegetables. The company was owned by one of my mother's uncles in Shreveport. There were very few Jews in Louisiana, and especially in Shreveport, but somehow one of my mother's uncles ended up there during the great depression. He acquired an open cart, successfully sold fruits and vegetables in the street, and was able, subsequently, to provide employment to his family and other families as well. I have one brother, Bob, an older brother, who like our father became an expert salesman. He sells primarily cleaning materials and lives together with his family in Baton Rouge, Louisiana.

Did your parents experience anti-Semitism?

My parents never discussed anti-Semitism with us. If discrimination affected us, we accepted it as a fact of life. We lived in a completely segregated society when I grew up. Blacks were not allowed to attend schools with whites, nor to eat in the same restaurants, play in the same sport facilities, or use the same water fountains. The restrictions on Jews were much more subtle. My brother and I were the only two Jews in the elementary school that we went to in Shreveport. Our family was much less well-off economically than the other Jews in Shreveport and we lived in a different section of the town, a neighborhood for people with very modest economic resources. I can remember occasions when my brother and I had to run home from school in order to avoid getting beat up. The majority of the kids in our school were Baptists, but there was a minority of Catholics and every now and then the Baptists would beat up the Catholics, who then beat up the Jews, my brother and I. But, these were isolated incidents.

What was your name originally?

Bachalor, which I think means a wine goblet in Russian.

Your wife is also an astronomer. Did you ever work together?

Yes, we did. In fact you can see on my bookshelf a picture of my wife, our two sons, and me; that picture was taken just after we made a discovery together in the early 1970s. We had exclusive use for a summer of the 1 m telescope in Mitzpeh Ramon in the Negev desert of Israel. The telescope was just being brought into operation and our task was to convert the facility into a working scientific observatory. I had to develop the darkroom; we worked together on the first instruments.

X-ray astronomy was in one of its golden ages of discovery; Riccardo Giacconi and his group had found a number of pulsing neutron stars in binary systems (two stars going around each other) that emitted X-rays. But no one could find the optical counterparts of any of these X-ray binary systems. So, we didn't really know what they looked like, how far away they were, what type of ordinary stars were going around the neutron stars.

Since the observatory had only one operating instrument, a camera, the only thing that we could do was to take pictures. We took repeated images of the directions on the sky where the X-ray binaries were located. Neta noticed that one of the stars in our observing program had a period equal to the binary period of the X-ray binary, Her X-1. Bingo, we had made an important discovery. What luck! With rudimentary observing equipment, we succeeded ahead of all the astronomers in the rest of the word who were trying to identify the X-ray binaries with sophisticated equipment and clever techniques. No one else tried to look at the slow light variations, analogous to the phases of the moon, due to the binary motions of the X-ray star and its companion (Hz-Herc).

We collaborated together on a number of projects over the years, including studies of quasars and of globular clusters. We have not collaborated scientifically in the last decade or so, because Neta's interest has become more focused on cosmology and my interest has become more focused on neutrinos. But we talk about astronomy a lot.

Did your children mind that when they were small?

I think they may have thought that our conversations were limited. Both Neta and I have a scientific view of the world. However, when our kids were growing up, they probably didn't know enough to protest. We were the only family they had ever lived in. They just took things as they were.

Please, tell us something more about your family.

I consider myself lucky in life because, after an enormous effort on my part, Neta agreed to marry me. I am still feeling lucky every day. I have always admired her smile, her intelligence, her practicality, her logic, and her ability to get things done. We share the same cultural, intellectual, and ethical values. Our kids are great; they are a source of wonderful pleasure. Now, they are more like our good friends than our children.

We have three children. Our oldest son, Safi, is 34; he has a Ph.D. in theoretical physics and is the CEO of a bio-tech company. Our middle

The Bahcall family: Dan, Orli, Safi, John and Neta, 2003 (courtesy of J. Bahcall).

son, Dan, is 31; he has a Ph.D. in cognitive psychology and is involved
with non-profit organizations that support environmentally-friendly activities
and other socially desirable projects. Our daughter, Orli, is 26 and is finishing
her Ph.D. in epidemiology at Imperial College in London.

How did you first meet with Neta?

Yuval Ne'eman arranged for me to give some lectures on nuclear astrophysics
in Israel. He wanted to start astronomy in that country. So, in 1965, I
came to the Weizmann Institute. The first day I was there, I went to
the basement of the physics building where they had a van der Graff
accelerator. I was looking for a friend of mine, Gabi Goldring, a nuclear
physicist I had met earlier at CalTech. Gabi was not in the lab, but I
saw this beautiful young woman with a wondrous smile, working on some
apparatus. When I finally found Gabi, I asked him to introduce me to
the smiling young woman who, it turned out, was his student. Gabi asked
Neta to show me the laboratory and she did. I invited her for a coffee,
I invited her to go for a walk, I invited her for lunch, I invited her to
a movie. Over the next several days, I asked her out many times but she
always said that she had too much work to do on her experiment. I was
to be at the Weizmann Institute only for 10 days; I began to feel desperate.
Finally, I called Neta at home and she said, "OK, my mother has tickets
for the opera tomorrow, but she can't go. Do you want to go with me?"

Of course, I said yes and we went to the opera. We were attracted to each other, but it was a great struggle to get her to agree to marry me.

What was the greatest challenge in your life?

The most difficult and challenging situation I ever faced, and the problem I worked hardest to solve, was to persuade Neta to marry me. If you ask me what was the best idea I had in my whole life, it was the following. After Neta and I met in Israel, we continued corresponding for about half a year. I wrote asking her whether she would like to come to Caltech to visit. She replied: you should come to Israel and settle here. I wrote back saying: there is no astrophysics in Israel; I can't work there. She said she would never leave Israel. Then I got the best idea I have ever had. I scraped together nearly all of the money I had saved up and bought a roundtrip ticket from Israel to the U.S. with an open return date. I sent the ticket to Neta, with a note saying I hoped she would use it. That was in December of 1965. At that time no one left Israel because nobody had money to travel. Neta said for her going to the U.S. was almost like going to the moon. So it was a rather dramatic thing to do.

I remember driving to the post office to mail the ticket with a friend of mine, a physicist named Joe Dothan, who had known Neta from the time that he was her graduate student instructor. He assured me that there was no chance that she would come — but she did.

What do you do when you are not doing science?

My hobby is literature; I like to read novels. Most everything I read recreationally nowadays is in Hebrew; I read primarily modern Israeli novels. I like the challenge of reading in a language that is not my native tongue. But, until I was in my middle or late forties, my recreational reading was almost entirely English language novels.

When did you learn the Hebrew language?

After we were married. First, I learned to speak conversationally, but later I taught myself to read.

Would you like to add anything?

No, I think we've covered a lot more than I expected.

Rudolf Mößbauer, 1995 (photograph by I. Hargittai).

12

RUDOLF MÖßBAUER

R udolf Mößbauer (b. 1929) is Professor of Physics at the Technical University Munich. He shared the Nobel Prize in Physics in 1961 "for his researches concerning the resonance absorption of gamma radiation and his discovery in this connection of the effect which bears his name". (The other recipient of the 1961 Physics Nobel Prize was R. Hofstadter, Stanford University, "for his pioneering studies of electron scattering in atomic nuclei and for his thereby achieved discoveries concerning the structure of the nucleons".) Our conversation with Rudolf Mößbauer was recorded on October 26, 1995, in Budapest.*

First, I would like to ask you about some comparisons. Let us start with the Max Planck Institutes versus university research in Germany. Why are the Max Planck Institutes so much stronger?

There is a political reason. Around 1968 there were student unrests in various countries, including Germany. In all the other countries, they regulated the problems in some sensible ways. In Germany new laws were introduced in 1976 that bind the universities in ways that even today most people do not realize. For example, the professors at the university are civil servants, like those in the post office or in the finance office. The same rules apply which is very bad because the university needs more liberty and more flexibility. If we would take a vote today at a German university, the majority

*István Hargittai conducted the interview. This interview was originally published in *The Chemical Intelligencer* 1997, 3(3), 6–13 © 1997, Springer-Verlag, New York, Inc.

of professors would vote for the present system because they have got used to it. Once you become full professor, you may be doing nothing for the rest of your life and nobody will bother you. It is very socialistic, and competition is reduced to zero.

The Max Planck Institutes are better off for several reasons. They have more financial assistance and more liberty. The Max Planck Society gets 45% of their income from the states, 45% from the federal government, and 10% from industry. This 10% gives them the liberty.

Is there a comparable organization in the U.S.?

They don't have a comparable organization. Their national laboratories have their German counterparts and that's the only common feature. In the U.S. the tendency has always been to do research at the universities and, in principle, this is better. When you are young, you are more active in research and when you get older you might become more active in education. These two areas are in harmony in the university.

Let's make another comparison: science today in the former West Germany and former East Germany.

In a West German laboratory, when I needed something, I always checked the catalogues first to see whether the stuff was available on the market. In former East Germany, due to lack of foreign currency they had to make everything themselves. This is why they had to have many times more scientists, and it's still a problem what to do with them. Even in the western part of Germany we have nowadays too many scientists, especially in many of the federal supported laboratories, now organized in the Helmholtz Society. There are altogether 16 of them, 13 of them in the former West Germany and 3 newly formed ones in the former East Germany. You cannot dismiss anybody, but with money frozen in and with salaries continuing to increase, the money available for research is gradually diminishing. In addition to the Max Planck Institutes and to the Helmholtz laboratories, there exist also the institutions of the so-called Blaue Liste. These are mostly remnants of former East German Institutes, which could not simply be closed down, but which now also drag on the financial resources.

Currently, I have 10 people in my research group at the Technical University in Munich, and none of them is from the former East Germany. This is not by design though. I have also tried to send some of my co-workers to become professors in the former East Germany, and it didn't work. The East Germans were blocking theese attempts. They have too many people locally. Why should they accept foreigners or semi-foreigners if they already have too many scientists of their own?

How do American and European research institutions compare?

I spent altogether nearly seven years at the California Institute of Technology (Caltech). I went there in 1960. At that time West Germany was so run down that whenever you got an offer to go, you did. I was leaving Germany to stay in the U.S. for good. I didn't want to come back. The reason why I came back was the smog in Los Angeles, so I returned, which in a sense turned out to be a mistake. Not so much returning to Germany but going to a university. I can't blame this on the Max Planck Society because I had had many offers from them. However, I felt, as a young man, that I might be able to fight the government and the administration, where the actual problems had been. Today I know it's useless. Coming back to the comparison of science in the United States and Europe, there is one thing there which we don't have in Germany and that is the private universities. Caltech is an example. These private places are setting the standards for the state-operated universities. For instance, Berkeley is a state university, part of the University of California System, another is UCLA and Irvine, and so on. These California state campuses are on that very high level only because they have to compete with the private places in California like Stanford and Caltech.

You were at Caltech when Linus Pauling was there and received the Nobel Peace Prize and then quit.

He was chased away because of his anti-nuclear movement. Actually he was not chased by his colleagues, the professors, but by the Board of Trustees and the rich donors on whom the financial health of the Institute depended. A rich oilman, for instance, who used to contribute one million dollars annually to the Institute threatened to withdraw if Pauling stayed. Pauling was finally fixed up in a nice way in Santa Barbara, but he definitely had to leave Caltech, which was one of the unfortunate aspects of a private institution.

You got the Nobel Prize in physics when you were very young. How did you survive it?

It wasn't so easy but my advantage was that although I had done the work in Germany, I got the prize when I was already in the United States. In Germany a Nobel laureate is a very special person. In the United States the number of Nobel prize winners is so high that I was just one of them. At Caltech we were very well protected. I had a bad flu at the time of the announcement, lying in bed, and I got a call from Caltech telling me that they were taking my telephone off so I wouldn't be bothered, and two sheriff cars appeared in front of my house and no reporter could get close.

Is European science similar to the American in strength?

In many aspects Europe is stronger now. Of course, we have many countries together, for instance, Italy and France are especially strong, and in high-energy physics we have the CERN Laboratory in Geneva and there is nothing like that in the United States. I was director of the Laue-Langevin Institute in Grenoble for five years, a joint German-French-British research center. Our main competitors were Brookhaven and Oak Ridge in the United States and Chalk River in Canada. I spent four months at these laboratories before I started my own directorship at Grenoble. The reason I went there was to find out what *not* to do. When I came to Grenoble I could stop a lot of things that were already in progress about which I'd learned that we shouldn't be doing. I felt that if after five years we could be on the same level as the Americans, I would be happy. Now what happened was that we were on top after only two years. There were several reasons for that. Firstly, the Institute had a unique arsenal of instruments, largely based on the work of my predecessor, Professor Maier-Leibnitz. Secondly, the Institute was operated as a user's Institute, where the experimental proposals originated in the entire scientific community associated with the Institute. And thirdly, the Institute operated on an international scale; this, in fact, is a general big advantage in Europe and we make too little use of it. The reason we were so good was that we were German and French there and we also pulled in the British while I was there. The three countries have very different mentalities. If you can put them to work together, you can do miracles which the Americans cannot. It doesn't matter that the countries are small, what matters is that they think differently.

What is your main current interest?

Neutrinos and developing new detectors in my Munich laboratory. They will have much higher resolution in energy than the present ones.

During the past years you have been actively involved in the GALLEX project. Could you tell us about it?

We measure solar neutrinos in this experiment, which is carried out by a huge international collaboration. The neutrino is a particle which carries no charge and which is subject to weak interactions only. Only the gravitation is still very much weaker, but it matters only on astronomical scales. Our body, for instance, is attracted by the entire earth and it is only for this reason, that gravitation appears significant to us. On nuclear scales, gravitation is negligibly small, even when compared to the so-called weak interaction. This latter interaction is so weak, that a neutrino entering the Sun on one side and leaving it on the other side hardly feels its existence. Neutrinos, in particular, are formed inside the Sun, as they are in other stars, when four protons are fused into a helium nucleus. The helium nucleus is a little lighter than the four protons, and this mass difference is converted into the energy which the Sun radiates. The generated neutrinos then leave the Sun, some coming to the Earth where we then can measure a tiny fraction of them.

If the neutrinos can travel through the Sun, how do we know whether they have travelled through the Sun or were generated in the center of the Sun?

Gran Sasso with its top almost hidden in the clouds, 1996 (photograph by I. Hargittai).

They can only stem from the Sun. The other stars produce similar amounts or even more neutrinos, but they are so far away that the solid angle which the Earth forms to them is so small that we cannot observe them. Only the Sun is sufficiently close so that we can measure its neutrinos.

How do you detect them?

It is simple in principle and complicated in practice. You use a big volume of a proper material. Take an ordinary radioactive decay, which goes from a higher energy state to a lower energy state, and you take the inverse. You shine the solar neutrinos on a nucleus and occasionally you get from the ground state of this nucleus with atomic charge Z to a higher state in a nucleus with atomic charge $Z + 1$. You need a minimum neutrino energy for this transition, giving rise to an energy threshold, but if there is more energy available, then that's all right. Gallium is especially favorable for this experiment because, among other things, relatively little energy is needed to initiate the transition to the neighboring element, which is germanium. The energy threshold, therefore, is low and thus one can detect the major part of the solar neutrino flux. Only a few of the incident solar neutrinos perform the transition from the ground state in gallium into the higher state in germanium. The measurement is then being done by looking at the inverse transition back, germanium to gallium. The number of such back-decays is representative of the number of initial excitations and thus of the number of incident solar neutrinos. This is how their number is being determined. Because we are dealing with weak interactions, most neutrinos go unnoticed through the detector, and there is only occasionally an interaction, about one per day.

For how much gallium?

Thirty tons. We measure one nucleus out of 10^{30} nuclei. This sensitivity is very much higher than in usual analytical chemistry. It's quite complicated chemistry not to lose the few nuclei that have been generated. This is done largely by radiochemists in this project.

Who is funding it?

The purchase of the gallium, DM 22 million, was funded by German sources; half of it came from the Federal Ministry of Science and the other half from the Krupp Foundation. In principle, this money can be recovered after the experiment, by reselling the gallium.

Chemical apparatus for the determination of the number of Solar neutrinos causing nuclear reactions at the GALLEX (Gallium Experiment) in the Gran Sasso National Laboratories, 1996 (photograph by I. Hargittai).

Was it a loan?

Originally, the gallium was given to us through a complicated financial arrangement. It would be easy to sell the gallium in its pure form, but we are using it as gallium chloride. The conversion may cost several millions DM.

Of course, in addition to purchasing the gallium, a lot of money had to be invested for setting up and running the experiments. These expenses were shared by the various European countries participating in the project.

Is it worth it?

Definitely. For the first time we learn that the Sun works the way we thought it does. We used to suppose, and now we know it, that the energy of the Sun comes from nuclear fusion. Of course, the same is true for all the other stars as well. It is basic research and there are no immediate applications for every-day life, but there is an enormous amount of side products.

Mossbauer spectroscopy. Is it easy to understand?

It is easy. In normal spectroscopy you observe an excitation or a de-excitation of an energy level, i.e. you go from a lower energy level to a higher level or vice versa. In other words, you observe the absorption or the emission of radiation. The energy involved in such a transition is separated into two parts: the major part of it goes into the radiation, optical line or gamma-line, or whatever it is, but there is also a small amount that goes into the recoiling nucleus.

In optical spectroscopy, the recoil energy does not matter and the entire transition energy, i.e. the energy difference between the two involved energy states, is available for absorption or emission. The situation is different in nuclear physics, where the transitions do not occur between electronic states of the atom giving rise to optical transitions usually appearing in the visible range of the spectrum, but between nuclear states giving rise to gamma transitions. Such gamma transitions have a million times higher energy than an optical transition, but are otherwise identical to optical transitions. The energy of the recoiling nucleus, although very small compared to the gamma energy, can no longer be neglected. For instance, if you want to do the inverse process to the emission, which, because identical energy levels are involved, is called resonance absorption, then it does not work. It doesn't work because already in the emission process the emitted gamma quantum is a little short in energy, a very small part of the transition energy being lost to the recoiling nucleus. The inverse process of absorption is likewise short in energy, because in order to perform the absorption you need the full energy of the transition, but a small part of the energy of the incident gamma radiation is again converted into the energy of the recoiling nucleus, thus leaving only a somewhat reduced energy for the absorption line. By consequence, the emission line is shifted to somewhat lower energies, the absorption line to somewhat higher energies. For most gamma lines, these energy shifts due to recoil are so large such as to prevent the observation of the inverse processes of emission and resonance absorption.

In the case of the Mossbauer effect, I developed a scheme binding the nuclei in crystals where you don't get the recoil energy shifts but you get the full energy of the transition into the line. Let me give you an analogy. If you are in a boat and you throw a stone, then the major part of your energy goes to the stone but the boat also recoils a little, and you lose a little energy. If you do the same experiment in winter, when the boat is frozen in the ice, then the recoil energy is not taken up by the boat but by the entire lake, whence it becomes negligibly small. Therefore, the entire energy goes into the thrown stone. This is what I do by putting the nuclei into crystals. The crystal as a whole is taking up the energy from the individual nucleus, and therefore there is no loss in recoil energy.

The emission line and the resonance absorption line are no longer displaced relative to each other, and the inverse processes of emission and resonance absorption now become possible.

Why is this of interest to the chemist? The levels between which emission and resonance absorption are performed are usually split. These splittings are a consequence of the chemical interactions between the various nuclei in a sample. You therefore no longer have a single gamma transition between these levels, but a number of them differing in energy. By measuring the slight differences between the various transition energies, you learn a lot about chemistry or biology. These energy shifts may be a million times smaller than the energies of the actual gamma transitions, but they may nevertheless be measurable and then provide very valuable information about the interactions between the atoms in the sample.

Was this a discovery by design or by serendipity?

At the time I did it first, in 1957, I measured the size of the effect directly via the so-called cross section. I wanted to look for the resonance transition. For this purpose I used a radioactive source and a resonance absorber engaging the same gamma transition which was emitted by the source. Changing the temperature of the source or of the absorber, I imagined the way the absorption might change. I finally found a very tiny effect, a change in the radiation intensity of about 10^{-4}, very difficult to measure, even today. But the effect had the wrong dependence on temperature, the wrong sign. I was worried about that. First I thought that I had made a mistake because there were temperature changes up and down, and different combinations of sources and absorbers. Later on, however, I determined that there was no mistake in sign, so I was puzzled. That was the start of it. Then there was a theory by Willis Lamb, the Nobel Prize-winner, that was of no use what he had made it for, but I found it to be of high interest for my area, and I could use it to explain the phenomena I had observed. It was only several months later, in 1958, that I discovered that by using Doppler shift methods I could detect much more easily the same phenomenon. This consisted of applying a relative velocity between source and absorber, thereby introducing relative shifts between the very narrow emission and resonance absorption lines. This Doppler shift method is still used today to measure the tiny differences in energy between the various lines, and it is in this way, that one can gain the information about chemical interaction effects in solid matter.

When did the applications come about?

Right after the Doppler shifts experiments in 1958 all hell broke lose. My prime discovery was during my Ph.D. studies in 1957 at Heidelberg, and I made the Doppler shift experiments right after I'd got my Ph.D. in 1958. The situation was still very difficult in Germany at that time, and I wanted to gain some time for experiments of my own. We didn't have money and we didn't have equipment. I knew that in the U.S. they were much better off. After the Doppler shift experiment I was fully aware of the potentialities of my discovery. I knew that if it leaked out, then I would no longer have any chance of my own. So I went to my former thesis supervisor, Professor Maier-Leibnitz, and asked him to suggest a journal in which I could publish my findings but where they would not be read by anybody. His advice was to publish in the *Naturwissenschaften*, which turned out to be a grave mistake. Within one week I got 250 requests for reprints. Of course, this was pre-Xerox time. Today, I would simply publish in the *Proceedings* of some remote Academy.

But you didn't lose anyway.

No, I didn't, although I'd sent preprints of the first paper even to my main competitors, Moon in Birmingham and Metzger at the Franklin Institute in Philadelphia. My paper was written in German but Metzger was of Swiss origin and therefore could read it and Moon got help from Peierls, who happened to be also in Birmingham. Fortunately, though, none of them thought about the Doppler shift measuring method, though they had been employing this method over many years. I myself hadn't been thinking about it at the beginning. It was only in 1958 that I realized that I hadn't yet done the most important experiment, and I was sure that the two were already doing it. I thought, in fact, that they were trying to gain time in writing me letters full of uninteresting things. I found out only later, that there had been no danger whatsoever.

You, as other Nobelists, participate in nominating future laureates. Is the application side decisive of a discovery?

No, not at all.

How about the fullerenes?

I think this is just a question of time. The people involved are high on my list.

What about your activities as a public figure?

I used to be very active but not anymore. I came back from the United States to Germany in late 1964 under very stern conditions. At that time we created a Department of Physics at the Technical University in Munich. For German universities, this was a completely new thing. It had been the custom at German universities to have institutes with a professor at the top. I made it a condition for my return, to create 20 chairs for physics professors, incorporated into a single department with a strong chairman at the top. A lot of additional changes were introduced at that time. I myself was chairman of this department for four years, followed by several others. In 1972, I left for Grenoble for 5 years. In 1976, new university laws were created in Germany in the aftermath of the 1968 student unrest. The state administration took the opportunity to eliminate the department altogether, and it was then that I decided to no longer get involved in political university issues.

I still speak up on public issues, such as nuclear power and on other issues of general interest. Yet, scientists and their opinion matter very little in German politics today. I shall give you a typical example. Some time ago the British Queen came to Germany, and the Chancellor of the country gave a big reception in a castle. Some thousand people were invited. I wouldn't have gone there anyway because of the long distance, but I found it most illustrative that not a single German scientist was invited. Yet, the leading tennis player was there. This is quite symptomatic.

I conclude by making a remark about present conditions in Germany. The spectacular recovery of the country after the war was, in my opinion a consequence of the complete destruction of all government institutions. People were free to do whatever they felt was necessary. In the meantime, institutions have been rebuilt, many ministries, offices, laws and regulations have appeared, and for this reason things no longer work. Brussels, or rather, the European Community, is of some help here. Most of the deregulation that has happened in Germany was forced upon the country by Brussels. Much of the present situation is directly related to the immovability of present-day politics. But the country is also lacking the spirit of getting involved with new things. Germany has always been known for its engagement in new things, and its prosperity has depended decisively on this. In this respect also, the country is beginning to show a disturbing deficit.

Arno A. Penzias, 2001 (photograph by I. Hargittai).

13

ARNO A. PENZIAS

A rno A. Penzias (b. 1933 in Munich, Germany) is a Venture Partner at New Enterprise Associates in Menlo Park, California. He shared half of the Nobel Prize in Physics in 1978 with Robert W. Wilson "for their discovery of cosmic microwave background radiation". The other half of that year's physics prize went to Pyotr L. Kapitsa "for his basic inventions and discoveries in the area of low-temperature physics". Arno Penzias received his B.Sc. degree from the City College of New York and his Master's and Ph.D. (1962) degrees from Columbia University. His positions included Vice President and Chief Scientist of AT&T Bell Laboratories and later of Lucent Technologies, Bell Labs Innovations. He has been with New Enterprise Associates since 1998. Dr. Penzias has been a member of the National Academy of Sciences of the U.S.A. and other learned societies. He has received numerous honorary degrees, and has been on the board of many companies. Between 1967 and 1985 he was associated with Princeton University on a part-time basis. His books include *Ideas and Information: Managing in a High-Tech World* (W.W. Norton, 1989, and numerous foreign editions) and *Digital Harmony: Business, Technology, and Life after Paperwork* (Harper Collins, 1996, and foreign editions). We recorded this conversation during the Nobel Prize Centennial in Stockholm, on December 11, 2001.*

---

*István Hargittai conducted the interview.

Would you please summarize your prize-winning work?

Having come to Bell Labs I decided to work in radio astronomy, using the available equipment that it was uniquely suited for. The absolute measurement of radiation from the Milky Way seemed to be something that we could carry out. This radiation was quite evident at long wavelengths. As the wavelengths get shorter, however, it gets weaker and weaker. The thought was that we could make a contribution by making this background measurement at about 20 to 30 centimeters. This would help understanding the mechanism by which the Galaxy radiates and illuminating some of the electrodynamics of the Milky Way. That was the plan.

The actual measurements were very difficult and first we wanted to make sure that our equipment produced real information rather than some artifact. We decided to make the first measurements at a wavelength of 7 centimeters, where the Galactic background should be infinitesimally small. If we got zero there, we would have more confidence to believe our other measurements. To do this I then created a quite elaborate source of noise with which to compare the temperature of the sky. We started with the 4 kelvins of liquid helium. The problem with liquid helium is that it boils in the tank where the standard, this black body, is and that would interfere with the measurement. By using a pump, we lowered the temperature of liquid helium to the superfluid point, that is, about 2 kelvins. Thus we had a noise source of about 2 kelvins, which I thought would still be higher than our zero but not all that much higher. We expected zero from the sky with a very small contribution from the antenna (about half a kelvin), which we could correct for and a contribution from the atmosphere (about 2 kelvins) for which we also had to correct for.

In our first experiment, we had first switched the input of our radiometer to our standard and we had calibrated our temperature. Then we expected that the sky would be slightly colder. Instead, it was enormously hotter, by almost 4 kelvins. It was a huge amount and it was totally unexpected. This then became the object of our search for the next year. We had to eliminate all possible natural sources and get our results right at this frequency because otherwise we could not do the experiment we wanted to do in the first place. We cleaned the antenna, did seasonal experiments, we did experiments day and night, and so forth. To our surprise, when we cleaned the antenna — there were lots of pigeon droppings around — there was a difference of only half a kelvin. We also knew that prior experimenters using that same antenna had, in fact, hints of excess noise in their system.

They had very low noise receivers and they couldn't separate the sky component from the receiver component, but the total amount was higher than expected. They thought that it was just a measurement error. The same thing actually happened at two different wavelengths.

When we did our experiment and we couldn't find an explanation for these results, I thought that we should publish them as a section in another paper. This being the second refereed paper in my life would not strike my career with something, which is foolishly wrong. Then, early in 1965 I learned about the work at Princeton. In 1964, in Russia, Doroshkevich and Novikov had realized that relic radiation would have a blackbody spectrum. That same year, in 1964, Fred Hoyle abandoned the steady state theory for the reason that there is too much helium in the stars. In a steady state Universe, the only helium any galaxy contained is that which is produced during its lifetime if the continuous creation of matter starts with hydrogen. The huge "excess" amount of helium was the first experimental evidence for the Big Bang. Hoyle was a big opponent of the Big Bang and yet he had found the first evidence to support it. He was able to kill his own theory. As a result of Hoyle's work, Doroshkevich and Novikov suggested a new look at Gamow's theory.

Gamow had prepared his theory to understand the origin of the elements, but it was a failure because they could not get stable nuclei produced past mass 4 in the Big Bang. One of the consequences of the Gamow theory is leftover heat. Gamow extrapolated this heat knowing that it would expand adiabatically because the energy density above 10,000 kelvins is much greater than the energy density of matter. As the Universe expands the energy density of matter goes down as $1/r^3$, whereas in radiation it goes down as $1/r^4$. Once you get down below 10,000 kelvins, the radiation decouples from matter (because the latter is no longer ionized) so the radiation continues to cool adiabatically, as described by a simple relationship: the temperature is proportional to the reciprocal of the radius (which, in turn is proportional to the cube root of the matter density).

In a theoretical tour de force, Gamow was able to calculate the matter density of the Universe at a temperature of one billion degrees (at which point neutrons were no longer created) from the amount of primordial helium we see in the world today. Having a relationship between radiation and matter density at one temperature, Gamow was then able to track the temperature and density as the Universe expanded. Using this relation, Gamow was able to draw some interesting conclusions about the formation

George Gamow lecturing (courtesy of Igor Gamow, University of Colorado at Boulder).

of the galaxies. This was a marvelous thing. Later, Alpher and Herman used Gamow's formula to the present day value of the radiation temperature.

When we published our results, Alpher and Herman had long been out of this picture. By then the Gamow theory was considered a failure because it could not explain the origin of the elements except for the lightest ones. In the mid-1960s, the Russians were looking at it again as an alternative to Hoyle's steady state theory. As a result of this revisiting the Gamow theory, Doroshkevich and Novikov suggested that there should be a 3 degrees temperature left in the Universe, and that the best place to look for it would be Bell Laboratories. They actually found a paper, which had been written by an engineer named Ohm at Bell Laboratories using that same antenna. We were working at 7 centimeters, 4000 megahertz; Ohm and his co-workers had been working at 2400 megahertz as part of an earlier satellite experiment, the Echo Baloon. They were measuring the noise temperature of their receiving system very carefully and all the components of that temperature. One of these components was the noise contribution of the atmosphere (which Ohm called the sky temperature). Because the atmosphere absorbs radiation, it also emits radiation, by simple thermodynamic balance. Unfortunately, the Russian workers misread the original publication from Bell Labs and interpreted the results as if there was no difference between the apparent noise temperature of the sky and the noise temperature due to absorption by the atmosphere. They concluded that there was no cosmic component to the temperature and hence the Gamow theory should be revised.

When we completed our experiments showing the relict heat in the Universe, I suggested to the Princeton theorists that we publish a joint paper with them, but they declined. We then decided to publish our separate papers back to back, and we had to wait for their paper to get accepted because of the objections by one of the referees of their paper. In the meantime people started referring to our work.

Was there any reaction from Gamow?

Gamow was very happy although this was not a direct proof of his theory. He used the following analogy. I lose a nickel and you find a nickel. I can't prove that it's my nickel except that I lost a nickel, just there. People loved it. Gamow was very gracious. Alpher, on the other hand, was very upset. He claimed that he had predicted the microwave background. I looked at his paper and told him that he did not predict an observable microwave phenomenon. He predicted the existence of a background energy. In a later article in *Physics Today*, for example, he and Bob Herman had, in fact, written that the cosmic background would be undetectable because it is masked by starlight and cosmic rays, both of which had similar energy densities.

How great was Gamow as a scientist?

Personally, I think that he was a better scientist than Galileo, at least as far as their respective contributions to cosmology are concerned. Galileo's "proof" of the motion of the Earth around the Sun was based on the tides. The Earth undergoes a compound motion from its orbital motion and its rotation, so the side away from the Sun moves faster than the side facing the Sun. He made an agreement with the Catholic Church not to write a popular pamphlet until he had incontrovertible scientific proof of the motion of the Earth. Galileo's proof was based on the tides, but for his theory to work he needed one tide a day with a 24-hour cycle. People had known for centuries that there are two tides a day and Kepler demonstrated their relation to the Moon's phases. Some people criticize Gamow for the rough approximations he used, but as I've tried to show with my Galileo example, he wasn't the only one to do so.

When did you realize the importance of your discovery?

When *The New York Times* put it onto its first page even prior to the publication of our paper. I was suspicious because so many of the early radio-astronomical theories had been simply wrong. It made me feel more comfortable when other people offered other explanations for our observations. By that time we had made our second measurement confirming our observations.

Would you care to comment on the Star Wars concept?

It's flawed from the beginning to end.

Didn't it help to bring down the Soviet Union?

Yes, but consider the risk. If the Soviet Union was brought down by the threat of an anti-missile system, that threat risked a nuclear war. Ronald Reagan bet that they would back down, but if he had lost his bet, our entire species would have been destroyed. He bet the entire planet, and no one has the moral right to bet the entire planet on that kind of confrontation. I would rather live with the Russians than risk a nuclear war. Nuclear stalemate between us and the USSR was based upon a set of treaties set up to insure that neither side could win by launching a pre-emptive strike on the other. If successful, Reagan's "Strategic Defense Initiative" would have allowed the side possessing an anti-missile system to launch such a pre-emptive strike.

How about Hitler?

I'm a Jew from Nazi Germany, remember? If the difference was that some part of the world would live under the most horrible regime, even if I were in that part of the world, I would opt for it rather than risk nuclear war. I don't think we have the right to take a risk with the entire planet.

How about the Manhattan Project?

It was very different. It did not threaten the end of the world. There were no hydrogen bombs in those days. Getting an atomic bomb to stop Hitler is fine. Getting an atomic bomb to stop the Japanese was fine.

Would you then be in favor of a limited nuclear war?

Sure. But there were other means to attack the Soviet Union, like the war in Afganistan. When Andrei Sakharov came to the United States, I was the second person to see him (after his doctor). For some reason he picked me. I went to Newton, Massachusetts, to his stepdaughter's house and we sat down for a whole afternoon talking about this. He was coming to argue against the SDI [Strategic Defense Initiative]. I said that it was immoral for Reagan to do this, because the Russians could also change the rules of the game; the Russians at some point might launch their nuclear weapons. Sakharov said, no, they wouldn't; he said, they would back down, and Reagan would win. Even though he was fighting against SDI, he thought that the odds were very high that Reagan would win. I still felt that it was immoral. It was a 90 percent bet and a bet with the future of the planet is immoral. There were other means of bringing down the Soviet Union. A Russian scientist I knew called the USSR a low-temperature society with an infinitesimally small specific heat. A slight perturbation would destroy it. A few people in Russia went to the authorities and declared their intention to emigrate and the whole system went nuts.

The collapse of the Soviet Union had been predicted since its formation, yet when it happened it came as a big surprise, especially for the experts.

The Soviet Union was spending about one third of its gross national product on the military. The Americans knew that the Russians could no longer spend to keep them up.

Don't you then think that in this sense SDI helped?

SDI worked in that direction, but there were other things, which we could've done without risking nuclear war. All they did with SDI was they saved 10 years at best. The Soviet Union was totally going downhill.

Was building the hydrogen bomb by the United States a mistake?

I'm not sure what I would've done. I would've taken advantage of the United States' military supremacy to do something. There was a reasonable way of living with hydrogen bombs, which was the mutually-assured destruction. That kept the peace for 50 years. The SDI was a clear violation of the existing system in which it was worst to shoot first. Once you have a shield over your missiles or over your cities then you can shoot first.

Once you have a shield, you have an incentive to shoot, which destabilizes the world.

But it has been argued that SDI would not work in the first place.

You can't be sure and the people who believed it would work would also be the ones who would decide to shoot first.

So you were against it?

Of course. It was destabilizing the world. Ronald Reagan wanted to make sure that he won the Cold War before he was out of office. For this he was ready to risk a war. He could've done his part to make sure that in a few years the Soviet Union was going to be finished. There were plenty of things he could do to make it happen. Of course, this doesn't have to do much with this conversation.

But your opinion about SDI does count because you are an astrophysicist and a Nobel laureate at that, so your words carry a lot of weight.

I'm not speaking as an astrophysicist; I'm looking at it as a technologist. I was head of research at Bell Labs. I don't know about nuclear weapons. My perspective is a systems perspective and how people think and how systems work.

Turning to more recent things, your departure from Bell Labs caused quite a stir in 1998.

This is what happened. I was an officer of the company and there is a rule that at age 65 you must leave. It would've been retirement, but I didn't wait until I was 65. We had a whole bunch of high-level people at Bell Labs, me included, who had our training decades ago and people at the top were needed who understood modern technology and we ought to make room for them. Back in 1992 I suggested that officers should step down at the age 60. My boss was hysterical about it. He never wanted to leave. So my suggestion never got through. I volunteered to step down at age 60 and become a researcher again, but nobody would hear of it. By age 62 I had finished the work I had wanted to do about making Bell Laboratories more organized. There is a book about this re-organization, *Engines of Tomorrow*. At that point I decided to learn about how research is done in small companies. I moved to California and rather than going

Arno and Sherry Penzias in Stockholm, 2001, during the Nobel Prize Centennial (photograph by I. Hargittai).

to meetings and reading, I decided to learn by doing it. I became an advisor to small companies and that's what I've been doing for the last 5 years.

Did you have a lot of interest in Lucent Technologies? Its stocks have gone down recently.

Yes, I did have some interest, but at the time of my retirement I did reduce my interest considerably. I went to California just at the time when Lucent was formed and I then kept an advisory role with Lucent, which stopped in 1998. But my actual departure was in 1995. It's a lot easier when you're coming for a job if you are doing it before your retirement age than if you had just reached retirement age at your company.

Did this move work out for you?

Beyond my highest expectations.

Could we get back to your early history? In your Nobel biography, you described a frightening experience of being on a train for deportation to Poland but then you returned to Munich.

We were Jewish and Polish citizens and lived in Munich, Germany. In October 1938 the Polish government abruptly put an expiration date on all passports. Then it instructed its consulates not to renew any. This was a way of exporting their Jews. The Germans were getting ready to deport all the Polish Jews under their control, and the Poles didn't want them back. I was born in Germany, but my grandfather was born in Lemberg (Lvov), which was Poland when he was born [today it is in the Ukraine]. In around 1905, he moved from Lemberg to Munich. My father was also born in Munich, but we were Polish citizens. Originally, grandfather was an Austrian citizen as Lemberg used to be a part of Austria-Hungary, but when it became Poland, grandfather could not keep his Austrian citizenship and he had to become a Polish citizen even though he was living in Munich.

In October 1938 we had only a couple of days to leave Germany; for one night we were put in jail; and on October 31 we were loaded on a train. However, by the time we got to the Polish border, it was after the deadline. The people who could cross the Polish border in time were put to an open field by the Poles and about half of them froze to death. We were fortunately late and went back to Munich. My father was given a few months to find some other way out of Germany or to be put in a concentration camp.

After the deportation, the son of a couple that died in that field in Poland, shot a German official in Paris. Because of that there was this organized riot called *Kristallnacht*. So the *Kristallnacht* was an indirect consequence of the deportation. *Kristallnacht* awoke the conscience of the British, though no one else, to take in refugees, but only children, no adults. My parents had applied for affidavits, which was the first English word I learned. With an affidavit my parents could get a visa to the United States. My father was the first in the family who got an exit permit. At that point our parents put us on a train to England. My brother was 5 and I was 6 years old. Eventually, we all ended up in America.

How much has Jewishness been part of your life?

I am a religious person, not observant, but religious. It is important to me emotionally, but I'm not heavily constrained by ritual. I feel, as many others who survived do, some obligation to represent the people who did not survive. I would like to speak for them as well as myself because they can't speak for themselves.

What did your parents do in Munich?

My mother took care of us and my father had a one-man leather business. He was a salesman. He did not have language skills and this is also why

Arno Penzias in Stockholm, 2001
(photograph by I. Hargittai).

we did not leave Germany before we were forced to. One of the most
telling reasons why people did not leave came to me when I visited Anne
Frank's house in Amsterdam. I realized that for a long time it was very
easy to believe that you could tough it out. All the rules appeared to
be rational. Every rule looked as if it was the final rule that you had to
obey. After the Nazi occupation, Jews in Amsterdam were restricted in
their use of public transportation to the hours of 10 a.m. to 3 p.m., or
something like it, off the rush hours, *except* to go to work. Jews were
restricted in their shopping to the low-volume hours, *except* drug stores.
They did not want to interfere with the Jews going to work or they did
not want this poor Jew not to get an aspirin when he had a headache.
Could you then imagine that they would throw children into ovens just
to save a few cents worth of Cyclon B? The initial steps did not indicate
that because they indicated that there were limits of how far they wanted
to go. It was terrible but it did not imply that it was going to be any
worse. People thought that they would figure it out. People did not know
about the extermination camps at that point. Everything appeared so rational.

Think about the question of who is a Jew. The Nuremberg Laws were logical. Who is a Jew? Two Jewish grandparents? If you marry a Jew, or say that you're a Jew? That's convergent. You can decide unambiguously on anybody you meet whether he is Jewish or not. It was well thought out. This is what made it especially horrible. It was worse than cold blood. This is also why people did not leave. The other thing is, where would you go? My father's friend finally got an affidavit for us in America, but it was a hassle on the part of those who signed such an affidavit. But when somebody went through it, he saved lives.

How do you feel when you are in Germany today?

It's all right, but I still can't bring myself to go to Austria. I was there once, in the mid-1980s, and found that there was hatred around. In Germany they at least try to deal with the past somehow. What is most difficult for me is to go to a synagogue in any place that was part of the Third Reich. I only did that once. After my visit to the Anne Frank House, I have a hard time going back to Amsterdam. I get nightmares about the bridges, about opening the bridges during the occupation and trapping the Jews on the island.

I suggest to conclude our conversation on a scientific topic. You worked on the chemistry of interstellar space.

Arno Penzias and István Hargittai in Stockholm, 2001, during the Nobel Prize Centennial (photograph by M. Hargittai).

After the discovery of the cosmic microwave background radiation, I wanted to continue to expand the scope of my research. In 1972, I joined Princeton as an unpaid associate. I supervised Ph.D. students and did a range of experiments with them on a number of subjects. After a while I started working on interstellar molecules. I concentrated more on isotopes than on chemical composition, because of my background in physics. Using our millimeter technology to measure rotational spectra, we changed the rules of the observational game. Before us, only a couple of odd lines were known, such as the inversion transition of ammonia or some weird states of water. There were very few ways of looking at molecules. Once you get to millimeter spectrum, where everything has rotational spectra, it becomes very easy. In the first days, we found a bunch of molecules, starting with carbon monoxide. Then we did some isotope work. Now, hundreds of people are doing similar work. For my own interest, from a cosmological perspective, we used this to study deuterium. I was the first to measure deuterium in interstellar space. Suppose, you are a farmer in Egypt. As your plough goes down a furrow, you find this trap door, one which leads obviously to a tomb. What do you do? You would do well to cover up the door and you go to archeology school, because the first archeologist down the steps discovers the tomb. You have to be an archeologist to get credit for the discovery. Having just only published a couple of papers by the time of our cosmic background discovery, I had to get my astronomer's license anyway. What gave me my astronomer's license was not so much that we did the background radiation, but that we discovered the interstellar molecules.

Robert W. Wilson, 2002 (photograph by I. Hargittai).

14

ROBERT W. WILSON

R obert W. Wilson (b. 1936 in Houston, Texas) retired from his original workplace, Bell Laboratories in 1994 and has since been a part-time associate of the Harvard Smithsonian Center for Astrophysics. He shared half of the Nobel Prize in Physics for 1978 with Arno A. Penzias "for their discovery of cosmic microwave background radiation". The other half of that year's physics prize went to Pyotr L. Kapitsa "for his basic inventions and discoveries in the area of low-temperature physics". Robert Wilson graduated from Rice University with a B.A. degree with "honors in Physics" in 1957 and received his Ph.D. degree from Caltech in 1962. He is a member of the American Academy of Arts and Sciences and has received the Henry Draper Award (1977) and the Herschel Medal (1977). We recorded our conversation in Dr. Wilson's home in Holmdel, New Jersey, on March 1, 2002 and below are edited excerpts from that conversation.*

You shared the 1978 Nobel Prize in Physics with Arno Penzias, but the other half of that year's physics prize went to the great Russian physicist, Pyotr Kapitsa in 1978.

In some ways it was appropriate because the availability of good cryogenics made our experiment possible. If we hadn't had easy access to cryogenics, we mightn't have done what we did.

---

*István Hargittai conducted the interview.

Penzias and you shared the prize for your joint work, but the two of you gave two different talks at the Nobel session in December 1978.

I spoke about our discovery and Arno spoke about the formation of chemical elements. After our original discovery, we got into the formation of stars. We discovered carbon monoxide in interstellar space and discovered giant interstellar molecular clouds. Arno expected to learn more about element formation. When the Big Bang happened, it started only with hydrogen, helium, and almost nothing else. The heavier elements were made somewhere else, it could have been either in the first generation of stars or as the galaxies developed. It appears that it was in the first generation of stars. We could examine the galactic evolution by looking in places where star formation is much more rapid, like the galactic center. In our own neighborhood it is much less rapid than in the galactic center.

Originally the main obstacle in accepting the Big Bang hypothesis was that it could not explain the formation of heavier than the first few light elements. Then your discovery supported the Big Bang theory although it still did not explain the formation of the heavier elements.

That's correct.

How did you resolve this problem?

It had actually been resolved already by a paper by Willy Fowler. Fred Hoyle described element formation in the stars, which is the way it happens. They had access to good nuclear physics information through Willy Fowler. They understood those two or three processes by which the chemical elements are formed. This still did not make Hoyle change his mind about the Big Bang although he himself recognized the shortcomings of the Steady State theory.

The Big Bang was advanced by George Gamow. Did you know him?

I only met him after our discovery was announced. There was a meeting in New York of the American Physical Society and Gamow was there. But I did not get to know him well. I read his popular science books as a child. However, at that meeting, at first I did not even remember that he had predicted what we had discovered. Arno and I talked with him and our discovery came up and its relationship to his prediction. Arno recalls him saying something like, "If I lose a nickel and you find a nickel,

I can't prove that it's my nickel, but I lost mine near where you found yours."

Very gracious.

Yes. To me it was very satisfying for him to know that this all worked out well.

If you had the opportunity to talk to him now, what would you tell him?

I would like to know more what he was thinking about when he was working through the theory, what his approach was. I have talked with Alpher and Herman; they seemed to have a rather tense feeling of having been left behind. They were different from Gamow, trying to get their due. It was an unfortunate thing, they did this early and no one paid a whole lot of attention. Then when the discovery came, what they had done became the real theory. I think everyone agrees that the paper by Alpher, Follin, and Herman[1] was the first proper calculation of element formation in the early Universe. But it did not describe the temperature of the early Universe. By the time the actual discovery came and people started to take it seriously, they had moved on to other things whereas Peebles, who was right at the beginning of his career, really jumped in and has made a lot of contributions to the theory.[2]

You are saying that people did not quite take seriously Alpher and Herman's work. Do you think they themselves had realized that it was so important when they had actually come to it?

Their prediction of the background, they probably did not take it very seriously. One reason why I don't think they took it too seriously is that they did not ask anybody to measure it and thus test their prediction. I'm convinced that the first measurement made by the Princeton group was made with an equipment that was very like a World War II equipment. If they, in the mid-1950s, had wanted to make a measurement with a better equipment, they could have. Everything that would have been needed to do it was available.

How straightforward was this question about the temperature in the interstellar space? Today we understand its importance, but when it first came up, didn't it sound a little esoteric?

Obviously, it is not an equilibrium situation. You get away from the Earth and there is this microwave background, but then there is also the starlight and dust radiation. You have to separate all the foreground things from the background in order to make the measurement. It was lucky for us that we started out at a wavelength where the Galaxy was not very strong and the Earth atmosphere was nice and transparent, and it was possible to see the excess.

What was your error estimate?

It was three point one, plus and minus one kelvin.

Did Bell Labs gain anything from your work, apart from prestige?

That actual discovery did not make much difference to anything. Other parts of our work were used. We made careful measurements of the brightness of astronomical radio sources for scientific reasons, but our data also helped in calibrating Earth stations for satellite communications. The contracts were written regulating the payment according to the signal to noise ratio of the station. We measured something up in the sky and that was used commercially. There were other applications as well, related to satellite communications. We probably paid our way and that was the comments of other people too, at Bell Labs. You could, of course, ask the question, "Why would they hire two radio-astronomers?" I can only speculate, but I think there were two things. One that we had experience with things like pointing antennas and seeing through the Earth's atmosphere and that would be useful in the satellite communications program. The other factor may have been that the people who had put together the receiving system were proud of it and they were happy to see some science done with it.

When you received the Nobel Prize, do you think there were people who felt left out?

People who probably felt they were left out were Bob Dicke, and maybe Jim Peebles, who did the calculations. Gamow was dead by then, but Alpher and Herman felt left out. We could argue for all those people one way or another that they were part of the project. The actual discovery process involved not only our measurements, but getting the hint from the Princeton people that radiation remaining from the Big Bang might be the explanation, and that was Dicke and Peebles; Dicke had had the idea. He was a very good physicist. During the Second World War he invented the standard microwave radiometer that actually measured the brightness of things. He clearly understood that the heat left over from the Big Bang would be in a form of microwaves that could be measured.

How long will that heat still be around?

It will be around forever, just getting less and less. Leaving out this new notion about acceleration, it goes down with one divided by the time.

Very slow.

It is, but originally, of course, it was very fast, it was about three thousand kelvin at a couple of thousand years of age of the Universe. It's three kelvin now. If the Universe is now 15 billion years old, this temperature will be 1.5 kelvin when the Universe will be 30 billion years old. There's plenty of time to make measurements.

If the experimental error is one kelvin, then it covers a tremendous interval in time.

That was our first measurement, today it is a hundredth of a kelvin. Have you seen the spectrum that the COBE satellite produced? When they drew the spectrum, they drew the error bars 400 times larger than they are. If you were to reduce them down to what they are, it would be less than the line width they used to draw the spectrum. It's fantastically accurate. Still in a hundred years they're still not going to see any difference.

When did you start the work that eventually led to your discovery?

I went to Bell Labs in 1963. Arno had joined them the previous year and we got together in May 1964 and we soon made the first measurements that showed the excess heat.

Were you partners?

Yes. We both reported to Al Crawford, who was the Head of the Department and above him there was a series of people. Ours was a very good partnership, we complemented each other very well. It lasted for over 20 years. It started breaking apart when Arno started getting higher up in management in the Company and had less and less time for astronomy. But we were still doing things together, so it was gradual.

Were you partners only or friends as well?

In the early days we spent a lot of time together, observing and so forth. We talked about things other than science too.

Did you know about his background?

I knew a fair amount, the story of his getting out of Germany. He told me about it. We did not frequently get together in the evenings socially. We probably had them over a few times and they had us over a few times. His family was fairly strictly kosher and that provided some barrier to social interactions. So we could not have them for dinner in our home. We did have dinner in their home though. I have no reason not to eat kosher food.

Was it easy to get your discovery accepted?

I was surprised to see how easy it was to get accepted. It was a paradigm change although it may not have been too much of a paradigm change because no one really took the Steady State model too seriously. When we submitted our report about our discovery to the *Astrophysical Journal*, I think that Chandrasekhar did not even send it out for review, he read it himself and decided to print it.

You are no longer at Bell Labs.

I took an early retirement package and left in 1994 and I took a job at the Harvard Smithsonian Center for Astrophysics. I am there half the time and the other half I am at home. I am working on a new telescope called the sub-millimeter array telescope. It will be eight six-meter diameter telescopes working together to measure at wavelengths short of a millimeter down to two tenths of a millimeter. The main interest is star forming regions. We'll be able to see a lot of the molecules, especially at higher temperature, and we'll also be able to see dust with it.

The millimeter and sub-millimeter transitions are largely rotational transitions. Have you exploited this information for determining molecular structure?

There are parallel operations; since the conditions of formation are very different in space than the laboratory, there has been a problem of doing microwave spectroscopy in order to know what the frequencies and structure and so forth are. The fact that these things were out there has called certain people to do the microwave spectroscopy, which might not have been done otherwise.

Have they discovered new molecules that had not been known before?

There was a molecule (called exogen, at first) with a strong transition that no one understood what it was. It turned out to be HCO^+. In other cases, people had done blind surveys and then tried to identify them. So the motivation for doing microwave spectroscopy is more identification than structure elucidation. However, once you identify several lines, deducing the structure will help you in the identification of the rest of the spectrum. Our interest has been in understanding the conditions under which these molecules form. It's mostly gas-phase chemistry, ion-molecule interactions. We are talking about very low densities here; 10^4 atoms per cubic centimeter would be a relatively high density for us. The collisions are very rare under these conditions and it may take hundreds of years to reach equilibrium.

So the non-equilibrium situations appear as if they were frozen?

They could be. They may even last millions of years.

It would be a different kind of chemistry than the one we are used to.

Yes. It's not only collisions and excitation, but there are photons around. They heat or excite the molecules. But the collisions are not frequent.

From the time of your Nobel Prize-winning discovery till now, what was your most important scientific contribution?

Probably the discovery of carbon monoxide in the outer space. It made a big difference to our understanding of star forming and to our understanding of what happens to a gas in the Galaxy. People used to think that heavy stars don't last very long, so they have to be forming, but what we found was that it's the gaseous clouds where it all occurs.

What was there before the Big Bang?

I don't know. I'm not a theoretician. I've heard people say with a straight face that perhaps the Universe is a quantum fluctuation, but that's not my field. I can't propose a scientific explanation of how it all got started.

If you say that God did it then you have to put as much information into God as you get out about the Universe. As scientific theories go, that does not help. You want a theory in which you put a little bit in and you get a lot out. In science you have a formula and it'll describe a lot of things. If you suppose that God created the Universe, then the next question is where was God and that doesn't help you with much else.

For most people it probably doesn't matter, but how do you feel about other scientists who believe in creation?

I don't argue with them. I'm amazed at how much emphasis the world places on religion when there is very little evidence for any of it. I'm amazed that sophisticated people in these days are still believing the things, which only come down as hearsay from thousands of years ago. I'm very puzzled by how people think that way. There are two purposes for religion. One is that the priests' class controls the rest of the society. The other is that religion provides a moral background and gets people to think about how to organize society in a very productive way and that could be very

Robert Wilson and Arno Penzias in Stockholm, 2001, during the Nobel Prize Centennial (photograph by I. Hargittai).

useful. Of course, religion can go very wrong and you can have the Spanish Inquisition, or some very recent things and we would be better off without them. But I understand the trappings of religion, the celebration of Christmas, and other things are quite satisfying, even if I don't believe in the basic reason for doing it. I think you must almost have a segmental mind. Politicians probably feel that they have to display being religious. I sometimes wonder what's going through their minds.

Yet the general perception is that many scientists are religious.

I don't go out of my way either to say that I don't believe. There is social pressure and it's easier to let it ride. But if someone asks me the question, I'll tell them how I feel about it. Because of my science, in some sense, I am in the religion business.

Another sensitive question that you are supposed to be close to is Star Wars. What do you think about it?

I still don't think we can do it. I've generally felt that such things as SDI [Strategic Defense Initiative] would be a bad idea because, certainly in the days of the Cold War, that would just lead to additional weapons being made by the other side. It's always easier to overwhelm a defense than to build a perfect defense.

Wasn't it an important contribution to bringing down the Soviet Union because it could not match the expenditures?

Probably so. I'm sure that had something to do with it, but they had plenty of paranoia within the Soviet Union even without our doing much. They were already spending a large fraction of their gross national product on what would seem to be an unnecessary defense. If the Soviets had coldly analyzed SDI, they would have said that it was fine for the United States to spend a lot of extra money on it. Today, SDI may have a more defensible reason for existence in that there may be some states willing to launch a missile against us. But it would be simpler to send a bomb on a ship, for example, or some other way, and blow it up than to send a missile. What worries me more now is the nuclear material that is in the former Soviet Union, and that may be getting out. We ought to support the scientists there and try to get all the dangerous stuff somewhere safe before it gets into the wrong hands.

I would like to ask you about your origins. You write that your grandparents moved to Texas after the Civil War. Where did they come from?

I can't trace back my ancestors very far. The Wilsons came from Georgia, but I don't know how they got there. I have skin that is very sensitive to ultraviolet light so I'm sure that I'm from pretty far north in Europe. I know a little more about my mother's family. They were also in the south and there was a French influence there. My mother was a housewife and my father was educated as a chemical engineer at Rice University, and he worked for a company in the oil industry. He was involved in the oil drilling technologies and used some electronics and he got me interested that way. Back in the 1940s people worked five and a half days. On Saturday morning I would go with him to the company shop and I would patter around the shop, play with the electronics stuff. He used to fix radios and I learned how to do that. During my high school years I fixed radios and television sets. I did my undergraduate studies at Rice and for my Ph.D. in physics I went to Caltech. When I got there and started looking around, I was beginning to feel a little discouraged about the projects I saw. I had thought that I wanted to do low temperature physics, but that group was more than full, and there weren't any other things that were appealing to me until I met David Dewhirst from England who was also staying in the Atheneum, the faculty club of Caltech, as I was. We did a number of things together, like walking along the Grand Canyon. He did not have much experience with cars and I helped him learn to drive and so forth. David was working in the Astronomy Department and he told me about the radio astronomy effort that was started. John Bolton and Gordon Stanley were the senior people in the group and that seemed to be a perfect match to me. My doing my first experiments with John Bolton was probably my formative experience. I remember him saying, when I started to write up my experiment, "It's, of course, important to do the right thing, but it's much more important to say what you did, correctly." In other words, you should be completely honest about what you have done in writing up the paper. I have stuck with this. The papers that came out of that group were very short, two or three typewritten pages.

Your present family?

I have been married for 40 years, we have three children. My oldest son should have become a scientist; he has the curiosity and he has a keen observation. He likes to get to the bottom of things until he understands them. He was at Swarthmore College at the time of my Nobel Prize, and he dropped his physics course shortly afterwards. He became a philosophy major for a while. He may have felt that he did not like competing or something. Later he became a successful computer scientist.

How did the Nobel Prize change your life?

It brought me into public life. I just returned from a trip to Hawaii where I gave two public lectures. Many fewer people came for the second than showed up for the first.

Who are your heroes?

I've admired people who had the courage to change things and were successful at it.

References

1. Alpher, R. A.; Follin, J. W.; Herman, R. C. *Phys. Rev.* **1953**, *92*, 1347.
2. Peebles, P. J. E. *Physical Cosmology*. Princeton University Press, Princeton, 1971.

Owen Chamberlain, 1999 (photograph by I. Hargittai).

15

OWEN CHAMBERLAIN

Owen Chamberlain (b. 1920 in San Francisco) is Professor Emeritus at the University of California at Berkeley. He shared the Nobel Prize in Physics in 1959 with Emilio Segrè "for their discovery of the antiproton". Owen Chamberlain got his B.S. in 1941 from Dartmouth College and his Ph.D. in 1949 from the University of Chicago. He worked for the Manhattan Project in Berkeley and Los Alamos between 1942–1946. He has been at the University of California at Berkeley since 1948. He is member of the National Academy of Sciences of the U.S.A. and has received many other distinctions and honors. We visited Dr. Chamberlain in his home on May 11, 1999, and asked him about the origin of his interest in science, his experience with the Manhattan Project, and the discovery of the antiproton. In spite of his failing health he was kind and patient and his wife, Senta Pugh, was very helpful in making this visit possible. Here are some edited excerpts of what Dr. Chamberlain told us.*

I was about 12 and we lived in Philadelphia. My father was a radiologist, and there was another radiologist in New York who would come to our house. She brought her husband with her, an amateur magician who gave me problems. He would ask me, "In how many jumps can a frog get out of the well?" They were simple numbers games but he made me very much interested in them. He told me that you cannot make a perpetual

---

*István Hargittai conducted the interview.

motion machine. I didn't know that this man called himself a physicist but he sparked my interest. I didn't know enough then to connect this man with physics but I knew enough to connect myself with his problems.

Following my undergraduate studies at Dartmouth, in 1941, I enrolled in graduate studies at Berkeley. I had had just one semester of graduate work when the attack on Pearl Harbor put an end to it. I got quickly involved with the Manhattan Project. First I heard the rumor that Professor Lawrence was looking for people to help with some secret program and I volunteered. He assigned me to be a helper to Professor Segrè, and working for Professor Segrè was more valuable than going to graduate school. We tested some of the conditions that might effect the final bomb design, such as the spontaneous fission rate for nuclei. In the middle of 1943, I moved to Los Alamos with Oppenheimer and Hans Bethe and many of the scientists who had escaped from Nazi Europe. Szilard was one of them but I was not so much in touch with him as I was in touch with people who were in touch with Szilard.

I had a lot of interaction with Segrè and with Fermi. We had temporary quarters. Once I walked into a little shop where Segrè was having a conversation with someone in Italian and he introduced me to Fermi. That was our first meeting. Fermi was sitting in a corner of this small room, not saying much, and my mouth just dropped open. Later I learned that it took two years at the University of Rome before the janitors accepted Fermi as somebody of importance. In contrast, E. O. Lawrence's appearance was very impressive, whenever he entered a room everybody was aware of him.

One day, about six months after I'd arrived in Los Alamos, Oppi (that is, Oppenheimer) sent me to visit a laboratory in Ohio. When I came back I had to tell him that they were falling behind schedule in their work. When I was coming back to Los Alamos I climbed on the train in Chicago and saw Enrico Fermi in the compartment next to mine, traveling all alone. For a day and a half I had Fermi at my disposal, he was answering all my questions about physics. It was a rare opportunity and I enjoyed the benefit from it and it also gave Fermi a chance to get to know me. It was a very opportune time for me to have this contact because I had trouble at that time differentiating between Maxwell's equations in polar coordinates and in Cartesian coordinates. After the war I became Fermi's student at the University of Chicago.

There was an interesting twist in our work on the spontaneous fission rate back in Los Alamos. When we set up our measurements we asked the theoreticians, Bethe was one of them, whether there would be any problem with cosmic ray neutrons. They did some calculations and said that there should be no interference from them. However, when we compared our measurements in Los Alamos with those that we had done in Berkeley, the impact of the cosmic neutrons about doubled the spontaneous fission rate. There was, of course, a significant difference in altitude between Berkeley and Los Alamos. So for the neutrons producing fission, the cosmic neutrons proved to be important. This observation forced us to choose a more rapid way of compacting the components of the bomb. This was important for the bomb that was to be used in Hiroshima. The Nagasaki bomb used a different mechanism, plutonium implosion.

The Hiroshima bomb was not tested but the plutonium implosion bomb was. I was present at the test. It was just dawn, at 5 a.m. or so. I was at the three cabins where we had some equipment. We were listening to the countdown on the radio. When the explosion gave rise to this fire ball, my co-worker Clyde Wiegand and I were looking down at the ground to avoid direct contact with the light. I felt intense heat at the back of my head. It was a mushroom-like cloud; it had a stem and then it spread out. I had no idea how dangerous it was; we just didn't think about it at that stage. I just thought, "If Fermi thinks we don't get burned, I'll go."

After the test, there was a lot of discussion whether the bomb should be demonstrated for the Japanese first rather than dropped on a Japanese city. But I was not one that was busy sending letters to the President telling him how we should handle this. However, four days after the explosion I had to go near the center of the activity, which we called the virtual center. The desert looked amazingly unaffected so I took it as evidence that we shouldn't sign any letters. I thought that such a demonstration would not be effective, and would not change anything.

When the war was over I stayed for another six months or so in Los Alamos before going to Chicago. I wanted to be sure not to get drafted. During this period of time we set up the Los Alamos University. A Stanford professor, Leonard Schiff taught statistical mechanics and Fermi gave a course on electromagnetic theory. Hans Bethe was also among the teachers. My background was insufficient for some of the courses I'd signed up for but most people had more success. I remember how Hans Bethe was aiming at Fermi as he was giving his lectures so it was not surprising that the level was too high for me. I had to drop out but I was not intimidated.

It took me three years to get my doctorate with Fermi. He thought I was ready to go after two years but it took me an extra year to correct for some mistakes.

Teller was also at the University of Chicago when I was there. There was always a great deal of negative feeling between me and Teller. We all knew though that Teller was an extraordinary physicist because he developed ways of reasoning, which were correct but looked quite different from the usual theory. There we had a very interesting course. In the morning Teller would describe the work he planned for the hydrogen bomb. Then in the afternoon Fermi lectured trying to demonstrate that it wouldn't work. I attended all those lectures but I wasn't getting a full understanding of what they were talking about. Finally Teller came through and took the steps that led him to the right answer.

Teller played the piano very nicely. For a couple of months I lived above his apartment in Los Alamos, and I liked to listen to his playing the piano.

From left to right: Emilio Segrè, Clyde Wiegand, Edward Lofgren, Owen Chamberlain and Tom Ypsilantis at the Lawrence Berkeley National Laboratory in 1955 at the time of the discovery of the antiproton (courtesy of the Lawrence Berkeley National Laboratory).

But our political differences showed up as soon as we got to know each other.

I was learning physics primarily from Fermi and Segrè and many of my attitudes in physics clearly stem from them. When we were in Chicago we invited Fermi for lunch in our home. He liked to talk to his students and we were the beneficiaries.

From Chicago I went to Berkeley. We were doing scattering experiments, measuring proton-proton scattering at 300 MeV. I was working with Segrè at that time. Panofsky, one of the most respected physicists, was doing proton-proton scattering at 30 MeV, so our experiment was at ten times his energy. Clyde Wiegand and I were responsible for the apparatus and Segrè kept all the contacts with the outside world.

We were looking for the antiproton rather explicitly. The question was how you decided that you'd seen it? We found some negative particles, and we measured each particle that went through our system with a pathway we wanted it to go. We determined their time of flight with what was essentially a mass spectrometer. We made two measurements of the momentum and two measurements of the velocity. We established that there were particles that were negative and had a mass within five percent of the proton mass. We could even have pushed it to within one percent with a more complicated setup. The question was how much evidence do you want to collect before you can say that you have the antiproton. You have to make certain choices. We entitled the piece for the *Physical Review*, "Observation of Antiprotons".[1]

We were then inundated with congratulations. Then the Nobel Prize came. Although I was satisfied with the quality of the experiment because it was the best it could be at that time, and finding the antiproton was valuable in science, I was surprised because I felt that if we hadn't found it, somebody else would've in the next six months. I remember also the reaction to the announcement of the Nobel Prize by one of our children. She said, "Does that mean that one of Dad's experiments worked?" I was delighted, of course, by the Nobel Prize but would've been more satisfied if Clyde Wiegand had been included as a third recipient.

Reference

1. Chamberlain, O.; Segrè, E.; Wiegand, C. E.; Ypsilantis, T. "Observation of Antiprotons." *Phys. Rev.* **1955**, *100*, 947.

Mark Oliphant, 1999 (photograph by I. Hargittai).

16

MARCUS LAURENCE ELWIN OLIPHANT

On July 22, 1999, we visited Sir Mark Oliphant (1901 in Kent Town, near Adelaide, South Australia — 2000 in Canberra), as he was known in his home in Canberra, Australia, where he lived together with his daughter, Vivian Wilson. Oliphant was co-discoverer of tritium and helium-3 with Ernest Rutherford. Later, Oliphant was a principal co-worker in establishing efficient radar for the defense of the Allies in World War II; he was also a leading British participant of the Manhattan Project, founder of the Australian Academy of Science, and co-founder of the Australian National University. At 98 he was still a commanding presence and we came away from this visit deeply impressed by his wit and care and interest.

There is a book about Mark Oliphant's life and times,[1] which we consulted for this account and which we used as the source for three quotations. Otherwise, this writing is based on our visit with him and uses direct quotations from our conversation.*

Mark Oliphant received his B.S. degree from the University of Adelaide in 1921 and stayed on there to do research. In 1925 he attended a lecture by Lord Rutherford and decided to go to Cambridge and work for Rutherford.

---

*István Hargittai and Magdolna Hargittai conducted the interview. This interview was originally published in *The Chemical Intelligencer* 2000, *6*(3), 50–54 © 2000, Springer-Verlag, New York, Inc.

In 1927 Oliphant won an 1851 Exhibition Scholarship[2] and sailed to England together with his wife. Oliphant earned his Ph.D. at the Cavendish Laboratory in 1929 and stayed on for another 8 years. He became Lecturer in Physics and Fellow of St. John's College in 1934 and the same year, also Assistant Director of Research of the Cavendish Laboratory. In 1933–1934 he built an accelerator, and with deuterons from heavy water (donated by G. N. Lewis) as bombarding agent, helium-3 and tritium were discovered. Oliphant, Paul Harteck,[3] and Rutherford reported this discovery in 1934.[4]

Oliphant was elected Fellow of the Royal Society in 1937 and moved to Birmingham as Professor of Physics the same year. He established his own laboratory there and stayed until the early 1950s, when he returned to his native Australia.

The citation upon his election to the Royal Society summarized his achievements in physics to that point[5]:

> (Oliphant is) distinguished for his experimental researches on the action of positive ions on surfaces and for his contributions to our knowledge of transmutations. Has been active in the design of high voltage apparatus for the production of swift positive ions and has taken a responsible part in experiments which show that two new isotopes, hydrogen three and helium three, were produced by the bombardment of deuterium by deuterons. He has made an accurate study of the modes of transmutation of lithium, beryllium and boron by the action of protons and deuterons, and determined the masses of the light elements.

Oliphant actively participated in defense projects throughout World War II. First, he did fundamental work in the radar program, in particular in developing the magnetron. Then, he joined the atomic bomb project as a leading member of the British team and spent long periods of time on various American sites of the Manhattan Project, primarily at Berkeley, working with E. O. Lawrence and his team. His contribution was recognized by a 1946 recommendation for the highest award the United States Government can bestow on a foreign citizen, the Medal of Freedom with Gold Palm. The citation accompanying the recommendation read[6]:

> Doctor Marcus Laurence Elwin Oliphant, Australian Citizen, during the period of active hostilities in World War II, rendered exceptionally meritorious service in the field of scientific research and development. He brilliantly conceived, developed and perfected the cavity magnetron, an important factor in the entire radar program of the United States, and, in his further collaboration in scientific research, he made outstanding contributions in the

development of the atomic bomb. Through his exceptional scientific knowledge and resourcefulness, Doctor Oliphant contributed immeasurably to the success of the allied war effort.

The award was actually never made because when the United States requested the consent of Australia, the Australian government was "unable to agree to the acceptance of the proposed award". By an ironic twist of events then, in 1951, the United States denied Oliphant an entry visa on the grounds of an unsubstantiated FBI accusation of a wartime indiscretion. This was in McCarthy's time.

When Oliphant returned to Australia in 1954, he founded the Australian Academy of Science and served as its first president. He was one of the co-founders of the Australian National University. He served as Director of the Research School of Physical Sciences until 1963, as Head of the Department of Particle Physics until 1964, and as Research Fellow in charge of the Physics of Ionised Gases Unit until 1966, when he became Professor Emeritus.

The Nobel laureate physicist Luis Alvarez succinctly characterized Oliphant in 1980 in the following way[7]:

> Mark may not have carved any new niche in the Hall of Fame after he left the Cavendish. When he got out on his own playing field, I guess he just failed to deliver. But when he talked physics, he still talked good sense, and he still exudes a sort of magnetism which makes you feel that you are in the presence of a great man.

Oliphant was mobilized out of his retirement in 1970 when he was appointed to be Governor of South Australia. He retired from that office when his term was over, in 1976.

What follows are excerpts from our conversation.

Recalling his most important discovery:

We were doing experiments with all the possible projectiles in order to produce transformations in elements. It was natural to try to use heavy hydrogen and, indeed, the results were very interesting. The experiments with heavy water brought about two discoveries, one was helium-3 and the other was tritium.

On his interactions with Rutherford in connection with these discoveries:

Rutherford was the greatest influence on me and on so many other people at that time in Cambridge. He was my scientific father in every sense of

Lord Rutherford, drawn by "W. R." in 1925 (courtesy of Mark Oliphant).

the word. Rutherford didn't like his associates keeping long hours in the lab. He thought it was silly to overdo it. But this didn't mean that Rutherford stopped working at any time. One day we went home without having understood the results of an experiment and our telephone rang in the night at 3 a.m. My wife told me that the Professor wanted to speak to me. Rutherford said, "I've got it. Those short-range particles are helium of mass three." I asked him about his reasons and he said, "Reasons! Reasons! I feel it in my water!"

Was he an approachable person?

He was very approachable. He was always ready to talk to anybody about anything under the sun and he was also a tremendous influence that way. His booming voice could always be heard at the high table in college. The Rutherfords were very good in a social way too. They had Sunday afternoon tea parties to which they invited the students. They had a cottage in the country to which they invited students to spend the weekend. It was always a great experience. [Oliphant wrote a book on Rutherford, see Ref. 8.]

Is it true that Rutherford did not believe that nuclear power was possible?

He hated the idea. He knew it was possible but didn't want it to be possible.

What do you like to remember best about your career?

Going to Cambridge, the entry into the holy of holies.

Did you try to establish something like the Cavendish Laboratory in Birmingham?

Of course. One always tries to follow up one's youthful experiences. I had moments of satisfaction but they were few and far between, compared with the Cambridge experience.

Would you care to say something about your experience in working on the atomic bomb?

I was mainly in contact with the California people at Berkeley, with Ernest O. Lawrence and his colleagues. Lawrence was the biggest influence, mainly

Lord Rutherford demonstrates the deuterium reactions at the Royal Institution in London in 1934. Oliphant is standing in front of the bench (courtesy of Mark Oliphant).

because of his enthusiasm for physics. The bomb was just a side issue as far as physics was concerned. At that time it was not a moral dilemma working on the bomb because if you didn't do it, somebody else would've. There's no use in shutting your eyes to these things. Lawrence's Berkeley lab was like Cambridge in England. It was a place where everybody interested in nuclear physics went. I don't remember a great deal about the war period. In that time one was interested in practical results and not in thinking about physics. To my mind it wasn't a very fruitful period for physics and new ideas were rare but it gave a push for physics and for one, money was available and that's always important.

Did anything ever interest you outside science?

I don't think of things that way. I enjoy music, I enjoy the arts but I don't set them apart deliberately from my other interests.

Have you had any hobbies?

I had lots of hobbies when I was young. I tried everything. Biology intrigued me. Also, I always liked making things. I did silver smithing, made jewelry, particularly using the Australian stone up in northern South Australia, colorful stones, the opal is the best known among them. I have a small workshop here with all the machinery I needed, including a fine lathe.

Do you see students nowadays?

I see students when I'm invited to talk to them. I tell them to be aware of physics. That it's a never-ending search for truth. In most parts of science, like chemistry, one can determine the structure and properties of a given substance. In physics it's much more difficult because the physical properties of the substances are intrinsic to them, they belong to them alone.

At the beginning of your career, you were taking both chemistry and physics. What tipped the balance?

Chemistry is very interesting but it is not exciting like physics, one hasn't got these fundamental challenges that one has in physics. Chemistry is a matter of care, observation, and hard work. Physics begins with hard work and very often leads nowhere but it is always exciting.

In 1970 you became the Governor of South Australia. Was it a political appointment? What were you doing?

It was an honorific post. I was fiddling. I always was a fiddler. I didn't have any staunch project, which I followed through my life, like so many people do.

Would you like to see Australia become a republic?

I can't see any point in it. It's rumbling along quite well as it is, why change? There must be a reason for changing. I don't see any for the moment.

Are you religious?

Yes but not in an organized way. I have a feeling about things we don't understand, that are rather special. I've always listened and accepted very little.

Once you gave a lecture "A Physicist's View of God".

I think the theoretical physicist is the closest to God, the theoretical physicist who thinks about the fundamental realities.

Mark Oliphant with his daughter (courtesy of Mark Oliphant).

Do you think we should be looking for a unifying principle in nature?

I don't think so. I think all knowledge is important and it's just as well that we have a very wide and chaotic approach to knowledge, that people of all sorts are seeking fresh information in every corner, that's as it should be. Nature is very complicated. The more one knows it, the more complicated it becomes. If it's the other way around, then one isn't thinking.

What do you do these days?

Nothing. I read mostly, *Physics World* and other magazines. I watch TV occasionally, the news. One gets an enormous amount of information, not very deep but that's what people demand. They don't want to have to think. I still find great interest in my work and in thinking. That's the important thing to keep me going. I think it's a great mistake to retire. There's, of course, the official retirement from a paid position, for me it was in 1966, but I never really retired. I don't think a genuine physicist ever retires.

On other physicists:

J. J. Thomson

J. J. had a great influence on me. I was interested in electricity in gases and he was a major contributor to that. He was always happy to talk about physics but nothing else.

James Chadwick

I worked more with Chadwick, the neutron man, than with Rutherford. Rutherford had such a string of activities then. He was President of the Royal Society and had other sorts of activities. Chadwick had a very great influence on the research students of my time. Also, one hesitated to ask a question of Rutherford because one might have got more than one bargained for. In this sense, it was difficult to extract information from Rutherford. Chadwick was more careful when you asked him a question. He would think about his answer very carefully whereas Rutherford would have just spelled out what was on his mind at the moment. He always wanted to tell you what he wanted to say and not what you wanted him to say. Rutherford thought too fast and about too many jumbles of subjects. But still he had a very great influence on all his students. People took him seriously. Chadwick, on the other hand, was a simpler person. One could understand his thinking much better than one could understand Rutherford's.

Peter Kapitsa

He was a great character and a man of tremendous influence in England. He used to visit his Russia every year, and in the end they told him that he wasn't going back to England again. After that, I visited him several times in Russia. There were very interesting physical scientists there, always worth listening to. I think Kapitsa was worried at the beginning but he soon settled down to the life in Russia. He was such a versatile man. He couldn't be kept under very long.

On speaking freely with Russian colleagues in the Soviet Union:

You could speak freely with them if you were alone. I learned rapidly not to inquire too much to avoid certain agonies.

Niels Bohr

Niels Bohr had a very practical attitude to physics. He had the greatest influence. His public lectures were boring but he was far from boring when one was talking with him personally. I went to Copenhagen several times but only to meetings and did not spend much time there.

Robert Oppenheimer

I had a tremendous affection and feeling for him. He was a very great man but he didn't sell himself well.

Leo Szilard

He was a brilliant person, not to everybody's taste.

Was he to yours?

Yes and no. There were times when he irritated me by his certainties, but that was one of his charming characteristics, that he spoke up for himself. He utterly believed himself. He was a very able man. He was very important for the Manhattan Project at the beginning, in generating interest, in that he was the most important. Then he became just one of us. He was an original thinker although you didn't have to think very far to be an original thinker at that time, there was so much going on.

Edward Teller

Nobody could help but meet Teller. He was a rowdy person. You could always hear him. He had a loud voice and he made sure you could hear it.

Did you have any arguments with him?

Many but that was his life and that was very useful too because he questioned everything. You couldn't have an argument with Teller — you just had to listen to Teller. He was a very cocky man, very fond of himself. He was a great one for leading a team and getting the most out of something.

Was it correct to blame Teller for what happened to Oppenheimer?

No, what happened was inevitable.

References and Notes

1. Cockburn, S.; Ellyard, D. *Oliphant: The Life and Times of Sir Mark Oliphant.* Axion Books, Adelaide, 1981.
2. The 1851 Exhibition scholarship was for postgraduate research, awarded annually to a few students of the British Empire. It was financed by the surplus income from the Great London Exhibition of 1851. Rutherford himself had been an 1851 Exhibitioner from New Zealand.
3. Paul Harteck (1902–1985) studied in Vienna, Berlin and Breslau. He worked with Fritz Haber at the Kaiser Wilhelm Institute for Physical Chemistry (1928–1933), and in Cambridge with Lord Rutherford (1933–1934). From 1934, he was Director of the Institute for Physical Chemistry of Hamburg University. Harteck was one of the ten German nuclear scientists suspected of participation in the German atomic bomb program who were detained in Farm Hall near Cambridge at the end of World War II. In 1951 Harteck became Research Professor at the Rensselaer Polytechnic Institute in Troy, New York [*Operation Epsilon: The Farm Hall Transcripts.* Institute of Physics, Bristol, 1993].
4. Oliphant, M.L.E.; Harteck, P.; Rutherford, E. "Transmutation effects observed with heavy hydrogen." *Proc. Roy. Soc. A* **1934**, *144*, 692; *Nature* **1934**, *133*, 413.
5. As quoted in Ref. 1 from the records of the Royal Society (London).

6. The Medal of Freedom citation was uncovered by one of the authors of Ref. 1 in 1980 while researching Oliphant's wartime career in the United States, Ref. 1, pp. 196–197.
7. Ref. 1, p. 250.
8. Oliphant, M. *Rutherford: Recollections of the Cambridge Days.* Elsevier, Amsterdam, 1972.

Norman F. Ramsey, 2002 (photograph by M. Hargittai).

17

NORMAN F. RAMSEY

Norman F. Ramsey (b. 1915 in Washington, D.C.) is Higgins Professor Emeritus of Physics at Harvard University. He was co-recipient of the Nobel Prize in Physics with Hans G. Dehmelt and Wolfgang Paul in 1989. Ramsey received half of the prize "for the invention of the separated oscillatory fields method and its use in the hydrogen maser and other atomic clocks" and the other half was shared by Dehmelt and Paul "for the development of the ion trap technique."

Norman Ramsey received his A.B. and M.A. degrees from Columbia University and Cambridge University, England. He did his Ph.D. work under the direction of I. I. Rabi at Columbia University in the new field of magnetic resonance and participated in the discovery of the deuteron quadrupole moment. He worked for short periods of time at the Carnegie Institution in Washington, D.C., and at the University of Illinois. During World War II he worked at the MIT Radiation Laboratory, where he headed the group that developed radar at 3 cm wavelength. Later he was a radar consultant to the Secretary of War in Washington and participated in the Manhattan Project in Los Alamos. After the war he was active in establishing Brookhaven National Laboratory and was executive secretary of the Laboratory and chairman of its first Physics Department. He has been at Harvard University since 1947.

He is a member of the National Academy of Sciences of the U.S.A., the American Philosophical Society, the American Academy of Arts and Sciences, and many other learned societies. He is a foreign associate of the French Academy of Sciences. His many distinctions include the Presidential Certificate of Merit (1947), the E. O. Lawrence Award (1960), the Davisson-Germer Prize (1974), the Compton Medal (1986), the Oersted

Medal (1988), and the National Medal of Science (1988). He has received honorary degrees from many universities. We recorded our conversation in Dr. Ramsey's office at Harvard University in February 5, 2002.*

Please, tell us something about your family background.

My mother was a mathematics instructor at Kansas University; her parents came to the U.S. from Germany. My father's parents came from Scotland. He graduated from West Point and was an officer in the Army Ordnance Corps. He was often assigned to new posts and thus I constantly had to change schools. One of the results of these changes was that I graduated from high school very early at the age of 15.

What made you interested in science?

It was an article on the quantum theory of the atom that triggered my interest in science relatively early. But at that time I did not think that physics could be a profession. My parents wanted me to go to West Point just as my father did but I was too young to be admitted there. At that time my father was again assigned to a new position in New York City. So I entered Columbia College in 1931 as an engineering major. It was not long before I realized that the engineering curriculum would not give me a deep enough knowledge to understand Nature. Therefore, I shifted to mathematics. I graduated from Columbia College in 1935 and that was when I realized that physics was a viable profession and that physics was the field that I was most interested in.

What is the scientific result you are most proud of?

There are quite a few that I am proud of. There are the methods for making much more accurate measurements of radio-frequency and higher frequency spectroscopy; the separated oscillatory field method, which was partly what I got the Nobel Prize for (the other part was for the hydrogen maser). These were all primarily developed for the measurement of the fundamental properties of atoms. However they also have important applications, particularly for the atomic clocks. The applications are far greater than I originally anticipated.

*Magdolna Hargittai conducted the interview.

Would you care to tell us how the development of the separated oscillatory field method came about?

Of course. I had just recently come to Harvard and I was developing a molecular beam magnetic resonance experiment. A decade before this my thesis supervisor, I. I. Rabi, at Columbia, had invented the first magnetic resonance experiment and I had the good fortune to work with him at the time when he invented it. I wrote the first Ph.D. thesis ever written on any kind of magnetic resonance. Later at Harvard in 1949 I wanted to set up a much better magnetic resonance apparatus. I realized that to do this I needed a longer magnetic resonance region; by the Heisenberg uncertainty principle, the longer the time you measure the better the accuracy.

So we made the apparatus longer but it was impossible to get the magnetic field sufficiently uniform to take advantage of that, and I was struggling with this. Then I had the idea that maybe if I just put the oscillatory field at the beginning and the end, but not in the middle, and had the two phases correlated, so that the oscillatory fields run up and down together, then I could get the same accuracy, in fact a factor of two better accuracy than with having the oscillatory field in the middle, and the resonance would not be bothered by the magnetic field being irregular. In other words, the energy levels would go up and down with the magnetic field but I would be able to measure accurately the average value. This was very valuable for that purpose. Later I realized that there were other advantages to the method as well. It overcame the first-order Doppler effect and we could have the two oscillatory fields many wavelengths apart; so it brought us much greater accuracy not only by the factor of two or five that I was first aiming for but factors of hundreds. This made it possible to make accurate atomic clocks.

For any clock what you need is a regular motion, which you can observe. In case of the pendulum clock, it is counting the number of swings of the pendulum, and the swings depend on the temperature and other factors, so it is not terribly accurate. The best pendulum clock at that time before we did this experiment, was slightly better than the best magnetic resonance method, so this method was not used for atomic clocks. With the method of separated oscillatory fields that I invented I realized that I could make the apparatus many wavelengths long and therefore get much, much greater accuracy than what one could get with the pendulum. Since then this method gives the best time; even at present, the definition of the second is based on measurements by the method I invented. It is usually operated with

Norman Ramsey adjusting the molecular beam apparatus on which the method of separated oscillatory fields was developed (courtesy of N. Ramsey).

cesium but one can use other elements as well; what they use now is cesium, which has the advantage that cesium is heavier and thus moves slower and that gives another factor of two benefit. If the atom moves slower, it is in the apparatus longer and by the Heisenberg uncertainty principle the resonance is narrower and the measurement is more accurate. It is also easier to detect.

Originally when I worked out the method, the atomic clock was not in my mind. I invented it to overcome the effect of our irregular magnetic field and was hoping that, if we made an apparatus three times longer than before, we'll get a gain of a factor of three, which would have been important. It was only later that I realized that we could do much better than that because we can go to higher frequencies and then we can work with a shorter apparatus; so this evolved in time. Of course, for making atomic clocks, there was still a lot to be done; the clock involves much engineering work but I was not involved with that. The reason for the original apparatus was not to make an atomic clock but for the molecular beam studies we wanted to do that time: we did basic research. It was Zacharias at MIT, a very good friend of mine and also Perry and Essen in England who developed means that made it feasible to make atomic clocks that could run for a month, rather than a day, the way our apparatus ran at that time. Today they can run for ten years. It was around 1954 or 1955 when these cesium clocks became the standard for time. They are extremely accurate; we may say that such a clock would gain or lose less than a second in three million years. Zacharias worked with a commercial company and they made a commercial clock with a lifetime of half a year or more. At the same time Perry and Essen developed one in England and it was immediately used for the British time standard. In 1967, the 13th General Conference on Weights and Measures first defined the International System (SI) unit of time, the second, in terms of atomic time instead of the motion of the Earth, as was done previously. Accordingly, a second was defined as "the duration of 9,192,631,770 cycles of microwave light absorbed or emitted by the hyperfine transition of cesium-133 atoms in their ground state undisturbed by external fields". Of course, they are also used for standards of frequency. Since frequency and time are just reciprocals of each other, if you develop a good atomic clock you also have a good frequency measure.

Joe Taylor and Russell Hulse, who discovered the binary pulsars, said that the binary pulsars could also be used as the best ever clocks.

I know of them very well. In fact the binary pulsars were one of the great applications of atomic clocks; their discovery of these binary pulsars and the measurements they made, they all depended on the time measurement and that is measured in terms of the cesium clock. In one of their papers they say that one of the binary pulsars may even be more stable than a cesium clock. Some signals that come from the pulsars, the so-called millisecond pulsars, that have a thousand pulses a second, are very regular and extremely stable, and are comparable to what you can get with the atomic clocks. The only way to say that a clock is better is to have a number of countries make a number of them and then see that all agree with each other better than any previous clocks do. So far this has been done only with the cesium clock and the hydrogen maser. Even better clocks based on laser cooling and higher frequencies are now being developed, but the international time signals are still based on the cesium beams and hydrogen masers. It was these early accurate clocks that made Joe Taylor's beautiful experiment possible. They always measured the ratios of the frequencies of the pulsars versus the frequency of an atomic clock.

The hydrogen maser was one of the topics mentioned in your Nobel quotation. Would you mind telling something about it?

It is very simple to describe. It is essentially a bottle of about 10 cm in diameter, made of quartz, and coated with Teflon on the inside to reduce the effect of collisions on the wall. A beam of atomic hydrogen is produced and enters the cavity. With an inhomogeneous magnetic field, only those atoms that are in their high-energy, excited state are selected. We fill the bottle with atoms in their high-energy state and if we surround that with a cavity that is tuned to the resonance frequency, we achieve a spontaneous oscillation and the system becomes a hydrogen maser. The word "maser" stands for *m*icrowave *a*mplification by *s*timulated *e*mission of *r*adiation, and it was first invented by Charlie Townes in the nineteen fifties, and was at a much higher frequency. Also, it was done not with an atom but with a molecule, ammonia. In fact, people originally thought that the difference in energies in atoms were too small for a maser to oscillate. Yes, the energies were small, but we found that if the lines were sufficiently narrow and sufficiently precise, the noise was small and we would get oscillation. With the hydrogen maser we had this nice, regular signal, so we just had to count the number of cycles and that gives us the time — or the rate at which the oscillations come out gives the frequency. Incidentally, a maser is very similar to and the forerunner of the much better known laser, with the difference that while in lasers visible light is used, in the masers we use longer wavelength microwaves. The masers are even more stable than the cesium atomic clocks for short periods of time. Their frequency stability is about 1 part in 10^{16} for a few hours up to a day; while the cesium clocks are stable for longer times. At present the best broadcast-time signals are derived from low noise hydrogen masers periodically tuned for long-term stability to cesium beam tubes.

We invented the hydrogen maser in the nineteen sixties, together with my then most-recent graduate student, Dan Kleppner, who is now a professor at MIT. He is the one who later started experimental research on Bose-Einstein Condensation (BEC), which is one of the results of our interest in the hydrogen maser. This is one of the interesting things in science, if you do research on something, you may discover something quite unexpected and unrelated. I, for example, wanted to do better spectroscopy for learning the nuclear properties, and that way we realized that we worked out the means for the best atomic clocks. The same way, to make better atomic clocks, people were anxious to make the atoms move more slowly because the more slowly they go, the longer they are in the field and the longer you can observe them. There are also different ways to trap them; we trapped them in a bottle but you can also do it with a laser beam but they have to move very slowly to do that. Then laser cooling was developed, the first idea was due to Dehmelt and Wineland and to Haensch and Schawlow. So here the incentive was to develop good atomic clocks but it turns out that one of the most interesting things was that with laser cooling you get Bose-Einstein Condensation, a new form of matter; you can get phase-coherent atoms. Just like the laser, which gives phase-coherent radiation, in which the oscillations go only up and down together and not randomly as in the case of sunlight. You can do the same thing with atoms, you can have a phase-coherent beam of atoms, their quantum mechanical wave oscillates up and down. The latest Nobel Prize was given for this Bose-Einstein Condensation, but originally the incentive for laser cooling was for atomic clocks, and interestingly it happened only later that the laser cooling worked well for atomic clocks.

I would like to go back in time. After graduating from Columbia in 1935, you went to spend some time in Cambridge, England. As far as physics is concerned that was one of the most exciting places to be at that time, with an incredible concentration of great names there. Could you, as an undergraduate, benefit from their presence?

Very much so. I got my bachelor's degree in Columbia in mathematics, but I was very much interested in physics, so they gave me a scholarship to go to Cambridge University. There, since I already had my undergraduate degree, I had a choice; either I could start to do research but in that case I would never have had advanced graduate courses because there you got that as an undergraduate. At that time their undergraduate courses were equivalent with the first two years of graduate studies here in the U.S. While here when you are an undergraduate, you take all sorts of courses, in England if you are a physics major undergraduate, you take only physics courses. Therefore, I decided to take the undergraduate courses, and I did that in two years instead of the usual three. I studied with Lord Rutherford; I even took a course from J. J. Thomson, who was the discoverer of the electron! I also took a course from Dirac and Cockroft; it was a great group.

What kind of teachers were they?

They varied a lot. Dirac, who was a great scientist, one of the founders of quantum mechanics, was terribly shy and a poor undergraduate teacher. At that time he basically read from his book. He also discouraged people to do their thesis with him; when they asked him to be their advisor, and asked him to suggest a problem for them to work on, he sometimes used to say: if I knew of a problem I would sit down and solve it myself! Which he would. Of course, he produced at least one very good Ph.D. student, Abdus Salam, who got the Nobel Prize later, in 1979. Even worse than Dirac as a teacher was Eddington, who had essentially memorized his beautifully written book on relativity and repeated it back in class. Cockroft was a wonderful, clear lecturer.

With Rutherford I had the following intriguing experience. He was a great name and I wanted to take his course, but initially I was very disappointed. I remember even writing to my parents after living in Cambridge for a while that Rutherford lived up fully to my ideal of a typical British lord; he dressed in a peculiar cutaway tweed jacket I had never before seen on anybody else and always with a bow tie. With his courses I was rather disappointed. He asked numerous questions at seminars, some of which were pretty trivial and even the undergraduates could see the answer. In class he derived Rutherford's scattering law, I was terribly disappointed. His derivation just fell apart; he could not get to the end so finally he said, go and work it out yourselves! So as a previous mathematics concentrator, I was very disappointed. But then the following summer I had to write a paper for my tutor, Maurice Goldhaber, who was a refugee from Germany. I did not have any books at hand, so I looked up my notes from Rutherford's lecture and I found the answers to my questions in Rutherford's notes. I also found the notes to be very interesting. If you looked at them from a different point of view, not that of a mathematician, but from the point of view of ideas that he threw out. For example, when he talked about his experiment on Coulomb scattering from nuclei, which shows the existence of the nucleus, and its atomic structure, he asked why don't you worry about being scattered by the electrons? Well, he said, it is like walking in the woods and being scattered by the mosquitoes. Of course, I had the feeling that if he had ever lived in New Jersey or New Hampshire, he would have known that you could be scattered by the mosquitoes; but this is something different. But this idea of his was good. So eventually I went back to his course for the second time and it was excellent. I concentrated on his ideas and his approach, and I found his ideas stimulating. Likewise, I discovered that although he did ask many foolish questions, he also asked some very good astute questions, so you could always eliminate the stupid ones. In fact, most of the research in the lab even then originated from Rutherford's questions. For example, eight or ten years before Chadwick discovered the neutron, Rutherford had speculated about the possibility of there existing what he called an atom of zero charge.

So I had a very valuable time in Cambridge. Maurice Goldhaber asked me to write a term paper on magnetic moments. This is how I learned about I. I. Rabi's work at Columbia, and was very much interested in it. When I came back to Columbia, I asked Rabi if I could work with him and he agreed. After my time in Cambridge, I started graduate work at Columbia by doing only research; I only took one course, by Fermi. I told Rabi that I wanted to do work on molecular beams but he was very discouraged about that field at that time. He did not think that it was a good choice for graduate work because all the exciting discoveries that could possibly be done in this field had already been done. Then, about two months after I started, he invented the magnetic resonance method and immediately we shifted our experiments to magnetic resonance. Soon afterwards we discovered the quadrupole moment of the deuteron and did some other work as well, so I got my Ph.D. in just two years.

What kind of a person was I. I. Rabi?

He was a great scientist, very creative, with great ideas. He also had very high standards for the quality of the work that people did with him, but he also wanted us to do interesting things. He let the graduate students work on their problems for a while and let them alone. But then he would come back and ask them about their results, and if he felt like it, he sometimes

At the celebration of Isidor Rabi's 60th birthday in 1962. Front row, from left to right: J. R. Zacharias, I. I. Rabi, J. Schwinger and G. Breit. Second row, from left to right: N. F. Ramsey, C. H. Townes, V. Hughes, E. M. Purcell and W. Nierenberg (courtesy of N. Ramsey).

said, "Why are you wasting your time on such an uninteresting problem?" He did the same thing with other people in the Physics Department, which annoyed many people. But it also made them all think. The graduate students quickly learned that they had to argue with him that it was really a good thing that they were doing or else, they better do something else instead. He was also very good in giving independence to his graduate students. He would suggest a problem and let the student work on it; while the student was building the apparatus he might not show up much; he was not interested then. But when the student started to get data, Rabi was back again and keenly interested and worked with him. Rabi really trained people to be independent and this was a much better attitude than when a supervisor did most of the work for the student and thus the student did not learn any independence at all. The associates of Rabi had this great characteristic of being independent and many of them eventually won Nobel Prizes.

You participated in much war-related work during World War II. You started first working at the MIT Radiation Laboratory. Is not that name a misnomer?

Of course, it is. In fact I am an expert on this subject because I am a co-inventor of that name. The Radiation Lab started in the fall of 1940, when the Nazis had invaded France and everything was falling apart in Europe. The British sent a delegation over to urge the U.S. to help on research. The British had had this great success with radar; they developed a so-called "chain system", meaning that they put big tower radars all along the coast of Britain which detected the German bombers as they came in so they could direct the defending fighters onto the tails of the bombers. Radar was a key element of the success of the Battle of Britain. But that system worked during the day when the fighter pilots could see the bombers but the British were worried about night-fighters for which they could not use that long wavelength method and they wanted to develop higher-frequency equipment to go onto the fighter planes for that. John Randall, Mark Oliphant, and Henry Boot developed the magnetron at 10 centimeters, but they wanted to go further and asked us to participate. They sent over a delegation which met with some of the members of the National Defense Research Council, that Vanevar Bush helped form somewhat earlier. They also brought with them a few of their 10-cm magnetrons. Our scientists — and they were very senior people, such as Rabi and Lawrence, for example, — were very impressed, and so they agreed to participate in the further work. The U.S. scientists decided to start a small group at MIT to work on the development of radar. At that time I had just started to work at the University of Illinois as an associate, which was a very low rank. The chairman of our department at Illinois, Wheeler Loomis, suggested me to go and work in this group. I thought I would be soon drafted anyway, so after staying in Illinois for about five or six weeks, I moved to Boston.

There were only a few of us at the beginning, maybe 8 people. We had a meeting about what this new laboratory should be called. The British did not use the word "radar" at that time; they called it "radio location", which is a fairly informative term. We thought that if we called it Radio Location Laboratory there would be a dead giveaway of the purpose of the lab. I don't think that any of us heard the word radar that time; later we realized that the Navy was also doing work on it and they called it radar, but we were not aware of it. In fact, radar was independently discovered by at least four different countries including England, U.S., France and Germany. Ernest Lawrence was one of the people who were urging people to go and work on the new project, and we knew that he directed a Radiation Lab in California, which was working on particle radiation. So we thought, perhaps we should call our lab also Radiation Laboratory, for several reasons. Partly, it would be a tribute to Lawrence and it was technically correct, since we were also working with radiation, just not nuclear but electromagnetic. Finally, and importantly, the name would also be misleading for the Germans, who might think that we were wasting our time with something as impractical as a nuclear bomb.

Did you know Mark Oliphant?

Oh, I knew him very well. He was one of the lecturers at the Cavendish Laboratory in Cambridge when I was a student there; he was in fact a very good lecturer. I also met him often in connection with the radar work. Once I was sent to England to see their work on radar and to tell them about our work. This was exactly when the invasion of Britain was expected, so it did not make my wife very happy. But then it turned out that the Germans attacked Russia instead, although it was their ally at that time. I visited Oliphant's laboratory at that time and later after the War was over.

What was your goal with the radar work here, in the U.S.?

There were different things. When I first arrived, we discussed what components of the project were the ones that we should worry about. One of them was the magnetron. Rabi was the head of the group of the magnetron and I was his deputy. Incidentally, just to show our ignorance, at that time there was no group on the transmission lines which later turned out to be one of the key things. Soon afterwards, Rabi became deputy head of the whole laboratory, and I became head of the magnetron group. Later it was decided that a so-called Advanced Development Group would be useful, of which I became head. We jokingly said that we were fighting World War III rather than World War II. Most of the time I spent in this group. Our major goal was the development of a 3-cm magnetron, in contrast to the 10-cm one that the British used. In connection with this work we also had to work on systems, because for the 3-cm microwaves, we really had to change the hardware. For 10-cm microwaves the rest of the lab could use coaxial cables, but at 3 cm the cables became too small to carry so much power, so we realized that we had to use waveguides. Eventually we got the system working on the rooftop of MIT, picking up planes.

Then we formed three subgroups: one for developing the 3-cm radars for submarine detection, another for night-fighter aircraft interception, and one for navigation over land, maybe even bombing over land. We had two planes working with us, one for the night fighters and one for the sea work. With the submarines, even if they run on batteries, periodically they have to come up with their periscopes to the surface to charge their batteries for which they needed air. The goal was then to detect these very small objects with our radar. Our first test flight was not very successful. We wanted to see something that was about the same size as these submarine conning towers, so we just looked for trash, little floating objects, like soda cans on the water. We asked the pilot to find such objects, so he flew above some but we did not see them. Then we asked the pilot to pick us a small boat, which we did not see either. Then we asked the pilot to pick a larger boat, which, again, we could not detect. Beginning to get quite desperate, we asked him to find a larger boat, to which he answered, in disgust, "That was the Queen Mary we just flew over!" Apparently, in our eagerness we made too many mis-adjustments on our system and this blocked the signal completely. In a relatively short time we fixed the problems and we could pick up very nice signals. We saw that the system would be good for detecting submarine conning towers. We also took some pictures over Cape Cod and Nantucket and it was obvious that these 3-cm pictures

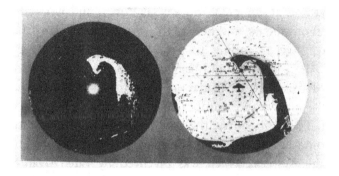

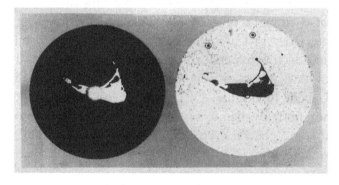

The radar scope photographs of Cape Cod (top) and Nantucket Island (bottom), taken by Ramsey's group in the early 1940s, with the first 3 cm (or x-Band) radar they developed (courtesy of N. Ramsey).

were much better than the 10-cm ones. We also found we could do some navigation over land with 3-cm radar.

From MIT I went to work in Washington, at the Army Air Corps. At that time they were not very good in making decisions about what type of equipment to buy so I was sent there with the awkward title of Expert Consultant to the Secretary of War. It was awkward because there was the joke going around that time about what an "expert" is: the "ex" means that he is a "has been" and spurt is "a little drip". Edward Bowles, a professor of engineering from MIT was heading that office. My task was to help the Army Air Force decide what it should purchase; I was doing this for about a year. One of the most fruitful of my actions there happened about 9 months after I got there. On a Friday afternoon a young second lieutenant, the lowest officer rank, came frantically to my office and told me that he was given the task to put together a draft for a five-year procurement program for radar for the Army Air Force, by Monday morning. He did not know anything about radar. So we sat down, and worked all weekend; I had reports about radars, and he had the procurement plans about airplanes. He would say, "We are going to get so and so many more B18s," and I would say, "All of them should have a tail warning radar and at least every tenth should have a 10-cm radar as a guide plane." Then he reported plans to purchase a certain number of night fighters and I would suggest radar for each. Then, of course, there would be many B29s, and I would say, every B29 should have a 3-cm bombing system, and one in ten should have a so-called eagle system, which was invented by Luis Alvarez, and had a very high-precision with a radar antenna extending the full wing span. So we picked out a number of each of these, and eventually it turned out that the whole radar procurement cost would be 1 or 2 billion dollars. He then submitted this material to his superior and as far as I know, with small modifications, this plan went through, so this was the first 2 billion dollars that I spent in my life. Of course, if you wanted to confirm this, you would not find my name, or the lieutenant's because ours was the first draft and some general eventually signed it, but basically this was the plan that was adapted.

From Washington you went to Los Alamos. Who was the one who recruited you?

It was Oppenheimer. There was a good restaurant, called Watergate Restaurant, and it was at the location in Washington that later became famous but at that time the Watergate apartments had not yet been built. Oppenheimer met with Ken Bainbridge and myself at that restaurant. This was very early, in 1943, and the project had not even moved to Los Alamos yet. I thought that it was perhaps time for me to go; I had spent about a year in Washington and done a lot. When I arrived there fresh a year before and made a proposal for a radar the staff said, "Oh, it takes about 50 signatures to get approval for this, and then it takes a training program, and all this takes forever, so maybe we should not even start." At that time I said, OK, I'll collect the signatures, and indeed, I did and we got it through. But after I'd been in Washington for about a year I discovered that when the people from MIT came down with a newly developed radar, I heard myself saying, "It really takes 50 signatures to get this through…" That was the point I realized I was perhaps good for Washington for one year and after that I started to become like the rest of them, so I was quite prepared to go.

But I had quite a problem with leaving. Oppenheimer said to me, don't do anything, I'll discuss this with General Groves; he reports directly to the Secretary of War and he never lost a battle yet. In the meanwhile there is some testing to do; they wanted to start testing the ballistics of some dummy bombs falling and he asked if I would supervise these tests. Sure, I said, I would do it, it was not to take more than a couple of weeks before I would be transferred. But the months went by and nothing happened. Eventually I learned why.

Bowles called me and asked if I would tell Groves and Oppenheimer that I wanted to stay in Washington. I told him that I would rather go. Finally I understood what the problem was. There were two people who reported directly to the Secretary of War, one was General Groves and the other Bowles, the supervisor of the radar work. So what happened was the following: Groves wrote a letter to the Secretary of War, Stimson, that I should be released, who then after a few weeks forwarded it to Bowles. Then it took again some time for Bowles to answer that he wanted to keep me. Then when the letter got back to Stimson he was reviewing the troops in Africa, so another month went by and this went on back and forth for a while. This whole thing became a test between two people who never lost a battle. Of course, when I told Bowles that I wanted to go, this became a question of saving face for him. So he suggested the following: I should stay as an expert consultant of the Secretary of War and be on his payroll but I should go to Los Alamos and work as a regular staff member there. So this is what happened. Therefore, during the whole of my time in Los Alamos, I operated as group leader but officially I was a government employee and not employed by the University of California, which supported the Los Alamos program. An amusing side-result of this was that at the end of the war all of the University of California employees were given a Navy E award of excellence; everybody got it except me because I was not an employee.

Would you care to tell me something about Oppenheimer?

He was a great leader, very understandable, very smart, very fast, very eloquent, very persuasive, in one respect perhaps overly so. I learned how to overcome that problem perfectly well. During the first couple of meetings I had with him, as a group leader I wanted something for my group and I would go and talk to him about it. It would be a great conversation, he would be very persuasive, but when I left I realized that I had not gotten what I went for. What I did later was that I always made some notes about what I went to see him about and during the conversation I always tried to get back to my original question — which he did not mind and we got along very well. Partly I reported to him and partly to Captain Parsons, who was the head of the Los Alamos Ordnance group.

What was your work at Los Alamos?

It was classified from the rest of the lab. Since earlier I had spent a year with the Army Air Force, I was asked to be in charge of the Los Alamos work related to the Air Force. There was, for example, a Los Alamos colloquium every week about what was being done in other groups. I attended them, and could ask questions, but I was never asked to give a talk; my group was not supposed to be informative. One of the secrets of the lab was how rapidly it had progressed. The security officers did not want too many people to know about our work. Of course, things do leak out but as far as I know there was no leakage from our group. Fuchs did not know what we were doing. By now the work is fully declassified.

One of our tasks was to test bomb shapes and to see that they do not fall head over heels. We were involved with the modification of airplanes that had to be done to carry the bombs. Originally, when we first started, both the uranium and the plutonium bombs were going to be very simple narrow shapes. What one had to do to make an atomic bomb was to get the two subcritical pieces together fast enough so that they then generate a lot of energy before they blow themselves apart. It was planned to do this by having a subcritical amount of uranium or plutonium at the receiving end and have another subcritical piece fired from a gun into the receiving end to accomplish the task. That would have been a very straightforward job. Very shortly after I arrived at Los Alamos the discovery was made that there was a certain amount of spontaneous fission from plutonium. So when you shot the two parts together, there would always be neutrons present, and before the two were close enough together for a huge energy release, they would produce enough fission energy to blow themselves apart. So that method could not be used with plutonium. The implosion method was used instead, because it was much faster, but it was also much more difficult. I was concerned with both types. With the first version, which we called a sewer-type bomb, it looked like a sewer pipe welded together to each end. The implosion bomb had to be highly symmetrical, and it had to fit inside the bomb bay. It was not particularly stable and the first ones fell head over heel. I was responsible for improving that. We wanted to have the bomb as large as possible but on the other hand it had to fit into the airplane. That size was determined at my end; I knew that the most suitable war plane for us to get was a B29 and that it could take up to about 60 inch diameter outside for everything and we had to fit inside that. When we made the first models for those, we had two problems: it was the heaviest thing that the Air Force had ever used and they did not have suitable bomb-release mechanisms. The first ones that the Air Force provided did not release the bombs and they stayed in the bomb-bay. After several tries, we finally devised our own mechanism utilizing a British bomb-hook. Then when they were dropped, they did not fall properly, and just rotated all the way down. So we had to make them stable. We did this by adding drag plates to the bomb fins.

Because of my experience with the Air Force, I knew that with the Army you cannot easily transport things overseas; you cannot just take equipment that were not put in the "table of equipment" before. I had learned, however, that you could put things in this general form of "kits". So before we knew what we needed at all, I put in a general "kit for assembly". It turned out very useful because we could specify later what that included. It eventually included 15 whole buildings, structures for vehicles, and so on. So eventually we ended up very well equipped.

I was at the bomb assembly and Trinity Test in Alamagordo primarily to obtain information that might be valuable for us later. At the test I was lying next to Rabi. I was awed that so much spectacular energy could originate from such a small plutonium sphere. I was hopeful for a quick end to a terrible war but was deeply disturbed about prospective casualties and worried about future implications. That afternoon I drove Oppenheimer and Rabi back to Los Alamos. We have often been asked about what we discussed during that long drive, but we were so overwhelmed that none of us could recall.

The next day I left by plane for Tinian Island in the Pacific where the 509th Composite Group was based. The Los Alamos team there was headed by Navy Captain W. S. Parsons and I was his deputy and Chief Scientist. The Los Alamos team carried out a number of tests at Tinian, assembled the two bombs and participated in the analyses of the results.

Did you know Edward Teller?

I knew him well, beginning before World War II. My first job after receiving my Ph.D. was as a Research Fellow at the Carnegie Institution. The Institution had a nuclear physics laboratory headed by Merle Tuve, as part of its

Department of Terrestrial Magnetism in Washington, D.C. At that time
Teller and Gamow were members of the faculty of George Washington
University, so we frequently attended the same physics seminars. My wife
and I were also good friends of Edward and his wife at Los Alamos.

Teller is a brilliant and creative scientist who has made important contributions to both chemistry and physics, especially molecular spectroscopy
and nuclear physics. However, he sometimes has strong obsessions. He
felt so strongly that his early proposals for a high priority hydrogen atomic
bomb development were so much in the best interests of the United States
that he concluded that Oppenheimer's opposition clearly indicated disloyalty.
Teller testified strongly against Oppenheimer before the AEC [Atomic Energy
Commission] board that eventually recommended the removal of Oppenheimer's clearance. Actually there were many good justifications for Oppenheimer's opposition which were fully consistent with his being loyal. I was
one of many scientists including Rabi and Fermi who testified in support
of Oppenheimer. I believe that Teller's strong feelings at the time of the
AEC hearing were probably intensified by the then ongoing Russian
occupation of Hungary, which deeply disturbed Teller, a Hungarian by
birth.

*What is your opinion of the hydrogen bomb project? Was it an important
one?*

It was an important one, but I am not sure if it was a good one. First
of all, the bomb that was being discussed by the AEC General Advisory
Committee (GAC) of which Oppenheimer was chairman was very different
from the one that eventually got built. It could not be carried in any
existing airplane. What Teller wanted at that time was a crash program.
The other argument was that developing it would be more harmful to
the U.S. than beneficial, because there were big targets in the U.S. The
GAC felt that it was not practical or beneficial to the U.S. They thought
that we should concentrate on other things. Later there was another invention,
made by Ulam, that could lead to a practical usable bomb and that was
pushed by more people. I don't know what Oppenheimer's opinion would
have been about that very different project. I am not sure that the hydrogen
bomb is all that important since without it the fission bombs can produce
more than enough damage.

*As I understand, you participated with Rabi and Zacharias in establishing
the Brookhaven National Laboratory.*

Yes, that is correct. During World War II I shifted from being a member of the University of Illinois staff to a member of the Columbia University staff, so after the war I went back to Columbia. We were quite annoyed by the following. We thought that Columbia had been "done in" by Arthur Compton, who was the head of the Physics Department at the University of Chicago in the following way. Before the war Enrico Fermi was a professor at Columbia and he did the early work on a potential atomic bomb and was developing plans for the first nuclear reactor there. Then it was decided that the government should consolidate the work just like the radar work, which was concentrated at MIT. Compton, head of the nuclear research portion of the National Defense Research Committee, decided that the main center of this activity should be Chicago. Thus at the end of the war, Columbia no longer had that work and no longer had Fermi. Chicago, on the other hand, had both Fermi and the Argonne National Laboratory. Rabi and I felt that we also needed something like Argonne at Columbia. Although I was mostly concerned with starting the molecular beam work going at Columbia, we thought that we should also have a nuclear reactor built there. Eventually we realized that it would be too big an undertaking for Columbia alone. Then we proposed that we have a New York area group to do this. We had a meeting that included Bell Labs, Princeton and Yale. I was elected to be the executive secretary of the group that organized this. We sent a letter to Groves, head of the Manhattan District, about this idea. Soon thereafter our previous close collaborator in the molecular beam work, Zacharias, who by then was professor of nuclear science and engineering at MIT, heard of our proposal and made a similar proposal for the Boston area. That letter went also to Groves. Groves then said the following: he was interested in both proposals, but he was managing what was probably the dying phase of the Manhattan District, which would be replaced by something else soon. He could see himself in that phase setting up one new laboratory in the country but not two and certainly not two in the Northeastern section. So he suggested that if we got together and agreed on a single proposal, he would probably support it; if not, — he said — he was a busy man, don't bother him any more. So we had a couple of very tough meetings and eventually we agreed that we would collaborate. Each person, of course, wanted the lab to be close to him. Rabi was very clever at these things. He suggested we should also include Cornell and Johns Hopkins. I think he did this because it would move the center of gravity closer to the New York area. Also at that time there was not much plane transportation and we had to rely on trains, but even the train service was pretty bad from Boston to Cornell. Eventually we agreed to cooperate, but for a while Rabi and Zacharias were not as good friends as they used to be before these tough negotiations.

I was asked to be the chairman of the site committee. The criteria were that the laboratory should be within a one-hour drive from the Grand Central Terminal. I picked up my first speeding ticket when I wanted to prove that one of the sites was within that limit. We found a couple of sites but then the Manhattan District said that they wanted only property which they already owned. That restricted the choice; so there were two in New Jersey (one at Sandy Hook, which was an Army base and another one at Camp Kilmore which had been a recruiting station), and two on Long Island (one at Brookhaven and another one near New Rochelle). Then the Manhattan District ruled out New Rochelle because it was too close to the city, and Sandy Hook was ruled out because the soil was not good. So Brookhaven quickly became the favored site. It was a rather decaying, unattractive place at that time because it had been a prisoner-of-war camp immediately before. Kistiakowsky, a professor at Harvard, said that the site selection was decided on the basis of "equalization of disappointment".

After the site was selected, I became the first Chairman of the Brookhaven Physics Department on a part-time basis shared with my physics professorship at Columbia. A year later I resigned both positions to become a physics professor at Harvard University.

How did you become involved with Fermilab?

In 1963 I was Chairman of a Joint High Energy Physics Panel of PSAC and AEC. We recommended the future high-energy physics program for the United States, including an 800 GeV proton accelerator, colliding electron beams at SLAC, and colliding protons at 35 GeV. In 1971, at the suggestion of the National Academy of Sciences, the leading research universities formed a new consortium, Universities Research Association (URA), to construct and manage the proposed high-energy proton accelerator and its laboratory, eventually called Fermilab. I was elected President of URA and we appointed Robert Wilson to be the first Laboratory Director. He was an excellent director who advanced the schedule of the original planning group by two years, increased the planned energy and completed the construction $6.5 million under the budget. We did have one difficult year when he announced his intent to complete the project three, instead of two, years ahead of the original schedule but with failed operation during the first part of that

Normal Ramsey, his first wife, Elinor, and their four daughters (from left to right): Patricia, Winifred, Margaret, and Janet around 1955 (courtesy of N. Ramsey).

extra saved year and only intermittent operation during the remainder of that year. Subsequently, the operation became reliable and the energy increased to 1 TeV. Under the second Laboratory Director, Leon Lederman, the accelerator has been upgraded to provide colliding beams and thereby much higher energy in the center of mass system. A number of important scientific advances have been made at the lab, including discoveries of upsilon particles (due to bound states of the b or bottom quarks) and the top quark.

Please, tell us about your family life.

My family life for the most part has been happy and enjoyable. My wives and children have been very supportive of me and my work and I have enjoyed trying to be supportive of them. My first wonderful wife, Elinor, and I had five daughters, including a set of triplets, of which one survived for only two days. On the whole we had an enjoyable time growing up together. I was very busy with my research and teaching, so most, but not all, of the care of the children was provided by Elinor. We spent many happy weekends, vacations and sabbatical leaves together, mostly hiking in the summer and skiing in the winter. We particularly enjoyed New Mexico,

Lake O'Hara in the Canadian Rockies, Christmas in our cabin in Ripton (Vermont), Les Houches in France, and two sabbatical leaves at Oxford during which we had summer trips to Europe and skiing trips to Austria.

Our greatest family tragedy was the death of Elinor in 1983 from a long siege of ovarian cancer. I was terribly depressed for the next year and a half, but had the good fortune on an Appalachian Mountain Club hike to meet Ellie Welch, a widow with three fine children. Ellie and I were married in 1985 and have had a delightful time together including trips to Europe, Asia, Alaska, Canada, Colorado, Utah, California, and Sweden for the awarding of the Nobel Prize. We particularly enjoyed a four-week trek in northern India, from the Moslem-dominated Srinigar (Kashmir) to the Buddhist-dominated Leh (Ladakh). Unfortunately, on the day we returned to Srinigar, the leader of the Moslem independence movement was in an automobile accident. He was recognized by the Indian police and put in jail, which in turn led to an immediate strike that closed down all of Srinigar. This complicated our departure and led to subsequent elimination of this beautiful trek. Unfortunately, the arrest led to an escalation in violence, which still continues with deaths of thousands of Hindus and Moslems.

I officially retired from Harvard in 1986, but I have an office there as well as at our home in Brookline, Massachusetts. I continue studying new advances in physics, doing some theoretical research and writing on both new physics topics and history of physics. I am no longer teaching regular classes, but in response to special invitations, I continue to give lectures at scientific meetings and university colloquia throughout the world. Occasionally I lead a regular physics class for a single day at Harvard or elsewhere. I am still very busy but hope soon to have more time for my wife, children, grandchildren and great-grandchildren.

Are you religious?

No, I would say no. There are many things we do not understand about the origin of the Universe, but it seems to run without an external god-like intervention. Quantum mechanics and evolution by natural selection are remarkable processes. However, I am very tolerant of most opinions on this subject that are held by others.

Has the Nobel Prize changed your life?

In my case, I would say it has not much changed my life but it has made it possible for me to continue it the same way for longer. The Nobel

Prize is time-consuming and this is thus harmful for all winners. For some there are other harmful effects. Some winners think that after the Prize they have to do only better experiments than before and eventually they do not do much at all. In some other cases they stop trying. In my case it was an advantage that the Nobel Foundation was somewhat slow in giving me the prize. Some people when congratulating me said that it is great, only it is a pity it had not happened earlier. My response was that I am actually happy it did not happen earlier, because I received it about three years after I officially retired from Harvard. I still had a visiting affiliation with other institutions but in my case the Prize helped me to keep active in physics much, much longer. Not in the form of any direct experimental work or analysis; I was mostly advising people and was lecturing. I have had a large number of students during my career and they continued doing well later in their lives; I had altogether about 86 graduate students. People are working very vigorously nowadays on atomic clocks and I am very much concerned with that. The topics change somewhat but I am still very much interested in them, although I am already 86 years old. I am still interested in high-energy physics; when I was president of the University Research Association, which is operating Fermilab, I was in close contact with that.

What was the greatest challenge in your life?

Norman Ramsey, his second wife, Ellie, and Sheldon Glashow around 1995 (courtesy of N. Ramsey).

It is hard to define; almost everything was a major challenge. Probably a major unsuccessful challenge was the search for parity non-conservation. Purcell and I had the first idea that parity might not be conserved and we did an experiment on parity, testing it at a time when everybody thought that this was something one should not waste time on. This was about 5 to 6 years prior to the Lee and Yang paper. In fact the only paper they quoted was one of our papers, even if they slightly misquoted it by confusing it with our later experimental limit whereas the paper they referred to was the first one, which pointed out that parity symmetry was not obvious and should be tested experimentally. We did an experiment in which we were looking for an electric dipole moment as a test of parity when everybody believed that it was a somewhat stupid experiment because parity had to be conserved. We worked on that for about 5 or 6 years, and did a very good experiment, but the problem was that we were looking at the nuclear strong forces. At that time people did not differentiate much between strong and weak forces. We did a very, very sensitive experiment but what we tested was the nuclear strong forces and there, as far as we still know, parity is not violated. When Yang first reported on their theoretical work at an MIT colloquium, I immediately tried to arrange an experiment to do a parity search with weak forces, and arranged with Louis Roberts at Oak Ridge Laboratory, the only person who has ever aligned a nucleus in large amounts, which he did with Cobalt-60. Unfortunately, soon after we made our arrangements the Oak Ridge management postponed our experiment.

Was your experiment the same as the one Madame Wu did?

Yes, it was. I corresponded with Yang about this but I did not know then that Madame Wu was even considering an experiment.

What gave you the idea originally to look for parity violation?

That is an interesting question. Shortly after I came to Harvard, I was giving a graduate course on molecular beams. Ed Purcell, the co-inventor of NMR, who was a professor at Harvard that time, sat in the course because he was interested in the topic. I discovered the following: it was fun having Purcell in the class because it led us to have interesting discussions. But I also learned one hazard of having Purcell in the class; if I was talking about a subject that I did not really fully understand, I could count on Ed asking an astute question that not only would convince me that I did not understand but would also convince him and the whole class.

In the class I was about to give the well-known proof: if parity is conserved there cannot be an electric dipole moment. All one knows about the orientation of the nucleus is its angular momentum, as if something is spinning, and if it is spinning in one direction, the only way you can tell what's up and what's down is that you grab it with your right hand and your fingers go in the direction of the spinning, then your thumb points in the up direction. If you do it with your left hand you get the opposite answer. Therefore, if you have an electric dipole, that is a violation of parity. I knew this and I knew how to give its proof. But then I had the vision that Purcell would ask "but what's the evidence for nuclear forces of parity being a good assumption?" So I thought I better figure out before he asks. I looked at all the experiments and I could not find any evidence. At that time, as I said before, we did not talk much about strong and weak forces, only simply nuclear forces.

I thought that the best way to defend myself against a Purcell attack would be a counterattack. So a couple of days before class I went to Purcell and told him about this. He said, oh, there must be lots of evidence, so he went to look for evidence and he could not find any either. Eventually we wrote up a paper saying that we would do an experiment in which we would look for electric dipole moment as a test for parity [*Phys. Rev.* **1950**, *78*, 807]. Which we later did, with a graduate student, Jim Smith [*Phys. Rev.* **1957**, *108*, 120]. We improved the sensitivity by a factor of about ten million, and there was no electric dipole moment.

But interestingly, I got quite comfortable with the idea of parity not being conserved. It was at that time that Frank Yang gave a talk at MIT saying that he and Lee were speculating that some particle physics experiments might be interpreted as a failure of parity in the weak interaction. Since I had enjoyed winning many theoretical arguments on parity during our electric dipole experiments, I was very sympathetic to the idea of parity non-conservation and immediately wanted to do a weak force parity test with polarized beta radioactive nuclei to see if more electrons came out in one direction than another. In the discussion period following the lecture, I proposed doing the experiment on Cobalt-60, the only nucleus that had been significantly polarized in macroscopic quantities. Yang agreed that this would be a good experiment during the discussion and in subsequent correspondence. I telephoned Louis Roberts at Oak Ridge, and suggested we jointly look for parity non-conservation in the decay, with Louis providing the polarized Cobalt-60 and I the electron detection equipment. We agreed to do the experiment as soon as possible. Unfortunately before we could start the experiment Louis made an interesting, but not world-shaking discovery about the angular distribution of the neutrons in fission and the theoretical advisory committee at Oak Ridge urged that the likelihood of parity being non-conserved was so small that Louis should postpone our parity experiment until he had completed his fission studies. By the time I was told of his decision and could locate another source for polarized Cobalt-60, I learned that Madame Wu was already well started on the same experiment. Missing out on this discovery was my greatest physics disappointment, which was only slightly compensated by my winning $50 from Dick Feynman, who was so sure that parity would be conserved that he made a bet with me at fifty to one odds that parity would be conserved in radioactive beta decay.

Soon after Lee and Yang and Madame Wu published their papers on parity non-conservation, a number of theoretical physicists, including Lee, Yang, Landau, and Schwinger, published theoretical papers claiming that there could still be no electric dipole moment because of time reversal (T) symmetry. I then wrote a theoretical paper [*Phys. Rev.* **109**, *225* (1958)] pointing out that just like parity, time reversal was an assumed symmetry that had to be tested experimentally. I, and others, therefore, continued electric dipole tests of ever increasing sensitivity. So far no electric dipole moment for an elementary particle has yet been found. However, failures of CP and T symmetry have been found in the decay of the long-lived neutral kaon. Some of the theories to account for the neutral kaon experiments have already been eliminated by the upper limit we have established on the electric dipole moment and there is great interest in the ongoing improved electric dipole experiments as further tests of remaining theories.

I have heard that you were at one time also involved with chemical shifts. How did that come about?

That is a rather amusing story. It turned out that the most frequently-quoted paper of mine is about chemical shifts. This came about the following way: I mentioned earlier that we could measure the nuclear magnetic moment much more accurately with the separated oscillatory field method. But there is one disappointment; the nuclear magnetic moment is inside the molecule, so when one puts on an external magnetic field, the electrons, which are outside the nucleus circulate and produce a magnetic field to shield it and thus disturb its measurement. So the value that we get for the nuclear magnetic moment is not quite correct. Willis Lamb had developed a theory of magnetic shielding for atoms, which is quite simple since the motion of the electrons due to the magnetic field are simple circles centered on the nucleus. However, for molecules it is not that simple. So I figured out how to calculate the general theory for this shift, which is different in different molecules. Independently Felix Bloch, Bloembergen, and others observed chemical shifts, so I included "chemical shifts" in the titles of my later papers on magnetic shielding. My theoretical chemical shifts had the characteristics that they were temperature-independent. One day Bloch came to my office and said: Now, I've got you! He showed me curves, which he got with ethyl alcohol, CH_3CH_2OH, and he said that the OH signal is temperature-dependent! It was a bit of a blow to me and I could not immediately explain the temperature dependence. I am not much of a chemist, so I just left the formula on the blackboard. A couple of days later the representative of the Office of Naval Research, Urner Liddel, came around to see how I was doing. He looked at the blackboard, saw the formula and said, what are you doing with it, are you studying or are you drinking it? I said I was studying. He said his Ph.D. thesis was also on alcohol and he had discovered that there was molecular association in alcohol, through the hydrogen bonds of OH and that was temperature-dependent. It was immediately clear to me that this temperature-dependent change in the form of the molecule would lead to a temperature-dependent chemical shift, compatible with my theory. So Liddel and I did a few calculations to see how this agreed with our theory. We wrote it up the same day and mailed it to the *Journal of Chemical Physics* that afternoon; the fastest paper I ever wrote [Liddel, U.; Ramsey, N. F. *J. Chem. Phys.* **1951**, *19*, 1608].

There is a side story to this. The same people who put together the *Citation Index*, also put out a special article on the Nobel Prize, basically self-congratulatory, showing how you could tell a person should have gotten the Nobel Prize based on the *Citation Index*. When I got the Nobel Prize, they planned an article on me attributing my Nobel Prize to my most cited chemical shift paper. They sent me a draft of the article and I had to point out to them that, although my most quoted paper was on chemical shifts, it had nothing to do with what I got the Nobel Prize for.

David E. Pritchard, 2002 (photograph by I. Hargittai).

18

DAVID E. PRITCHARD

D avid E. Pritchard (b. 1941 in New York City) is Cecil and Ida Green Professor of Physics and a principal investigator in the Center for Ultracold Atoms and the Atomic, Molecular, and Optical (AMO) Physics Group in the Research Laboratory of Electronics of the Massachusetts Institute of Technology. He earned his B.S. degree at the California Institute of Technology in 1962 and his Ph.D. degree at Harvard University in Cambridge, Massachusetts, in 1968. His thesis title was "Differential Spin Exchange Scattering: Sodium on Cesium". He has been at MIT since 1968. He is a member of the National Academy of Sciences of the U.S.A., Fellow of the American Academy of Arts and Sciences, and is a fellow of several learned societies. We recorded our conversation in Dr. Pritchard's office at the MIT on March 11, 2002.*

Last December in Stockholm, it was a moving moment during Wolfgang Ketterle's (see next interview) Nobel lecture [2001] when he asked you to stand up and acknowledged your contribution to his prize-winning research. When Ketterle was your postdoc at MIT and he was to become independent, staying at MIT as an Assistant Professor, you magnanimously gave him one of your research areas, which then happened to be the field where he did his prize-winning discovery. Is this a correct description of what happened?

---

*István Hargittai conducted the interview.

That's correct, except that my promise to him that he could have the experiment to himself if he chose to remain at MIT was made prior to his accepting this position. Would you like me to talk about it?

Yes, please. It's not common that people have Nobel-level projects and even rarer that they give them away. I may sound provocative, but did you recognize its importance at the time?

Yes, we had made the Dark Spot magneto-optic trap (MOT) that could collect a dense enough sample of cold sodium atoms that we should be able to get evaporative cooling to occur if we transferred the atoms to a magnetic trap. This was very exciting, as it seemed to me that with a little luck we could get runaway evaporation. I had estimated some collision cross sections that showed this might lead to Bose-Einstein Condensation (BEC) or to some unanticipated piece of physics that would prevent it.

But at that time I had three projects in my group. We had a precision mass measurement experiment with single ions, which was and continues to be the most accurate mass measurement apparatus in the world, and our atom interferometer was unique and was just starting to give all sorts of new results. So I had a choice: give one of three projects — albeit the most exciting — to Wolfgang and get a wonderful junior colleague, or keep the three projects at the expense of creating another strong competitor in the cold atom field — which already had a half dozen strong groups in it. I knew that Wolfgang could develop my ideas to get to BEC faster and better than I could, even if I didn't have to manage two other experiments. Finally, I thought that being the youngest AMO professor at MIT for 20 years was long enough. So it wasn't a hard decision, and in the end I got a Nobel Prize medal anyhow.

Yes, I heard about that. Can you tell me how and when Wolfgang gave it to you?

As soon as we were both back from Stockholm, Wolfgang asked if I would like to see his medal (which I hadn't seen there). When I said yes, he quickly picked up his briefcase and led me into my office, putting several things on the table in my office. I thought this a bit strange because as a matter of courtesy, I should have stayed in his office to see his medal. But I passed this off as Wolfgang's usual energetic approach to things.

To my surprise, he put three medals on the table. "Dave, they actually gave each of us three medals — the real gold medal, and two replica bronze

Wolfgang Ketterle and his advisors at the Nobel Prize award ceremonies in Stockholm, 2001. From left to right: Herbert Walther, Wolfgang Ketterle, David Pritchard, and Jürgen Wolfrum (courtesy of D. Pritchard).

medals plated with the same karat gold as used on the solid gold medal. Can you tell which is the real one?" I often tell the story of how Archimedes developed his principle to determine non-destructively whether the King's new crown was solid gold. His method relies on the fact that gold is the densest metal commonly available, being about 50% more dense than lead. So I was quickly able to pick out the gold medal even though the bronze ones were slightly thicker so that their weight is not that much less.

I had it in the palm of my hand and passed it to Wolfgang so that he could verify that I had correctly identified the gold one. Instead of taking it however, he wrapped his hand over mine, closing my fingers over the medal, and said, "You keep holding on to it — it's yours. I'm giving you this one. I'll give one of the replicas to the department and keep the other for myself." My mother will be glad to hear that I remembered to thank him for this. But I was too stunned to express my true feelings to him until the next day.

In the meantime, we decided that it was appropriate for me to carry the medal in my computer case, as it is no more valuable than the computer, at least monetarily. Later I mused on how Wolfgang's method of presentation had achieved three objectives at the same time — another example of his unmatched ability to find solutions that solve several problems at once. In case I had known about the replicas, it made sure I knew I was getting the gold medal. It also made me realize that he was not giving up his only Nobel medal. Finally, by getting it in my hand and holding the hand shut, he made it more difficult for me to decline the gift.

The gift of this medal has had two tremendously beneficial side effects. The medal has been enriched by being given. Furthermore, it is possible for me to show it and talk about its doubly positive reflection of Wolfgang (i.e. both winning it and giving it) without it seeming to be in bad taste as it would if Wolfgang himself were to exhibit it. I have passed it around at several department gatherings and at a gathering of the alumni of my research group. Many of those who hold it are deeply touched and thank me profusely for risking it being passed around. I doubt that any other Nobel medal has generated so much happiness in so many people.

That's wonderful and obviously doesn't happen often. But I suppose giving exciting experiments away is fairly unusual also.

Well, there is lots of precedent for giving away experiments in our group. Dan Kleppner led the way when he gave Bill Phillips the machine he'd constructed to do his thesis in — it housed the first successful atom slowing experiments at NIST. Then I gave my thesis machine to Stuart Novick at Wesleyan University, and later gave my Ph.D. student Brian Stewart my experiment on molecular energy transfer and the large double monochrometer that went with it; he has produced new results with it for 15 years. This year I'm giving my atom interferometer to my current postdoc, Alex Cronin, and next year I plan to give the mass spectrometer to Ed Myers at Florida State. I find it easier to stop doing old things and move on to new things if I know that my still state of the art experiments will go forward — often faster due to the infusion of new blood.

Now can you tell me about measuring atomic masses?

Yes, a key point is that you can determine the energy of the nucleus from its mass by using $E = mc^2$. We measure mass to one part in 10^{10} or maybe a little bit better. If you measure both ^{14}N and ^{15}N, you can find the energy that's liberated in the n-gamma reaction on ^{14}N. That gamma ray is used to calibrate the gamma ray energy spectrum. By measuring the mass difference to five parts in 10^{11}, we measured the gamma ray energy to one part in 10^7 which was a hundred times better than the existing error in that field. So we made a huge improvement in X-ray metrology. We have also made a key link in a new route to measure the fine structure constant.

Aren't there similar studies in the Siegbahn Institute in Stockholm?

Yes. They measured the mass of Cs to one part in 10^9; concurrently we improved it by an additional order of magnitude. They only did one measurement to one part in 10^9 but we have done about 15 to one part in 10^{10}. That Cs mass is needed to measure h/m [h is Planck's constant] in a precise way which leads to a precise measurement of the fine structure constant. We (and Stockholm) did the mass and Ted Haensch measured the wavelength and Steve Chu has been measuring h/m. This route to the fine structure constant motivated us to measure its mass and Haensch to measure its wavelength. All three measurements are necessary for this method of determining the fine structure constant, which has the further advantage that it is based on quite simple physics so that this result might test understanding in other fields where the fine structure constant is measured.

What was your involvement in the Bose-Einstein Condensation experiments?

I have been interested in using BEC's for atom optics and atom interferometry, and have been collaborating with Wolfgang to build two new BEC machines. Recently my group started an atom interferometer experiment, the first one I have done on the BEC without Wolfgang as a co-author. In fact, we did it with the (much improved) apparatus I gave him. It is a new type of atom interferometer to measure h/m.

Oh, maybe you mean my past involvement. So let me go back a ways. When I got tenure, I was doing a lot of chemical physics — atom-molecule collisions and spectroscopy of van der Waals molecules in several different ways. In the late 1970s, I reinvented Ioffe's trap formally proposed for plasma confinement — it is often called the Ioffe-Pritchard trap when used for atoms. I had worked out some of the things that we could do with that trap, including a way to cool atoms below the Doppler cooling limit, that presaged the main idea of Sysiphus cooling. But I hadn't pursued trapped atom experiments because there wasn't a source of the necessary cold atoms, and traps for neutral atoms can only hold atoms below 1 kelvin, even

William D. Phillips (b. 1948, Nobel Prize in Physics 1997) at the Nobel Prize Centennial in Stockholm, 2001 (photograph by I. Hargittai).

if you work really hard on the magnetic field. I didn't start working on those ideas until Bill Phillips started demonstrating that it was possible to slow atoms down with radiation pressure. In the meantime, I had started work on light forces on atoms and demonstrated the Kapitza-Dirac effect (with atoms, not electrons which was done only recently) and Bragg scattering of atoms from standing light waves, and later from nanofabricated diffraction gratings. These experiments started the field of atom optics (indeed, I published the first review article with that name) which has since become an active sub-field of atomic physics.

Our work on diffraction gratings for atoms enabled my group to build the first atom interferometer — the first apparatus based on atom optics. We put three of these gratings in a row and made a three-grating atom interferometer, where the atoms split in the first grating, are directed back together by the second grating, and recombine at the third. We added a thin metal foil so that each atom's wave goes on both sides of the foil. To perform the atom interferometer experiment, you expose only the portion of the atom wave on one side of the foil to an interaction, then measure the phase shift of the interference fringes at the third grating. Many of the ideas of atom optics and atom interferometry have found interesting application or generalization in BEC's and much of my current collaboration with Wolfgang focuses in that direction.

In our cold atom work we demonstrated the first Ioffe-Pritchard trap for atoms. We made a big one in 1987, and trapped 10^{11} atoms, which is as many atoms as anyone traps these days. But the trap was quite large and it had superconducting coils, so it was not possible to change the trap parameters to compress the atoms in the trap. We built it to be sure we could trap atoms, but in doing that we made the trapping volume so large that we didn't see any collisions between the atoms, so we couldn't do evaporative cooling. We settled for doing some observations on Doppler cooling and radio frequency spectroscopy of trapped atoms.

I also had a lot of ideas about using light forces to trap atoms and some of those ideas went into a joint paper with Carl Wieman [Nobel Prize, 2001] that we wrote in 1985. That work culminated in my group's invention of the magneto-optic trap (MOT). This is a six-beam trap that we demonstrated with Steve Chu [Steven Chu, Nobel Prize, 1997] at Bell Labs in 1987. The MOT completely revolutionized the cold atom field because everyone could quickly get cold atoms. We'd shown in the original experiment at Bell Labs that we could capture the atoms in the room temperature vacuum container. Then Carl Wieman showed how cheaply you could get millikelvin atoms this way, and the slow atom field took off. At one point 11 of 14 assistant professors in AMO physics in the U.S. were basing their careers on that trap and I think all but one got tenure.

By 1990 I had decided that our cold atom experiment based on the big trap we demonstrated in 1987 had run its course, so I was looking for new directions when Wolfgang came as a postdoc in 1990. The status of the cold atom field at that point was that the cooling had proceeded quickly beyond the Doppler limit. Ashkin, Bjorkolm, and Chu demonstrated optical molasses and amazingly they reached the Doppler cooling limit with ease. Then Bill Phillips made some careful measurements and showed that it was in fact six times colder than the theory would predict. That set off real intellectual excitement in the atom cooling business and Cohen-Tannoudji [Claude Cohen-Tannoudji, Nobel Prize, 1997] explained what was going on.

We knew at that time that Bose-Einstein Condensation was the holy grail in the field. We also knew how to get the final part of the way to Bose-Einstein Condensation — that we could get the very, very low temperatures that are required for condensation by using evaporative cooling, which had been demonstrated by Harold Hess in the MIT hydrogen cooling group of Kleppner and Greytak. Evaporation occurs spontaneously when the hottest molecules escape from the trap taking away their excess energy. Those left behind must then re-thermalize by collisions, both establishing equilibrium at a lower temperature and producing more hot atoms to evaporate further. It is a good trade-off: you throw away 10 or 15 percent of the atoms to lower the temperature of the remaining atoms by a factor of ten. This process cools the atoms — or your coffee — rapidly when they are hot, but to continue it you have to weaken the trap because at the lower temperature the energetic atoms no longer have enough energy to escape. (The fact that you can't reduce the intermolecular attraction of the water molecules in your coffee explains why it doesn't cool far below room temperature.) Unfortunately losing atoms, weakening the trap, and slowing the atoms by lowering the temperature all work to lower the collision rate, possibly preventing evaporation of atoms from continuing at a satisfactory pace. Fortunately, in reviewing our work on radio frequency spectroscopy of trapped atoms, Kris Helmerson and I had suggested using radio waves to evaporate atoms. In this scheme you don't have to weaken the trap when you evaporate, you just lower the radio frequency as the atoms cool. The advantage is that when you cool atoms while maintaining a stiff trap, they cluster closer and closer to the bottom of the trap, increasing the density fast enough so that the collision rate actually increases, allowing the evaporation to proceed faster. Thus you predict a "runaway evaporation" that seemed able to carry you low enough in temperature to reach BEC, provided only that some bad collisions that can kill the atoms don't overwhelm the good elastic collisions.

But we had no way to cool alkali atoms with enough density for evaporation to begin. Both the magneto-optic trap and our techniques cooling atoms below the Doppler limit fail when the atoms get dense enough to scatter the trapping and cooling light. Thad Walker and Carl Wieman showed that the scattered light bounces off one atom and hits another, making a force between the atoms that pushes them apart. Alan Gallagher and I showed that the excited state atoms have a huge cross section for collisions that expel both colliding atoms out of the trap. The scattering cross section for resonant light scattering is about three orders of magnitude larger than the cross section for the atoms to collide with themselves, stopping the cooling well short of the density needed for evaporation. Actually, the gap is more likely only one order of magnitude than three because with good vacuum the atoms can rattle around the trap for a hundred bounces before they collide and evaporation will still work. If you have a very good vacuum in your trap, it's even less as Randy Hulet showed.

It was this perspective that led Wolfgang and me to write a paper stressing that useful cooling meant increasing phase-space density, not just decreasing temperature. Essentially, what is the condition of getting Bose-Einstein Condensation? The answer is, you want to have one particle, one boson, in every little cube half of a de Broglie wavelength on each side. That's a quantum phase-space density of 1. If you add any more atoms to the system at that temperature, they will all go to the ground state, and you get a macroscopic population of the ground state, which is the Bose-Einstein condensate. In a normal gas, the phase-space density is 10^{-12}. If you cool and trap with light, you can increase phase-space density by a million, so the phase-space density is now 10^{-6} but you need 10^0 for Bose-Einstein Condensation. A condensate can have a density of 10^7, which emphasizes the amazing nature of the route to BEC — an increase in the phase-space density of an atomic vapor by 19 orders of magnitude!

With the quest for density as a motivation, Wolfgang and I developed a variant of the magneto-optic trap that enabled us to get about a hundred times more density in a trap. The idea takes advantage of the fact that alkali atoms have two hyperfine levels. The MOT works on atoms in the upper hyperfine level. Often the atoms spontaneously fall to the lower hyperfine level which interacts only weakly with the trapping light, necessitating use of a repumper laser beam to excite them back into the level that is trapped. We got the very simple idea of blocking the repumper light for those atoms that were in the very center of the trap. Atoms away from the trap center would be strongly trapped as usual. But when they were cooled and trapped in the center of the trap, they would not be re-excited by the repumper laser beam, but would collect in the untrapped level. They would just sit there and occasionally, from stray light, would be re-excited and re-trapped, cooled down again and put into the middle. If they should drift out of the center, they would be subjected to the full repumper laser and would quickly be pushed to the dark center of the trap. This way we achieved a huge increase in the density, trapping a hundred times more than enough atoms to absorb light resonant with the *lower* hyperfine level. At this density, trapped atoms would have to make only 10 trips across the trap before hitting another atom. At that point we realized that if we transferred the atoms into a magnetic trap and turn off the lasers to avoid heating due to photons bouncing off the atoms, there would be enough collisions to allow evaporative cooling. Depending on how lucky we were in terms of the elastic collisions that enable the evaporative cooling versus the inelastic collisions that would cause heating or atom loss, we could go all the way to the Bose-Einstein

Condensation. That was the situation about 6 months before Wolfgang came to the end of his postdoc.

Much as I would have liked to keep Wolfgang working on this experiment with me, it was time for him to be independent. I decided that it was important to keep Wolfgang at MIT as a colleague since I was 50 years old and we badly needed a junior colleague in our field. I could testify that he was comparable to the top echelon of people who had come through our group — people like Eric Cornell [Nobel Prize, 2001], Carl Wieman, Bill Phillips, and several other extremely talented people — so I was able to convince the rest of the faculty to hire him.

It is unusual in the United States to become Faculty where one had been a postdoc.

It is, and it is essentially due to the shadow problem. Wolfgang was not known in the cold atom field — before he came to work for me, he had worked in applied physics. If I had kept working in the field, he not only would have been competing with me in real terms, but his reputation would have suffered because some referees would have said that he was

Eric A. Cornell (b. 1961, Nobel Prize in Physics 2001) in Stockholm, 2001 (photograph by I. Hargittai).

Carl E. Wieman (b. 1951, Nobel Prize in Physics 2001) in Uppsala, 2001 (photograph by I. Hargittai).

not doing anything different from me. That would have jeopardized his chances for tenure. My idea was that we would offer him an assistant professorship and I would step aside with respect to cold atom research. That meant that I could work harder on the mass measurement and atom interferometer experiments. It seemed natural to give my cold atom apparatus to Wolfgang in order to give him the best start possible. What I got out of this deal was that I had a new younger colleague and that I could see my ideas for BEC carried forward. I gave him the apparatus, two senior students, and two grants, but I also told him that from that point on, when the laser tube broke, it was his responsibility. I knew that Wolfgang would put one hundred percent of his effort into that project and I knew that he would go ahead with it faster as an untenured Assistant Professor than I would as a senior professor with two other projects that were giving world-class data. I also told him that I would no longer be involved in the project and I would not put my name on any of the papers until he got tenure. In fact the Department Head made me sign a letter to this effect. Nonetheless, Wolfgang wanted to put my name on the first BEC paper, but I didn't allow this.

At some point, of course, he will have to come up with his own project.

He already has and it didn't take him long. Wolfgang is very creative and doesn't second-guess himself. What impressed me the most was the decision he made right after he got BEC. The situation was that Wolfgang and my former Ph.D. student Eric Cornell of the Joint Institute for Laboratory of Astrophysics (JILA) were basically racing. The JILA people also were trying to transfer atoms from a dark spot into the magnetic trap and trying to make evaporation work. To get the best chance of attaining runaway evaporation, both Eric and Wolfgang selected a spherical quadrupole trap — the strongest magnetic trap.

The downside is that a spherical quadrupole has a small hole right in the middle, so when the atoms get cooled into a small ball, they leak out of the hole. It's a virtual hole, the spin of the atoms can't follow the sudden reversal of magnetic field direction at the center of the trap, and they jump into a state that's not confined by the trap anymore. Once runaway evaporation was achieved they had to figure out how to plug up that hole. Wolfgang got the idea of using a green laser beam to repel the atoms from that region, and Eric got the idea of shaking the whole trap in such a way that the zero point in the field, which killed the atoms, went into a circle that lay outside the volume of the confined atoms. In spite of the shaking, the atoms stay confined in the middle. It was called the time orbiting potential trap or TOP trap. That enabled Eric to win the race. The problem with Wolfgang's laser plugged trap was that it was hard to align and that jitter would stir the atoms, which heated them. Once he solved that problem, and he got the BEC, he had about 100 times more atoms and he could produce them about ten times quicker than the Colorado group.

When he had gotten the biggest BEC in the world, Wolfgang made a courageous scientific decision. Remember, he was still an untenured Assistant Professor and this was the first science he had ever done that generated significant media attention — his phone was ringing off the hook with speaking invitations. At that point he said that the Ioffe-Pritchard trap is a better trap to do experiments in than this trap with the green laser plug, and he abandoned the spherical quadrupole trap. He figured out a clever variant of the magnetic coils called the cloverleaf trap. In the Ioffe trap there are coils at the ends and there are four current-carrying bars in the middle; it's a trap which is rather cluttered. In the BEC business clutter is a real problem because you have to get the six laser beams in there for the magneto-optic trap, you have to get some probe laser beams in, and you don't want the laser beams to hit the wires in the trap because the light scattering would ruin your camera or whatever sensitive optics you are using for diagnostics. Wolfgang decided that he didn't have to put those bars in; he could put the clover leaf coils at the end and they would still make that desired magnetic field configuration. That solved yet another problem because every time you have something like a water-cooled coil inside your vacuum system, if you ever make the mistake to forget the cooling water or the cooling water plugs up, then the coil with all its current in it will melt and it'll cause a disaster. It has happened once here and that kept them back for several weeks while they took everything apart and cleaned the whole system. With his new configuration, Wolfgang was able to position the actual physical coils outside the vacuum system; he had a much better optical access, and he didn't have anything inside the vacuum system. That apparatus has been in operation for five years without a need for opening it up.

I go through this in detail because it demonstrates the kind of intellectual creativity that Wolfgang has. It isn't the ability to create a brand new field that no one had thought about before, it's the ability to get in a complicated area where there are a number of considerations and to somehow come up with a solution that solves three or four of the major problems in one stroke. I have never worked with anyone like that. I have worked with some excellent people (e.g. six Nobel Prize winners), but Wolfgang has the unique ability to look at the whole field of play and think about it for a couple of days and come up with just one idea that solves several problems at once. He applied the same ability to renovating the lab. We needed to create a conference room and were planning to knock down three or four walls when Wolfgang solved the problem with just knocking down one wall. This is a salient feature of his intellect. It underlies the way that he goes about giving talks. I don't know if you've heard him speak?

Yes, twice, in Stockholm and in Uppsala. These were identical talks, his Nobel lecture.

He thinks very carefully about the condition of the audience and what's important in the work that the audience should know and how to communicate that in a very clear way. Then he puts that all together.

Can you envision a situation 15 or 20 years from now that he forfeits one of his projects in favor of one of his former postdocs to help him start his independent research career? Will he come up with his own questions in due time?

I don't know. I don't know for certain, but I think he is likely to exercise this type of self-sacrifice if it would be best for MIT.

In a way you sound as if you were still his mentor.

He is currently more famous than I am but, yes, if anybody can sit him down and say, Wolfgang, it's time for you to stop doing slow atoms, you really ought to do something new, I would probably be that person.

Your mentor was Daniel Kleppner. Would you care to tell me something about your relationship?

I have always admired Dan as a scientific and human role model, and have learned many things from him. Perhaps he has been so valuable to me because of the great contrast in terms of personality and in terms of interest too. I'll mention two characteristics in comparison with myself. Dan is very English; he is rather circumspect in the advice that he will

Daniel Kleppner at MIT, 2002
(photograph by M. Hargittai).

give. He will say, "Well, you ought to weigh the opinion of all the people concerned", when he means that some particular action would anger quite a few of the Europeans. If I were in the same place, I would say, "X and Y and most of their European colleagues would be pissed if you did that." The other difference is that Dan is very focused on hydrogen, on fundamental physics, for example on doing BEC with hydrogen. I remember when his postdocs were pushing him to do Rydberg physics with alkali atoms, which were much more accessible whereas Dan wanted to do hydrogen. Finally his postdocs prevailed and they did a brilliant work in the spectroscopy of Rydberg atoms and then in cavity QED, a field they really started. In contrast, I'm not constrained by fundamental physics. For instance, atom optics and atom trapping involve lots of creative quantum engineering. The magneto-optic trap is a very ingenious, creative thing that my student Eric Raab and I came up with, but it's not fundamental physics, just an elegant configuration of laser light, polarizers and magnetic fields that exploits the laws of quantum mechanics of atoms and radiation pressure in a very clever way. Its scientific value is that it does something quite remarkable, and that it is so easy to set up and do interesting science with.

What was your connection with the Colorado Nobel laureates, Carl Wieman and Eric Cornell?

Carl Wieman was an undergraduate thesis student here at MIT. Carl and I worked on opposite sides of the same lab when I was an Assistant Professor. We were very good friends and Carl was a regular crewmember on our sailboat for a couple of years. We spent many hours together on the boat, some of them talking physics, of course. He paid me a typical backhanded compliment when he said, "Dave, you're so smart you used to win arguments with me even when you were wrong." The other Nobel laureate, Eric Cornell was my Ph.D. student. He worked on the ion trap experiment here, then worked for Carl as a postdoc.

Was your trip to Stockholm last December your first participation in a Nobel Prize award ceremony?

It was. I was invited before, but I didn't go.

Who invited you?

The Nobel Committee of Physics invited me in 1997 when Chu, Cohen-Tannoudji, and Phillips received the prize for the slow atom work.

Why didn't you go?

I was very busy and I also figured that I'd go when the BEC prize was given. It was clear to me when the first prize went for the slow atom discovery in 1997, which was two years after the BEC discovery, that there would be a BEC prize. I suspect that the BEC discovery pressured them to recognize the slow atom results.

Was it an unambiguous combination to give the Nobel Prize for BEC to Cornell, Ketterle and Wieman?

It was clearly those guys who made the big breakthrough. It could have also gone to Kleppner, being a pioneer, but Eric and Carl were hard to separate, and Wolfgang was a must. So the committee did the obvious — how could they justify two winners from MIT when the first BEC was made at JILA?

How did Kleppner take it?

I didn't see any evidence that he took it hard personally and I think the three that were chosen was a natural combination. Dan still has a good chance of winning a Nobel Prize for his Rydberg atom and cavity QED work. He pioneered that field. It may well be that the next award in atomic physics will involve atoms and non-classical radiation unless something else really big comes up, like quantum computing.

How about yourself?

It's an outside shot, but I could still be considered for atom optics and atom interferometry. That has become a significant theme in atomic physics and arguably the most noteworthy advance in the first half of the 1990s — I've been told that atom interferometry led all the physics citations in the year we made the first interferometer.

Isn't it an unusual situation that Dan Kleppner, you, and Wolfgang Ketterle, three generations in the same line of research, work in the same place?

That's ideally what you try to do if you are going to have a tradition of excellence in atomic physics.

Isn't it more the European approach?

To do it in a top U.S. university you must work hard to prevent the shadow of the older person from obscuring the light of the younger person. For instance, Dan was an Assistant Professor with Norman Ramsey [Nobel Prize, 1989] at Harvard. Because Dan kept working predominantly on the hydrogen maser, which he co-invented with Ramsey, some in the referee community could not rave about his great scientific originality, so he didn't get tenure at Harvard. I had that lesson in front of me when, after being a graduate student with Dan (and Norman Ramsey) I got an assistant professorship here at MIT. (As I like to kid Norman, this gave me the best of two worlds — an MIT thesis and a Harvard degree.) If I had done the same kind of hydrogen trapping and fundamental constants experiments as Dan was doing, I would have had the same shadow problem. It turned out to be an easy thing for me to avoid because I had proposed and then started doing atomic collisions experiments as a graduate student. MIT had hired Dan to build up AMO physics at MIT and in those days you built up by appointing your best students. That was very much the American style prior to 1970. So I stayed on here as an Assistant Professor and I had the collisions machine that I'd built for my thesis, but I didn't have any dowry. These days you get half a million dollars as a start-up package. In those days you were just apprenticed in the lab with a senior professor who had lots of money from the government. I was in that position vis-a-vis Dan. After a couple of years I realized that I wasn't going to get tenure doing collisions. At that point the dye laser had just been invented, and I got the second single mode dye laser that Spectra Physics ever made and quickly did some novel experiments that got me tenure.

What was Laszlo Tisza's (see elsewhere in this volume) contribution to the intellectual preparation for the BEC?

There had been a debate about whether BEC was an artifact in the mathematics or represented reality. Then London had suggested in 1939 that helium-4, not the liquid helium but the superfluid helium, was a BEC. Tisza did the two-fluid model of that. Helium has turned out to be BEC, but a very strongly interacting one; it's only about 10 percent pure BEC, whereas the atomic one is 99.9 percent pure.

What is your background?

My Pritchard ancestors were Welsh miners and my grandfather was the first generation born in this country. They lived in upstate New York. My mother's father was related to Ethan Allen who was a patriot in the American Revolution — his private army, the Green Mountain Boys, captured Fort Ticonderoga. Then they brought the fort's cannons over the ice to a hill next to the harbor here and drove the British fleet out of Boston. Being from Vermont, technically he was not an American at that time. Most of my other ancestors on my mother's side came from French Canada. I was raised on the East Coast of the United States; I spent my formative years, age 4 to 11, in Milton, Massachusetts.

Is it where Buckminster Fuller was born?

That may well have been. It's also where Quincy Adams and John Adams lived; it's an old New England town. Then we moved to New Jersey where I went to high school. I went to Caltech when I was 16 and graduated at 20. I came to Harvard to graduate school and have stayed around here ever since. My parents were the first generation to have gone to college. My dad had gone to MIT and he graduated at the depths of the Depression and stayed on and got his Ph.D. He was a concert-level pianist, and he was good with languages. He won an MIT fellowship and went to the Sorbonne and Goettingen for a year — each of those two places for half a year. He took courses with Franck and Brillouin and toyed with the idea of becoming a physicist, but then, being from a working-class family, he decided to be an engineer and wisely chose to do electronics rather than power engineering. I learned a lot of science from him at the dinner table. My Mom tells me that I was a demanding student. Once when my Dad admitted that he couldn't answer one of my questions I told her he was dumb and asked where I could find smarter people to answer my questions. She suggested his teachers at MIT, so that's one reason I'm deeply happy to have finally made it here.

You have been an excellent mentor.

At MIT we have a mentoring system. So even when Wolfgang was no longer my postdoc and became an independent Assistant Professor, I was designated officially to be his mentor.

You are not any more, I suppose.

Not officially anyway. Once Wolfgang was promoted from Assistant Professor to tenured full professor (in one fell swoop), at that point we started collaborating again with a more equal relationship. But we always talked, and I hope always will. When I call my wife saying that I am coming home, she asks whether Wolfgang is still at work. She knows that I walk by Wolfgang's office on my way out to my car, and if he is still in the office, there is a good chance that I will be half an hour later than I said.

You could have become a Nobel laureate, but did not and you don't sound bitter about it, which is very rare in similar circumstances in my experience.

The Nobel Prize would really crimp my style, which is that I like to investigate anything, for instance the UFO Alien phenomena. Right now I am working on electronic education. Sometimes it may be advantageous to have the authority of a Nobel Prize, but many times people would take me too seriously. In the Alien business it would just have been tremendous news that a Nobel Prize-winner was looking at the phenomenon. The assumption is that because you're looking at it, you believe in it. People come up to me and say, "You are a famous scientist, do you really believe in Aliens?" My response is, "I don't believe in Aliens, I believe in the scientific method,

I believe it's a tenable hypothesis that there are Aliens but it's an issue that has to be decided on the basis of objective (i.e. scientific) examination of experimental evidence."

It is assumed that if there are other places with intelligent life in the Universe, they are probably older than our civilization.

The idea is that they wouldn't be phase-locked to us temporally, and some of them would be older. We're talking about a 13-billion year Universe in which planets, like the Earth, could've existed 5 billion years after you had some supernova to make heavy elements. If there is other intelligent life, you would expect them to be many millions of years separated from us as a matter of random occurrence. So some would be technologically ahead of us, and they would be the ones capable of contacting us if they wanted.

Somehow this implies an optimism that a place like the Earth would not blow itself up in the course of million years of intelligent life.

I'm concerned about that and would expect it to happen on a much shorter time scale than a million years if it does. The question, "Can we find communications, artifacts, or visitors from other intelligent civilizations?" requires only *some* high-tech civilizations to survive and importantly to stay interested in other intelligences and/or in cosmic exploration. With respect to the Aliens, my research question is, is there physical evidence that Aliens are visiting the Earth?

Is there any evidence?

I didn't find any existing claims that met scientific scrutiny, and I couldn't find any physical evidence myself. Folklorists find many more consistent story elements in abduction reports than in normal human oral tradition, i.e. in urban myths, fairy tales, etc. One of these consistencies is that implants are put in prepubescent human beings by the Aliens. I've managed to get hold of one of these alleged implants from a person who could document telling his abduction experience about 10 years before these stories reached the public media, and had a 10-year old doctor's report of a small object being under his skin. He had described how the implant was put into him on a ship and that he could watch the procedure on a big television screen that showed that it had "wires" sticking out of it. He ultimately pried it out of his skin and brought it to me here (at MIT). The alleged implant had microscopic filaments sticking out of it, so I thought it would be interesting to do elemental analysis. I looked at it by myself without any help by students, because I didn't want anybody to get a bad reputation (except me). It was made of elements common in living things. Ultimately, with the help of a dermatopathologist from Massachusetts General Hospital, we determined that it was human damage material — possibly from an ingrown hair — with cotton fibers from the man's underwear that got stuck in it and that looked like wires. This explanation was consistent with everything that I could see and measure about this physical object. It strongly suggested a mundane origin, but the guy who brought it to me is still convinced that it was put into him by Aliens when he was 8 years old. He has unpleasant flashbacks to this alleged experience like victims of post-traumatic stress disorder.

The remaining question is what happened to him, or inside his head, that generated these memories? At the present time I place the abduction stories more likely in the category of spiritual experiences than encounters with physical Aliens. More scientific study of spiritual experiences is warranted in my opinion because in their various forms (encounters with religious beings, past lives, near death experiences, out of body experiences, Alien abductions, etc.) they have been experienced by a significant fraction of normal adults who have real anxiety about these deeply-affecting memories. But it's probably not a job for physicists.

In a way this is also a mentoring activity for a broader constituency than just your students. May we return to your mentoring your students?

As part of my mentoring I talk with my students about what experiments we should do. If you're a good scientist, you can think of more than one experiment to do. How do you figure out which one to do? You look at the scientific payoff, estimate the probability that it'll pay off for you (before one of your competitors), and then you divide by the effort. You have to maximize that ratio. Incidentally, if you can apply this approach to the Aliens question, you realize that discovering that there are other intelligent beings, that they are present on the Earth, and then discovering that they are doing experiments on human beings, that'd be equivalent to two or three Copernican revolutions. So you have a huge payoff multiplied by what realistically has to be a very small probability of success. So it's infinity times zero, how do you evaluate that? You put in some effort and when it isn't paying out or looking promising, you go elsewhere and do something else. Now I'm channelling that effort (and more) into a web-based tutoring program (www.myCyberTUTOR.com) and into developing interactive content with it for introductory physics. I hope this will make a big impact on the world, and a more positive one than discovering Aliens.

How much effort did you put into it?

A few hours a month for a couple of years. In addition, I organized a conference on the subject with Dr. John Mack, out of which came what is still the definitive book on the Alien abduction experience, called *Alien Discussions* (ed. Andrea Pritchard *et al.*). It was a very interesting conference because conversations between sociologists, historians, psychologists, medical doctors, scientists, folklorists, and skeptics tend to be more wide-ranging than those between physicists. Everything was put into the book, which you can read, including the post-presentation discussions. But there aren't any conclusions because, obviously, opinions differed. There were people there who think that Aliens are real and others who think that all this is absolutely false, and there were others in between.

Was there any religious component to the meeting?

There were ordained ministers and at least one religious scholar there too. Some feel threatened; others would say that if God made Aliens they would be governed by His laws too.

How do you look at religion?

I am pretty much agnostic. I think that there is a spiritual dimension to human existence. I don't meditate, but I've been to a couple of weekend Shaman schools, and I have gotten some insights into some human issues in my life via that route. It's divorced from my scientific side. I have no place in my life for organized religion. My parents drove me to and from Sunday school, but didn't go to church.

American scientists find it more difficult to speak about their being atheists if they are than European scientists.

We have some puritanical religious roots here. Religion was incredibly important to the founders of this country; that was one reason they came here in the first place.

Let's return to mentoring, again, if we can. I wonder if you ever thought of putting together a scientific genealogy in your line of research. There seems to emerge a distinct line of Ramsey-Kleppner-Pritchard-Ketterle.

That's right. The line started with Stern, then Rabi, who was Ramsey's advisor. You can also include Eric Cornell, Bill Phillips (who was Dan's student and my postdoc here), and Carl Wieman (Dan's undergrad thesis student and my crewmember).

Whom would you expect to be the most likely to open a new field?

Eric Cornell is the most widely-ranging, creative person of the lot. He is also the youngest, which is another advantage. His Nobel Prize has affected his life the least, and he is intrinsically convinced that taking chances will pay off. Wolfgang might take a project like LIGO (Laser Interferometric Gravitational Observatory) and turn it into a Nobel Prize winner.

Will Eric survive the Nobel Prize?

He already is practiced at this. He was consciously quite happy to let Carl Wieman give the lion's share of talks on their work while Eric stayed home working on his experiments and enjoying his family.

How do you compare the creative atmospheres at MIT, which has a formidable reputation, and the Colorado lab?

Norman Ramsey and David Pritchard (courtesy of D. Pritchard).

The Joint Institute for Laboratory of Astrophysics is a very high-quality collection of people who have a special relationship in that institute. It's also a place that has succeeded in attracting a great many excellent people. They have a very fortunate situation in that NIST puts money into it and there are permanent NIST positions like the one that Eric Cornell has. He doesn't have to teach if he doesn't want. He can concentrate on research full time. Bill Phillips has a similar position at NIST in Gaithersburg.

MIT is a private institution whereas NIST is a government organization. Why would the government invest in such research that could be done at a private institution?

MIT is not doing it without government money. Most of the money to do research here comes from the government. Work being done in this hall is funded by NSF, the Office of Naval Research, the Army, the Department of Defense, NASA, and the Packard Foundation. MIT puts up money to start a new faculty member, or in case of Wolfgang to keep him here. MIT's philosophy is to start with young people, to back young people and give them resources; young people who, hopefully, prove out. Then, what tends to happen is that institutions that have big endowments (and a larger ratio of endowment per faculty than MIT does), like the Scripps Oceanographic Institute or Harvard, will offer an MIT person in his early forties a big incentive and steal him or her away.

Can you survive that challenge?

So far. Our strength is that we train outstanding young scientists. Incidentally, in addition to the four Nobel laureates whom I have mentored, I have also mentored two national thesis award winners, and a couple of others have also come from our group. Speaking about our group — Kleppner, Ketterle, and myself — we have an exceptional AMO group (ranked first by *U.S. News and World Report*). We have created the Center for Ultracold Atoms with physics faculty members from Harvard, and this super community is very stimulating for students, faculty, and visitors alike. I really hope and expect to be able to attract and mentor talented and inspired young people like Vladan Vuletic and Ike Chuang into this group to replace Dan and me over the next decade.

Wolfgang Ketterle, 2002 (photograph by M. Hargittai).

19

WOLFGANG KETTERLE

Wolfgang Ketterle (b. 1957 in Heidelberg, Germany) is John D. MacArthur Professor of Physics at the Massachusetts Institute of Technology (MIT) in Cambridge, Massachusetts. He received his Diploma in Physics (equivalent to a Master's degree) from the Technical University of Munich in 1982 and received his Ph.D. degree from the Ludwig Maximilians University in Munich in 1986. He did postdoctoral work at the Max Planck Institute for Quantum Optics in Munich-Garching, at the University of Heidelberg, and at MIT and joined the physics faculty of MIT in 1993. He shared the Nobel Prize in Physics for 2001 with Eric A. Cornell (b. 1961) of the Joint Institute for Laboratory of Astrophysics (JILA) and the National Institute of Standards and Technology (NIST) in Boulder, Colorado, and Carl E. Wieman (b. 1951) of JILA and the University of Colorado at Boulder "for the achievement of Bose-Einstein Condensation in dilute gases of alkali atoms, and for early fundamental studies of the properties of the condensates."

Wolfgang Ketterle received the I. I. Rabi Prize of the American Physical Society (1997), the Fritz London Prize in Low Temperature Physics (1999), and the Franklin medal in Physics (2000), among other awards. He is a Fellow of the American Academy of Arts and Sciences, a member of the European Academy of Arts and Sciences, and other learned societies. We recorded our conversation on March 11, 2002, in Dr. Ketterle's office at MIT.*

---

*Magdolna Hargittai conducted the interview.

First I would like to ask you about your family background.

I was born in 1957 in Germany, I have an older brother and a younger sister. My mother ran a small business selling first-aid products. She inherited this business from her father-in-law. However, for most of the time she stayed at home bringing up the children. My father was the administrative director in an oil and coal company. My brother is a tax consultant and certified accountant, and my sister is a high school teacher. I am the first scientist in the family.

How did you become interested in physics?

I liked to construct things, with Lego, for example. I fixed things at home, like lamps and small appliances. I took things apart and then put them back together. I had an electronics kit and a chemistry set and I did experiments in the basement and I had fun with it. When I first thought about a profession, I considered becoming an architect. Later, I got more interested in mathematics, and finally concluded that physics was a good combination of abstract concepts and the real world. I still like to construct things. I built major pieces of equipment for my experiments. What is different from being an engineer is that we build unique apparatus to explore new science.

When was your first encounter with the problem whose research then led you to the Nobel Prize-winning project?

The first encounter with ultra-cold atoms came when I was working in physical chemistry, but I wanted to change my field. I wanted to get into a more fundamental area. I looked around, talked to people, studied conference proceedings to find out what the hot topics were. Cold atoms was a hot topic. At that point I applied for my third postdoctoral position, which brought me into collaboration with Dave Pritchard at MIT, and in the summer of 1990, I became his postdoc.

Earlier, I had done my Ph.D. and some postdoctoral research in molecular spectroscopy, but then I was at crossroads. I wanted to do either something more applied, or something more fundamental. At that point, I decided to go into a more applied area, into physical chemistry, and to pursue combustion research. This was my second postdoctoral appointment.

Please, give a summary of the Bose-Einstein Condensation (BEC).

Bose-Einstein condensates can be regarded as the coldest matter that man has ever produced. It may even be the coldest matter in the Universe.

Matter at those temperatures has very special properties. It behaves like no ordinary matter. The special properties of this very special matter can be best described by using a comparison with the optical laser. If you compare a light bulb to a laser beam, the difference is very striking. The light bulb emits light in all directions, that is, many electromagnetic waves go from it in all directions. In the laser beam the light is directional; it's just one single wave. There is a similar difference between an ordinary gas and the Bose condensate. In an ordinary gas, the atoms move around independently, in all directions. In the Bose condensate, all the atoms march in lockstep; it just forms one big wave.

Why is it called a different state of matter? What is it closer to, to an ordinary gas or to a solid?

It's a very good question. Honestly, it's a little bit arbitrary to classify the states of matter into gaseous, liquid and solid. Some people add plasmas as a fourth state of matter, and now, as a fifth state, the Bose condensate. In our everyday life, the classification of matter into gases, liquids and solids, makes a lot of sense. Physicists have seen such a remarkable variety of forms of matter, and it is arbitrary, which forms of matter we should call a new state of matter. To classify the Bose condensate, I would adopt what Fritz London had suggested and regard superfluids as a new state of matter. In this state of matter, all the particles act coherently. This state of matter would include super-fluids, liquid helium three and liquid helium four, it would include superconductors, and it may even include the optical laser as a special case, and now it includes the gaseous Bose-Einstein condensates. So this fifth state of matter would include all the forms of matter in which all the particles act coherently, that is, constitute one wave. We should not say that the Bose-Einstein condensate was the first representative of this kind, because it belongs to the class of superfluids and superconductors. Considering its density, it's a very dilute gas. If we want to distinguish between the forms of matter on the basis of density and consider low-density matter a gas and high-density matter a solid, then the Bose condensate is a gas. It's a superfluid gas.

What are its other physical properties, like viscosity?

This is a subtle and interesting question. The viscosity of the Bose condensate can vanish under certain circumstances and this is the property of superfluidity. If you stir it at high velocity, it just behaves as an ordinary gas.

I've read that Laszlo Tisza (see next interview) had a paper in the late 1930s, in which he suggested that there is a connection between superfluidity and the Bose-Einstein condensate [Tisza, L. "Transport Phenomena in Helium II," Nature 1938, 141, 913].

There were two papers in 1938 in *Nature*, one by Fritz London [*Nature* **1938**, *141*, 643] and another one, a few weeks later, by Laszlo Tisza. These were the first papers connecting the recently observed superfluidity in liquid helium to Bose-Einstein Condensation.

Tisza was a professor at MIT, now retired, but we met with him in Budapest in 1997 when he was 90 years old and he was alert and very interested in science.

He's still around. He will be the featured speaker of our seminar on ultra-cold atoms next month. It's absolutely amazing that there is someone who was involved in one of the first two papers in the field and 60 years later he is still there. He can walk into my office and tell me about those days with vivid memories. I see him every few months. He is already 95 years old, but there are people who are beyond age.

How do you observe the Bose-Einstein condensate?

We observe it by shining light on it and then we observe the scattered light or we observe the absorbed light.

By the naked eye?

Almost by the naked eye because it's visible light and we record it with an electronic camera. It's like having a TV camera and observing the result on a screen. We use an electronic camera to achieve good spatial and temporal resolution.

Is the Bose-Einstein condensate a quantum matter?

It's a new form of quantum matter. It's one of the few examples in physics where the laws of quantum mechanics manifest themselves on a macroscopic scale. It's a macroscopic system with a size comparable to the thickness of the human hair, sometimes with an extension of even a millimeter, and it's completely governed by the laws of quantum mechanics (note: the human hair is about a tenth of a millimeter).

Is it unique, so far, in this respect?

It shares such properties with other superfluids and superconductors, but is distinguished by its very low density.

How do you achieve these extreme cold temperatures?

You have to build special refrigerators. We use laser cooling and evaporative cooling. In principle, it is very simple to obtain Bose-Einstein Condensation. Take a gas of bosonic atoms and cool it to a very low temperature. The challenge was to develop cooling methods. We reduce the temperature of a gas from about room temperature down to micro- and nano-kelvin. This is a temperature drop by a factor of a billion. There is no single scheme to achieve this and we had to apply different schemes of refrigeration. We used several methods of laser cooling, and then evaporative cooling.

How can you avoid condensation of the gas at such low temperatures?

We achieve that by keeping the gas at extremely dilute densities. Even at room temperature, the most stable state of sodium is not to form a dilute gas, rather, to condense into a solid form, a piece of metal. The condensation process into a solid requires that several atoms meet each other and form little clusters, little droplets. This process is very, very slow at low densities. It is the low density that enables us to have a gas in equilibrium as a meta-stable form of matter, but the most stable form of sodium would be to form a piece of metal.

When you joined David Pritchard's group in 1990 at what stage of the project were they?

When I came, the Pritchard group was just in the process of building up a new experiment for laser-cooled atoms. The goal was to study cold collisions and the formation of cold molecules.

How then did the Bose-Einstein Condensation come up?

Soon after I joined the group, I was drawn to consider the current limitations of temperature and density. Dave Pritchard was a pioneer of the field, he started it in the 1980s and he and I resonated on those questions. We were brain-storming on new ways to get colder and denser, and we published two theoretical papers on the limitations of cooling and trapping particles.

It was on our mind to push for the limit. On the other hand, we wanted to build up a laser cooling experiment, which would provide an intense source of ultra-cold atoms to study cold collisions and the formation of molecules. This was our experimental agenda. When we were building up this experiment, we tried to build it very well and we tried to get around certain limitations. In this process Dave and I had discussions about the limitations of laser cooling and we tried to push laser cooling to higher densities. Then we had an idea, which is now called the dark spot trap, a modification of the standard light trap, which avoids certain density-limiting processes in laser cooling. We put the experiment together and it worked incredibly well. In the summer of 1992, we had a unique combination of high density and high atom number. Even before we wrote up these results and published them, we started discussions about the next step. What can we do with this very special sample of dense atoms? One suggestion was that we would now be in a very good situation to pursue our original research agenda to study cold collisions and the formation of cold molecules. This is something that Dave had worked out in great detail and he had very good ideas. Several groups became successful in working along those lines in the following years. But there was then this other suggestion, which was put forward mainly by me and the graduate students. We thought that we could go for the Bose-Einstein Condensation because the density of our sample was very favorable to start the second stage of cooling, the evaporative cooling.

It may be a short excursion, but I should say that the challenge to getting to BEC was the combination of two cooling schemes. One was laser cooling, which works at low density and evaporative cooling, which works at high density. For many years, people used to say that there is a gap, laser cooling stops before evaporative cooling can be started. With this new idea of the dark-light traps, we felt we had closed the gap. We felt that with a little bit more of engineering, we could start the second stage of cooling, the evaporative cooling.

After we had realized the dark spot trap, there were weeks of intense discussions. For me, it was a moment of absolute greatness on Dave's part, when he gave up the ideas he had worked out on cold collisions, and supported the idea of getting evaporative cooling to work. He encouraged me to go for something that was fairly speculative and risky. We immediately ordered equipment for the next stage of cooling and only after we had placed those orders, we wrote our paper on the dark spot trap. When we pursued evaporative cooling, we first did not state Bose condensation as our goal because we were still six orders of magnitude away from the desired density. We did not want to attract too much attention by stating such distant goals. Instead, we wanted to stay down to earth and focus on the next step, which was the merging of the two cooling schemes. We knew what the merging of the two cooling schemes might enable us to do.

Did Dave Pritchard stop his participation?

He was still involved. We were still one team and Dave was still my postdoctoral advisor. It was only in the spring of 1993 that MIT considered making me an offer for a faculty position. Earlier, I had applied for jobs both in Germany and in the United States and I had offers from Berkeley and later received another offer from Chicago. When an offer at MIT was being discussed, Dave made this surprising suggestion that if I would stay at MIT, he would get out of the field of laser cooling and ultra-cold atoms, because he wanted to avoid the problem for me as a younger person to be in his shadow.

I liked MIT and I appreciated Dave Pritchard's mentorship, and Dan Kleppner's too, so I decided to stay. I still can't fully understand what Dave did for me, it is unprecedented. Dave handed me over his lab and the laser cooling apparatus, and told me just to go and run with it. He told me that I could still ask him for any advice, but I was independent to make all decisions, and we won't publish the results together. He gave me full independence, but he also gave me a head start. It was unbelievably generous. When people start as an Assistant Professor and build up their own lab, if they are fortunate, after two years they may have built up some complicated machinery and may publish their first paper. I was an Assistant Professor for two years and I was the co-discoverer of the Bose-Einstein Condensation. I know that I couldn't have done it without the support of Dave Pritchard. In fact, when we did accomplish the Bose Einstein Condensation, I invited Dave to be a co-author of our first publication, because I regarded it the culmination of our previous collaboration, but he decided to stay off it.

Did he move on with the original idea of forming cold molecules?

Actually, he didn't. He was working on three experiments, ion trapping, atom interferometry, and laser cooling. He completely stepped out of laser cooling. For the next few years he did not do any work on cold atoms. He focused on his other two experiments. Five years later, after I'd been established, we got engaged in a collaboration, now as two equal partners, joining my expertise in Bose-Einstein Condensation with his expertise in atom interferometry, and we have been very successful in our joint projects.

It seems as if he had just stepped out of the most exciting part of it. Did you talk with him in retrospect whether he had any regrets?

He once used this expression, "I gave him the key to the family car because I knew he could drive it faster than I could've driven it." I have never heard any word of regret from him. He has always taken a fatherly pride in my accomplishment. He knows that he has done a lot for MIT and he has done a lot for atomic physics, and he has done a lot for me. The leader of the first group that realized Bose-Einstein Condensation, Eric Cornell was his student and the leader of the other group was his postdoc. So he influenced the field in a major way.

He must be an exceptional person, because the Nobel Prize is the ultimate recognition in science and having no regret seems to be superhuman. Was your goal winning the Nobel Prize?

[After a long silence] You see me hesitating what to say because the Nobel Prize was never my goal. My goal has been excellent science, which I have enjoyed doing. Just to be in the midst of exciting results, living through such exciting times was ample reward. The Nobel Prize recognizes that and it needs a combination of luck and other things. I would never

Dan Kleppner, Wolfgang Ketterle, Thomas Greytak, and Dave Pritchard with their experimental apparatus at MIT (courtesy of W. Ketterle).

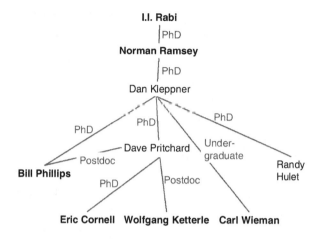

Wolfgang Ketterle's scientific "family tree". The names of Nobel laureates are printed in bold (prepared by and courtesy of W. Ketterle).

recommend to any scientist to aim for the Nobel Prize and I did not have it in my mind either. There is more Nobel-Prize-worthy work than can be rewarded and the quality of my work would be the same if I hadn't gotten the Nobel Prize for it.

As I understand, your relationship with Dr. Pritchard was part of a longer chain of similar relationships. Recently I talked with Norman Ramsey who told me about his former pupil Dan Kleppner, who in turn was at one time Dave Pritchard's mentor. Did you have any interaction with Dr. Kleppner?

Of course, I am very close to Dan. The pursuit of Bose-Einstein Condensation of diluted atomic gases started with working on polarized atomic hydrogen in the late 1970s. The leading groups were a group in Amsterdam and a group at MIT led by Tom Greytak and Dan Kleppner. It was their efforts that put Bose-Einstein Condensation into the minds of researchers. For a whole decade this work focused on atomic hydrogen gas. In the end of the 1980s, with the advent of laser cooling, there were different atoms that could now be used to pursue ultra-cold atomic physics. But the original work on atomic hydrogen directly effected or even triggered the search for Bose-Einstein Condensation in the alkali atoms. I knew all the preceding work, I had read all the papers, and it was an enormous source of inspiration. For me, there was even more direct inspiration because Dan Kleppner is here, his office is in this hallway, and we often talked to each other.

It must have been a difficult decision for the Nobel Committee to choose the three persons to award the prize for Bose-Einstein Condensation.

It must have been because if we consider the people with direct contribution to the discovery they had to choose between the Boulder group, my group here at MIT, and the group of Tom Greytak and Dan Kleppner. These groups were the major contributors. There must have been a very serious discussion about how to select three people from these groups. I don't want to comment on whether they made the right choice or not, but the person who would've deserved it and did not get it is Dan Kleppner. He worked on it for twenty years, he is a pioneer, but the Nobel Committee has made its decision to recognize the spectacular breakthrough and not the early beginning. They have done this also in the past.

Did you ever discuss this with him?

No, I didn't. But I know that he is such a modest person that he would never claim anything for himself.

How did you decide on using sodium?

The alkali atoms can be easily laser cooled. So all the standard laser cooling methods had been developed with alkali atoms. Other elements require more complicated laser equipment or are harder to evaporate. The alkali metals are easy to evaporate and produce a usable vapor pressure at rather low temperatures. Besides, the alkali atoms have one unpaired electron, which is easy to excite. With infrared or visible lasers the alkali atoms can be brought into their first excited state whereas other atoms, which have more tightly-bound electrons, require ultraviolet lasers and it could be a major effort to build such lasers. You first try out the easy candidates, and in our case, these were the alkali metals. I picked sodium, because sodium has been the workhorse in atomic physics for decades and Dave used to work with sodium and I took over some of his equipment. The group in Boulder was setting up new labs and they decided to use first cesium and then rubidium, because they can be excited by diode lasers, which have become available. The choice of element mainly reflected the available technology.

As I understand it, Kleppner's group succeeded in achieving Bose-Einstein Condensation with hydrogen not very long after your successful experiments with sodium.

Yes. This is a heroic story of an effort, which continued for over twenty years, with a lot of persistence and focus, and which finally succeeded. It's a great example.

Please, don't think that my main interest is in the Nobel Prize, but one can't help wondering about the contrast of one group having worked for twenty years on the same problem that the other group then solves in the course of a few years.

It is this contrast and the surprises, which makes science exciting. In 1978, the group of Greytak and Kleppner picked the only system that in those days offered the possibility and that was atomic hydrogen. Their success in 1998 showed that they were right, that it was possible indeed. It was a fantastic experiment. However, while they were pursuing this work, some new development came along that provided an easier approach to the same scientific goal, and that was what I used.

Let me add here something that fits very well into this point. When I was hired as an Assistant Professor in 1993, I had the following expectation. I knew about atomic hydrogen, and they were within one order of magnitude from the temperature necessary for BEC. At that time, with laser cooling, I was still away by six orders of magnitude. It seemed absolutely clear to me that in the next few years hydrogen condensation would be pushed

Wolfgang Ketterle receiving the Nobel Prize from the King of Sweden in the Stockholm Music Hall, 2001 (photograph by M. Hargittai).

over the edge and those people would succeed. I felt that I was working on something that was further away from the same goal, and it was also unknown whether it would work or not. I was optimistic though, I thought that in the long run, I would get it to work, and then I would have a new system to explore the rich physics of Bose condensation. I never had any expectation that I would reach the goal earlier than the hydrogen people. They were very close, while I expected to take about a decade or longer to accomplish Bose condensation in alkali atoms. But I felt that focusing on such a big goal would bring out the best in me and I would work at the frontier of the field and every obstacle that I would find and eliminate would be interesting science. It was to my great surprise that there were no further obstacles. We built the apparatus and got it to work and accomplished six orders of magnitude increase in phase-space density in a very short time and achieved Bose condensation.

It was a piece of luck, or you may say that Nature was kind to us. We know now that the sodium atom and also the rubidium atom, which was Bose condensed in Boulder, have almost ideal properties. In hindsight, the achievement of BEC only required the complete dedication of some research groups, who put all their resources to Bose-condense rubidium or sodium, and finally it worked. There were little tricks and combinations of techniques, and modifications to existing techniques, but no major new idea was needed. The combination of laser cooling and evaporative cooling was the crucial breakthrough.

You have said that there was a gap between the two techniques of cooling. How did you bridge the gap?

When I was a postdoc in the early 1990s, I considered combining the two but I was told that there was a gap. Most people were skeptical or felt that it was impossible. In collaboration with Dave, I pushed laser cooling to higher densities and this helped closing the gap. In hindsight I should say that the gap wasn't really there because evaporative cooling works even at low density, but it is extremely slow. If you have a sufficiently good vacuum, you can wait for the long time that is needed for atoms to collide, and this drives the evaporation process. In the end, the gap was closed with different technologies; our contribution was the dark light traps, which were crucial for us and also for the Boulder group. But it was just one way to close the gap.

How large is your research group?

There are about 20 people, 12 graduate students, 5 postdocs, 3 undergrads. Several of these people are jointly supervised by Dave Pritchard. When we built our first Bose-Einstein Condensation experiment, it was only me and three graduate students. So things have been quite dynamic. In the early days, there was just one big project and we did everything together. When we had vacuum problems, we worked together on the vacuum. There were evenings when we all had to take turns in tightening the bolts because it was physically exhausting.

Do you discuss problems on a one-to-one basis or do you prefer group discussions?

We usually have four to five people working on one experiment; they form a team, and I try to avoid individual discussions so as not to leave out the others in the team. I meet every week with each lab. We also have a weekly plenary meeting where all the labs are together. Of course, the larger the group, the less I can get involved in the details of every project. I try now to leave the management of the groups and the day-to-day activities to the people in the labs. It's different from the time when there was only one lab.

What are your current projects?

We are exploring the possibilities of ultra-cold gases for further research. There is a vast research agenda opening up. Bose condensation has given us atomic physicists access to novel samples of atoms with unprecedented properties, and more recently those cooling techniques have been applied also to ultra-cold fermionic gases. We have been involved in several interesting directions of what I would call ultra-dilute condensed matter physics. We have carried out experiments on vortices, superfluidity, and phonons in a Bose condensate. In those experiments, the gaseous Bose condensates share properties with solids; this refers back to an earlier question of yours. We try to explore this aspect. We have another experiment in which we place the Bose condensate on a micro-fabricated surface with wires, which magnetically guide the atoms. More colloquially these are called atom chips. With these experiments we are working on guided matter waves and atom interferometry on a chip. This is in collaboration with Dave Pritchard. Then in another collaboration with Dave, we try to use falling condensates for precision atom interferometry. One goal is the more precise measurement of the fine structure constant. In summary, my current research program is a wonderful combination of cold fermions and cold bosons, condensed matter physics, atom optics, and precision measurements.

As I understand, the fermionic atoms can't make a Bose-Einstein condensate. What happens to them in these ultra-cold situations?

Let me start with a more general picture. At higher temperatures when the de Broglie wavelength is much shorter than the spacing, all gases behave classically. When you cool down gases, the atomic matter waves start to overlap when the de Broglie wave of atoms is comparable to the spacing between the atoms. Then the gas no longer behaves classically, it develops new properties. That's also when the difference between bosons and fermions becomes manifest. When you reach this point, you either form a Bose condensate with all its special properties, or you form what is called a degenerate Fermi sea, which also has properties very unlike a classical gas. So there are lots of things to investigate. The simple properties of the Fermi gas reflect the Pauli exclusion principle, two fermions cannot be at the same place at the same time. There are profound implications mainly in the suppression of light scattering and particle scattering. This is something worth exploring, but these are single particle effects whose theory has been worked out. What is even more exciting is to study interacting fermions. When fermions interact, there is a possibility that fermions form pairs, and pairs of fermions have bosonic symmetry. Those fermion pairs can undergo a transition to superfluidity, which you may call the Bose condensation of fermion pairs. This is a current frontier of many-body physics. This would be truly exciting and open up years of truly thrilling research. We don't even know if it can be accomplished, but this is the most demanding goal with cold fermions.

You have told me about basic research. What are the possibilities for application of Bose-Einstein Condensation?

Bose-Einstein condensates may find applications in precision measurements. When you cool atoms to a standstill, when they are not running away any-more, you can make more precise measurements. This has been one of the major motivations behind laser cooling and Bose-Einstein Condensation. I would expect that in the near future Bose-Einstein condensates may improve the precision of atomic clocks and may also be used in the next generation of atom interferometers, which can be called matter wave sensors, used for sensing gravitation and rotation. Bose-Einstein Condensation is a quantum object; it's quantum mechanics on a mesoscopic scale and this may contribute to nanotechnology indirectly and maybe even in a direct way. Finally, when we learn something about coherence in a Bose condensate, it may contribute conceptually to the field of quantum computation. There are ideas to build the next generation of computers based on quantum mechanical coherence. I personally doubt whether we can use the Bose condensates to construct the quantum computer, that the Bose condensate will ever be used as a practical computer, but the system is so pure and so well controlled that we may elucidate some of the concepts and contribute to advances in this field as well.

What would be the significance of having an atom laser rather than a light laser?

The atom laser is a different kind of laser. The optical laser emits light, whereas the atom laser emits atoms. I can state very generally: if you have a demanding job with light, you don't use light bulbs, you use a laser beam. If you have a demanding job with atoms in the future, you may want to use an atom laser beam, or atoms extracted from a Bose-Einstein condensate. They have properties that are superior to ordinary atoms. It's not that the atom laser would replace the light laser; they are two different things. For me, the laser represents absolute control of light because the photons are put into one single wave, into a single mode of a cavity. Similarly, atom lasers and Bose condensation means that we are controlling atoms at the quantum level. We can prepare a sample of atoms in which the atoms are represented by just one wave function. Atom lasers may be superior devices if you want to deposit atoms on a surface, if you want to focus atoms to a pinpoint, maybe for a new kind of atom microscopy.

Is this the line of research you are currently pursuing?

We currently exploit the properties of atom lasers in atom interferometry for precision measurements.

Could you single out the scientific result that you are most proud of?

It is hard to decide between two. The two I am most proud of were the accomplishment of Bose-Einstein Condensation in 1995 and then, about a year later, when we saw the interference of the two condensates. These were absolutely exciting moments and special contributions to physics.

Is the Nobel Prize changing your life?

Fortunately, not a whole lot. I still find time to do research and to have discussions with my group; I still have time for my family. People told me after the Nobel Prize had been announced that other people would take control of my life and I won't be able to get anything done for a whole year. I was scared and concerned, but it did not happen. It takes only a fraction of my time to answer inquiries, to give talks and to give interviews. I can still dedicate most of my time to the things I care about.

Are you still as much interested in science as before?

Yes, I am.

You are an experimentalist. Is there a serious divide between experimentalists and theorists?

Yes and no. There is a divide in certain areas where both experiment and theory are very specialized and it is impossible for one person to master both. I certainly promote closeness between experiment and theory. I was trained as a theoretical physicist, I did my Master's thesis and first publication in theoretical physics. My present work is purely experimental, but sometimes we stumble over some concepts in the course of our experiments and then we write papers that are purely theoretical. So I still practice the combination of the two. There are aspects of the theory of the Bose condensates that are quite sophisticated and require special mathematical tools and special knowledge. In those cases, there is a divide.

The separation of basic science and applied research seems to be somewhat blurred in your field.

Not really. My area is mainly basic science, but sometimes it is only one step from basic science into applied science. If you consider atomic physics, the laser was developed as purely basic science, but then it has found lots of applications. The development of the atomic clock in which Dan Kleppner was a pioneer happened just as fundamental physics. Later on it formed the heart of the global positioning system. Bose-Einstein Condensation may play a role in gravitational and rotational sensing. This may be relevant for geological explorations and navigation, but we are too early in the game to be sure that it will ever materialize. In my heart, I'm a fundamental physicist and I try to do what I'm curious about, what I consider to be the frontiers of science. At the same time I find it satisfying that in this pursuit we're sufficiently close to applications, that we may have a broader impact than just impacting pure science and new knowledge.

I would like to ask you a little more about your mentors. You have mentioned Dave Pritchard. Were there others?

I should mention my mathematics teacher in high school, Mr. Strobel. He challenged me and instilled in me a playful approach to solving problems. His approach was to play with numbers, make a drawing, just get a feel of the problem, rather than using a recipe to solve the problem. That's an important approach in science. He also encouraged me to take part in some special mathematics competitions. I had to solve problems where at first I had no idea how to solve them, but I had to get into it and develop ideas. Then I have been privileged that all the advisors I have had in my career have been special and I could learn unique things from each of them, Professor Wolfgang Götze for my Diploma work, Professor Herbert Walther for my Ph.D. thesis, and Professor Jürgen Wolfrum and Professor Dave Pritchard as my postdoctoral advisors.

Please, say a few words about your family.

My wife is a part-time teacher's aide in preschool. We separated a month before the Nobel Prize was announced. I have two sons, 16 and 9 years old and a daughter, who is 13 years old. They have diverse interests and I'm not pushing them in any direction but would like to help them to be able to choose from a wide range of opportunities.

Wolfgang Ketterle with his daughter, Johanna, and two sons, Holger and Jonas in Stockholm, 2001 (photograph by I. Hargittai).

What are your ambitions?

I would like to help my children in growing up. As a scientist, although I know I have accomplished a lot, I'm still maturing, acquiring a broader view, and I still feel challenged by doing science. At the same time I can't expect to be once again part of such a discovery as the Bose-Einstein Condensation. Such a discovery requires unusual circumstances. You have to be in the right place at the right time, working with the right people, and you also need a little bit of luck. This is a combination, which is singular and unique, and I have been privileged that I have been part of it once. So I'm not putting myself under pressure for achieving something similar. I just want to do good science and be a responsible scientist, both in research and education.

What do you teach?

I teach undergraduate and graduate courses, but this term I got a term off as a consequence of the Nobel Prize.

You mentioned challenge, what has been the greatest challenge in your life?

Life is just one ongoing challenge. I like to do something demanding, something that brings the best out of me. In science, when we set our eye on the Bose-Einstein Condensation, we went for two years without publishing any major paper and we could not know whether our experiment would ever work. This was the biggest challenge and the biggest effort I've undertaken in science.

Was it easy to get support for such an uncertain goal and for going on without publishing for two years?

It is important to have bigger goals in science and bigger goals take a while. In our field that usually means a couple of years. Grants are often given for three years, one usually has about five years for tenure and promotion, and the graduate students usually stay in graduate school for five years. Thus you can go for a big goal and accomplish it within one such cycle. I could build the Bose condensation experiment and achieve Bose condensation within the lifetime of a graduate student. As for funding, I could ensure continued funding even before the big success came. Of course, I took a risk in 1993. Since we didn't go for any smaller fish, we did not publish any less important papers, we just kept our focus on one big goal. In 1995, I got nervous. I needed renewal of my grant of the Office of Naval Research, but the program officer, Peter Reynolds, was extremely encouraging and told me not to worry, just keep the focus. He had confidence in me; it is true that in the American system an assistant professor must do something special to earn tenure. This often involves identifying some special problem that would distinguish you from the rest and that may involve taking a risk. I also felt a tremendous freedom as an assistant professor. I was a new hire and nobody expected any results from me in the first or the second year. I felt that I had time to do something major. I was building up my group and people knew that, so they did not expect me to achieve anything big very soon. It was in this period that I felt I could put my whole focus on something speculative.

I have looked up your group pictures over the years and the group was always all-male except the last one where one female appeared. Any comment?

Now there are two female graduate students. But they are in the minority. Certain areas of science, and physics is one of them, are male-dominated. I can't think of a good explanation except, perhaps, tradition. In France and Italy, there are many more female physicists than in Germany or the United States.

Do you have heroes?

In physics, of course, one is inspired by famous scientists like Einstein, Schrödinger, Heisenberg, and Newton. For me though they are too historical to be my direct heroes. I've felt more motivated by living persons, especially by my mentors. I learned from them and they are my examples to follow.

What is your present status in America?

I'm a German citizen and a permanent resident in the U.S. I've been here for 12 years, I am very happy to be here and have no plans to leave. But I still feel as a German and my family is a German family. We speak German at home. My children speak English in the school and sometimes they speak English among themselves at home, but they are aware of their German background and their German heritage.

Are the times of great discoveries in physics over?

Physics is as exciting as ever. If you look at the recent accomplishments in physics in cosmology, in condensed matter physics, in particle physics, in atomic physics, it's exciting. Now is as exciting a time to do physics as ever. Maybe it's just the perspective when you look backwards that you know how important the work was and what further development it triggered. It may only appear to you that the earlier work was more important than the work you're currently engaged in.

Do you think that physics in Germany will ever reach the height it had in the 1920s and early 1930s?

Physics in Germany is in good shape. There are excellent groups in all areas of physics in Germany. German researchers have made excellent progress and have won several Nobel Prizes. The United States has built an enormous tradition and has been very successful in physics since the Second World War. I expect that the U.S. will continue its leading role in physics, but many other countries in Europe and Japan are making major contributions.

Have you seen the play Copenhagen?

I have seen it in Germany a while ago.

What do you think of Heisenberg?

I was fascinated by the play and by all the characters. For me, the main message was the message of uncertainty. Uncertainty is well defined in quantum mechanics where we know how to write down the uncertainty relationship and deal with it. This play extended uncertainty to the uncertainty of memories and the uncertainty of events. Certain events have taken place in some form, but it will always be uncertain what really happened, because the only record that was left was in the memory of humans. This is not a scientific record. Even the humans, Bohr and Heisenberg, to the best of their knowledge, may not be able to give a completely unbiased account of what had happened. Therefore, what happened in Copenhagen will remain uncertain.

What is your opinion, apart from the play, of Heisenberg and his role during World War II?

[After a long pause] My opinion is that it's a mix of things. The Farm Hall transcripts seem to indicate that he had made some mistakes about the feasibility of nuclear weapons, but quickly corrected it when he learned that the Americans had built an atomic bomb. I also believe that he was not pursuing it so vigorously for ethical reasons. If somebody makes a decision of not pursuing something, it may be because that person is skeptical, it may be because he thinks it would not be ethical, and it may be because he feels that the goal is impossible to accomplish. The decision may be based on all reasons, and the person may not know himself what the main reason was. On the other hand, if I think something is impossible but I am fanatic enough to do it, I may work hard and may make the impossible possible. So maybe there is some truth in all the different opinions about Heisenberg. We will never know for sure.

You were born long after World War II and I wonder how you relate to the period of national socialism in German history?

I was born in 1957 and what happened earlier in the 20th century has a profound impact. Germany is responsible for what happened. As Germans, we are aware of what happened on German soil. If you live in a country, you also live in the tradition of this country and there are things to be learned from its history, from the mistakes that must never be repeated.

What do you do when you are not doing science?

Currently I split my time between science and my family. As for hobbies, I like outdoor activities, jogging, bicycling, photography.

Is there anything that you think I did not ask about but you would care to bring up?

I think you've covered everything. Maybe I should stress that I see my work in the tradition of the whole field and especially in the tradition of MIT. The Bose-Einstein Condensation emerged on the basis of laser cooling and evaporative cooling. My two next-door neighbors were the pioneers, Dave Pritchard of laser cooling and Dan Kleppner, together with Tom Greytak and Harold Hess, of evaporative cooling. I was privileged to take the work of my two colleagues and carry it to the next level, and I have an enormous respect for them. Of course, the Nobel Prize distinguishes something that is spectacular, but you stand on the shoulders of others and then some major breakthrough happens. The distinction for this major breakthrough is not just for the person who gets it but for the whole field. This Nobel Prize means a lot to all the researchers of the whole area, which has gained some kind of glamor from it.

Laszlo Tisza, 1997 (photograph by I. Hargittai).

20

LASZLO TISZA

Laszlo Tisza (b. 1907 in Budapest) is Professor Emeritus of Physics of the Massachusetts Institute of Technology (MIT) in Cambridge, Massachusetts. He grew up in Budapest where his father owned a book store. Already before university, he distinguished himself in competitions in mathematics and physics and met Edward Teller (see next interview) when they shared the first prize in a physics competition in 1925. He started his university studies in Budapest first majoring in mathematics. After two years he continued his studies in Germany, first in Göttingen, then, at Teller's urging, in Leipzig. Tisza did his first research work in studying molecular properties employing quantum mechanics. His doctoral project was the application of group theory and he gradually became a physicist. Upon return to Hungary he completed his Ph.D., but got involved in the communist movement and was sentenced to a year in prison. After he had served his time he joined Lev Landau in Kharkov in the Soviet Union in 1934. He acquired yet another degree there, approximately equivalent to a Ph.D. He returned to Hungary from the Soviet Union in 1937, but stayed for a short while only. Teller once again helped his friend when Tisza arrived in the United States in 1941 but had no job. Tisza worked as Teller's assistant for a short while before he found his employment at MIT where he has stayed ever since. We recorded our conversation on October 11, 1997, in Budapest, during Professor Tisza's brief visit in Budapest.*

---

*István Hargittai conducted the interview.

How did your career begin?

I entered the University of Budapest in 1925 as something like a *Wunderkind* in mathematics. Although there were severe restrictions on Jews, I got automatically into the University because I had won the High School Competition in Mathematics. I was extremely lucky. It used to be a ritual that Hungarian students stood at the doors of the University and asked for ID papers and the Jewish students were beaten up every morning. Only from 1925 to 1927, the two years I was a student of the University of Budapest, this was absent.

I felt myself very much at home in mathematics at the University. Eventually, however, I found that I was not made to become a professional mathematician. I wanted to test myself so I went to Göttingen, Germany, for a Spring semester and even gave two seminars on my own research in mathematics. It was in Göttingen that I heard for the first time about quantum theory. Although I was always interested in mathematics, I was also asking myself the question, "What is mathematics good for?" I knew mathematics was about truth, but truth about what? Engineering didn't interest me, and quantum mechanics suddenly provided an interesting application.

I enjoyed learning quantum mechanics in Göttingen. I attended Max Born's course and he was a very good lecturer and so was Walter Heitler. However, I wanted to get back into research. It's hard to imagine today that in 1928 the first excitement about quantum mechanics was over and the applications have hardly started yet. So it was not easy to find research problems. When, in 1930 in Leipzig, I asked Werner Heisenberg for a problem, he said, "It is not so easy, atomic spectroscopy is finished, molecular spectroscopy is still alive but I'm not interested in it, and Felix Bloch finished solid state theory." Nuclear physics, of course, has not started yet. Heisenberg finally suggested to me to generalize Bloch's theory of electric conductivity, which Bloch had worked out for the three-dimensional case, to a thin layer of metal.

But I'm getting ahead of myself. When I was looking for advice still back in Göttingen, Edward Teller came for a visit. I had known him since we had both been winners of Eötvös contests in Budapest. He told me to go to Leipzig and this is what I did for the next term. During that term we were in constant collaboration.

Teller pulled me into the problem of molecular structure and rotational spectra. In methane and methyl halides there was a fictitious moment of inertia appearing from the rotational transitions. It appeared because of a deficiency of the mathematical relationship between the spectra and the moment of inertia. The rotational spacing gives the moment of inertia but methane had two bands, one around five microns and the other around seven microns, and one gave a higher value for the moment of inertia than the other. I thought it might be because methane might have a pointed pyramidal structure but Teller knew this was impossible because methane was a symmetric tetrahedron. He had a chemical training although he didn't graduate in chemistry.

Teller brought up various ideas and we discussed them and gradually it crystallized that what happens is that the vibration is degenerate in symmetric molecules and if this is so, it will have an internal angular momentum which interacts with the total angular momentum and this gives a falsification of the connection between the rotational spacing and the moment of inertia. We published a joint paper and later there was a follow-up paper in which Teller gave the definitive theory. There are two constants, the moment of inertia and the coupling constant of the two angular momenta. The observed spacing is a function of these two.

I returned from Leipzig to Budapest in the Summer of 1930. At that point my father said, "That's it, you were long enough in Germany." My father was a bookseller, *Tisza Testvérek* (Tisza Brothers) was a well-known book store in *Fö utca* (Main Street) in Budapest. It was started in 1903 and the communist government expropriated it in the early 1950s, but they left my father to manage it. Then, in the late 1950s, my father joined me in America.

The next work in physics, which became my doctoral thesis in Budapest, was the application of group theory for unsymmetric polyatomic molecules, with selection rules for infrared and Raman spectra. It also involved excited states. Eugene Wigner had written a paper on the spectra of polyatomic molecules but his treatment was quite classical. When my thesis was completed, the Professor of Physics, Ortvay, sent my manuscript to Wigner, who was in Göttingen at that time, to evaluate it. I had known Wigner from Budapest.

Edward Teller[1] mentioned in our conversation with him that you were imprisoned in Budapest.

It was in 1932 and it was after I had completed my thesis. It was a foolish thing. That was Depression time. I had an old school friend who was a fervent Marxist and he convinced me that it was the way to go. I got involved with it through my friends. They said, "Once this is your conviction, why don't you help us," and I foolishly said, OK. Then I did some perfunctory connection jobs, like taking a manuscript from a political leader to the printers and that sort of things. My friend was arrested and so was I. I was in prison for about a year.

Teller was very helpful. He visited me in prison and brought me a copy of my Hungarian thesis, and told me that he was leaving for Germany and I had two days to translate my thesis into German. I did the translation, working fervently from early morning till late night for two days and Teller took the manuscript to Germany. He submitted it to the *Zeitschrift für Physik* where it appeared in 1933.[2]

If Teller was your friend and he helped you then he must have been very open-minded because you were a Marxist at that time.

He was very open-minded and he had other leftist friends too, but he didn't sympathize with communist ideas even then.

Then you went to Kharkov in the Soviet Union.

Yes, and it was also with Teller's help, and he arranged a position for me with Lev Landau there. I spent two years in Kharkov, 1935 to 1937.

Did you meet Hans Hellmann[3] there?

I met him. My connection with Landau started in 1934 when he arranged a small international meeting in Kharkov. Niels Bohr was there and a number of other people. I was there and Hans Hellmann from Germany. After the meeting I wanted to visit Moscow and Leningrad and I spent a day or two in Moscow as a guest of the Hellmanns. He was later deported and killed.

My appointment in Kharkov was surprisingly informal. My work with Landau was on relativistic quantum mechanics.

You left the Soviet Union in 1937, quite a dangerous time, a big wave of political purges.

It was a very dangerous time. In the winter of 1937, our Institute was split in two parts, one was Landau and his friends and the other was a rabid party group. Landau's collaborator, Koretz, was arrested and Landau felt uncomfortable. By then Pyotr Kapitza had been forced to stay in Moscow, following one of his visits from Cambridge, England, and he invited Landau to work in Moscow. I remember the meeting in Kharkov where Landau stood up and announced, "Next Monday I'm leaving for Moscow."

After that there wasn't much reason for me to stay. A number of my friends were arrested and I felt increasingly uncomfortable. Besides, they were planning to turn the Institute into war work and didn't want any foreigners around. There were other job offers, from Sverdlovsk, Kiev, Odessa, but nothing worked out and I decided to leave. At that point I needed an exit visa which failed to arrive. So I wrote to the Hungarian Consulate and asked for their help, and I got my exit visa. I returned home but I also knew that there was no future for me in Hungary.

Teller then helped me again. He talked to Szilard and Szilard arranged an invitation for me to Paris. Szilard had friends all over. I went to Paris where I had a scholarship at the College de France and worked mainly with Fritz London. I stayed in Paris until the Germans came in June, 1940. First I went to the South of France, then, in early 1941, to America. I landed a job at MIT.

Did you ever have any problem in America because of your political past?

There were some problems. There were some interviews with the FBI and I told them the story, and they believed me. This was before the McCarthy period. I know Teller likes to tell the story of my conversion from Communism but this is not something I want to be associated with or would like to advertise for the American scientific community.

At MIT, first they appointed me to teach physics, to replace the people who went away to do war-related research. During those four years I gave all courses in physics at MIT and I really learned physics.

Under Landau's influence I had got very interested in thermodynamics. At MIT Slater, who was the Head of the Physics Department, used to give a course on thermodynamics. He was also leaving and asked me to take over his course too. From then on I gave a course on thermodynamics for many years and I developed a quite popular course and a textbook too.[4]

*****

Following the Nobel Prize in Physics in 2001 (see the previous interview in this volume), it was natural for me to ask Professor Tisza about his involvement in the two-fluid concept.[5] He published a short note "Transport Phenomena in Helium II".[5] When I asked him about it in the fall of 2002, he gave me his lecture notes about the subject, prepared

Laszlo Tisza with Wolfgang Ketterle, 2001 (courtesy of W. Ketterle)

for the centenary meeting of the Eötvös Physical Society in Budapest in October 1991. With his permission, I am quoting from these notes. They had appeared in full in Hungarian[6] and Professor Tisza plans to publish them eventually in English in their entirety, including the figures.

... I was quite reluctant to retell this old story, an intimate blend of success and failure. Moreover, the problem was complicated by controversies involving the "triangle Landau-London-Tisza". Being the last survivor, I wanted to avoid even the appearance of making a partisan point. However, my old friend Nicholas Kurti, a witness of the early events, convinced me that recording my thought processes and my interactions with Lev Landau and Fritz London would not be without interest.

I worked in Landau's group in Kharkov from 1935 to 1937. Somewhat surprisingly, the subject of liquid helium was never mentioned during that time. True, superfluidity was discovered only in January 1938, but liquid helium was already known to have unusual properties of quantum origin, which were being examined in depth by Fritz London, at that time in Oxford. These properties were in line with Landau's interest and I am still puzzled why he did not take note.

Although I never discussed helium with him, Landau is nevertheless part of my story. He was the only one among his generation of quantum physicists to be genuinely interested in thermodynamics. Shortly after my joining his group he published a number of papers on second order phase transitions. I was greatly impressed by the fact that thermodynamics was alive and well and offered the possibility of applying conceptual considerations to empirical problems and the mathematics did not involve tedious calculations. I decided to try my hand in this field, provided I would find a research topic, which did not seem easy at the time.

This was to change shortly, as in September 1937 I found myself in Paris in close touch with Fritz London. He did not have the universality and the incredibly sharp mind of Landau, but he was eager to discuss his insights and he introduced me to his phenomenological theory of superconductivity. It was a great experience. The discovery of superfluidity next January by Kapitza, and independently by Allen and Misener, was exciting for both of us. London looked at it as an analogy to superconductivity and I sensed the emergence of the type of research project I have been hoping for. While waiting for the weekly issues of *Nature* to come out with additional surprises, London filled me in on the remarkable properties of helium with which he had been involved for some years. He was originally stimulated by Francis Simon's beautiful thermodynamic analysis, which implicated quantum effects.

Instead of crystallizing with the lowering of the temperature, helium undergoes a so-called second order phase transition into another liquid phase helium II. It was evident to the experimenters that something dramatic was happening: as the temperature is reduced by pumping, helium I boils vigorously, but at the transition temperature the boiling stops, and junctures tight to helium I become leaky to helium II. It took some time before the first of these effects was traced to an anomalous internal convection, and the annoying "superleak" turned into the discovery of "superfluidity" ...

A glance at the phase diagram tells a great deal to the experienced thermodynamicist. The liquid solidifies only under 25 to 30 atmospheres of pressure. Simon concluded from the asymptotically horizontal direction of the phase separation line that helium II has the same vanishing entropy as the crystal, and he attributed the disruption of crystalline localization to the quantum mechanical zero point energy. London has followed up these qualitative considerations with a number of calculations on the delocalization of the helium atoms in the liquid, as distinct, say, from liquid hydrogen. However, "delocalization" did not imply "disordering". It was an utterly novel instance of melting, which proceeded as a mechanical decompression without latent heat, from an "order in coordinate space" to an "order in momentum space", the whole phenomenon was a "macroscopic quantum effect". The expressions in quotes were conceived by London in an attempt to give an intuitive feel of the "fourth phase of matter", the zero entropy fluid.

London came to think of liquid helium more as a gas than a crystal and as our discussions turned into an informal collaboration, he suggested that I explore an analogy with the Fermi gas. While I was following up this false start, he suddenly presented me with his own version of the Bose-Einstein Condensation. This phenomenon suggesting the condensation of an ideal gas into its lowest quantum state was believed to be spurious as Uhlenbeck pointed out a fallacy in Einstein's argument. However, London showed that the fallacy was in the criticism and that Einstein was right.

In order to underline the similarity and contrast between the ordinary van der Waals and the Einstein condensations, London framed the concept of a "condensation in momentum space" to characterize the latter, in full harmony with his earlier conception of an "order in momentum space". He immediately connected the kink in the specific heat of the Bose-Einstein gas with the λ-point anomaly of helium. Although the resemblance was rather tenuous, I shared London's confidence that he was on to something.

We both thought that the new approach would help also to clarify the puzzle of superfluidity. However, there was a bold step from the foregoing equilibrium considerations to a hydrodynamic context, and I was more willing to go out on a limb than the more careful London.

My first step was to examine the concept of viscosity in liquids and gases. It was instructive to compare helium and hydrogen in their gaseous and liquid states. Mere inspection leads us to distinguish between what I called "dynamic" and "kinetic" viscosities. The former is typical of liquids, excepting only helium: the flow process comes about as an activation process overcoming intermolecular potential energies. The resulting viscosity is large, with a large negative temperature coefficient. By contrast, kinetic viscosity comes about from the transport of transverse momentum in gases. ... Liquid helium I has no dynamic viscosity. This is explained by the fact that the zero-point motion that prevents crystallization also overcomes the potential energy barrier against flow.

By contrast, liquid helium I has a kinetic viscosity, which is of the same order of magnitude and has similar temperature dependence as gaseous helium and even hydrogen. This viscosity is continuous at the λ-point, and although it appears to drop in helium II, it is utterly inconsistent with the superfluidity manifest in capillary flow.

The situation was particularly dramatic in the thermomechanic or fountain effect according to which superfluid flow is readily produced by thermal radiation.

In view of the above-mentioned discrepancy between the methods of the measurement of viscosity, my conclusion was the following: this is not a kinetic coefficient of an unusual value, but the breakdown of the viscosity concept. There is no Navier-Stokes equation with a viscosity parameter! I concluded there must be a "mixture" of a superfluid and a viscous, or normal, component. A very narrow capillary (acting as an "entropy filter") was permeable only to superfluid flow, but not to the normal fluid. Such a division into interpenetrating, but distinct entities was provided by the Bose-Einstein Condensation, but the gas pressure was now more an *osmotic pressure*, and the solute and the solvent transformed into each other as heat was removed or added from or to the system.

I had this idea one evening and spent the rest of the night thinking through the whole array of experimental facts which indeed were in qualitative agreement. Thus the very high heat current (not proportional to the temperature gradient) was explained in terms of an internal convection in which the normal component carried an energy proportional to the temperature, rather than to the temperature difference.

When I presented all this to London the next morning, he was unimpressed, and his attitude was duplicated by every theorist I tried to convince for a number of years. It is true that my ideas did not qualify as a "theory" by any admitted standard.

Counting on the usefulness of phenomenological framework for experimenters, I published a note in *Nature*, which received much attention. I made the minor prediction that the thermomechanical effect ought to have an inverse: a superfluid transfer from vessel A to B should lead to heating of A and cooling of B. This was readily verified.

However, London persisted in his opposition to the idea that two velocity fields could persist in a liquid, and leaned toward the suggestion of his brother H. London that the fountain effect was a surface effect.

I had no hope to be able to derive my model from basic quantum mechanics, and I wondered whether that would be even possible. It occurred to me, however, that the postulate of a two-fluid hydrodynamics in the bulk liquid leads to a dramatic experimental prediction.

I knew that the complicated non-linearity of the hydrodynamic equations is linearized to give sound propagation, and I concluded that my internal convection should give rise to a second type of wave propagation that would be excited by periodical heating to produce a "temperature wave". I established a semi-quantitative derivation based on a simplistic mechanical model. In Eulerian hydrodynamics one replaces the volume element by a mass point, I replaced it by *two* mass points. In November 1938, I published the following formulas for the propagation velocity of the temperature waves:

$$v_T = \sqrt{\frac{\rho_s}{\rho} \frac{dp_n}{d\rho_n}} \tag{1}$$

$$v_T = \sqrt{\frac{kT}{m}\left[1 - \left[\frac{T}{T_0}\right]^5\right]} \tag{2}$$

where

$$\rho = \rho_s + \rho_n$$

and p_n is the "osmotic pressure".

Equation (1) followed from my argument, while Equation (2) was inferred from phenomenological assumptions relating my hydrodynamic parameters to the measured specific heat. These equations were meant to hold only to about 1 K. Below that, the specific heat would be mainly due to phonons, and I assumed — erroneously — that the phonons should be associated with the superfluid, and I expected no temperature waves.

I presented a preliminary version of these results already in July at a low-temperature conference in London … [However], the suggested experimental test did not come about.

The sequel is history and I can sum up the main points briefly. In 1941 Landau published an impressive quantum hydrodynamic theory of helium and predicted the propagation of two sound waves. After unsuccessful attempts to detect the second, E. M. Lifshitz recalculated the theory and found that the second sound is in fact a temperature wave, which then was observed by Peshkov. At this point London approached me and suggested that Peshkov's measurements do not differentiate between Landau and myself, except for adjusting a free constant of my result. However, as the measurement was extended to lower temperatures, the situation changed, and the measurements verified Landau's prediction of a rising velocity of second sound due to the contribution of phonons to the normal fluid. Although I had been

Laszlo Tisza in front of the Bose-Einstein Condensation experiment at MIT, 2001 (courtesy of W. Ketterle).

aware of the presence of phonons, I assigned them to the superfluid, which was an outright mistake.

Afterthoughts

What is the moral of this story for the modern student? Landau's two fluid hydrodynamics has a clear precedence over my earlier attempt. Nevertheless, his overall approach has a feature, which I consider objectionable. In his first paper Landau gives the impression that his theory is a deduction from quantum mechanical principles. However, this program is feasible only to the extent that he bases the discussion around the energy spectrum $\varepsilon(p)$, where p is the linear momentum of the excitation. Yet, the nature of super-fluidity depends also on the structure of the ground state. Since this problem involves a strongly interacting many-body system, the deductive approach is of not much help. By contrast, the somewhat complementary method of a direct interpretation of the experiments as demonstrated by Simon, London, and by myself proved effective. London's order in momentum space and its manifestation in macroscopic effects is of consequence not only for the student of helium, but constitutes a refreshingly positive aspect of the Heisenberg principle, instead of the usual epistemological limitation.

Insofar as my own contribution is concerned, I believe this is mainly to provide a bridge from London to Landau's two-fluid hydrodynamics. Maybe more specifically also, having advanced the concepts of kinetic and dynamic viscosities. I am mystified why this was rejected out of hand by Lifshitz.[7] The same attitude surfaces in Lifshitz' assertion that helium becomes a quantum liquid only at the λ-point, and he notices nothing special about helium I.

If history has a lesson, it is that the "winner take all" attitude deprives one of the pleasures of being the heir to the best of different traditions, even while avoiding their intolerance against each other. London in his *Superfluids*, Vol. II, made real strides toward such a synthesis; the book is a labor of love. Its one flaw is to be too thorough and the rich detail catering to the specialist tends to hide its conceptual sweep. At one time London expressed to me his disappointment that people would not understand that superfluidity is not only for the specialist, but is also a part of fundamental physics. Maybe it was not possible to do justice to both aspects in one book, and he had no time for a second.

References and Notes

1. See, Teller interview in this volume.
2. Tisza, L. "Zur Deutung der Spektren mehratomiger Moleküle", *Zeitschrift für Physik* **1933**, *82*, 48–72.
3. Hans Hellmann (1903, Wilhelmshaven, Germany — 1938 perished in Soviet captivity) got his doctorate in Stuttgart in 1929. He taught physics in Germany until in 1934 he became Professor of Theoretical Physics at the Karpov Institute of Physical Chemistry in Moscow. Hellmann was the author of the first quantum chemistry book that was translated into Russian (Gel'man, G. *Kvantovaya Khimiya*, Glavnaya Redaktsiya Tekhniko-Teoreticheskoi Literaturi, Moscow, Leningrad, 1937). A somewhat abridged German version has appeared (Hellmann, H. *Einführung in die Quantenchemie*, Deuticke, Leipzig, 1937). In 1938, Hellmann was charged by the Soviet authorities as a German spy, and soon after he perished. See, Kovner, M. A. "Hans Hellmann of the Hellmann-Feynman Theorem." *Chem.*

Intell. **1996**, *2*(1), 54–55. Michail Kovner was a doctoral student of Hans Hellmann. Hellmann's name is best known today for the Hellmann-Feynman Theorem according to which, for the exact wave function the energy gradient is equal to the expectation value of the derivative of the Hamiltonian.

4. Tisza, L. *Generalized Thermodynamics.* MIT Press, Cambridge, MA, 1966.
5. Tisza, L. *Nature* **1938**, *141*, 913 (May 21).
6. Tisza, L. *Fizikai Szemle* **1992**, *42*, 303–306.
7. Lifshitz, E. M. and Andronikashvili, E. L. *A Supplement to "Helium"*, translated from Russian. Consultants Bureau, New York, 1959, 54.

Edward Teller, 1996 (photograph by I. and M. Hargittai).

21

EDWARD TELLER

Edward Teller (1908, Budapest — 2003 Stanford, California) was a Senior Fellow at the Hoover Institution, Stanford University and University Emeritus of the University of California at the time of our conversation. He made major contributions to nuclear chemistry and atomic, nuclear, and solid state physics. He was also involved with the Manhattan Project, the hydrogen bomb, and what is popularly known as Star Wars. He started his education in chemical engineering but eventually became a world famous physicist. He was recipient of, among others, the Enrico Fermi Medal and the Albert Einstein Award.

We spent a memorable afternoon with Edward and Micike Teller in their home in Stanford, California, on February 24, 1996. Dr. Teller was just convalescing after an illness, but especially when he spoke about science, and in particular about his views on the uncertainty principle, he was most vigorous and captivating and kept our mini-audience of two in awe. Although the social chat was in Hungarian, the entire conversation recorded here was in English.*

Could we start by your telling us something about your family background?

I'm not terribly eager to do so. My father was a lawyer in Budapest. He came from Érsekújvár which is now in Slovakia. My mother came from

---
*István Hargittai and Magdolna Hargittai conducted the interview. This interview was originally published in *The Chemical Intelligencer* **1997**, *3*(1), 14–23 © 1997, Springer-Verlag, New York, Inc.

Lugos which is now in Romania. They met in Budapest and lived in Budapest, and I was born in 1908. Otherwise I cannot tell you anything of great interest.

Around the turn of the century there was a great concentration of intellectual Jewish families in Budapest.

Hungary was a mildly anti-Semitic country at that time, when I was young, and I lived there the first 18 years of my life. I say mildly because I call Hitler not mildly. Hungary never duplicated Hitler until he came in. Those last months were truly terrible. I felt anti-Semitism in school but only something very disagreeable but not threatening.

You asked about Hungarians. Many came to this country and I won't talk about them, except I want to mention four of them, very famous, whom I had actually met through our families, before I left Hungary in 1926. In the order of their ages, Theodore von Kármán, whom I only met in the United States, Leo Szilard, Eugene Wigner, and Johnny von Neumann. All of us have made real contributions to the American war effort.

If we may jump to another subject, did American science benefit from competition with the Soviet Union?

No. It was not competition with the Soviet Union. It should have been, but among the scientists there was a strong tendency not to compete with the Soviet Union, not to go into practical matters. Of course, Theodore von Karman had nothing to do with all that. His work was mostly in connection with the Second World War. His main merit is that we entered World War II if not with airplanes, at least with good plans for airplanes. He had excellent relations, not with the Air Force, which did not exist at that time, but with the branch of the Army that handled airplanes.

Szilard played a decisive role in getting nuclear explosives started, but that was in no way competition with the Soviet Union. It was the fear that the Nazis would develop the atomic bomb. In those early discussions the Soviets did not play any role. After the War, Szilard was a strong exponent for working with the Soviets, rather than competing with them, and indeed for abstaining from further development of nuclear explosives. In this I differed from the great majority of scientists in this country, and I differed from Szilard in a very definite way. However, we have been and remained friends. It was not the case in all other similar relations. Eugene Wigner and Johnny von Neumann were for working on the hydrogen bomb and competing with the Russians but played a relatively minor part in it.

But you played a major role.

There I was almost alone, but not completely.

You have been called the Father of the Hydrogen Bomb. Is this an accurate description?

There were quite a few people working with me, but among the people who worked on the hydrogen bomb, I was the only one who had some little access in Washington. Not because of merit but because of history, or accident, if you please; of the scientists that were listened to in Washington, I was practically the only one who advocated that we go ahead with the hydrogen bomb. In that respect Eugene and Johnny were on my side, and Szilard was on the opposite side. But Eugene and Johnny did not strongly expose themselves. In a way, I don't want to be called the father of anything but I was the relevant advocate, and in this sense, I had an influence because without my advocacy things might have gone very differently.

Some people who have criticized you maintained that your opinion was colored by your history back in Hungary. Have you seen such criticism?

Often. Let me tell you, when Hungary went communist in 1919, I was 11 years old. The communists were in Hungary for four months. Our family didn't like them at all. My father, as an advocate lawyer, had no job. Then there was the counter-revolution, and there were shootings in the streets which I, from quite a bit of a distance, witnessed. I was certainly anti-communist but at the age of 11, I was not exposed to the problems of communism in a direct way. I was exposed in the school to anti-Semitism.

In 1926 I went to Germany to study. This was just two weeks before I turned 18, after I'd completed Gymnasium and attended the Budapest Technical University for a few weeks. I just finished the first term there in chemical engineering. My father did not approve of mathematics, and we compromised on chemical engineering.

At that time, I can honestly say that I had an open mind about communism. In 1926 the horrors of the Stalin regime had not yet become obvious. In Europe, at that time, communism was considered the way

Edward Teller and István Hargittai during the interview (photograph by M. Hargittai).

of the future, and particularly so after 1928–1929, after the big, worldwide depression that was clearly the end of capitalism. The only people who had new ideas, so it was said, were the communists. Now, I was not a communist, not at all, but I was genuinely open minded. I can best illustrate this to you by telling you about my friends. After studying for two years in Karlsruhe, I went to Leipzig to work with Heisenberg. There I was exposed to a small, interactive, international community. Politics was not in the center of interest, but I had two very close friends — one, a clear anti-communist, the other, a devoted communist. The anti-communist was Karl Friedrich von Weizsäcker, the elder brother of the man who later became President of the West German Republic. He was an excellent physicist, later philosopher, a close friend, and we studied together in Leipzig, in Göttingen, and later in Copenhagen where we lived in the same house. I disagreed with him about communism. He was clearly anti-communist and he was at that time also anti-Nazi but not as clearly. Of the two evils he considered Nazism the lesser evil.

Another good friend at that time, with whom I also published a paper, was Lev Landau. He was a devoted communist. He considered capitalism as something ridiculously wrong, and talked about it a lot.

I want to mention to you another friend, a Hungarian, László Tisza. He and I worked on molecular structure. We had a molecule with threefold symmetry. We deduced the moment of inertia from the spectra. We found that in case of degenerate vibrations the spacing is strongly influenced by the interaction of vibrations and rotations. This interaction can falsify the data from which you would deduce the moment of inertia.

Tisza was very strongly pro-communist, not a member of the party but he worked with them. When he went back to Hungary, he was arrested. I continued to work with him and visited him in prison in Budapest. Eventually he was let go but his chances for a job were precisely zero. I then wrote to Lev Landau who got a job for him in Kharkov. In a few years he came back from the Soviet Union. He was completely disillusioned and told me about the horrible treatment of Landau, who had been arrested. This was about the time I came to the United States.

I have been an anti-communist, but not on the basis of what I had experienced in Hungary. It was after the conversations with Tisza that I just described, when he returned from the Soviet Union. This was in 1936. I can also tell you that the capstone of all this was a book, of which you know.

Darkness at Noon *by Arthur Koestler?*

Darkness at Noon. I read *Darkness at Noon* in the first month when I was in Los Alamos. That was in the spring of 1943. The book could not have been published much before that. By that time I did not need much convincing. It is an excellent book because throughout the book Koestler holds you in suspense of which side is right. Then he comes out very strongly against Stalin in the last 50 or 100 pages.

Did you know Koestler?

Johnny von Neumann knew Koestler. On one of my visits to Princeton, Johnny took me along and I met Koestler, I talked with him for an hour. I remember Koestler much more from his books than from our personal meeting.

So the statement that I am an inveterate Hungarian anti-communist is just not so. I am an anti-communist essentially of the school of Koestler.

Who used to be communist before.

He was but he changed his mind, slowly and painfully. I compare myself with him divided by a factor of hundred.

In what sense?

Well, I was not a communist, but I considered communism as a possibility. For Koestler to decide that communism was wrong, was a very difficult decision. For me it wasn't. But it was a decision. He did it very much more on the basis of personal experience. I can say that for me, as it was for Koestler, at the same period, it was a problem. Only, for Koestler it was a hundred times bigger problem. But the statement that I had come from a country which was communist and this was the reason for my anti-communism is simply not true.

For me though, it was not a vital problem because I was not a politician, I was a physicist. But I had to be a politician, as all of us had to be at that time to survive, to know what to do. I started from a doubtful position, and I had lost my doubts a hundred percent by 1943.

Europe at that time was clearly going either Nazi or communist and of the two, communism was probably better. When Hitler came in 1933, I left Germany for England and Copenhagen, and arrived in the United States in 1935. It was not before 1934–1935 that I started to recognize that democracy was a real alternative. Democracy made a mess of economics and there was the worldwide depression of 1928–1929. There didn't seem to be a solution. That there was a solution, and that the choice was not between communism and Nazism, became a little clearer to me before that time but began to become really clear in 1933–1935. The whole process took years, and I landed on this side, but not, as some people say, entirely prematurely.

Eugene Wigner and Leo Szilard were in disagreement about politics because Szilard was for the communists and Eugene not. Later Szilard was with the Pugwash group. Once I took him to dinner in Washington. We never were on a first-name basis. We called each other Teller and Szilard or Teller Úr and Szilard Úr. We were good friends. He said to me, Teller, you are mistaken. What the Soviets in Russia are doing is very good. You should go to Russia, spend two weeks there, and your mind will change. And I said no, I won't go. And Szilard said, why not? Szilard was a very tolerant person. He allowed people to disagree with him but not without reason, so he asked, why not? So I told him the reason. I said, look, my mother and sister are in Hungary. I go to Russia and I'll be blackmailed.

I won't go. Szilard said, you are wrong, they never would blackmail you but I'll do something about it. The funny part of the story, and the nice part, is that he did. At the next meeting of the Pugwash group which was in Austria, he went to the Russian leader and asked him, why don't they let Teller's mother and sister come out? The Russian said "I have nothing to do with that. They are in Hungary; it is an independent country," but the same day the Hungarian representative came to Szilard. They had a conversation and in a couple of months my mother and sister had the permission to come out. This was in 1959.

Did you then go to the Soviet Union?

No, I didn't. But I went there after the Soviets were gone. In my opinion the communists knew darn well that Szilard was an effective spokesman for them, and he was. Had they not let my mother and sister come out, they would have proved me right, but they wanted to keep Szilard's high opinion, rather than lose it, and they let my mother and sister come.

Wigner was a very polite man. Do you know the story about Eugene Wigner when he got into trouble at a garage. They didn't repair his automobile well. Eugene got very mad. You know what he said? He was very upset. He said, "Go to hell, please."

You said that you were listened to in Washington in the 1940s and later. But even in President Reagan's time you had great influence with the Star Wars.

Sure.

How was it possible for a physicist to maintain this level of influence for decades?

My influence was never that great.

But people think it was.

I have to be blamed for it, and in this blame there is some exaggeration. Exaggeration that for me is not entirely unwelcome. To be said that I am more important than I am is not factual but it is not an insult.

Let me give you an example. After the war I was not named to the General Advisory Committee on Nuclear Energy. They gave me a job on which I worked seriously for years, reporting to them on the safety of reactors. It was a job away from politics. But I had been involved in arguments about the atomic bomb with Szilard before even Oppenheimer heard of it. Then the question of the hydrogen bomb came up in 1949. People got interested in what was going on in Los Alamos and I was the only known person in Los Alamos who was favoring the development of the bomb. An important senator, Head of the Joint Committee of Supervising Atomic Energy, Senator Brian McMahon, a Democratic politician, was involved in this. He sent for me, obviously for the reason that he wanted to have an opinion from Los Alamos which differed from the general opinion, which was, to stop the hydrogen bomb. I went to see him in the fall of 1949. The first thing I heard from him was that he had seen the report of the General Advisory Committee, and it made him sick. He had already made up his mind without listening to me. But he had no physics facts.

The most outstanding person for strong defense was Earnest Lawrence, but he knew nothing about the details of the hydrogen bomb. In the fall of 1949, right after the nuclear explosion, he and Luis Alvarez came to me in Los Alamos and asked me about the hydrogen bomb. I was known to have worked on it, and I was known to have been with atomic energy from the beginning. So I had a voice.

You didn't ask me, but I'll tell you about something. In 1945 the democracies had to disarm. Public opinion would not allow them to do otherwise. Stalin did not disarm. He was going to win, except for the scientific component in the defense of the West. The scientists themselves were generally opposed to the military. Had they had their way, we now would be all communists.

May I ask you about the Star Wars?

Now, that's a big jump. You have not asked me but I'll tell you anyway, about the accomplishment I am most proud of. It's not the hydrogen bomb. It's the establishment of Livermore. To have a second laboratory so that these serious questions of defense should not be considered from the point of view of only one group of people, those of Los Alamos. Secrecy restricted discussion to a small group, which I claim in 1949–1950 was about to make a very big mistake. They were against the hydrogen bomb because they wanted to be popular with the rest of the scientists. That was the opinion of Oppenheimer and the rest and that was the way Los Alamos had to go. I argued for the second laboratory so that questions

President Kennedy visits the Lawrence Berkeley National Laboratory in 1962. From left to right: Norris Bradbury, John Foster, Edwin McMillan, Glenn Seaborg, John Kennedy, Edward Teller, Robert McNamara, and Harold Brown (courtesy of Lawrence Berkeley National Laboratory).

of such importance could be considered at least on the basis of a debate. With the definite help of Earnest Lawrence, we did go ahead with the Lawrence-Livermore National Laboratory (LLNL).

Reagan became governor of California in 1967. Early that year I went to see him, and told him: here is a laboratory in your state, you should visit us and see what we are doing. He came, and at the end of 1967 he heard in great detail about the possibility of defense against missiles. It was quite clear from that discussion that it was a subject new to him but interesting to him. He is a man well known for his excellent property of shooting from the hip. My opinion is that he heard about this possibility in a serious way in 1967, and he did shoot from the hip in 1983. Throughout the years he talked with many other people, and was very slow to make up his mind. But when he made up his mind, he did so firmly, and, furthermore, he knew my position, and I was then in the position to be listened to by him. I think that I did have an influence on a variety of people, on Earnest Lawrence, on McMahon, on Louis Strauss, Chairman of the Atomic Energy Commission, on Nelson Rockefeller, and on Ronald Reagan.

Quite a list. You had a debate with Linus Pauling on television.

I'll tell you one story about it, and I won't forbid you to publish it although I am writing my autobiography, and I don't like to be scooped. But it's a good story.

We had a debate on the question of whether we should stop nuclear testing. The debate was broadcast in California on KQED and was rather generally listened to. A few years later, a very pretty girl ran after me in New York, at the airport. May I have, please, your autograph, Professor Pauling? I made a very ugly face and I said, no. That was the only time I was impolite to a pretty girl.

You had only one debate.

Certainly only one public debate.

Did you have any other interaction with him?

I had some. Positive. A few years before the television debate, he was denied a passport and some of us in Chicago wrote a letter to the State Department protesting against this, and I signed the letter. At that time I got a very pleasant letter, thanking me, from Linus Pauling.

There is yet another thing that I'd like to tell you, as chemists, about Linus Pauling and me. You know that Linus Pauling is rightly credited with the simple quantum mechanical explanation of the Kekulé idea that benzene has two formulas. That much you know. You may or may not know that in the Soviet Union this work of Pauling was declared sacrilegious.

A little before that time I and my friends wrote a beautiful paper confirming Pauling's theory. My contribution is the spectroscopic evidence for the various chemical bondings. Pauling's theory is that benzene should be described by one Kekulé formula plus another Kekulé formula; a linear superposition of the two. If there is *this plus that*, where is also a wave function when it's *this minus that*? I answered this question about 1940. In the spectrum of benzene the lowest transition is a forbidden transition in the near ultraviolet. My specialty was the vibrational structure. I am oversimplifying but in the lowest transition of benzene you start from the nonvibrating molecule, you absorb, and the only lines you find are when a particular vibration of benzene is excited with one quantum. In a decent, permitted transition, that should be forbidden. The point is that the transition between the two Kekulé formulas (plus and minus), which in classical physics corresponds to a jump from one Kekulé formula to another, has no dipole moment, and there is no absorption. But you distort the molecule, and there is a dipole moment between the two. This reasoning was worked out quantitatively in that paper: an excitation of the appropriate vibration produces a dipole moment. The ground state of benzene is the positive superposition of the Kekulé formulas, and the first excited state is the negative superposition of the same two, and that explains the spectrum.

I'm extremely proud of this because I claim that while to consider the superposition is the obvious thing to do, the point that you can find both superpositions, and that the spectroscopic behavior is the predicted one, really annihilates, to my mind, all possible objections to that interpretation.

Did you ever discuss this with Linus Pauling?

No.

Do you know if he was aware of your paper?

I don't know. Pauling was a chemist and our paper was about a physical detail which I'm honored when chemists listen to a paper like that.

You have several named effects, such as the Jahn-Teller effect, which is very important in chemistry.

Do you want to hear about the Jahn-Teller effect? Of course, it is not more chemistry than the other research I mentioned before. As you know I started as a chemist and that stigma stayed with me. This effect had something to do with Lev Landau. I had a German student in Göttingen, R. Renner, a very nice man, and he wrote a paper on degenerate electronic states in the linear carbon dioxide molecule. This man later had to serve the Nazis and work hard for his father's business after the war and never wrote another paper. But he was a good man. The problem I originally put to him was to take a transition of carbon dioxide where the transition dipole moment is perpendicular to the CO_2 axis. He made a good paper out of that, assuming that the excited, degenerate state of carbon dioxide is linear.

In the year 1934 both Landau and I were in Niels Bohr's Institute in Copenhagen and we had many discussions. He disagreed with Renner's paper, he disliked it. He said that if the molecule is in a degenerate electronic state, then its symmetry will be destroyed and the molecule will no longer be linear. Landau was wrong. I managed to convince him and he agreed with me. This was probably the only case when I won an argument with Landau.

A little later I went to London, and met Jahn. I told him about my discussion with Landau, and about the problem in which I was convinced that Landau was wrong. But it bothered me that he was usually not wrong. So maybe he is always right with the exception of linear molecules. Jahn was a good group-theorist, and we wrote this paper, the content of which you know, that if a molecule has an electronic state that is degenerate, then the symmetry of the molecule will be destroyed. That is the Jahn-Teller theorem. The Jahn-Teller theorem has a footnote: this is always true with the only exception of linear molecules. So the amusing story of the Jahn-Teller effect is that I first worked via my student, R. Renner — my name was not even on that paper, but it was a paper published in 1934 — that presented the only general exception to the Jahn-Teller effect.

Was Landau happy to see your footnote?

I hope so, but I never saw Landau again. On some occasions I have written this down and quoted him. It really should be the Landau-Jahn-Teller theorem because Landau was the first one who expressed it, unfortunately using the only exception where it was not valid.

What happened to Jahn?

I don't know. I think he was an Englishman who had studied in Germany but I'm not sure. He may have been German.

Let's move to another topic. My favorite Teller book is "The Pursuit of Simplicity". Nowadays there is a lot of discussion of complexity. Would you care to comment on this?

No. I forgot what I wrote in that book. But if you refresh my memory about some of my statements, I might tell you whether I still believe them.

My impression was that much of what you wrote about simplicity appears today in the discussions of complexity. Recently I reviewed Murray Gell-Mann's book The Quark and the Jaguar.

I just finished reading the book. It is not quite as horrible as I had expected it to be.

I am quoting from my review of Gell-Mann's book, referring to its last part: "It is here that Gell-Mann makes an attempt to chart, rather than just to speculate about, the possible future path for the human race and the rest of the biosphere. Curiously, and obviously not by mere coincidence, another book on simplicity and complexity considered such studies as leading 'to decide more effectively the course best suited to human needs' [Teller, E. The Pursuit of Simplicity; *Pepperdine University Press: Malibu, California, 1981; p. 11]" Later, the review states that "the book will be a great service in bridging the gulf between people of science and people of letters and arts. C. P. Snow complained about this gulf 35 years ago in* The Two Cultures, *and Teller also stressed the importance of a common understanding of ideas and their consequences in order to work together for the future [Teller, loc. cit., p. 12]." There is no reference to your* Simplicity *book in Gell-Mann's book.*

He may not have read my book. Well, I read his, and felt a little less uncomfortable reading it than I'd expected to. But let me tell you something I am thinking about now, a point Gell-Mann brings up but then does not finish the thought. He talks about the uncertainty principle. He talks about the idea that there are infinitely many worlds around us. God, instead of playing dice, has realized all the worlds, and we just happen to be living in one of them. I want to tell you about a discussion I had with Johnny von Neumann.

Before that let's review the story of Schrödinger's cat. The uncertainty principle says that the observation should have an effect on what you find. Schrödinger did not like quantum mechanics and proposed the following experiment to show how absurd it looks on a macroscopic scale. Here is a box with an alpha emitter and a counter, and the counter is connected to an apparatus that opens a container that has poison in it. There is a cat in this box with food and oxygen supply. It is so arranged that if within a certain period of time, say one hour, an alpha particle hits the Geiger counter, there is a 50% probability that the container opens and the cat dies. Now, Schrödinger says, I come along a week later. When the apparatus is opened, the act of observation brings about one of two situations, either the cat is found dead or it is found alive. Up to that time, prior to the observation, neither the dead cat nor the live cat represented full reality; it is the interaction of the observation with the system which brings about the reality of either alternative. The uncertainty principle says that it is my observation that transforms the situation into one in which the cat is alive or else into one in which the cat is dead. And this, says Schrödinger, I do not believe.

We discussed this with Johnny von Neumann, in 1946 or 1947. He expressed similar views, that is, that the uncertainty principle leads to impossibilities. I gave him an answer, and Johnny accepted it, but that answer is generally not accepted.

The crucial point is: what is an observation? Not only Johnny von Neumann said that he didn't understand this but another Hungarian, Eugene Wigner, said that he didn't understand this either. Eugene's argument is that an observation involves an observer. I cannot talk about the uncertainty principle and an observer unless I describe the observer, and I don't know how to describe the observer. What is an observation? What happens when I notice that the cat is dead or the cat is alive? If I don't describe that, the whole thing is fuzzy.

Well, I gave Johnny an answer, and I am very proud that Johnny accepted it. I claim that in every case in which Heisenberg applied the uncertainty principle, there is an irreversible process involved. The importance in the observation is *not the presence of the observer but the irreversible process*, because when the alpha particle is emitted, it can go in this direction, or that direction, or yet another direction. We have a superposition of its going in different directions. The important thing in the analysis of a measurement is that there should be in it an irreversible process. So that from the result of the measurement I cannot get back to a conclusion about what was the state before the measurement.

Would you care to give an example?

The working of the Geiger counter. That's an irreversible process. After it has worked, the result after the measurement is not equivalent with the result before, because from the result after the measurement I cannot get back to the result before. And from the fact that before there was a possibility for interference, I cannot conclude that afterwards there is still a possibility for interference. Now, Gell-Mann has a similar statement but I want to go farther than he does. You may remember that he talks about tracks of alpha particles that are a billion years old and are preserved in some stones about which I then find out the lifetime in the evolution of the stones. And they were there, independent of whether there was an observer or not. That sounds a little like what I am saying.

I want to repeat the same thing in two more ways which point in the same direction. The one is religious. I'll talk to you about God. I don't want to convert you to any religion but God has the property that we have all talked about him, and to the extent we talked about him, we know something about him. You know Einstein's famous statement.

God doesn't play dice?

Yes. I can imagine that he governs the world in any way, but he does not play dice. I claim that Einstein is wrong, and I go so far as to say, God does play dice. Or, if he doesn't, there is an important point here because of the way I describe it. In 19th century physics God may have existed or not, but assuming that God exists, he was unemployed. He made the world 15 billion years ago together with cause and effect; out of that everything followed and there was nothing more for him to do. In the decisions of how experiments come out, whether God exists or not, at least there's a job for him.

I'll now repeat the thing in a different way. Not religious, solely physical. Gell-Mann even says it explicitly and repeats what people have said in the last century, that the world is running down. The entropy is increasing, more and more disorder, predictably nothing interesting can happen. My statement is, maybe he's right, but maybe he isn't, and we don't know that. Because I don't like — I can't say it's wrong — but I don't like the coexistence of infinitely many universes of which the choice is made. I gave you now a long speech to the effect that the choice is not made by an observer; rather, *the choice is being made whether there is an observer there or not.* According to Gell-Mann, it may have been made a billion years ago. And if I choose to be around a billion years later, I may now qualify as an observer, but what if I have not looked? There is an infinite number of processes nobody had looked at, but anybody could look at them.

What I say is this. In quantum mechanics I look at a probability distribution that is predictable and calculable until I interfere with it, or until something interferes with it. What interferes with it is not I but an irreversible process, a chaos process where something relatively small, for instance, one alpha particle grows into the life or death of a cat. This is a process not describable by entropy. The transformation of all these probabilities vanishing, and one of them becoming real, is not included in the discussion of the increase of entropy. That does not prove that the world isn't running down, but that does prove that the running down of the world is not the whole story.

I am not constructing a religion or an opposite thereof. I am not even saying that in my cells by some chemistry a decision is made that corresponds to what I feel as my free will. I substitute all of this by the simple statement: I don't know. But I do know one thing, namely, that out of a set of probabilities, by your interference, or my interference, or Gell-Mann's interference, or God's interference, or a Geiger counter's interference, there develops and emerges one. And this is something not included in the discussion of the increase of entropy.

You are introducing an element of doubt.

An element of doubt which was unjustified and had no substance prior to quantum mechanics. Quantum mechanics, however, had to introduce that doubt in order to avoid direct contradiction between waves and particles. That doubt, therefore, is introduced, as Heisenberg says, except, where he says observer, I say, irreversible process. I even say that as a teacher, explaining things, he has advantages in calling it an observer. But if you look into it, in fact, what it means, it means an irreversible process — in me, or in the cat, or in something not alive, something in this world which is left out of the process of increasing entropy.

What comes out of a quantum state through the processes that are usually described by the uncertainty principle is mutually exclusive situations. They are not completely exclusive because taking probabilities of 10^{-100}, the cat may come out alive, and everything may be reversible. No doubt we have never seen it because we usually don't see things that have probabilities of 10^{-100}. The entropy argument is based on such improbabilities. I am saying that what was not known before quantum mechanics is that whenever I choose to look, I see that somehow a choice has been made.

If you want to believe in God, then I claim, to the extent you talk about God, to the extent you invoke him, permit him, please, to play with dice. But once he has played it, another act of creation, an unfinished act of creation, has been completed.

The uncertainty principle is an explicit conflict with the increase of entropy, because instead of making things more complex, things become more simple. The track of the alpha particle is observed in that stone, or the cat is alive.

Take, for example, the recording of electron scattering on a photographic plate.

Precisely.

You have mentioned the spectroscopic evidence of the Kekulé structures and the Jahn-Teller theorem. What else would you like to mention in this connection?

There was the paper with Tisza on the determination of the moment of inertia. Then there was our work on multilayer absorption, by three of us, another Hungarian, an American and myself — B.E.T., Brunauer, Emmett, and Teller. Steve Brunauer was a good Hungarian physical chemist in Washington. He came to me and said, we have this multilayer absorption and there must be some long-range forces. I said, nonsense. But he said, there are the facts, how do you explain them? I said, OK, you have, according to Langmuir, a layer at an appropriate pressure and it is completely occupied. On top of that layer there is another layer, not due to long-range forces but due to forces between the second layer and the first layer. The theory of multilayer absorption has served as a basis for good determinations of effective surface areas. What enters into it is only two parameters — the pressure and the heat of evaporation of the liquid or solid of the material that is being absorbed. If you measure the amount that is absorbed as a function of pressure, knowing the interaction between the absorbed particles, you can derive the area. This was in the late 1930s and I have written more about it a little later.

One of your books, The Dark Secrets of Physics, *names your daughter Wendy, as a contributor. What is her profession?*

My daughter is a computer scientist. She worked for a company and she did not quite like the work and she is now an independent consultant on computers. We cooperated on my notes when I was lecturing in Berkeley.

Micike and Edward Teller in their home in Stanford, 1996 (photograph by I. and M. Hargittai).

Wilson Talley, the other contributor, created order in our old notes and we put together that book.

We also have a son, Paul, who is a Professor of Philosophy at Davis.

How about your wife?

My wife is a young lady whom I have personally known and known well for more than 70 years. She studied mathematics, she helped me in one of my enterprises, the Hertz Foundation, where we selected students to work on applied science. She has some professional background.

Micike, what was your maiden name?

Micike Teller: Schütz-Harkányi. My father was Harkányi. He died. My mother then married Aladár Schütz.

Any hobbies?

Playing the piano.

What other languages do you speak?

I claim that I speak German with a Hungarian accent, English with a German accent, and Hungarian with an American accent.

Your Hungarian is perfect and you have a very good choice of words.

Mici and I usually disagree in Hungarian. Whenever I have a fight with Micike, it is in Hungarian, so I must be very careful about the choice of my words.

John A. Wheeler, 2001 (photograph by M. Hargittai).

22

JOHN ARCHIBALD
WHEELER

John Archibald Wheeler (b. 1911 in Jacksonville, Florida) is Professor Emeritus of Physics at Princeton University. He graduated from Johns Hopkins University and did postdoctoral work with Gregory Breit at New York University and with Niels Bohr in Copenhagen. He has been at Princeton University since 1938, except for a ten-year break, starting in 1976 at the University of Texas at Austin. He has honorary degrees from 18 universities. He has been a member of the National Academy of Sciences of the U.S.A. (1952), the American Academy of Arts and Sciences (1954), the American Philosophical Society (1962), foreign member of the Danish Royal Academy of Sciences (1971) and the Royal Society of London, England (1995) and many other societies. He has received numerous awards, among them the Enrico Fermi Award of the U.S. Atomic Energy Commission (1968), the National Medal of Science (1971), the Niels Bohr International Gold Medal (1982), the J. Robert Oppenheimer Memorial Prize (1984), the Franklin Medal of the American Philosophical Society (1989), and the Wolf Prize in Physics in 1997.

Together with Bohr, he worked out the theory of nuclear fission, participated in the Manhattan Project and the hydrogen bomb project and was the mentor of Richard Feynman and collaborated on investigations of the concept of action at a distance. He suggested, first, the existence of black holes. He has had seminal contributions in other areas of physics, such as quantum gravity and the theory of nuclear scattering. I recorded several conversations in May 2000 and in May 2001, and the material was finalized in the Spring of 2002.*

You have been involved with the most diverse areas of physics; the book The Story of Physics *calls you "one of the most versatile physicists of the 20th century." Looking back at your career which of your achievements would you single out as most memorable?*

Richard Feynman regarded the one that when you shake a particle here, the reason that it loses energy is because the faraway matter of the Universe interacts with it. I went to talk about that with Einstein. He was pleased with it. This brings together cosmology, the faraway, and radiation reaction, which is close up. Various people have applied that idea in other areas but nothing ever gave me that thrill. You see this arm and you see this leg? I would be glad to give them away, if in return, I could understand "How come the quantum?" "How come existence?" They do not seem to be the same question but the answer will be the same.

Which of your achievements you consider the most important in retrospect?

The most important? All those American troops on the island of Okinawa in the fall of 1945, who were ready to invade Japan — they knew that the Japanese were ready to die rather then to give in; I have had so many of them come up to me and say that those two bombs saved their lives. I am proud that I participated in the Manhattan Project and if anything, I often wonder how many more lives could have been saved, had we done the bomb a year or so sooner.

This question has a painful personal relevance to me. My younger brother, Joe, was killed in action in October of 1944 in Italy. Shortly before he was killed, I received a letter from him in which he wrote: "Hurry up!" Although only vaguely, he knew that I was involved with war work. By the time my brother got killed the uranium separation plant in Tennessee was already operational but only produced grams of uranium and we needed kilograms for the bomb. Similarly, only grams of plutonium were produced, again, instead of the kilograms that we needed. The idea of how to design the uranium bomb (a so-called "gun-type" bomb) was already more or less

---

*Magdolna Hargittai conducted the interview.

John Wheeler in the mid 1930s
(courtesy of J. Wheeler).

clear among the scientists working in Los Alamos that time but the idea of how to build the so-called "implosion-type" bomb was only starting to surface. This was eventually used for the plutonium bomb, dropped over Nagasaki. I have reflected many times on my own roles in the events of those days; could I have sped up the process? Even with a rude calculation, one can judge that had the war ended a year earlier — in mid 1944 instead of mid 1945 — possibly 15 million lives would have been saved. A heavy thought ...

Getting the H-bomb took special initiative on my part. I was not just simply a small cog in a big machine. We got the H-bomb only 8 months before the Soviets did. If it had been the other way around, it would have been too bad for the world.

Although I am proud of my participation in the H-bomb project, it was not an easy decision to make. After having spent years on the Manhattan Project, I was happy being back in academia again. I was full of ideas. I received a grant from the Guggenheim Foundation to go to Paris to work on topics that interested me that time. Earlier I had worked with Bohr on the liquid-droplet model of the atomic nuclei. Later the independent-particle model of the nuclei appeared to be also a possibility and I wanted to try to develop a unified way to describe the behavior of the nuclei. In the summer of 1949 my wife Janette and I went to Paris and I started to work there, with frequent visits to Bohr's laboratory in Copenhagen. In September of that year we heard that the Soviets succeeded in making their atomic bomb, and we suspected that they might have already got a head start in the development of their hydrogen bomb. Soon thereafter I received a telephone call from Washington, asking me to join the H-bomb project. This was an awful dilemma for me. I thought, together with Edward Teller, that this was an obvious threat to our country. At the same time, I just recently started to work in Paris and was enjoying it. I was struggling between my obvious patriotic duty and my love for physics. Janette and I had long discussions about this. Later, I went to visit Bohr again, and told him about all this. He asked: "Do you think that Europe would be free of Soviet control today had it not been for the atomic bomb of the West?" Finally, later in January 1950, I decided to join the H-bomb project.

Many scientists refused to work on the thermonuclear program. It must have caused some dissent with some of your colleagues. How did you handle that?

Yes, many of my colleagues did not understand why I joined the H-bomb project. That was sad but I can't recall getting into big political discussions because you could not change anybody's mind. Next door to us at Princeton lived the art historian, Erwin Panofsky, who had two sons, very promising ones; one of them was known as the dumb Panofsky and the other as the bright Panofsky. The bright Panofsky was No. 1 in his class and the dumb Panofsky was No. 2. We knew that Panofsky did not approve. The FBI men, who have come around to check up on our reliability by asking questions of neighbors, came to the Panofskys and asked them. Panofsky told them, "They are not subversives, they are mass murderers! We are the subversives." After he died there was a little meeting and I remember one person saying about him: "He hated children, grass and birds, he loved all dogs, a few friends, and words."

You wrote in your book that Edward Teller was a good friend of yours. He was especially strongly criticized for his actions. There were several other famous scientists who participated. Why was he considered to be the villain alone?

It was a break in the solidarity in the community of Jewish businesses. Alvarez was not Jewish, Lawrence was not Jewish but Teller was. This is like being a traitor to your country, a traitor to a group of people. That is a theory; I'd like to see that theory examined for or against it. I did not testify at the Oppenheimer hearing. On the day before Teller's congressional hearing he and I happened to stay at the same hotel in Washington. We talked about this long into the night and he was agonizing about what he should

A group of participants of the Theoretical Physics Conference at George Washington University in 1937. In front, Niels Bohr, second row, from left, Isidor Rabi and George Gamow, third row, Fritz Kalckar and John Wheeler, fourth row, Gregory Breit (courtesy of John Marlow, Princeton University).

do. He asked my advice and I told him, you should tell it the way you see it.

Teller is a temperamental person. During the war and later, during the H-bomb project, he made many enemies with his impatience and arrogant behavior. His campaigning for the new weapons laboratory did not help either. By the time he testified at the Oppenheimer hearings, almost all scientists disliked him. I felt differently. In my opinion he fought obstinately for what he believed in. I may have disagreed with his tactics but never with his goals.

This business about being divided on the H-bomb was similar to the division you see today on Star Wars. There is something very puzzling to me about the way communities stick together and people are opposed to Star Wars. The intellectual side of it has been spelled out many times but what's the emotional side of it, I do not know.

I think that in the first part of the next century we are going to have an enormous war bigger than any war we've ever had. I do not know how it is going to develop.

Would not a new world war lead to total destruction?

When you see pictures with people fighting with swords, you wonder why were not all killed because nobody wore armors. Do you ever look at *Encyclopedia Britannica*? There is an article about the "Whisky rebellion". After America won its independence from Britain, all the colonels wanted then was to get back to peace. But Alexander Hamilton felt that it was absolutely essential to have an army and the constitution had no real provision for an army. So he invented a cause for a war. He got congress to pass a tax on whisky. The farmers in the western part of Pennsylvania had no good way to get their grain out in the market; the only way to get real money for it was to convert it to whisky and sell the whisky. So by putting in a law against it he got a big fight going between the farmers and the revenue collectors. Well, the fight leads to guns and the guns to government action. So Hamilton called up the militia and put them on the Whisky rebellion.

What turned you originally to science?

I think my mother had an interest in science, although she did not have a college degree. I may have inherited my mathematical and scientific aptitude from her. I remember, when she went to the grocery store to buy groceries and the clerk would add up the numbers she could read the numbers upside down and add them in her head faster then he could.

Who are your heroes?

I have a stone on the island in Maine that my son and his wife brought back from the outskirts of Athens from a garden where Plato and Aristotle walked and talked. And I hope some day to find a machine one could put the rock into and the conversation will come out. Those were certainly heroes. Giordano Bruno burned alive just because he had views about the structure of the Universe. Copernicus, Galileo, Boyle. He took the ideas of Galileo and wrote up in the form that caught the attention of Newton. Voltaire made fun of him; Voltaire also ridiculed Leibniz, so I don't consider Voltaire a hero, but Leibniz certainly. Leibniz taught us that philosophy is too important to leave it to the philosophers. Then, of course, in modern times, Rutherford. I can recall being at a reception at the Royal Society at the time of a big international conference on physics in September of 1934 and I had no tuxedo, no black tie. There was Rutherford standing in the hall with a group of about 15 younger people around him and listening to every word he said.

I walked around in Rutherford's lab and I don't remember the exact wording he used but it was something like this: "If you are interested in cosmology, it is time for you to go." He wanted to get down to practical things. There are so many heroes. Bohr and Millikan. Millikan took the California Institute of Technology and raised it from nothing to an important school. I used to see him at meetings with a little black notebook he had and I understood that he listened to the talks and wrote down what he thought about the speaker, whether to try to get him for Caltech. He tried to get Einstein, he thought that Einstein would sign up but Flexner from the Institute for Advanced Study went to talk to Einstein and Einstein went to Princeton instead of Caltech.

There had been a debate between Robert Millikan and Arthur Compton about the nature of primary cosmic rays. Are they radiation or are they particles? Compton argued for particles of matter and Millikan for photons. So Compton worked and worked and worked measuring the intensities of the cosmic rays at different places and concluded that the Earth's magnetic

Albert Einstein, Hideki Yukawa (1907–1981, Nobel Prize in Physics 1949), John Wheeler, and Homi Bhaba in Princeton, 1953 (courtesy of J. Wheeler).

field had a big influence on them so they must be charged particles. The person who calculated the orbits in the magnetic field of the Earth was a Mexican; Manuel Sandoval Vallarta. My wife and I were once having dinner with the Vallartas in Mexico City and he said, come into the library after dinner, I want to show you something. He went to the shelf and opened up something and inside was a gold coin. He said that this gold coin was given to his grandfather by the Emperor Maximilian. Remember, Maximilian was executed. He told the members of the firing squad: I know that you are doing this as a matter of duty and not as hatred for me, so I want you to feel a good conscience about this. So he gave each of them a coin.

You have worked with some of the greatest minds of the 20th century, and especially when you were in your young formative years. Was this a decisive influence on your life? Have you realized this already at the time?

I worked with all these great people when I was young; oh, I was so lucky!

You knew Einstein personally and worked on topics very close to him. Would you mind telling me something about him; what kind of a person he was, what was his attitude towards his pupils/colleagues, what was his working method, etc.?

There were many young people working with him. I remember, Einstein often walked home from the Institute for Advanced Study to his house passing my house with Peter Bergman on one side and Valentine Bargmann on the other. Once our children's cat followed him home. He took the phone: tell your children that their cat is over in my house.

I think Einstein took it almost as a joke that he was so famous. I remember, once a neighborhood girl came to his door and rang the doorbell and said: my parents tell me that you are a great mathematician, could you give me some help with my mathematics in school? Almost everything that came up he made a joke out of it.

Would you care to comment on Einstein's "cosmological constant"?

Einstein's reaction to the cosmological constant and to what I call the mystery of the missing mass was that he told me that he had not taken it that seriously at the beginning, because it led to the idea of a Universe starting at a certain moment. Einstein's greatest hero was Spinoza. He was expelled from the Synagogue in Amsterdam, for denying the Bible account of creation.

One of your first mentors was Niels Bohr. Would you tell us something about him?

I first heard Bohr in Chicago at the World Fair although saying that I heard him is an exaggeration because he talked in such a soft voice. Later, I was fortunate enough to spend a postdoctoral year, in 1934–1935, in his institute in Copenhagen. That was the most exciting place to be at that time; most of the world's greatest physicists came to visit Bohr at one time or another. He had a very special working style. He liked people to be around him and liked to discuss everything, often pacing up and down in the room. He was also famous for his long walks with his colleagues, during which they discussed science.

Would you care to tell us about your work with Bohr in 1939 on the theory of fission?

When we worked on the theory of nuclear fission that was a great time! Bohr came to Princeton, where I was an assistant professor, in January 1939. He came to give lectures at the Institute for Advanced Study and to visit his longtime friend, Albert Einstein. The two of them could not agree on the meaning of quantum mechanics. As you know, Einstein could not embrace the probabilistic nature of quantum theory, he was of the opinion that "God does not play dice." He felt that the Universe is lead down and there is no probability aspect in it. On the other hand, Bohr believed that uncertainty and unpredictability are intrinsic parts of quantum theory. But when he came to Princeton in 1939, his mind was on the newly-discovered nuclear fission. As it turned out, he just heard about it before embarking his ship and this is what had occupied his mind all the way during the trip. Just before leaving for the U.S., Otto Frisch, a German refugee physicist, working in his institute, told him about what he and his aunt, Lise Meitner figured out during a Christmas Eve walk in the woods in Sweden, where Lise Meitner lived then. They were thinking about the strange results of Otto Hahn and Fritz Strassmann, who found that when they bombarded uranium with neutrons, barium was formed.

Niels Bohr
(courtesy of David Shoenberg, Cambridge, U.K.).

They suddenly realized that the uranium atom, bombarded by neutrons, must break into smaller pieces, which are the nuclei of other, smaller elements. Bohr got very excited by this idea and could not stop thinking about this topic on his ocean trip.

Already years earlier, Fermi produced fission in his laboratory in Rome but misinterpreted it by suggesting that they created elements that are heavier than uranium. Interestingly, a German chemist, Ida Noddack wrote a paper in 1934[1] in which she suggested that Fermi's experiment split the uranium but she was obviously too ahead of her time and nobody paid any attention to her. It is an interesting question; had this suggestion been given by a man rather than a woman, would it have been listened to? It remains a fact that physicists all over the world remained blind to fission during the middle 1930s. This, in retrospect, is a blessing. What could have happened to the world if scientists in Germany had noticed and followed up Ida Noddack's paper in the mid 1930s and produced the atomic bomb before the Allies?

When Bohr arrived in Princeton in 1939, he asked me if I was interested in working with him on the detailed theory of fission. I was interested in this topic as a basic science and happily said yes. This was an exciting time. The word fission was originally suggested by Otto Frisch, who borrowed it from cell biology where it is used to describe cell division. Bohr did not like this word at all. He said: "If fission is a noun, what is the verb? We cannot say that 'a nucleus fishes'." We even went to the library and looked through all sorts of dictionaries trying to find another word that he would like better. We did not succeed.

We had to understand this new nuclear phenomenon, fission. It was obvious that the nucleus of such a heavy element as uranium must undergo a considerable deformation before it splits. For that it needs energy. When the uranium is bombarded by neutrons, the neutron can provide this energy; we say that the nucleus is excited. This excitation then could initiate a vibration in the nucleus that could deform it. Our Hungarian friend, Eugene Wigner helped us out. He ate some oysters in downtown Princeton and got sick and was in the hospital on the campus. I went to see him at the hospital to get some help. The questions that Bohr and I were dealing with; were like a chemical reaction. Uranium breaking up is like carbon monoxide breaking up into carbon and oxygen. I remembered that he had worked in that field with Michael Polanyi. And he helped us and, eventually, getting also ideas from discussions with other colleagues, such as Placzek and Rosenfeld, Bohr and I saw how fission works. Bohr left Princeton in April of that year and during the following months I wrote the paper and we submitted it to *Physical Review* in June. It came out in the September 1, 1939 issue; by strange coincidence the same day when Germany invaded Poland.[2]

Have you seen the play Copenhagen? *What do you think about Heisenberg's sudden visit to Bohr during the war?*

Yes, I saw that play, there was also a reception in New York that I attended. That question came up when my wife and I talked with Margrethe Bohr. Bohr did not seem to realize what Heisenberg's reason was.

Why do you think Heisenberg stayed in Germany and continued working during the war?

Oh, to me this is simple. Loyalty to his country. If he is working for his country and if the uranium is working, and if it's important for the bomb they should work on it. Have you read the Farm Hall transcripts? I never asked him, I should have. But I remember dedicating the Heisenberg laboratory in Munich after the war, he was no longer living. There, in the coffee break, I heard two young Germans talking to each other. One was saying: Heisenberg should be criticized because he did not give us the bomb. And the other one said: Heisenberg should be praised that he did not give us the bomb.

He told me the following story: when the outcome of the war seemed to be clear and the Allied forces were approaching, Heisenberg left his research laboratory without authorization and decided to get home, to Munich to his wife and children. He was cycling home on his bike. But that time Hitler gave orders to the sentries to shoot everybody. Heisenberg was stopped by a sentry who was given orders to shoot whoever came without orders. The sentry asked him, "Where are your orders?" He did not have any orders but he reached into his pocket in which he had a pack of cigarettes; so he gave one cigarette to the sentry. Heisenberg realized that he had a new equation: one cigarette equals one life.

Questions like the divide between the quantum world and the classical world; I think it was questions like that that Bohr and Heisenberg could not agree on. I wanted to know what they disagreed about and listened to them. And that is when Bohr answered, "To be, to be, to be, what does it mean to be?" If you get to the point, you have to ask what does it mean, "to be" in the quantum world. That is what I am going to spend the summer on. As I said if I could give my arm or my leg to get that answer; I would give it. What is the underlying reason for quantum mechanics?

My friend over at the Institute, Edward Witten, thinks that string theory is the answer to all this. I am so glad he thinks so because this would mean that he would mark out all the applications and implications. He has a kind of religious faith in this string theory, he is a true believer.

Are you religious?

Some people don't call the Unitarian denomination a true religion; some people think it is too abstract for the fatherhood of God, the brotherhood of man, the leadership of Jesus, salvation by character. My wife and I helped to start the Unitarian church here, in Princeton. But some people would not call it a true religion.

What is your great goal in life?

Find out, how come existence?

Knowing all that you know today, if you would be starting your scientific career today, what would be the research area of your choice?

It would be quantum mechanics. I packed many books in boxes that come with me to Maine or the summer, that may help me to find out "how come the quantum?"

What do you think about the responsibility of a scientist?

Scientists are part of the community and they have to feel their part in the community. That gives an inner drive to the scientist to feel that he is trying to do something for the community. If the community does something for science giving the scientist encouragement, the scientist does something for the community.

What is your opinion of the Nobel Prize as an institution?

The Nobel Prize is a wonderful institution. Just have a look at these pictures down the hall that is wonderful. That committee does real work.

Would you care to say something about your family background and your present family?

My family background: in the hallway outside our apartment there is a beautiful portrait. In the days of the Great Depression — you probably don't remember the Depression but it was terrible. A man across the street from us killed himself because he did not see how he could support his family. People came knocking on the door, and asking for any work, mowing the grass, paint, anything, just to have some work.

John Wheeler with two of his granddaughters, Eleanor Ufford and Frances Ruml at Christmas of 1963 (courtesy of Letitia Ufford).

My father was a librarian. He wanted to make the library that he had taken over in Baltimore into a good library. It was in some old apartment houses, spread between them. He got an architect to do the plan, he got a price, and then he had to get approval to spend that much money on this. The city was not allowed to commit itself to that much and it had to be state-approved. So I remember my father went to the state legislation of Maryland and he talked to different legislators to get their approval. Finally they voted and we got the library. He was very much committed to what he was doing. He arranged for storefront windows in Baltimore to have books on display; he also made arrangements with a laundry to print information about the library in the cardboards that they put into shirts.

My wife and I married in 1935; it has been a wonderful marriage. By now we've been married for over 65 years. We have three children, Letitia, James, and Alison. My oldest daughter, Letitia, is writing a book on the 19th century Middle East; my youngest daughter, Alison, raises money for the Peddie School, and does lots of other things; and my son, James, is a pathologist who has recently retired from the Department of Pathology at the University of Pennsylvania.

I had too many books in one of my rooms and my daughter told me that they do not want to deal with them when I die so I should do something

Janette and John Wheeler during Christmas in Brookline, 1980s (courtesy of Letitia Ufford).

with them. So I asked my friends in Texas, what school there has good physics, good teachers but not a good library. They had a meeting and after that suggested the Abilene Christian University in Abilene, Texas. So thirty boxes of books went there.

References

1. Noddack, I. *Zeitschrift für Angewandte Chemie* **1934**, *47*, 653.
2. Bohr, N.; Wheeler, J. A. *Phys. Rev.* **1939**, *56*, 426.

Freeman J. Dyson, 2000 (photograph by M. Hargittai).

23

FREEMAN J. DYSON

Freeman J. Dyson (b. 1923 in Crowthorne, Berkshire, England) is Professor Emeritus of Physics at the Institute for Advanced Study in Princeton. He received a B.A. in mathematics at the University of Cambridge (U.K.) in 1945, spent some time as Professor of Physics at Cornell University, and became Professor of Physics at the Institute for Advanced Study in Princeton in 1953. He participated in the ORION space-ship program and the TRIGA reactor program and served as a consultant in several governmental agencies, such as the Space Agency, the Disarmament Agency, and the Defense Department. He has been elected Fellow of the Royal Society (England) in 1952 and to the National Academy of Sciences of the U.S.A. in 1964. He was the recipient of numerous awards, among them the Danny Heineman Prize by the American Institute of Physics in 1965, the Lorentz Medal of the Royal Netherlands Academy in 1966, the Hughes Medal of the Royal Society (London) in 1968, the Max Planck Medal of the German Physical Society in 1969, the Wolf Prize in Physics (Israel) in 1981, the Enrico Fermi Award in 1995, and the Templeton Prize (see on p. 479) in 2000. He holds honorary degrees from over 20 universities. He has written several books, *Disturbing the Universe* (Harper & Row, 1979), *Origins of Life* (Cambridge University Press, 1986, Second ed. 1999), *Infinite in All Directions* (Harper & Row, 1988), *From Eros to Gaia* (Pantheon Books, 1992), *Imagined Worlds* (Harvard University Press, 1997) and *The Sun, the Genome and the Internet* (Oxford University Press, 1999). Our conversation took place in Professor Dyson's office at the Institute for Advanced Study in Princeton on April 24, 2000.*

About 30 years ago, in Physics Today, *you prognosticated physical research to the end of the 20th century. We are now there; how good were your prognoses?*

I don't remember what I said but one thing I do remember, which came out right. I was saying that the cosmic rays would come back into fashion, that accelerators would reach their limits, and that we should look for the future of particle physics with going back to observing cosmic rays and, of course, that is what has happened. During the last few years most of the interesting discoveries have come from passive detectors underground. So that was one thing that came out right. But I don't remember what else I said.

You also mentioned that physicists should look for possibilities to help molecular biology. For example, the sequencing that takes so much time by wet chemistry, should be done by physical methods.

Yes, that is something that has not happened. Of course, it still may happen, but not yet. It is interesting that this took longer than I thought. Already 30 years have gone by and it is still wet chemistry.

What would be your prognosis for the next 30 years?

I would think that probably the single molecule sequencing will turn out to be practical and probably will be done mostly by physicists. Of course, you never can tell how long these things will take. I would think that it ought to take less than 30 years, certainly, so we have very fast sequencing where you do not have to separate the molecules. Of course, similar problems are in protein chemistry, there we still do not have a really rapid way of solving protein structures. There is a young man at the University of Washington Medical School, his name is John Sidles. He is a physicist working at a medical school and he has invented a device, which he thinks one day will do protein structures, more or less one molecule at a time. You don't have to make crystals and it is done by a combination of atomic force microscope with a magnetic resonance imager. It is a single atom scale

---

*Magdolna Hargittai conducted the interview.

magnetic resonance. It does not yet work, but I think there is a chance that it will and then that would do the same thing for protein structures that the single molecule sequencing does for DNA. That kind of thing is very likely, not the one that Sidles has invented, I think its time has gone by; I think he proposed it about five years ago and as far as I know nothing has happened. So maybe somebody else will invent something better. But it seems to me that there is great room for physics there. Many physicists are interested in such projects; there is a fellow called Bob Austin here, in Princeton, who does similar things. But these problems are, of course, difficult.

You suggest in several of your writings that computer science and genetic engineering could eventually be combined. You call it computer-assisted reproduction. Don't you find this marriage dangerous?

Yes, of course. Everything connected with biology is dangerous. It is much more dangerous than physics, but nevertheless I think that it is extremely promising as well. So one has to watch out for the dangers but it would be stupid not to pursue the promise. The promise is, of course, mostly in medicine, but no doubt it will also be applied to the breeding of animals and plants. This will certainly become even more of an art than it is today and it could be a really exciting art form to design plants and animals. If you think how much effort goes into growing different kinds of flowers, such as orchids and roses, this could be done better with a certain amount of computer assistance.

This seems to be all right with most people as long as it goes on with roses and orchids; but we are moving into more dangerous grounds when it is done with animals.

Exactly, that is right. The closer we get to humans the more dangerous it is. That is quite right. You should talk to Lee Silver. He lives here in Princeton. He is a biologist and has made a study of fertility clinics. Fertility clinics are, of course, the places where you produce babies and that is the place where the action is. It is now all over the world not only in the rich counties but in poor countries as well; this is a fast-growing branch of medicine. Giving people the chance to have babies is an enormous thing. Lee Silver is, of course, talking about the future, but also about what has already happened. It is very quickly going to be possible to manipulate the embryos, to put in genes that you want or take out genes that you don't want. And that is when it really begins to be a problem.

Just think about twins, having the same DNA, looking alike and still being totally different persons ...

Yes, in fact we have two of them here, I have twin grandsons who are now six years old and it is a very good demonstration because it is our own cloning experiment. They have exactly the same genes, they have exactly the same environment, they always live together, sleep together, do everything together and still are two different people. The fact is that the growth of the brain is not determined by the genes, it is a more or less random process as far as one can tell, the fine structure of the brain, the neuroconnections are produced randomly. And, it is this randomness in the neuroconnections, which produces a person, fortunately not the genes.

So you do not take a DNA sample from Einstein and produce another Einstein.

No, you clone the genes, you do not clone the person. Fortunately.

J. B. Conant said somewhere that the first part of the 20th century was the time of physics and the second part was the time of biology. What will be the dominant science in the 21st century?

I usually say neurology, but of course, I could be quite wrong. The real dominant science may be something that has not yet begun. So we don't know. From among the present existing sciences I think it is neurology that is the most likely because it deals directly with the brain and the brain is the big unsolved problem.

Let us talk about you. What turned you originally to mathematics?

Just that I was good at it. Already as a very small child I loved numbers. I used to scribble big numbers. I remember that I was about 4 or 5 years old and I was calculating the numbers of atoms in the Sun! It was a huge number with lots and lots of zeros and I was very proud of that. So it came naturally, by itself. There wasn't any question, it came from the inside not from outside.

Were you a child prodigy?

No. That was simply my interest, numbers were just a delight, something I was very happy with.

Freeman Dyson in 1951
(courtesy of F. Dyson).

You started your career as a mathematician.

Probably that period of my life was when I was the most creative. I did some really beautiful things as a pure mathematician. But the problem was that every one of those papers was read by five or six people. So I decided that it was not what I wanted to spend my life with. Of course, I was doing rather old-fashioned mathematics. Old-fashioned number theory and the kind of things that Erdös did, not quite as good as Erdös.

What is your Erdös number?

Two, and I am very proud of that. Most of my friends here have only five or six.

Do you consider yourself a mathematician, an applied mathematician or a theoretical physicist?

It is a difficult question because it is all of the above. I would say that the truest description is probably applied mathematician. Everything I do is really mathematics but for professional reasons I call myself a physicist. Being a physicist gives me a place in society, which is a little bit different. But it really does not matter.

Exactly that is my next question, why is there a need for such a distinction at all?

It is a question of how people like to organize themselves. Here at this Institute we have a School of Mathematics, which also does applied mathematics, and we have the School of Natural Sciences, which does mostly physics. I find that, although I am really an applied mathematician, I fit better in the physicists' school, for personal reasons. Of course, my interests are somewhat different from my talents. My talents are all in mathematics, but my interests go much wider. I am interested in biology but I never could do biology. I am interested in astronomy but I could not do astronomy either. But still I love to talk to astronomers and biologists, so I fit better into the natural science school, even though what I am doing is much more mathematics.

What do you consider as your most important contribution to science?

Well, it is hard to tell until a hundred years later. My son tells me that my most important contribution is this little book on the origin of life — and maybe he is right, I don't know, but it is his view. Perhaps in the long run that is what I will be remembered for. Certainly that is a more important problem, because everything I did in physics was very unimportant. It was beautiful, it was exciting, it was a much better playground for me because I just love to play around with mathematics, so actually I became established as a physicist by solving puzzles. These were very interesting puzzles but nothing I did was really important. Whereas if you take the origin of life, perhaps that is important.

I wold like to get back to this question later, but for the moment let's stay with your physics. You did the unification of quantum electrodynamics; was that not important?

No, I don't think so, it was just a tidying up. It was necessary and somebody had to do it, but I am sure if I had not done it someone else would have done it very soon. It was obvious; you had these three theories which all gave the same answers but looked different. It was obvious that they had to be put together and I was the person to do it because I happened to have the mathematical skills that were needed. So I would say that it was a good piece of work and it was timely and helped a lot of people, but looking at it from the point of view of history it did not change anything. It only meant that the subject maybe was pushed ahead two or three years. So maybe it was somewhat important but not very important.

Is quantum field theory a good way to describe Nature?

Yes, certainly it is good. It is not final, it is still growing but it certainly does very well. It still has no mathematical basis, there is no strict mathematics to justify it, it is essentially a recipe for cooking. You put in the right ingredients and you get the right answer, but there is no mathematically rigorous structure behind it. As far as the nuclear interactions go, the theory exists as a strong interaction theory and that is what is called quantum chromodynamics now. Quantum chromodynamics is the extension of quantum electrodynamics to include the strong forces. Quantum chromodynamics is a beautiful theory but nobody can actually calculate with it; it is a theory, which has not yet become a useful tool. You cannot calculate the quadrupole moment of the deuteron, these are physically measured things that we tried to calculate 50 years ago and they still have not been calculated. Although the theory is very likely correct, nobody knows how to do the mathematics.

Why?

It is not clear why. Either because we are not clever, or because the theory is somehow radically incomplete. We don't know.

Einstein said that the most incomprehensible fact about Nature is that it is comprehensible. Is it?

Yes, I think so. Haldane said about biology, he was writing in 1923 and he always expressed himself very clearly: "We are at present almost completely ignorant of biology, a fact which often escapes the notice of biologists, and renders them too presumptuous in their estimates of the present position of their science, too modest in their claims for its future" (from Haldane, *Daedalus*, page 50). I think that that is true for physics as well; what we have done so far is quite modest but it does not mean that we can't understand everything at the end. There is still a long, long way to go.

Are the laws of physics the same in the whole Universe?

That is an experimental question, which we are trying to find out. In fact I wrote a paper quite recently on this. Whether the laws of physics remain the same is easier to test by going back in time, than by going to large distances, because if you go to large distances you automatically go back in time. So you might as well use the evidence that is here.

The most precise evidence that exists is from the fossil fission reactors in Africa which were running about two billion years ago. They are at a place called Oklo. The French discovered these when they were mining for uranium. These reactors were actually running two billion years ago and by studying the remains you can determine very accurately the nuclear physics as it was two billion years ago. It agrees precisely with what we measure today. So you can say that the relative strength of the forces was exactly the same as it is now within something like one part in ten billion. That was billions of years ago, so that means that the laws of physics did not change even by a tiny fraction. So that is the most accurate evidence we have. But, of course, there is a lot of astronomical evidence as well. If you look at distant galaxies, you see the same spectra with the same atoms that we have here. So chemistry seems to be the same.

Mutation is common in biology. Does it seem that there is no mutation in physics then?

Well, there may be but we have not been able to observe it.

Symmetry is an important scientific term and is also in our everyday vocabulary. Do you anticipate any further use of this concept in scientific development?

Oh, yes. It's been, of course, a goldmine and there is no reason to think that it has finished. I think that the most remarkable thing is, which came to me as a great surprise, that most of the important symmetries are broken. Of course, we all knew about symmetry for a long time but we did not understand this until the last thirty years. That was a big surprise that the whole Universe seems to be full of these broken symmetries. This turned out to be very powerful, that even if there was no symmetry, you pretend that there was. Thus you can actually use ideas from group theory, from Lie algebras, and such things as if the world was symmetrical when it is not. That has been extraordinarily fruitful. All the important symmetries are broken, almost without exception.

Do you think the Universe infinite or will there be a Big Crunch?

That is also an observational question and we don't yet know. But it is fairly sure, in fact, that there won't be a Big Crunch. It seems that the density is way below what is needed to cause the Universe to recollapse. Of course, it is a question, how much density there is, but all the measurements of density indicate quite a low value, so it seems to be expanding forever. But it is not 100% sure.

Some time ago a paper suggested that the Universe is not only expanding but doing that at an accelerated rate.

That is a very interesting question. Of course, the newspapers always like to present these things as a question that has already been answered. That is not true. What we have is new tools for attacking the questions, and that is what is important. Today we have several new tools to answer these questions.

What are these new tools?

The two main ones are the distant supernovae, which can now be measured in large numbers. Of course, the essential tool is the software, so that you can actually program a telescope to observe thousands of galaxies in one night and compare the images automatically. You can now more or less guarantee that you discover supernovae, while in the old days it was a matter of chance. Now with the good electronic imaging and good software you can observe ten new supernovae every night when the sky is clear. So finally you have good statistics. That is one tool. The other one is the cosmic microwave background and the anisotropy, which gives a lot of information about the early universe. That again is indicating an accelerating universe, but not very strongly. Both these tools are extremely powerful, so it is a big development but it is not the end. So I would say that whether the universe is really accelerating or not we do not know but we will probably know in another ten years.

Today they call it "dark energy" but it is essentially the same thing as Einstein's so-called "cosmological constant", what he later called the "greatest blunder of his life". Now it appears that it was not such a blunder after all.

Well, we don't know. Of course, it was not a blunder, just an interesting hypothesis. Einstein had, to my mind, a very unscientific way of looking at things. He never seemed to be much bothered by the observations, for him theory was a kind of revealed truth. For him to say something wrong was a blunder, but very often saying something wrong is actually a step forward. It is much better to be wrong than to be vague. So this cosmological constant was a good idea, even if it was wrong.

It is interesting that he did not care much about experimental evidence.

Yes, especially at the end of his life, he was very much wrapped up with his calculations. I found it very sad because I was here when he was an old man. He never came to the seminars; he was not interested in all the new experiments that were going on. He was very isolated.

Did you know him personally?

No. There was never a chance to get to know him, which is a shame. There were other old people here who were quite different. Niels Bohr, for example, was always talking to the young people, he wanted to know what was happening. We knew him well. Einstein was here about the same time but he was only a remote figure. So all I could tell about him would be second-hand information. No doubt he was a great man, but I would say that in the last twenty years of his life he lost interest in what most of us considered to be exciting.

What was occupying him then?

His unified field theory, which was, of course, a dead end, as far as one can tell. Bohr was much more human. He was interested mostly in nuclear physics; he is the one who, as we all know, built up nuclear physics in Copenhagen. It has remained a very central school in nuclear physics and for him it was very exciting. He was never disillusioned about nuclear energy. But he was interested in all sorts of things, new particles, for example. He was rebuilding science. Everybody in Europe, after having survived the war, was enthusiastic about that. Bohr was basically a very happy person, he had so much to do just to rebuild science, and he was also active in the United Nations. He was one of the chief persons in the Atomic Energy Commission of the UN. So he was extremely active in all sorts of practical things. But Einstein was not, so there was a big contrast.

Did you see the play Copenhagen?

Yes, just about a week ago. It was extremely interesting. Of course, I knew all three of those people quite well. It is a good play. The thing that is good is the main point it makes, that history eludes us, history eludes our grasp, we don't know what these people said and there is no way of finding out. History is something that very often leaves questions unanswered. That is important. There is another example of that which is much more important and I think it is a good illustration. That is a question, which has been so much in the public mind: would the Japanese government have surrendered if we had not dropped the atomic bomb? One of my friends is a historian, called Robert Butow. He is an old man and he lives in Seattle at the University of Washington. He was in a very good position to study the Japanese surrender. He wrote a book, called *The Japanese Decision to Surrender*. He happened to be in Tokyo soon after the war as a young man; he was a student but was there as part of the U.S. army, I think. Anyway, he made it his business to interview all the Japanese he could find and examined all the documents, just to answer that question, while the memories were still fresh. He asked these Japanese politicians who had been involved in making decisions about the war. He asked all of them the question: would you have surrendered if we had not dropped the bomb? And the answer he got was always: we do not know. It was very evenly balanced. Lots of army generals, and navy admirals were absolutely determined to go on fighting no matter what; they were about half of the government. Then the civilians on the whole would have been wanting to surrender. So there is no way to tell which way it would have gone. I think that is the truth, that history just does not answer these questions.

Are you familiar with the Alvarez letter?

Yes, but I think it did not get there soon enough to make any difference. It is a very interesting story but in fact it did not influence the Japanese decision. The people around the emperor did not know about it soon enough and they made the decision independently. But, of course, what would have happened if the bomb had not been dropped, nobody knows. Anyhow, that is a digression, but I was reminded of it very strongly when I saw the *Copenhagen* play because it is a similar thing, in a way. Although we would love to know what Heisenberg really said, we just don't know.

When I saw the play I had the feeling that there was one person who is missing there and that is Elisabeth Heisenberg. She was just as strong a character as Margrethe Bohr and she was very important. Of course, she was not there in Copenhagen, but I think that her influence on Heisenberg was extremely strong. I remember meeting them after the war, Elisabeth and Heisenberg, and she was much more impressive. It was like Lady Macbeth and Macbeth. She was obviously the driving force. Another aspect that both Margrethe and Elisabeth were interested in was surviving, for them that was what mattered. I think that Heisenberg probably had some vague idea that he and Bohr together could do something, what it was, I don't know.

He must have had some sort of a dream. Clearly, he was driven by this idea that Germany should lead Europe; he was a very patriotic German. Part of that, of course, would have been to make peace with the occupied countries. I haven't really examined what little evidence there is but it was remarkable how much Elisabeth and Margrethe were similar; each of them had six children and both of them were aristocratic types; they behaved like royalty and somehow everybody treated them like royalty. You could see how Margrethe made sure that as far as possible her family would survive. That was always her dominating motive and that was certainly true of Elisabeth Heisenberg as well. Both of them, in a way, succeeded.

How do you judge Heisenberg for staying and working in Germany when many of his colleagues were forced out of Germany or thought it proper to leave?

I respect that very strongly. I think that he was brave, it was a very hard choice, and he knew that. I think that he deliberately took upon himself a big burden and I respect him very much for that. Anybody who stayed in Germany at that time deserves a lot of respect. Especially since he clearly was not a Nazi himself. It took a lot of courage and he knew he would be blamed by both sides. So I think he was a brave man but probably he would not have been so brave if he had not had a brave wife. As history worked out I think you could say that he did less harm than almost anybody else. He saved a number of lives, I don't think that he did any harm, and his nuclear reactor was certainly a joke.

Do you think that his mistake in his calculations contributed to the failure of the German nuclear program?

There were lots of reasons. Certainly the fact that he never bothered to do the calculations in three dimensions had a lot to do with it. He was always dealing with flat plates because they were easier to calculate and of course if you wanted to be serious you would have made lumps of uranium rather than flat plates. It is harder to calculate but works better. So in a way I think that he was not very serious. There were, of course, other reasons why it failed. For example, the fact that the supply of heavy water disappeared, the fact that they had the wrong capture cross section in carbon, and a number of experimental errors. So there were a lot of things combined and it was very unlikely that it would have been a success; it was just not a very serious effort.

Could it be intentional on his part?

It is hard to tell when you are intentionally non-serious. I made decisions like that, too. Somebody asks me to do something that I don't really want to do so I am rather half-hearted. Not that you deliberately decide to sabotage, just that you do not put into it maximum effort, if you don't fundamentally agree with what they are trying to do. So I think that something like this may have happened in his case, a sort of a mixture, maybe different motives at different times. After all this thing went on for six years. Towards the end of the war it was clearly not going anywhere.

Somewhere you wrote a moving story about how you forgave Edward Teller when you heard him playing the piano in your house. How do you consider now his conduct in the Oppenheimer hearings?

It was politically wrong because the fight against Oppenheimer was a sort of intraservice battle between the Air Force and the Army. It is so often true that there are all sorts of fights between different parts of the government and, particularly in this country, the Air Force and the Army have always hated each other. They were more hostile to each other than either of them was to the Russians. So the Air Force wanted big bombs and the Army wanted small bombs, and Oppenheimer supported the Army, so the Air Force decided to destroy him; that was roughly how it went. The fight was started essentially by the Air Force people, and Teller was manipulated by them and they used him as a weapon against Oppenheimer. It was stupid of him to get into that position.

On the other hand, if you look at what he actually said at the hearings, it was all true and he was perfectly sincere. He believed what he was saying and what he was saying was not actually very extreme; he never said that Oppenheimer was a spy, he never said that he was disloyal; he just said that he was complicated and unreliable, which is true. I think that everybody who examined Oppenheimer would agree with that. He said very strange things and often told things that were not true, for no reason one could understand. So when you were dealing with Oppenheimer you felt that this is somebody you don't really feel at ease with and what Teller said was that he would prefer to have the security of the country in other hands. That, I think, was a correct statement. So I don't blame Teller for saying what he said; I just blame him for getting involved in it in the first place.

There were others, like Lawrence and Alvarez, of the same opinion as Teller, yet Teller's image suffered a great deal from it, not the others'. For some time he was almost a public enemy in the eyes of many.

I don't know, the public always likes to have one enemy at a time. Partly it was because Teller called attention to himself, he is also a prima donna. He took pride in having invented the hydrogen bomb and it was easy to declare him a public enemy. Lawrence never had that kind of public profile, he was well known but he did not perform in public. Neither did Alvarez, he only became famous later because of the dinosaurs.

What was the intellectual atmosphere at Princeton when you first came here?

It was rather pleasant because there were no old people around at that time at all. I did not even know if there were any professors at the Institute. It was an unusual time. The School of Physics was just starting; there was Oppenheimer, who was the director, and there were about twelve young postdocs who were doing physics, and that was all. We were just entirely a young crowd, Oppenheimer was hardly ever here because at that time he was still in Washington most of the time. So we just enjoyed ourselves and carried on. It was a state of anarchy, which suited me very well. Now, looking back, it is strange because now we consider the professors so important.

Please, tell us about Richard Feynman.

I have written most of what I knew of Feynman. When I came to Cornell as a student in 1947, it was just tremendous luck that I happened to hit Feynman because I did not know that he existed. I came to work with Bethe but as soon as I arrived at Cornell I found out that there was this young fellow, Feynman, who was absolutely wonderful. Immediately I sort of fell in love with him, he was such an exciting person and also he worked very hard. He was doing amazing calculations, which nobody understood. So that was immediately a challenge. I found out very soon that I would learn more from him than I would learn from Bethe. In addition to that, he loved to talk. He was a compulsive talker, he was lonely, he just lost his wife then, and he loved to have a listener. I was in the happy position of not having anything else to do so I could just be a resonator for his talking. We talked a lot about physics, a lot about Los Alamos, about his wife, everything. He was such a genuine person; he talked about almost any subject with a combination of seriousness and fun. He was obviously grieving about his wife but still could tell jokes about her. It was a wonderful time to be there but it was short, I was there just for one year. After that we never lived in the same place, he went to California and I took, in fact, his job at Cornell.

Hans Bethe?

Bethe is again a hero; in a very different way. I owe a tremendous lot to Bethe; he invited me to America, he taught me for a year. Actually he taught me how to do the old style of physics; Feynman taught me the new style. But the old style was also very important and I learned essentially everything from Bethe that I needed to be a physicist. He was extremely helpful. I was lucky twice over, they used to say that Bethe was like a battleship and Feynman was like a torpedoboat. That was true. Bethe is still going strong, it is wonderful how he continues. He came to New York recently when they had these reminiscences one afternoon after the premiere of Copenhagen. Bethe was there, and Wheeler, and others. They both were close friends with Bohr and Heisenberg. He is 94 now.

Have you had any scientific interaction with Wheeler?

No, not really. I love to read his writings, it is poetry more than science. They are full of wonderful phrases but they do not affect me because I think mathematically and he does not. He has this very physical way of thinking, he always thinks in terms of images and I am thinking in terms of equations. So we don't really connect as far as the science goes.

What do you think of his ideas about the delayed choice experiment and time flowing back?

These are good experiments, I do not find anything surprising there because I happen to take a sort of minimalist view of quantum mechanics. I think that quantum mechanics is only a very partial description of the Universe, it only applies within a certain well-defined framework. When you are talking about situations when you can predict the future, quantum mechanics tells you what the probabilities are. So I don't find anything strange in these experiments. They all agree with quantum mechanics. Quantum mechanics to

Conference participants in Princeton, 1979, commemorating the hundredth anniversary of the death of geometer William Clifford (picture in the center). Some of the attendees are, sitting, from left to right: John Wheeler, Robert Dicke, and Eugene Wigner; from right to left: Cecile DeWitt, Brandon Carter, and Stephen Hawking. In the second row, second from the right is John Klauder, third from the right is Freeman Dyson, then (somewhat behind) Bryce DeWitt, and Charles Misner (courtesy of F. Dyson).

me is just a beautiful description of nature, only it does not apply everywhere, it does not apply to the past in particular.

There is this famous question about the necessity of the presence of an observer; Bohr, Wheeler, Wigner, Teller, and many others discussed it a lot. Where do you stand on this?

I dislike having any sort of philosophy about quantum mechanics, it seems to me that it is stupid to treat the subject as if it was a closed subject. Like all of science, it is full of open questions. So I think it is foolish to say that I believe this or I believe that. I think that quantum mechanics, as it stands, is just a wonderful tool to describe nature but it is not the only tool. It is a great tool, amazingly exact and beautiful. It does not include everything and why should it, so much of science does not include everything. If you ask about the observer, it is clearly not true, that you have to have an observer for things to happen. I would say that what we need is a distinction between past and future. Quantum mechanics only speaks about the future whereas most of science also talks about the past. For quantum mechanics it is absolutely essential to have a distinction between past and future. You can call that "observer" if you like but it is only a reference time. You can say that somebody could have been observing it at a point and you can call it an observer but it does not really have to be an observer.

Who is your greatest hero?

Oh, that is difficult, there are so many different kinds. In science, undoubtedly I would take Feynman. He is the one whom I consider as the greatest human being as well as a scientist. But I was lucky, I met so many wonderful people. I remember when I was quite young, during my first year in America, I went to Berkeley and just by chance there was a talk announced about non-violence. I happened to be interested in non-violence, even when I was a child I was very much influenced by Gandhi, and I was a pacifist. So I saw this talk announced by a fellow I never heard about before, it was Martin Luther King. I was absolutely pulled over by him, he gave a wonderful talk. Those days I wrote home to my parents in England every week and I wrote them that this was the man I would be willing to go to jail for any time; I was so much impressed by him. So he was also a hero, and still is (but I never had to go to jail for him).

Another person I could mention is Carleton Gajdusek. He went to jail some time ago. He is certainly a hero. When he was released from jail I was there with a lot of other people, standing at the door of the jail when he came out. It was a great time and then we spent the rest of the day celebrating.

Why do you consider him your hero?

First of all he is a great scientist. He did this very important work on kuru, and also another very interesting disease in Siberia that he is still studying (*Viliuisk encephalomyelitis*). He was planning to go back there to study it. He used to say that if all else fails he could still go to Siberia and take a job there. As a scientist he is certainly great. He is also a wonderfully unconventional person. I always enjoy heretics. He adopted 60 Melanesian kids. Four of his adopted kids were there when we greeted him at the jail. They honor and respect him. The fact is that he got persecuted only because this country has an affinity for witch-hunts. They call somebody a child-abuser, it is like calling somebody a communist, and then nobody would come to their help and they blow it up terribly. Gajdusek got into this witch-hunt. Whatever he may or may not have been doing, it was clearly not bad for the kids, they came out of it extremely well. He adopted them, paid for their education, so everything he did was certainly for their good. He has a tremendous sense of humor about this business. When he was in jail he started to write letters, and I got lots of letters from him. They were always full of jokes.

This whole witch-hunt started when two of his kids brought charges against him for sexual abuse. These kids had their own sexual traditions that were very different from ours. Carleton said that if anybody corrupted anybody it was they corrupting him and not the other way around. Undoubtedly, they were very free and easy with sex, so according to the strict interpretation of the laws you can call that child abuse, but the case should never have been brought to court. In fact, what probably happened was that two of the adopted kids who did not do as well as the others, decided that this was the way to blackmail Gajdusek and they hoped to get rich by blackmail. Anyway, it is a sad story, but on the other hand, it is not so sad because he managed to handle it so well. When he came out of the jail the first thing he said was: they broke Oscar Wilde but they did not break me. So he is definitely a hero. I also wrote a piece about him, called "A Hero of Our Time". So altogether, I think that Feynman, King, and Gajdusek would be a good set of heroes.

Who had the greatest influence on you?

I would say my parents more than anybody. Both my mother and father were strong characters and they both set the pattern for my life. My father was very much interested in science and my mother was very much interested in religion. She was a lawyer by profession but her mother was a faith healer, so the idea that there are more things in life than just science was in the family. They were solid, middle class people. My mother was much involved with birth control; she ran a birth control clinic.

My father had shelves of popular science books, although he was a musician. He had interest in science, he had books by Huxley, Haldane,

Eddington, and other really good writers of popular science. I probably learned more from these books then I learned in school.

Can you single out any one book that was very influential.

Perhaps the most influential was the book, *Men in Mathematics*. I still have the original copy. It is a wonderful book and it was not from my father's collection, it was a school prize. Certainly, it had a strong effect and made me want to be a real mathematician. It is a collection of biographies of mathematicians, it is full of myths, and historians are very angry with it because lots of things in it are wrong. The writer was a mathematician and not a historian, but they are all good stories. There was one chapter about Abel, such a romantic story describing him as a poor young man dying of starvation and tuberculosis and in spite of that doing this marvelous mathematics. I read a piece recently by a historian who actually examined the papers that Abel left behind and there were all kinds of household accounts and bills for theater tickets, so apparently he was not starving at all. But he did die of tuberculosis.

Are you religious?

I don't think so, really. But I feel it is very much a part of the human tradition. I am not very religious but then almost everybody has it to a certain extent.

I have read that physicists are among the most religious scientists and biologists are among the least.

It is probably true; and it is partly because physics is a less dogmatic science and biologists tend to be very dogmatic.

Scientists who are not religious in the U.S. do not like to be quoted on this. In Europe, in Britain, for example, this seems to be less of a taboo.

It is perhaps because passion is not so strong in Britain, everybody learns religion in school, there is no separation of church and state. Oddly enough it was Thomas Huxley, the great evolutionist — he was on the royal commission to set up public schools in England in about 1870 — who insisted that every child should learn the Bible. I think that was very wise, that part of the culture should be taught to children and it has been helpful.

Thus everybody has a certain background, whether you believe it or not does not matter. So the country starts from a certain familiarity with what religion is all about. Whereas in this country with the strict separation between church and state, people who are religious feel threatened by the public schools, and there is a sort of antagonism created by this situation. I think that it is a great pity. On the one hand you have the parents of children who feel threatened by science, and are strongly against science and biology in particular being taught to their children. On the other side there are the dogmatic biologists who say we know what is true and that is what we'll teach. I think that the situation is much more polarized in this country. Mostly just because of this historical difference.

Did you ever meet John von Neumann here at Princeton?

Yes. When I was here he was very active, he was then just building the computer. He had a great team of people and they were all ready to use the computer, they were mostly meteorologists. It was a lively group and I enjoyed talking with them but I never worked with them. I actually made friends more with the young people than with von Neumann himself. He was then, like Oppenheimer, most of the time in Washington and just presiding from a distance. I got to know Klari, his wife, after he died; she became a very close friend. She eventually married Eckart in California. She was one of the two witnesses at our wedding and a very sweet person. She told many stories about John but I can't tell you anything in detail.

Did you meet Leo Szilard?

Yes, with him I had much more real contact. Szilard, of course, was very political at that time, he was organizing international meetings. I was several times at Pugwash meetings with him, and I met him sometimes in Washington, sometimes in La Jolla. We talked a great deal. I admired him very much. I would not quite consider him a hero; hero is not the right description for him but I loved his Ten Commandments. I think they are much better than the ones in the Bible; they are really beautiful. Somehow we always end up being friends with the widows; we got to be very close friends with Trudy Szilard after he died. That was a strange marriage; they were devoted to each other but they lived separately.

I remember Szilard was organizing the Council for a Livable World and I think that that was a brilliant idea to create a political organization to get senators elected which is what it was doing and it still exists and is quite effective. He always liked to go right to the top. He wanted to settle the question of the arms race by talking with Mr. Khrushchev and Mr. Kennedy and getting them together. He almost succeeded. He was certainly not conventional. He liked to be homeless, he never had a home, just lived in hotel rooms. Only when he got old he went to live with Trudy in La Jolla. When I met him in Washington, he used to live at the Dupont Plaza Hotel. He would invite me to lunch and then I had to eat very fast because he would eat the food off your plate. I was a slow eater but he never could stand seeing all that food on my plate. Yes, he was a little bit weird.

What did you talk about mostly, science or politics?

Mostly politics. He was not so much talking about science. He had tried to do biology, he lost interest in physics, then he tried to make the transition to become a biologist but that did not really work. He never really became active as a biologist. I think that politics was his main activity during his later years. That he was very good at. The Council for a Livable World was entirely his creation. It took a lot of organizing, he traveled around the country giving speeches and collecting members, almost entirely by himself. He was not good at organizing a group, was totally hopeless at that, he wanted to do everything entirely by himself. This is what he ended up with and of course, this was not very efficient. Still, it did work.

Did you know Eugene Wigner?

I hardly knew him and it was a shame, we lived so close all these years. I was friendly with him but somehow we never got close.

His famous politeness, wasn't it sometimes irritating?

There is a professor here, whose wife is Japanese. She is a sociologist and writes about Japanese society. One of the papers she has written, and I think it is brilliant, is called "Politeness as a Tool of Repression". It is certainly true for the Japanese society and perhaps a little bit with Wigner, too.

You have written in Imagined Worlds *that particle physics in the U.S. is struggling since the failure of the superconducting supercollider. Has anything changed since?*

Yes, everything is much better since they gave up that plan. They have been doing pretty well. First of all the two big machines that are still running, at Fermilab and at Stanford, are doing well. They are still making interesting discoveries. So things are not so bad. The funding provided for physics continued more or less flat in spite of the supercollider, it did not really make all that much difference. Now, of course, we have the underground detectors that, to my mind, are the most exciting. They are mostly not in the U.S. but still, they are producing very good science. The most interesting is the one in Canada, which is coming on line now. These are really doing new physics. It is true that the United States has somewhat fallen behind because here we only were building accelerators and not underground detectors. Meanwhile the other countries went ahead, such as Japan, Canada, and Italy. That is where the new physics is. Some of the ideas for new accelerators seem to be sensible, so they are planning to build a new positron electron collider at Stanford. It is not quite as extravagant as the supercollider.

The supercollider project was stopped after considerable investments.

I was against if from the beginning for scientific reasons. I think that it was a badly designed machine. Of course, the politicians were also against it, because it was mismanaged and it overran the budget by a large amount. There were also geographical jealousies. Before the site was selected and there was still a chance for several states to have it, they all voted yes. Then after the decision had been made for Texas the non-Texan politicians all voted no. When something is so big, it becomes essentially a question of jobs. The politicians made decisions based on that and they didn't care about the science.

In 1972 you wrote about missed opportunities for mathematicians in physics. Have any of those problems you mentioned then been solved since?

Certainly. There has been a lot of coming together of mathematics and physics since that time. Mathematicians have become much more willing to learn from physicists. There have been spectacular examples such as the theory of four-dimensional manifolds, which is a wonderful piece of mathematics opening up a completely new chapter of topology. All this resulted from the Yang-Mills physics. Yang and Mills wrote down these equations for the gauge field theory, which was motivated by trying to understand physics. It turns out that the solutions of the Yang-Mills equations on four-dimensional manifolds actually determine the topology in a beautiful way. So the whole subject of topology got revolutionized as a direct result of physics. That kind of interchange happened after I wrote that piece.

There are many other examples. Knot theory also has learned a great deal from physics. Now the mathematicians can classify knots using ideas from field theory. So there has been a very fruitful exchange and mostly in the direction from physics to mathematics.

Was it, partly, your impact?

Oh, no! I don't think so. In fact Yang and Mills had already done this work in the fifties but nobody paid attention to it. Then came a new generation of mathematicians who discovered it.

A missed opportunity today is the quasicrystals. Mathematicians have not done anything with it.

How about Steinhardt's work?

He is good but he is not a mathematician. The joke is that these are really only interesting in one dimension from the mathematical point of view, and all the applications in physics are, of course, in two or three dimensions. So that is why I think it did not get across the barrier. What Steinhardt is doing is all three-dimensional and it is beautiful because these things really exist, so you can learn a tremendous lot from the physics in three dimensions. But if you look at what a quasicrystal means in one dimension there is an enormously greater freedom. This is because you do not have symmetry in one dimension. In three dimensions you are restricted by the rotational symmetry, it has to be a discrete rotational group, like the icosahedron, so you have very few possibilities in three dimensions. In two dimensions you have a little bit more, you have all the regular polygons, but that still is only one discrete series and that is all. In one dimension there is no symmetry.

What makes them a quasicrystal then?

The definition of a quasicrystal is that it is a distribution of mass points such that the Fourier transform is also a distribution of mass points. That is the abstract definition and that is all it is. In a three-dimensional case you see that when you look at the X-ray patterns and you see the beautiful distribution of mass points. It is equally true in one dimension. If you are interested in looking at the distribution of mass points such that their Fourier transform is also a discrete set of mass points, it is essentially quasiperiodic, almost periodic. But in the case of one dimension, there is no classification known, nobody knows how great a variety of these things there might be. That is an unsolved problem. Nobody is even thinking about it as far as I know. Penrose did it in two dimensions but I don't think he has ever thought about it in one dimension. In any case, it is a deep problem. You can tell how deep it is because if you could solve that problem you could prove the Riemann hypothesis. Which, of course, immediately shows that I am crazy even talking about it. Everybody knows that the Riemann hypothesis is the deepest problem in the whole of mathematics. But it is certainly true that if you could classify these quasicrystals, the Riemann hypothesis would fall out. I think that this is what mathematicians ought to be looking at.

Have you tried it?

In a very modest way. I am clearly not the person to do it. It needs some very, very deep mathematics. But as far as I can see that is the most hopeful way to the Riemann hypothesis. All the conventional mathematics failed; Riemann would be laughing if he knew that this was still unsolved.

Other missed opportunities?

I have not been thinking about this lately at all. Of course, physics has opened up now. There are very new sets of ideas in what they call quantum computing and that is very exciting. Quantum computing is still not understood really at all. It is any sort of processing of data, processing of information using quantum logic instead of just yes and no. The individual bits of information are not just ones and zeros but they are quantum spins. In principle these are vastly more powerful. There is actually a theorem which says that certain problems that are unsolvable with classical computers are solvable by quantum computers. So we know that in principle quantum computers are enormously more powerful than classical computers. The physicists are, of course, interested in trying to build one but that is very difficult. There is a tremendous effort going on now in trying to build quantum computers. It turns out to be very interesting because in order to do it you have to understand quantum mechanics and even simple operations of a quantum computer give you completely new insights into quantum mechanics. So this is very helpful because it is pushing ahead the whole understanding of quantum mechanics. But on the mathematical side, I think that the whole subject of quantum computing ought to be another branch of mathematics just in the same way as classical computing has contributed tremendously to mathematics. That is one of the good things that has happened. Quantum computing should contribute a great deal more. And this is something that is just beginning. So this is not even a missed opportunity, it is an opportunity that is just emerging.

There is a "Dyson-shell" often appearing in the science fiction literature. Does it have anything to do with you?

It has nothing to do with me, really. It was a misunderstanding of a proposal I made about forty years ago for looking for Aliens in the sky. That was real. I proposed looking for infrared radiation from Alien civilizations. It is a good way to look because it does not require them to communicate with us. If they have a large population, a large industry, they are compelled to get rid of waste heat and whether they do or do not want to communicate, if we look for the waste heat we could detect them. That was my proposal. In fact it is interesting that at about the same time or a little after this proposal of mine, there was an Air Force satellite which was taking pictures of the Earth at night with exactly this kind of infrared camera. If you look at the Earth at night at the waveband of about 2 microns everything you see is human, there is nothing natural at all. The Earth is full with all kinds of infrared emissions and every one of them comes from what humans are doing. So if you want to detect humans on the Earth that is the way to do it. This applies equally to the Aliens; this is the waveband, where anything that is doing a lot of industrial activity will have to radiate away the waste heat. I used the term "artificial biosphere" to describe what we are looking for and by that I meant simply a habitat where the Aliens might be living. That would radiate from the outside surface and so you would see heat radiation from the surface. I said that we should be trying to detect artificial biospheres. Then, of course, the science fiction people got hold of this idea and they thought that when I talked about an artificial biosphere that means a big round ball which, of course, was not intended.

Is this how you became famous among non-scientists?

Oh, yes, you never know what will make you famous.

Are you a great believer in extraterrestrial life?

I would not say that I am a great believer, that is not true. I think that it is very exciting to find out whether they are there or not.

You participated in the Orion project. Would you tell us something about it?

That is a huge subject because it was a very ambitious project. That was a bomb-propelled spaceship. In fact my son is writing a book about it. He has collected a lot of the documents, which I did not even know existed. He is a very good historian.

It was a very exciting project for me, I worked on it for a year and a half to get the thing started. The notion was to have a big spaceship, built like a submarine, heavy very rugged construction, not like an airplane. It would have been built by a ship company. It was to carry about a thousand nuclear bombs and they were just to be thrown out at the back and exploded one at a time. You were to ride up to the sky on top of these bombs. It was a great idea and we worked out all the details of that and technically it was very good. I also liked it as a form of unilateral disarmament, in that if we got permission to fly this thing we could get rid of the bombs with tremendous speed. That seemed to me very desirable. Every flight would have been about a thousand bombs so we could have used up the stockpile pretty fast. I don't think that there has ever been any method suggested to use up the stockpile that made so much sense. So it was not such a bad idea, but there was one fatal flaw. That was, of course, contamination, because it produced fallout all over the place. It was obvious from the beginning that it was a horribly dirty way of getting into space and, in fact, it only made sense if you could have used much cleaner bombs. This was in the 1950s and we thought at that time that we would have clean bombs and that was the whole idea. We were not going to use the existing dirty kind of bombs. Well, it turned out that there were not any clean bombs, the ideas for clean bombs did not actually work well. They could get them cleaned up by about a factor of ten but that was not good enough. So in the end it was clear that as long as the contamination problem was not solved the Orion would not fly. I gave up after about a year and a half; to me it was clear that it was not going to work. The project actually continued for about 7 years after I left, it was only theoretical and not practical.

What do you think will be the next step in this respect?

There are a lot of good things going on at the moment. That kind of a nuclear ship is clearly not the direction you want to go. Now we have all these miniaturized instruments and miniaturized computing systems and communication systems. The direction now is getting smaller and smaller and not larger and larger. This is certainly much better. The really good systems now are quite small. The thing I am very excited about is laser propulsion. It just started about two years ago. There is a professor, called Myrabo, he is a professor at Rensselaer Polytechnic. He has built a little model, which he calls the light craft demonstrator. It is a laser-propelled ship and it only weighs about two ounces. But it flies! So he has a laser beam going up into the sky and the little model flies about a hundred feet up and it works using the energy of the laser to fly. This is the way one would like to do it, because this way you leave all the expensive machinery on the ground. The ship is quite small and the laser beam is just a public highway. Of course, at the moment it is only a toy but if you build it full scale, it sounds quite reasonable. If you have, say, a thousand megawatt laser — we already have hundred megawatt lasers, so it is not a huge step from what we already have. That would then be enough to be a public highway for something that weighs a few tons. So you would have your little spaceships which would carry a couple of people, or just instruments, and using water as your propellant. You could have a tank of water instead of having a huge rocket. With the laser you could have this public highway and anybody with a spacecraft coming along should just pay the toll, and off they go! The dream is that this would really be a lot cheaper than the present systems. But it only makes sense if you have a constant stream of traffic. It is also clear that it would be total stupidity to try to build something like that at the moment. There is nowhere for the people to go! Clearly this whole thing has to grow. It is not something that you just build. It is like the airplane. At the moment we are in the same situation concerning space travel as the Wright brothers were in 1904 after the first flight of the first airplane. It took fifty years before we had a really worldwide air transportation system with lots of airports and lots of airplanes and constant flights. It took fifty years before it really became cheap. I think that it is roughly how it will be. I would say that it is very likely the same with this laser propulsion. There will be spaceports in orbit and spaceports on the Moon, spaceports on various other places, on asteroids, on Mars, and there will be constant flights so there will be lots of choices for the passengers to go. It may take a hundred years but that is the way to go. In addition to being much cheaper, it is also much cleaner than the previous methods. We are not putting chemical fuel into the atmosphere, only plain water.

Where would people go on these travels?

Many of the places you could go to would be only instrument stations. Maybe people would go and visit or just put in the instruments, they would not be colonies. Human colonialization will also come but it takes longer. That is a problem of biology much more than a problem of engineering. If you want to have a real colony, I think it will have to be with modified people. If I wanted to live on Mars I would not want to live in a spacesuit; I would want to be out in the open, so I would have to have different kinds of lungs, different kinds of skin probably to live on Mars comfortably. I think that that is connected with genetic engineering and I would not advise anybody to try to get people genetically engineered for Mars at the moment. It has to wait until there is somewhere to go and there will be people who are actually prepared to do that.

Do you think it will come?

I think so. If you had genetically-engineered people on Mars, that would not cause severe social problems here. If you think of genetically-engineered people on the Earth, it creates all kinds of terrible difficult problems because it is like having different races or even more so. So it is hard to see how it could happen on the Earth without conflict. But if the people are living on Mars, thirty million miles away, they can be different. That would not cause conflict.

Why would people want to go there?

Thomas Gold, astronomer, and Freeman Dyson (photograph by Barrett Gallagher, courtesy of F. Dyson).

In order to live there comfortably. People do like to move around, it is a fact of history. Even the Hungarians do that; they came from somewhere in Asia.

When people think of Alien life forms they often imagine them similar to us; carbon-based, in need of oxygen, similar temperature range, and so forth. I do not see why it ought to be so.

Clearly it does not have to be so. The best example of not being so is the *Black Cloud* by Fred Hoyle. It is a wonderful science fiction book. Hoyle is a good scientist but he is also a good writer. It is a story about a creature that lives in space and is composed of dust grains, so it is essentially a cloud of dust, except that it is alive. It eats sunlight and instead of having muscles and nerves it has electromagnetic coupling between the grains and the grains are magnetized and electrically charged. It has electric and magnetic forces enabling it to move around and communicate with itself. So it is an interesting idea to have a life form that is completely independent of liquids, it has no liquid component at all. Life does not necessarily involve water and it could be made of all kinds of other materials. What is also interesting is that it could be at a much lower temperature, in fact it might work much better at a very low temperature, 10 degrees absolute, so it could be at home in interstellar space. It would not have to be near a star. I think that is quite plausible and could be a much better adapted form for life in deep space. There are many possibilities.

Do you feel any special responsibility as a scientist?

Yes, in a way. I always find that in the real world our importance has nothing to do with science really, it has to do with the international contacts. Science is a wonderful example of international activity that works. And that is what really gives us such a strong position. I worked for the government at various times and what is noticeable is that you very often have scientists in the government doing things that have nothing to do with science. And the reason is simple that they have much better contacts internationally than the political people have. So it really helps a great deal, especially in the old days, when we were dealing so much with the Russians. Scientists had this inside communication system independent of the government. Szilard was, of course, a good example. We often got much better information than the people in the government had. It, of course, gave us a lot of responsibility too. The fact is that all the progress that has been made limiting the arms race, was almost entirely due to the initiatives of scientists. Without the scientists it would not have worked. The same is true, of course, in biology, with genetic engineering. Biologists all over the world are all personally connected and they are much better able to deal with this than the governments are. I think that this is our most important responsibility; that we operate an international communication system which is the best there is. And that is much more of a responsibility than the fact that we know how to make bombs.

Hardy was your teacher. He is famous for having said that he was proud that he never did anything useful. Is that something to be proud of?

Well, it all depends on what sort of a person you are. For him clearly that was correct. He was an artist at heart, for him beauty was all that mattered. He was also a very good writer, so I think that personally for him that made sense. It did not make sense for me. I always liked to do things that were useful. Of course, there is a big responsibility if you are coming close to the destructive uses of technology, in particular things like biological weapons. At the moment we are much involved with biological weapons, something biologists have to think very hard about. It is a big problem but the biologists have handled it very well. In fact we have an international treaty declaring biological weapons illegal. I think that is very important and that happened mostly because of one person, Matthew Meselson, and I think that he is a great hero. So if you ask for heroes he should be on the list. He was very largely responsible and he is, of course, a professional biologist, and he has spent much of his life on this biological weapon problem.

Do you anticipate the general public becoming more science friendly or the other way around, getting even more alienated from science?

That always fluctuates. I grew up in the 1930s in England and the public was much more strongly against science at that time than they are now. Of course, the memory of World War I was still very vivid then; it was a particularly horrible war because it was essentially a chemist's war, poison gases were used at the western front. There were these images of people dying in the trenches being poisoned with all sorts of chemicals. So that was the image of science that everybody grew up with and science was violently unpopular. Very few people were interested in it and they were definitely considered to be second class compared to classical scholars. I think that it is the effect of particular events that strike the imagination of people. Vietnam had the same effect on this country. It was seen by the public as a technological war and thus gave technology a very bad image. Physics, of course, had a very bad image because of the nuclear bombs and the nuclear power industry has never recovered from that. So generally if the public reacts against science, it is for a good reason.

Do you anticipate any changes in science literacy now that we are in the age of the World Wide Web?

I don't know whether it makes any difference. I love email because it brings my family so much closer together. I just got an email from my granddaughter this morning who is 5,000 km away in Washington State, and I can communicate directly with her, which is wonderful. This makes a huge difference at the personal level but I don't know whether it has anything to do with learning science; I don't think it does. Whether you learn it from books or from the World Wide Web, I don't think it makes much of a difference. The real point is that most people do not have a gift for it anyway. I have a lot of grandchildren and none of them is much interested in science. And I don't see why they should be.

I would like to ask you more about your family. There is a book about your son, The Starship and the Canoe, *and I learned from that that you had some disagreement with him a long time ago.*

Yes, he was horrible as a teenager. What is remarkable, is that Esther, my daughter, who is the internet expert, was just the opposite of George. They were completely, diametrically, opposite. Esther was always successful, establishment, always got along well, did everything according to prevailing rules and George was always the rebel who did everything opposite. As teenagers they had nothing to say to each other. Now, after 30 years, they are very close friends. Each of them changed; she has become much more of a rebel and he has become less of a rebel. They are very close to each other and learn a lot from each other. Again, without the internet it is hard to see how they could have developed such a friendship.

What is your relationship with your son now?

We are very close. This is also largely due to the email; he loves to use it and writes practically every day. He also published a book, called *Darwin among the Machines.* I was amazed. He builds boats as a profession but he also turned out to be a real scholar. The book is on how we communicate over the centuries and is beautifully written. There are so many people in it of whom I never heard. He has a wonderful gift of discovering interesting characters in various periods of history. Of course, Darwin is not Charles, it is Erasmus. So this is my son, and he is becoming a serious historian, I would think. He is working on the history of the Orion project now.

The Guardian *called your daughter the "First Lady of the Internet" and the most powerful woman in the computer world. Have you read that?*

I read a lot about her. We also see her a lot; she is very much a family person. She has no children of her own but she is very close to her sisters and nieces and nephews.

Is she a computer scientist or a mathematician?

No, neither. She went into it through writing. She was a journalist and she has a gift for writing. Which is also true of George, they both have become serious writers. She began as a journalist; she had a job with *Forbes Magazine*, which is a business magazine. She got interested in the industry and then she decided that it was more interesting to be a player than just a spectator. She got a job on Wall Street as a stock market analyst. Then she started specializing in computer companies. But she has never really learned much about computers, she is mainly interested in the people who buy and sell

Two of the Dyson children; Esther, in about 1965 and George, in about 1975 (photographs courtesy of F. Dyson).

companies and who are running the industry and the internet in particular. Now she is chairperson of the organization called ICANN, which is International Corporation for Assigned Names and Numbers. That is what makes her most famous at the moment. This is a tremendously political job, so she is being attacked viciously by both sides. That is the reason why she took the job, in fact, because nobody else wanted it. It is a thankless job but she believes strongly that it had to be done. There was a moneymaking operation that made billions of dollars yearly by selling internet addresses. It was all in the hands of an American company, which had a complete monopoly. This international organization is designed to break up the monopoly and, of course, they had huge fights and the battles are not yet over. They hope to have it organized within a year and at that point she is out according to the law. Then there will be an elected board of directors. What is difficult to establish is who will have the right to vote. A huge amount of money is involved. Being chairman is not a paid job; she does not get a penny for it, she is supposed to do it in her spare time. Anyway, she is the right kind of person to do this job.

She has her own company; she is a venture capitalist. She has a particular interest in Russia, which, again, is philanthropy because she does not expect to get rich in Russia. She actually spends a lot of money in Russia getting companies started.

What is your opinion of women in science?

There are two strong opinions about women in science. I know both sides. People, like my daughter, for example, who say, yes, women can do just as well as men, it is no big deal, all you have to do is do it well and you will be recognized. On the other side there are people who say that feminist science is different from masculine science in that women in science will be a revolutionary force and will make science into something different. That does not seem to me sensible. I don't see any evidence that it ever happens, certainly not in physics. There are good female physicists who are just about as good as male physicists. They may have some human gifts, which are important, but as physicists they are not different. I definitely agree with my daughter that women do very well if they are given the chance. Nowadays we are working very hard to get women here, in the Institute, for example. But the fraction does not change. We work extremely hard, put out all kinds of advertisements that we welcome women, but the percentage remains the same, about 10% of the men.

When you are trying so hard to find women to come here, do you lower your standards in your efforts to accomplish this?

No, we do not. If there are several candidates of equal stature, we probably prefer the woman. But that is not a big factor.

So where then is the problem?

I blame the Ph.D. system to a considerable degree for it. I have five daughters and none of them has a Ph.D. I am pleased about that because I think that the Ph.D. system is particularly bad for women. It is so slow and it uses up the best years of your life. Of course, I don't have a Ph.D. myself so I have strong feelings against it. I think it is bad for everybody but particularly bad for women. I know in the case of my daughters that they want to raise families, they want to have freedom, and they saw that these Ph.D. programs would not do for them. I think if there were not such rigid systems they might have become scientists.

Two of my daughters are medical doctors and that is much more favorable to women, and among the medical doctors they are about 50-50 with men. It is, of course, partly because women have a natural gift for healing but also the system is much less rigid and it is possible to get through medical school and have babies at the same time. They work hard but they don't have the same kind of pressures as Ph.D.s have. These two daughters of mine actually did a tour of graduate schools, they were interested in biology. At the end they decided that people in graduate schools were generally unhappy and people in medical schools were generally happy — so they chose that and they both are very happy with it.

Freeman and Imme Dyson in Arizona, 1985 (courtesy of F. Dyson).

There is one daughter who is a veterinarian and the last one is a Presbyterian minister and she is very good at it. My wife has been very busy with raising kids. She came from a medical family, so she probably would have chosen that as a profession if she had had the chance. She grew up in East Germany and that was a very bad time to get an education. She was about 15 years old when they moved west and her previous education was not recognized. So she decided not to pursue that, went traveling and eventually ended up in the U.S. She has spent most of her time with raising kids and now grandchildren; we have nine of them.

You went to school in England while your children went to school in the United States. How do you compare the educational systems in the two countries?

I was glad not to have to deal with the English private schools. I grew up in private schools and that was a strong class problem. If you are upper class you go to private schools and that is something I did not want for my children. All of them went to regular public schools here and I found that much better and we could have them at home. Of course, the schools here are not that good academically but that did not matter much. In the Preface of George's book, *Darwin among the Machines*, I learned something that I never knew. Our son grew up in Princeton, and he went to the Princeton high school, at least we thought that he went to the Princeton high school. Actually what happened was that he just quietly disappeared from the school and spent the day at the University library.

Did they not miss him?

Apparently, not. That says something about the school. So you can get an education here, in spite of the schools.

But you have to be motivated.

Yes, that is important. The system is very good if you are motivated. Of course, the Princeton schools are better than average. Even at the time when the University did not allow female students to enroll, they allowed girls from the high school to take classes. Our daughters all officially took classes at the University when they were in high school.

Knowing all that you know today, if you were to start your career, what would be your choice?

Probably it would not be different, but it is hard to tell, of course. I have always been interested in biology and I really wanted to be a biologist but I discovered very early that I am not talented in it. I went to the river and caught a crayfish and tried to dissect it and it was such a stinking mess that it was clear, that is not me. Probably the same thing would happen now. I am still interested in biology but when I look at what it really takes to be a biologist I don't think I have the skills for it. If I were young today, I would probably be much more skilled with computers. So probably I would get into biology through computing.

In summary, I would like to say that I had two careers, one as a scientist and one as a writer, and being a writer is just as exciting as being a scientist. In particular, if you are fifty years old, it is much easier. So at the age of fifty I made the decision more or less to switch. It is clear to me that over the age of fifty it is much harder to compete as a scientist but it is quite easy as a writer. Of course, as a writer, I have made a huge number of friends and widened my interests very considerably, so that was a good choice. It seems that my son has done the same thing; he has spent his life building boats, and now his horizons have become much wider because of his books.

Which of your two occupations do you consider more important, the scientist or the writer?

Well, we never know until a hundred year later. But I think I would like to be remembered more as a writer than a scientist. I think that I spoke to a much wider range of people that way. Just to be a footnote in a textbook is not so exciting and my physics does not amount to much more than that.

What is your opinion about the Nobel Prize as an institution?

I think it is very good. The public needs something like that. They love to have stars and it has worked out very well. I am surprised how well they do, occasionally they make mistakes but by and large, in perhaps 90% of the cases they are really well deserved. Very often they pick out people who otherwise would not be known at all and almost always good people. On the whole I am very pleased with it. It has enriched the life of science very much.

This is from the side of the public. How about the scientists' side?

I don't know. Of course, we are all different. Obviously, Crick and Watson had a passion for winning the Nobel Prize and it drove them along. I never thought of this as my aim in life, it never had much effect on me and I am not sorry not to have got one.

Have you been close to it?

No. But I was amused just the other day talking to Jocelyn Burnell, who is famous for not having gotten it. She is a wonderful person and I admire her tremendously. She is always asked if she was not sad for this and she always answers, why should I be sad, I made a career of not having the Nobel Prize. With her it is a clear injustice but she is still such a happy person, it does not do her any harm.

She is an exception; there are others whose life was ruined by not having gotten the Nobel Prize.

I don't know, I have not met such a person. Of course, I should not ask you for names ... There don't seem to be such people in Princeton as far as I know. Of course, Johnny Wheeler is a good example, I think he should have gotten it. There are several things he could have got it for; the theory of fission, he invented the S-matrix, he had done so many things. Black holes are probably the most important. Anyway, I never noticed him to be embittered by this at all.

What are you working on nowadays?

I am not really working. I am 76 and I don't really do science any more. Most of the time I am preparing lectures, I do a lot of traveling, I do public talks — what I call performances rather than lectures. I hope in the next few years to travel less and perhaps write a couple more books.

John C. Polkinghorne, 2000 (photograph by I. Hargittai).

24

JOHN C. POLKINGHORNE

John C. Polkinghorne (b. 1930 in Weston-super-Mare, England) is a Fellow (and former President) of Queens' College, Cambridge University, and a Canon Theologian of Liverpool Cathedral. He studied physics at Trinity College in Cambridge and received his Ph.D. in 1955. He was Fellow of Trinity College 1954–1986. He was Lecturer, Reader and finally Professor of Mathematical Physics in Cambridge and was elected Fellow of the Royal Society (London) in 1974. In 1979, he resigned his professorship and trained for the Anglican Priesthood for which he was ordained in 1982. Having served the Anglican Church in Cambridge, Bristol, and Blean (near Canterbury), in 1986, he was appointed Dean and Chaplain of Trinity Hall in Cambridge and in 1989, President of Queens' College, from which he retired in 1996. He has been a member of the General Synod of the Church of England and of the Medical Ethics Committee of the British Medical Association. He was appointed Knight Commander of the Order of the British Empire in 1997. In 2002, he won the prestigious Templeton Prize for Progress Toward Research or Discoveries about Spiritual Realities. It is the world's best known religion prize in the amount of about one million dollars. The Templeton Prize has been awarded annually since 1973. John Polkinghorne is a prolific writer. We recorded our conversation in his home in Cambridge, on March 16, 2000.*

---

*István Hargittai conducted the interview.

First I would like to ask you to introduce yourself as a physicist.

I worked in theoretical elementary particle physics, very much on the formal mathematical side. Initially I worked in quantum field theory. Then people got disillusioned with quantum field theory; it was the way to think about relativistic quantum mechanics, but it didn't seem to work very well. So people tried other ways of approaching the problem. This led to the so-called S-matrix program, the basis of which is, "Let's concentrate on things we can actually measure and not talk about fields, which include many un-measurable aspects." If you want to do that, you have to figure out what properties these observable things are going to have. I spent a lot of time working on the mathematical properties of the S-matrix elements. It was an interesting program, but in the end, it collapsed under its own weight. We found that the S-matrix elements have very remarkable properties, but they are too complicated, in fact, to make them easy to use. After that I moved on to a different type of activity, which was connected with very high-energy processes. In physics, if you deal with extreme regimes characterized with very high energy or very high momentum transfer, or better still, very high values of both, then these extremes often produce situations that are easier to analyze and you can make predictions about the shapes things will take that you couldn't make in more complex, ordinary regimes. Extremity can lead to simplicity. We studied high-energy behavior and in particular we studied deep inelastic scattering a great deal. That was a particularly interesting thing to do because in deep inelastic scattering experiments people bounce electrons off protons and the electrons bounce back, indicating the presence of some point-like constituents sitting inside the protons. These, in the end, turned out to be the celebrated quarks. We constructed a number of models for various high-energy processes, based on the idea of there being this point-like structure within protons. From this we derived some shapes and general characteristics that these processes would have and analyzed some of them. That was an interesting blend of activity because it drew my work closer to experiments; we actually analyzed experimental results in the course of doing that. The mathematics and the more empirical physics came together. That was the last phase of my work. I worked on it during the last few years of my working life as a theoretical physicist, in the late sixties, up to the middle-seventies. It was quite a nice note on which to go out.

How did you become a physicist?

I came to physics from mathematics. I was good at mathematics in school. I enjoyed it and it gave me a sense of power when I could solve all those equations. When I came to Cambridge University, it was natural for me to study mathematics. During my undergraduate years I became interested in the fact that you can use mathematics to understand the physical world. So when I came to do a Ph.D., I decided that I would do it in theoretical physics rather than in pure mathematics. It is a well followed path in Cambridge for mathematical people to move into theoretical physics. I did my Ph.D. with Abdus Salam, the famous Pakistani physicist, who later on won a Nobel Prize. Some people have a great gift of generating ideas; ideas come pouring out of them. Some of them are good and some of them are bad and often those sort of people don't know which is which. Salam was a little like that. He was a stimulating person to be in the neighborhood of, but you had to step back occasionally and ask yourself, "Is this a sensible idea to think about or should we think of something else?"

Did he give you a doctoral project?

He didn't actually. He more or less told me what was going on in the subject, where he thought the action was, and then said essentially, "Think of something you'd like to do in the area." And I did that. That's how I wrote my Ph.D. dissertation, which was on the problem of what's called renormalization theory. It was a last tidying up operation in what had been a powerful program that Feynman and Schwinger had put in place immediately after the war, producing modern quantum electrodynamics. There were still some loose ends to be tidied up, and I did one of those in my project.

How does it work, the interaction between the mentor and the graduate student in theoretical physics?

It varies. When I came to have my own research students — in my career, I had about 25 research students who did their Ph.D. work with me — I felt that they needed more guidance than I had received from Salam. I usually wanted to suggest to my students an apprentice piece that they could do, a problem that I thought was worth doing, but was almost certainly doable, so that they could cut their teeth on that. It is not easy to find problems like that which are interesting, and that's why theoretical physicists tend not to have large numbers of graduate students. But if you have some calculational technique that has been invented and developed in-house, then you can propose a variety of problems that it can be applied to. Some of them may be quite tricky to apply, while others may be reasonably straightforward. The best people could complete that sort of exercise within a year or so, learn some of the techniques by doing that, and if they were really good people, then they would suggest their own problems. It was very exciting to see graduate students being self-propelled that way. However, even the best students need a bit of guidance at the start. My most difficult research students were those who were quite clever, but weren't prepared to accept any guidance at the beginning; they felt they knew what they wanted to do. On the whole, their work turned out to be more difficult than it would have been if they'd been prepared to be a little humbler at the start.

What was your family background?

I was the first generation in my family to go to university. My father came from a very large family, he had nine brothers and one sister in North Cornwall. This is in the extreme west of England. My family comes from a region in Britain that is part of a Celtic fringe, in the far west, where the Celts were driven by the Anglo-Saxons. The Cornish used to have their own language and they are connected to the Welsh to the north and to the Breton people to the south. They think of themselves as being different. The Cornish people say when they cross the river that divides Cornwall from the rest of Britain, "We're going into England." My father was a very clever young boy by all accounts, but left school when he was 14 because the family couldn't afford for him to take the train to the local grammar school. He had quite a successful career; he was a head postmaster. He obviously would have gone to university if he lived a generation later. My mother came from a different background. Her father was very talented with horses; he was a head groom and he rode horses both at work and in shows and in hunter trials. So my mother came out of a background where people were in service; they were respected by their employers, but they were servants. My father came out of a humble but very independent-minded background. My mother influenced me intellectually. She was a woman who very much enjoyed reading, especially the classics, and she encouraged me to read from a very early age. The love of literature that she had instilled in me has stayed with me.

How was religion in your childhood?

Both my parents were religious people, but they weren't people who talked about religion very much, as perhaps English people tend not to be. We went, as a family, regularly to church and it was intuitively clear to me that religion was important; it was central to my parents' life, and I absorbed religion through the pores. I can't remember a time when I was not in some way part of the worshipping, believing community of the Church. I have never stood outside the Faith. It's been part of my life.

Was it ever a topic among youngsters as you were growing up?

To a small extent at school and to a much larger extent when I was an undergraduate. At university we talked about these things all the time. I was at Trinity College, Cambridge. Between school and university I did one year of national service in the army and then came to Cambridge in 1949. One way or another I have been connected with Cambridge continuously ever since. Obviously, I had some believing friends, and some of them were more pious than I was, and I had some unbelieving friends, and many more who were interested in talking about these things but I didn't know exactly what their opinions were. At that time I was set to becoming a physicist. I do not at all regret the decision, but it was not particularly thought out. I was just following a conventional academic career. I was not a great creative physicist, but I was a reasonably successful physicist. I enjoyed physics very much and I still retain a lively interest in my old subject.

How did the change come about in your career?

Quite early on as a professional physicist I felt that I wouldn't stay in the subject all my working life. The reason was quite simply this: although I don't think that you do your best work before you are 25 in these mathematically oriented subjects, yet most of us do our best work before we're 45. I had seen a number of senior friends who had been very active in the subject, as they moved into their late forties lose touch with physics as the subject moved away from them. They remained in senior, quite important positions, but they seemed to me somewhat miserable in those positions. They knew they weren't quite at the cutting edge in the way they had been before, and that's an uncomfortable situation to be in. If you have a responsibility for a lot of quite young people, who're working in your group, to give them proper leadership, you don't want to be too far from the action, although you don't have to be the very best researcher in the group.

So I'd long thought that I would not stay in physics all my life and this feeling was reinforced by the progress in theoretical particle physics, where I was working and which was changing very rapidly. I had lived through a very interesting time in the subject, essentially the whole period of discovering the so-called Standard Model, in which we, collectively, as a community, discovered that protons and neutrons are made of more fundamental particles, the quarks and the gluons. It took about 25 years to figure all that out. It was a development that was largely experimentally driven, with the theorists limping along behind the experimentalists, continually adjusting their ideas to make sense of what was being discovered. It was a very interesting period to live through, but by the middle-seventies the dust had settled. Once the Standard Model had been established, the subject started to change character. It became more mathematical than it had been, more speculative than it had been, and more difficult than it had been. All these things, combined with my age, encouraged me to think that I had done my little bit for physics. I wanted to stay in touch with the subject, but no longer in a technically proficient way. That's when I left physics, but I did not act so because I was in any way disillusioned with the subject. I resigned my chair in Cambridge in 1979 at the age of almost 49.

What did you do then?

I went to a seminary in Cambridge to train for the Anglican priesthood. It was a two-year course.

Did you stay without a job?

I resigned my university job, but I decided to work my way through the seminary by staying a Fellow of my college, which was still Trinity, and teaching for them. I did undergraduate college teaching, and they paid me a salary for that. Of course, it was not a professorial salary, but it was enough for me to live on. I lived like that while I was a theological student — a kind of double life. When I got tired being a student, as I did from time to time, I could walk up the road, and turn into a Don. It was an agreeable mixed life for me.

Was it a big cut in your salary?

Yes, it was.

How did your wife consider this?

Of course, it was a joint decision. We both thought that I should do this. As a young woman, she had worked as a statistician. When we had children, she stayed at home to bring up our family in the way people did mostly in Britain in those days. By 1979, our children were beginning to leave home. The two older children were at a university, essentially independent, and our younger son was in his last years at school. Ruth herself thought that she would do something new. She didn't want to go back to statistics and decided instead that she would train to be a nurse, which was a sharper change than my change. Thus we were both engaged in middle age adventures. Our financial needs were considerably lower than they had been earlier. We had paid our mortgage on the house, so we didn't have very big financial demands on us. The cut in my salary, while I was at seminary was not as much as when I actually became a parochial clergyman. That was more substantial. However, we're both people of a fairly modest style of life. We like to buy a few books, but we don't have expensive tastes. When Ruth became a qualified nurse, she earned a good salary. In any case we adjusted to our changed financial circumstances.

What happened to you when you completed the theological course?

I remained in Cambridge for another year because Ruth had to train for three years and that was not finished. That year when I was a deacon, I was a part-time, unpaid parish clergyman, but I was paid for another year of college teaching. When all that was finished and we were free to move, we left Cambridge. I became a full time parochial clergyman, a full time parish priest, working in a large working class parish in Bristol under the supervision of a senior priest. In the Church of England, we have a kind of apprentice system, learning on the job. I spent my days walking around the parish, knocking on doors, talking to old ladies, visiting people who were sick. And, of course, I preached and led services.

Was this your original idea of becoming a priest?

Yes, I thought that was what lay ahead of me. I would have a sharp change in my career and develop into a long-term parochial ministry. I spent two years in this large working class parish in Bristol learning how to do it, and then I spent two and a half years doing it on my own in a large village in Kent as the vicar of the village. In the course of all that, I learnt two things. I realized first of all that in the longer term the intellectual side of me really needed a bit more of fulfillment. I like being with people, but my academic side didn't get very exercised by ordinary parish life. The other thing is that I came to the conclusion that part of my Christian vocation, part of my calling, was to think and write about how the scientific view of the world and the religious view of the world relate to each other. Of course, that was a natural subject for me to be interested in. I did a little bit of writing about this when I was a vicar. I had some time to do that, but it was clear to me that I wouldn't be able to do an enormous amount of writing. Then, after about two and a half years being in Kent, quite out of the blue, came the opportunity to return to Cambridge to another college, Trinity Hall (rather than Trinity) to be the Dean of the College, which meant that I would run the worship of the college chapel and I would have the pastoral care of the college. That meant that I would have a priestly ministry, but I would also have an academic component in my life. I would look after the teaching of theology at the college and I would have a chance to think and write. Thus we came back to Cambridge and we have stayed there ever since.

I was at Trinity Hall for three years. I enjoyed it, and I would have been content to stay there for the rest of my working life. But, again, quite unsought and unexpected and out of the blue, I was one day asked if I would like to be considered to be the President of Queens' College by somebody I knew in Queens', an acquaintance rather than a friend. I said that I would certainly be interested in it if I was offered the job. Of course they were exploring quite a lot of different possible people. Being President was a purely secular job, it didn't have any priestly component to it, but the college had a chapel. I went for the interviews and finally, they offered me the job. I consulted some older friends in the church and they thought it was a good idea for a clergyman to assume a secular job if he was reasonably well qualified to do it. Finally I decided to take the position and we spent 7 years at Queens'. During that time I played a major role in the life of the college chapel. Queen's had a Dean, of course, but I also preached and celebrated in the college chapel in collaboration with him.

Is Queens' purely secular?

None of the historical colleges in Cambridge are purely secular for they were founded for purposes that include learning, research and religion. All the historic colleges in Cambridge have a chapel and their statutes require them to maintain worship in that chapel. Colleges are required to have a Dean, who is the parish priest of the college. Thus there is a religious element in their foundation, and for many centuries there was a certain monastic quality in the college life. Until the end of the 19th century attendance in the college chapel by students was compulsory. Needless to say that has changed, but there is still a religious dimension available within college life for those who want to have it. Not all students are religious, but I think that more are than among the comparable age group of the general population. Intellectual people like to talk about religion, discuss these serious matters, so there is quite a lot of activity in the colleges in this respect.

When somebody is appointed to the college does he or she have to take an oath in the chapel?

If you become a Fellow, you're admitted in the college chapel by the President and I would normally admit people using a formula in Latin that would admit them in the name of the Trinity. But it was possible, for example, for a Jewish person, to be admitted in the name of God, or for a person who is actively a nonbeliever, not to be admitted in any name. But it still takes place in the chapel because nobody has ever asked to have this ceremony to take place anywhere else. It's not a statutory requirement to have it in the chapel, but it's certainly a customary requirement.

You are now a priest and a physicist.

I see no contradiction between the two; they are complementary to each other rather than in conflict with each other. They represent different perspectives on reality. Science and religion have different ways of investigating the things that interest them. What they have in common is seeking to respond to the way things are and desire to search for truth. There is no incongruity in being both a physicist and a priest — it is not like being a vegetarian butcher!

Can the two be reconciled?

I think they can. I do a lot of thinking and writing and talking about how they can be held in mutual relationship with each other. It's quite a subtle business. It's rather easy to make crude statements, which have a certain element of truth in them, but which are too unsubtle to describe the situation fully. You can say that science is concerned with how things happen, meaning the processes, and religion is concerned with why things happen, meaning the purpose. However, the way you answer *how* questions and the way you answer *why* questions have to fit together. There has to be some consonance between the scientific view of the world and the theological view of the world. Seeking that consonance necessitates some exchange between them. For example, science can tell religion what the history of the Universe has been and what its structure is, and religion has to listen to that. Science does not need outside help to answer its own questions. If you believe that the world is God's creation, which, of course, I do, and if you believe that the Universe has had a long evolving history, starting very simple and becoming very complicated, that tells you something about how the Creator may be at work in that creation. God is patient and subtle, not in a hurry. God did not snap a divine finger to form things ready made. In fact, in a very famous phrase coined immediately after the publication of *The Origin of Species*, God maintains a creation that's allowed to *make itself*. That's theologically how I understand an evolving world. There's potentiality present in matter to produce life, to produce consciousness, and so on. That potentiality is explored by creation itself through the process of evolution. This way of thinking deepens and strengthens the theologian's understanding of what is going on in the world. What theology can do for science is not to tell science the answers to scientific questions but to take scientific insight and set it within a broader, more comprehensive matrix of understanding. For example, it's very striking to me that mathematics is the key to unlocking the secrets of the physical Universe; that mathematical beauty is the actual tool for scientific discovery in fundamental physics. That seems to be a very strange thing, because mathematics seems to be such an abstract subject. Eugene Wigner called it "the unreasonable effectiveness of mathematics", which, he said, is something we neither deserve nor understand. Well, I'd like to understand it if I can, but it seems to me that science itself can't help me to do so. Science is simply very happy and content that mathematics has this marvelous power and then gets ahead with exploiting it. My religious belief makes me want to look at this issue in a more profound way. It seems to me that the mathematical beauty and intellectual transparency of the physical world shows the signs of the mind. For me, that's consistent with my religious belief, that it is Mind with a capital M, the Mind of the Creator that lies behind the marvelous order of the physical world. Here is the true source of that sense of wonder that scientists feel at the marvelous structure revealed to their inquiries. In that way, I experience a consonance between science and religion. They are different things, to be sure, but the question is, how do they fit together? Not without puzzles, of course, but I feel that I understand more if I look at the reality of the world in which we live in a two-eyed way, with the eye of science and with the eye of theology, than I would look with either eye on its own.

I have talked with quite a few scientists and those of them who said they were religious, seemed to have either of two reasons for that, and I am admittedly simplifying the matter. They may have lost a member of their family and religion provided the only consolation for them. The other reason was when, say, after having studied molecular biology, for example, the scientist still sees that facets of our life and our world that are unexplainable, that cannot be reduced to a scientific explanation. Do you think this is a realistic experience concerning the sources for a scientist to be religious?

I think they represent aspects that are present in the search for the discovery of religious belief. Let me say first of all something about consolation. Of course, I believe that religious faith can support one in life and in death — one's own death, eventually, and the death of those close to one — but it can't really do that unless it's also *true*. I'm very unhappy about religion simply as consolation unless that consolation is founded in the way reality actually is, in the way that God is. We have to be careful of wishful thinking, however well motivated. It is a very dangerous thing, both in science and in religion. So I am wary of too much emphasis on consolation. Yet it seems to me absolutely clear that, though science is very remarkable and very exciting, it couldn't possibly, by itself, be enough to give us an understanding of the rich and many-layered world in which we live. Take music, for example. Physics will tell you that music is vibrations in the air, physiology will tell you that there is a certain pattern of neural response to the impact of airwaves on the eardrum. But the reality of music slips through the scientific net. Whatever music is, it's a very mysterious and profound experience that in my opinion cannot adequately be caught within any form of scientific description. Science has been very successful just because it has limited itself to the sort of experience that it's prepared to deal with and the sort of questions it is willing to ask.

Eugene Wigner said in his Nobel lecture that the great success of physics was due to a restriction of its objectives in that it only endeavors to explain the regularities in the behavior of objects rather than to explain Nature.

I don't recall that remark, but I would say that what we are seeking as human beings is understanding and that understanding is more than explanation. Understanding involves both a comprehensive view of wide scope, and also an achievement of intellectual satisfaction. You can explain things by various ad hoc empirical rules, but you don't have an understanding within science until you reach some deep underlying theory that makes those rules intelligible within science. The greatest difficulty, I think, that scientists are having with religious belief is that religious belief centers on the unique and the particular: unique experiences, unique events, unique persons. Scientists professionally have a horror of unique, they like the general. They tend not to like things that are in any degree special. But it's impossible to understand human life and human experience without being prepared to take seriously what's sometimes called the scandal of particularity, that a particular person or even a particular set of writings might be a source of illumination and understanding that you wouldn't be able to find anywhere else. That's often hard for a scientist to take on board, it seems to me.

That is for the physicist, but biologists find most interesting, beyond the underlying general principle, the individual properties, beyond, for example, the general human behavior, the behavior of the individual.

Yes, I can see that. For the moment though, biologists are more hostile to religion than physicists, but that's a function, to some extent, of the state of the subject. Biology has had a stunning and successful period with the development of molecular biology. That encourages mechanical self-confidence of the biologists, just like the physicists in the 18th century. But that self-confidence won't last forever. Certainly, the historical sciences, like evolutionary biology and like physical cosmology, have a unique history to understand and they have only a fragmentary access to the evidence of that history. That kind of intellectual activity is much more like theological inquiry than a lot of physics, where you have much more under control and a high degree of repeatability.

I've talked with Brian Josephson whose interest in the paranormal is very much off the main thrust of physics and he also uses music as an example in discussing his inquiry. Does it bother you that he uses the same example?

No, not at all. Music is a very important phenomenon. It's an experience that's very widely appreciated by many scientists. There is considerable empathy between scientists and music. When Josephson was an undergraduate I taught him. Of course, I couldn't really teach him anything, but he came to me for quantum mechanics.

He has been engaged with telepathy and he mentioned something in our discussion as if considerations of telepathy and religion were merging somehow.

I'm a little wary of that sort of thing. Frankly, I don't know what to make of the paranormal. I just want to say that I don't think that religious belief, certainly not Christian belief, depends in any critical way on a validation of paranormal experiences. If they are there, they would enlarge our concept of what we know about our world, but I don't think that anything depends upon their being there. I'm agnostic about paranormal phenomena. But if there's telepathy — and it's very hard to understand it in terms of conventional science — but if it exists, it would also help to understand and explain a number of other paranormal claims. If there is some telepathic power, it seems to me that it's quite possible that it is better preserved among nomadic people living in isolated circumstances (the Lapps, for example), than it would be among modern, urban people. If there are even more strikingly personal paranormal effects, then they are more likely to be revealed in extreme personal circumstances than in card-guessing games. It seems to me that in the 19th century, when people collected and tried to evaluate anecdotes, that may well have been a better investigative strategy than the sort of clinical, laboratory-based card-guessing of marginal statistical significance. That makes it very hard to investigate paranormal effects, for if these powers exist, you wouldn't expect them to be easily manifested in circumstances that are congenial to the conventional scientific techniques.

Some people consider quantum mechanics so difficult to understand that they suggest that divine reality is involved in it.

I certainly wouldn't claim that. That would be suggesting some way or other a return to the discredited idea that God is to be found principally in things that we don't understand.

Have you met this view?

I've met almost every view about quantum mechanics. There's a type of thinking where, although they don't put it as crudely as saying that quantum mechanics is very peculiar so anything goes, there is a sort of quantum hype moving in that direction. Quantum mechanics is very challenging because of this paradoxical fact; it was discovered in 1925, we contrive to get the sums right and it has been a wonderfully successful theory, but we still don't understand how the quantum world and the world of every day are related to each other. The measurement problem is still a big problem. There are many different suggestions of how to tackle it, but they all seem to have difficulties. That's very odd; 75 years of highly successfully doing the sums but we still don't fully understand why the sums work. It's a humbling confession for physicists to have to make.

Do you ever have debates?

Oh, yes, from time to time, and I'm quite happy to do that. The last big occasion was a year ago [1999] when there was a meeting of the American Association for the Advancement of Science at the Smithsonian Institution in Washington. It was a three-day conference on the subject, "Is the Universe Designed?" There was a whole variety of contributions, some saying yes and some saying no. One of the set pieces of this meeting was a debate between Steven Weinberg, the greatly distinguished theoretical physicist, and myself. We have been old friends for a long time. The debate was on that very question. Steven said no and I didn't say quite categorically yes, but I said that there are indications that very strongly encourage me in that view. Those occasions can be fruitful and I enjoy them. They need a certain willingness for the parties to approach each other.

Was he?

Not as much as I would've liked him to be, to be perfectly honest, but not as bad as someone like Peter Atkins, for example. He and I have debated several times and there is really no common ground between us. He just says yes and I just say no.

What's the common ground between you and Weinberg?

If somebody maintains the position that science tells you everything and that there's nothing worth knowing that hasn't a scientific answer, then you won't be able to make any headway. But Steven wouldn't say that. He is a humane person. He'd admit that questions of morality and the experience of value are important. The difference between us there is that he sees these values as being a humanly-created little island of meaning in the ocean of a Universe that is meaningless and hostile to us. I see our human encounters with value and with moral imperatives as being encounters with a reality that is all around us and comes from beyond us. Thus between Steven and me there's something more to argue about than simply saying yes or no. We can find an intermediate ground for encounter in this kind of human experience. Yet Steven is unwilling to look at religious experience and he is very hostile to it. I regret that, for there are interesting things here that I would like to talk with him about. But we can find some middle ground, because he isn't saying that science is everything and we can discuss what the extra might be. That's sufficient for a degree of interaction. But if somebody isn't willing to move at least that far, the conversation becomes frustrating.

Other debates?

Not very many formal occasions. There is a dining club in Cambridge, called the Triangle Club, which meets once a term. It discusses these issues in a very broad way. The three sides are science, philosophy, and theology. That's an informal setting, we're a group of friends really. The number of public occasions isn't very large, I'm sorry to say, but I never turn one down if I can possibly manage it.

What's your impression of the general tendency, are science and religion converging or diverging?

There's a very active conversation going on between science and theology at the moment. It's taking place within the religious world and predominantly within the Christian sector of the religious world. I'm going to a conference, for example, at the end of June [2000], which will be held in Castel Gondolfo, the Pope's summer residence. We'll discuss issues of quantum theory and how that might relate to the ideas about divine action in the world, and general considerations of that sort. All the participants there will be well disposed towards religion. Most, though not all, will be Christians and all will be non-dismissive in relation to religion. There's quite a lot of activity of that sort, often involving a small group of highly selected people although there are also less organized opportunities. Quite a lot of this is going on because the Templeton Foundation is providing money to allow these things to happen. However, they are happening not simply because the money is there, but because there is also a real interest in discussing these issues. Today these discussions are probably more helpful than they were 20 years ago.

You lump all Christians together. Does it mean that there is no substantial difference, say, between Catholics and Protestants?

Of course, it looks very monolithic from the outside, but once you get on the inside, there are lots of nuances and differences. Then there are a lot of people among the Christians who are not at all interested in science. There are also some who are quite hostile to science. I'm not simply talking about the creationist movement in North America, which is an un-intellectual movement. Even among the more intellectually reflective people, there are a variety of different responses.

Are these differences along denominational lines?

No, they are more along the lines of different theological methods.

Is there a difference between Christian and Jewish approach?

I think there is to some degree. On the whole, Christians have been more concerned about the relationship between science and religion. Of course, Jewish people are very intellectually able and have made enormous contributions to science, but the Jewish religion is more concerned with practice than with belief. That means that, for example Creedal Statements, the definitions of orthodoxy of belief, play a less important role in Judaism than in Christianity. Jewish belief is very pragmatically oriented, concerned with the right way to live. Christianity has always been very concerned with matters of belief. The Islamic community is different again. Of all the three Abrahamic faiths, Islam has a much higher doctrine of scripture than either Christianity or Judaism does. Jews will believe the Hebrew Bible to be inspired by God; Christians will believe the Bible to be inspired by God; our Moslem friends believe that the Koran was *dictated* by God. That's why they don't like it to be translated, rather, it should be read in the original Arabic. They are much more the people of the Book than either the Jews or the Christians are, and that obviously can produce some problems in the modern world.

I've run out of questions. Would there be a message?

My attitude is this. I think that a very valid human need, a very profound human activity, is the search for truth. We need that search to be on as broad a basis as possible and to take into account as wide a range of experience as possible. The truth has to be sifted. It has to be interpreted, for it doesn't just come to us raw any more than experiments come raw. Science has been an important component of the quest for truth through motivated belief, and I see religion, and theology in particular, which is the intellectual reflection on religion, as being a component of that search for motivated belief as well. There are also many other forms of human activities all seeking the truth. As far as I'm able, I would like to hold these things together and reconcile them with each other. It's not an easy task and it's not a completed task, but that's what I want to do. When I changed from being a full time professional physicist to being a priest, my life also changed in all sorts of ways, but the search for truth has been common to both sides of my experience. That's my message.

Benoit B. Mandelbrot, 2000 (photograph by I. Hargittai).

25

BENOIT B. MANDELBROT

B enoit B. Mandelbrot (b. 1924 in Warsaw, Poland) is Sterling Professor of Mathematical Sciences at Yale University, and IBM Fellow Emeritus at the IBM Thomas J. Watson Research Center, Yorktown Heights, New York. He is most famous for his pioneering work on fractals.

Mandelbrot's family emigrated to Paris in 1936 and survived the Second World War in Vichy France. He graduated as an engineer from the École Polytechnique in Paris in 1947, received his Master of Science degree from the California Institute of Technology (Caltech) in 1948 and his doctorate in mathematics from the Faculté des Sciences de Paris in 1952. Until 1958, he worked for the CNRS (National Center for Scientific Research) or academia. Then he worked at IBM in the United States until he retired in 1993. At Yale University he started as Abraham Robinson Professor in 1987 and became Sterling Professor of Mathematical Sciences in 1999. He is a member of the National Academy of Sciences of the U.S.A., a fellow of the American Academy of Arts and Sciences, and a Foreign member of the Norwegian Academy of Science and Letters. He also belongs to other learned societies and has received numerous awards and distinctions. For the Barnard Medal for Meritorious Service to Science (1985) his citation read, "In the great tradition of natural philosophers past you looked at the world around you on a broader canvas." For the Franklin Medal for Signal and Eminent Service in Science (1986) his citation read, "For outstanding contributions to mathematics and the creation of the field of fractal geometry, and important and illuminating applications of this new concept to many fields of science." For the Wolf Prize for Physics (Israel, 1993) his citation ended by stating, "He has changed our view of nature." In 2003,

Mandelbrot received the Japan Prize. Benoit Mandelbrot's web page is: http://www.math.yale.edu/mandelbrot. We recorded our conversation in Stockholm, during the "Symmetry 2000" Wenner-Gren symposium,[1] September 13–16, 2000 and what follows are edited excerpts from our conversations, together with last minute additions by Dr. Mandelbrot.[*]

My parents were from Lithuania but I was born in Warsaw on November 20, 1924. In 1936, we moved to Paris and late in 1939, we moved on to Tulle in central France. Tulle is in a poor, mountainous area, where outsiders are not immediately accepted, but those who are accepted are accepted thoroughly. My uncle, Szolem Mandelbrojt, had moved to France many years before. When he was a professor in Clermont-Ferrand, he found an architect who was flattered to design a house for a university professor. Eventually my uncle moved to Paris to the Collège de France, the top of the French university system. But he kept his connections with the Tulle area, which became part of Vichy France.

Most of the French people whom I came across during the war did not want to think about the events happening around them. In fact, a few were more interested in settling old accounts with the British. My family was lucky to be helped by the close friends of my uncle and also a Dominican and a Jesuit who had all kinds of connections. It also helped that I did particularly well in high school, where my grades were better than anybody else's had been for a long time.

Foreigners were in grave danger and being a foreign Jew made matters worse. There were narrow escapes but until 1942 my parents found ways of handling the situation and we felt reasonably safe. Then the Germans occupied Vichy France and the situation changed from precarious to far worse. There was no question of attending college and I studied mathematics from old books that had many more pictures and explanations than the books of the 1930s.

Most of my Jewish friends tried to face events by staying together. We did not. I was supposed to be a member of a resistance group, which, however, never did anything. My younger brother, who was perhaps less adventurous, was with me most of the time and snap decisions often affected our survival. We were lucky. To survive we often had to gamble, and I have stayed a calculating gambler ever since, at least in my career choices. I also acquired a great deal of self-confidence.

---

[*]István Hargittai conducted the interview.

It was best to move around. First, I became an apprentice toolmaker and proved to be very good at it. I had firm hands, which were useful for the trade. Later, I raised horses in a small country castle.

In between, a stay of a few months in a school that prepared students for the elite French universities called "Grandes Écoles" revealed that I had an overwhelming geometric intuition. I also proved to be very good in drawing. When the war ended, I passed all of my examinations and, in fact, became a kind of hero. The math professor could not resolve a triple integral in the big examination problem but I resolved it in record time by recognizing that in a better system of coordinates it became the volume of a sphere. When told, my professor could only say "of course, of course."

My uncle encouraged me to go to the extremely prestigious École Normale Supérieure in Paris. He argued that, even if I failed as a scientist, I could always become a respected high school teacher. My father was strongly against any civil service job. He thought that the Communists could come to power and I might flee again, perhaps ending up in Brazil where my diploma would be worth nothing. My mother had been a doctor in Warsaw but at age 50 (two years before Munich) abandoned her profession and became a housewife in Paris. My father was a scholarly man but much preferred professions independent of states and political change.

My father had no chance of winning against his brother, who was forceful, had a highly-respected job, and moved in the best circles in France, so I

In this 1930 photograph, taken in Warsaw, sitting are Mandelbrot's uncle, father, Arnaud Denjoy, Jacques Hadamard, and Paul Montel. His white-bearded grandfather presides. According to Mandelbrot those men, except for Denjoy, greatly influenced him (courtesy of B. Mandelbrot).

took my uncle's advice. For a moment I was immensely proud of having outwitted Hitler and belong to École Normale. But my uncle had warned me that Normale was weak in physics and was becoming dominated by a very formalistic group of brilliant mathematicians calling themselves Bourbaki. I am anything but a formalist. Reality dawned and two days later I moved over to École Polytechnique.

You must have had a strong mother.

She was one of the first Jewish women to beat the restrictive "numerus clausus" at the Warsaw Medical School. My father went ahead to Paris and was very busy getting settled down and as a doctor my mother earned enough to raise us in Warsaw. But she wanted the family to reunite and Poland was becoming unlivable. She expected much of me and I was slow in developing. She lived long enough to be reassured, I think, but without this extraordinary piece of luck of bumping into fractals, I might have amounted to nothing compared with her expectations.

Do I understand it correctly that your self-confidence originated more from early life experience than from later achievements in science?

Benoit Mandelbrot with his mother, 1931
(courtesy of B. Mandelbrot).

Definitely. By age twenty, my mother and early experiences had done a good job. When Nigel Gordon was preparing the film he made of my life (*Clouds are not Spheres*. Films in the Humanities and the Sciences, Princeton, New Jersey), he needed photographs of me as a child or young man. Putting them together, I was struck (so was also my secretary) at how decisive I looked even early on. Also, in every case, everybody recognizes me at first glance.

Your question reminds me of the day when an MIT dean I knew well stopped me in the corridor to report that everybody was calling me crazy because a rolling stone like me could gather no moss. But he felt that, to the contrary, my behavior was proving that I was the least crazy person he knew because I did not worry about who I was. I knew that independence would mean a harder life but I was prepared to pay for it.

My early work did not amount to much, but I always thought that it would become significant. So perhaps that dean was wrong and I was crazy, after all. My first contributions to the study of difficult real problems came in the 1960s when I let myself become attracted to some very difficult problems of finance and of turbulence. I was well inspired in picking those problems and it was fortunate that I never distinguished among problems as belonging to either a high or a low caste. I did not care what other people thought about the "intrinsic worth" of the problems I tackled.

There is something else I want to tell you. I had a peculiar education from beginning to end. For example, I started elementary school very late because, before I was born, my mother had lost a child in an epidemic and preferred to have me tutored by an uncle (not the one who became famous). As a result, I never learned the multiplication tables properly or the alphabet. Because of this schooling, combined with the conditions of the war, I missed many standard topics. Many people are surprised that with such gaping holes in my education, I could be a university educator.

For your Master's degree, you went to Caltech. There was an exceptional group of people there. One of them was Carleton Gajdusek who remembers your being an exceptional member of that exceptional group.

Gajdusek was by very far the most exceptional young man there. By the time we were together at Caltech, he was already a full M.D. whereas I was a Master degree student. We became instant friends. Nobody is indifferent to Gajdusek. A minority admires him, but a majority has always hated him with a vengeance.

Why?

For performing so much better than most people in so many different ways and not hiding this fact at all. As you know, we could discuss Carleton for days. He went on to a brilliant and extremely odd career.

He told me that you were close to his family.

Yes. His mother was a formidable woman and no doubt Carleton inherited much of his "extravagance" from her. She lived in Yonkers between New York City and Scarsdale where I live. I used to go to see her often. After she suffered several strokes and moved to a retirement home, she was sadly diminished and I was one of the few people who visited her.

Gajdusek told me about the sizzling intellectual life at Caltech.

Indeed. After one year of a socially-limited life in mathematics and aeronautics, I was told that a new postdoc had arrived with his wife from the Pasteur Institute in Paris. This was Elie Wollman. By and large, socially and intellectually, I began to live with Wollman's crowd around Max Delbruck. That is how I met Gajdusek who worked with fascinating people like Linus Pauling, Max Delbruck and George Beadle. Other members of the group were Gunther Stent, the geneticist and Jack Dunitz, the crystallographer. The post-war situation had brought together an extraordinary group of people, many of whom already had complicated lives behind them. François Jacob was not there, but somehow I associate his name with this group too. Not all the members of that group had been top students (although Gajdusek was). The Caltech students who became Nobel laureates were seldom Number 1 in their studies.

What did you do at Caltech?

My uncle kept telling me that it is important to have a "patron". I went to Caltech to study with a famous Hungarian. You will like this Hungarian connection! Unfortunately, Theodore von Kármán had just left for Paris although he came back for several visits during my stay. I greatly admired him and he was a genius with practical things but once he tore someone apart in front of me. The person was a fool, but Kármán acted in excess of what would have sufficed. He never married and his sister managed his household. She hated the German language and called herself de Kármán. When he was a professor in Aachen, Germany, they lived in Belgium and he commuted every morning across the border; in this way she lived in a French speaking country. Then they went to Caltech but as soon as the war ended, they returned to Europe and lived in an elegant hotel in Paris.

My two years at Caltech were wonderful and inspiring but professionally did not bring what I had hoped. What I wanted to do did not exist yet and I had to invent it for myself, which did not happen until many years later with fractal geometry.

What did you do after Caltech?

I returned to France for a year of service in the French Air Force. I could not study much but had amusing adventures, met interesting people, and discovered the opera. All this somewhat compensated for overall boredom. When I got out of the army the privilege of being an alumnus of Polytechnique was drawing me to join the French bureaucracy. Again, my father worked to prevent this from happening. He looked for job offers in a newspaper and found one from the Dutch company Philips. He liked very much the idea that if there was a revolution in France, that big international company would send me elsewhere. Philips was developing color television and wanted someone with my interests and training. So I went to work for Philips as a consultant. The other staff members did not know much mathematics, and I already knew all kinds of tricks.

At the same time, I started on my doctoral thesis. My uncle and other friends thought that my attitude was irresponsible because I floated around and did not seek a patron. Finally someone agreed to write a report on my work. We did not like each other at all. I would have chosen Paul Lévy but he did not have the right to direct Ph.D. students. The regulations were incoherent therefore flexible. For me, this loose situation turned out to be far better than supervision by a maniac with an agenda, who would have forced me to accept a standard topic.

My Ph.D. thesis was very poorly written and no decent U.S. university would have accepted it. But in my case it turned out that not having a patron was not all bad. In fact some calculations from that very old text remain useful.

When did you get your doctorate?

Late in 1952. By then, my father had died. My thesis was titled "Games of Communication", but really combined the two topics of statistical thermodynamics and power-law distributions. My ideas had what they call "long legs", since they developed in due time into something important. One can say that they really concerned fractals but of course without the word, which I did not coin until 1975, and without a clear concept of where I was going.

And after the doctorate?

My first teaching job was at the University of Geneva, Switzerland. Then France experienced a great sudden need for applied mathematicians and found that few people had the proper background so I moved to the University of Lille.

But I soon realized that my problems with the French educational system were more profound than not having a patron. The French complain that the Germans are better organized, but France remains the tightly-knit, extremely well-organized country fathered by Mazarini and Buonaparte. To survive as a maverick is almost impossible, which is perhaps why there is no French equivalent to the word "maverick" and why there are no "middle-level" mavericks in France. Two who rose above mere mortals were Louis Pasteur and Pierre Curie. Pasteur defied categorization by being a chemist and curing human illnesses and doing many other things at the same time. He did not let himself be pressed by the traditional expectations and he was carrying out his work amid constant attacks by the media. He was also badly treated by the French universities and was saved only when outsiders endowed an Institute for him. Curie was jobless until he received the Nobel Prize. René Descartes was an earlier maverick who chose to work in Holland and after his death his works were put in the "Index of forbidden books". And French musicians are still baffled by Hector Berlioz.

You still like to be in France.

I can no longer stand the French university scene but I love the country and have many friends. We still have a very small apartment in Paris — my mother's former apartment with my mother-in-law's furniture. But I no longer try to work in France.

Where do you feel yourself an insider?

Nowhere. The situation was worse in France than it is in the U.S.A. I had said "No" to École Normale and later to an offer that looked glamorous but I knew would prevent me from bringing fractal geometry together. Each "no" was perceived as insulting.

In general, you don't change from being an outsider to an insider, but you may become accepted if you are useful. This is what happened to Pasteur, Curie, Descartes, and Berlioz, who of course are now hailed in France as major figures.

You joined IBM in 1958. Why did you leave academia?

First by chance and then by necessity. By chance, because they invited me, accompanied by my wife and our baby, for a summer visit. In successive stages, this visit extended to one year, then three, then thirty-five. By necessity, because no better place was available and IBM worked out for me. The safe thing at that time would have been to keep my academic job in France. I was too little-known to create resentment. I might have lived peacefully and obscurely. I was extraordinarily lucky to escape that sad fate by going to IBM and then to Yale.

It took a while for my wife to feel comfortable in America. She was born in Paris, but her parents had also come from Lithuania, more or less.

I wonder why IBM as a host institution was the right environment for your unorthodox activities.

It is often said that scientific creation presupposes three elements: the right person, the right place, and the right time. This is largely so but, at least, in my case, those three elements depended on one another. In due time, IBM became a major scientific power and remained one until 1993. But when I arrived, it was starting out of nowhere and had a loose disorganized feel that for me was just right. For many years, the Bell Laboratories were better and better known and we followed their example. But Bell was overly organized for my taste; few people on its staff were oddballs and not part of a well-defined group.

At the time most scientists perceived taking a job at IBM a gamble. For me the main advantage was freedom. The availability of computers came second. You must be surprised, therefore I must elaborate. IBM did not market graphic devices, therefore it provided them to us with extreme reluctance. The computing center was meant to help design future products, not to carry out wild scientific experiments. Many other institutions had bigger computers and their graphics were far better. But they were used with little imagination and produced little of interest. To the contrary, the prevailing loose and free-wheeling mood adopted the IBM of my time into the right place and the right time for me and a number of other mavericks. They provided ample computer time during the nights and weekends. But they mostly provided an environment where I could befriend and cajole competent persons into custom building special instruments we needed. They connected parts that were lying around. Those instruments and their software were clunky but adequate for me and my very few associates. Otherwise my skills at visual thinking could not have been utilized.

That is, we did not do well because expensive tools were available, or because money was available to buy them and beat the competition. In fact, I needed a very limited budget and there was no competition of any kind. We did well because that straight-laced bureaucratic giant included some persons in positions of authority who realized that a few free-wheeling characters hardly showed on the big budget. Ralph Gomory, who was my boss at IBM for twenty years, eventually moved up to be number three in the corporation and always supported me. He reported to the Chairman of IBM who reported to God. I did mostly science, but also I did several things that saved money for IBM. First I was a member of the IBM staff. In 1974, they made me an IBM Fellow. That is what made IBM an ideal place for me and why I did not return to academia until much later when events forced me to retire.

The IBM Research Center was not planned to be what it became. What it became is due to fortunate circumstances that sound like a Greek tragedy, the fact that Thomas Watson, Jr. literally overthrew his father. The father's company was great at producing mechanical devices that worked forever. But in the fifties, electronics arrived and the son faced a dilemma. Should he introduce the changes incrementally or start a brand new company within the shell of the old one? He decided to start a new company and it boomed. When I came to IBM, hardly anyone could imagine that a few years later, IBM would overtake RCA, a giant company manufacturing a broad range of products.

As Director of Research, Watson Junior hired Emmanuel Piore, who was born in Lithuania, got his Ph.D. around 1932 and had no real job until the war. Then he proved to be an extraordinary operator, but an operator with a vision and a soul. He organized the National Science Foundation and the Office of Naval Research as a fund-granting agency for pure research, and then built IBM Research out of nothing into a very large community. This started just when Sputnik went up and jobs had become so plentiful that the best group of scientists IBM could get included many "rejects" of the system. Some were justified rejects, many had career problems (not having a patron, or worse), and there were a few oddballs, like myself. IBM simply decided that something might come out of us and that the geographical isolation of Yorktown made it desirable to use us to build a small scientific community. The brilliance of Piore's vision was shown after the fact, when many of us became members of science academies and won prizes. Piore was prepared to take gambles and hired people prepared to take gambles as well. The same broad policy was followed by his most notable successor, who was Ralph Gomory.

How did you get to the idea of working on fractals?

There was a gradual process that began by exclusion. I had not liked the Caltech or Philips engineering and decided to return to science and obtain a Ph.D. One possibility was theoretical physics. However there was no role model in France and besides I did not want to work on anything related to the atomic bomb. I wanted to do geometry of the classical kind, but my uncle had convinced me that this topic was at a dead end. I hated algebra, so straightforward mathematics was out. I was an ambitious and rebellious idealist looking all around for some domain in a terrible mess to which I might try to bring mathematical order. All the while, my uncle was joking that my ambition was to become Kepler and I was late by several centuries.

But I did not think so because of something I had noticed. As the disciplines become organized, there is an increasing accumulation of observations that fit nowhere. I was free to look for potential treasure in dumps hidden in dark corners. Eventually, and against all odds I managed to clear several terrible messes and my romantic dream was completely fulfilled. For a person to be entitled to say so is truly overwhelming. An old man saying so is necessarily filled with awe. The process took a very long time and it is only recently that I began to see a system in my wild search. I realized that the fields I had excluded can be said to be ruled by smoothness, and my life work was to provide a theory of roughness.

When you gave the world this new word "fractal", did you think long about it?

A text I wrote in 1975 became the first book on fractals, but its original title was very clumsy. I gave the draft to Marcel-Paul Schützenberger, a dear friend and an extraordinarily brilliant man, who knew French thoroughly.

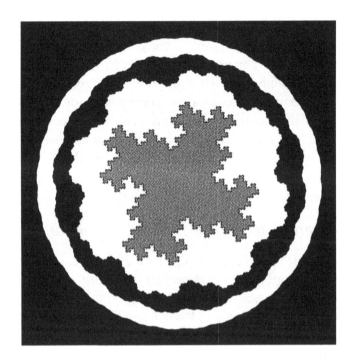

A sequence of fractal curves of simultaneously increasing roughness and fractal dimension. Mandelbrot explains: Fractal geometry has freed dimension from belonging to mathematical esoterica and made it into a very effective first quantitative measure of the "degree of roughness", a notion from time immemorial that had eluded earlier efforts. The "shading" near the center shows what is meant by a Peano motion of fractal dimension 2. Fractal dimension can also be negative. If so, it identifies and quantifies, for the first time, an important intuition that would have seemed irreducibly elusive, that of "degree of emptiness". Courtesy of B. Mandelbrot.

He was a genius, but too much of a dilettante and so disorganized that he is already forgotten. The first thing he told me is that if I would try to peddle this **** (he used a French five-letter word), he would protest to my publisher because he knew I could write in proper French, therefore must do so. I re-read my work and realized that "Marco" (as we all called him) was right. That draft was full of anglicisms. Someone had to help restore my previous command of the language.

The second thing Marco said was that I had brought to life something that did not exist before, so I was entitled to give it any name. Roughness is a concept connected with fracture so I looked up a Latin dictionary and was reminded that the word "fracture" comes from the Latin adjective *fractus, fracta, fractum*, which means "like a broken stone". The Romans did not care much for abstraction; they used very concrete words. The proper neologism was the Latin word "fractum" but it rhymed with quantum. This made it pretentious, in bad taste. Next I hit on "fractal", which sounded good in French and English — and later in many other languages in which my book has been translated. It is now in every dictionary.

Of course, there were various linguistic difficulties. In Russian, should the last letter of "fractal" be hard or soft (fractal versus fractal') and what should be the gender? A committee decided that it should be the soft fractal' and it should be feminine. If you compare new scientific terms, "fractal" is less problematic than "chaos". Every old word continues to be burdened by its previous meanings but a new word has none. Of course, my word, fractal, has already attracted new meanings.

So what is the mathematical definition of fractals?

There is no single mathematical definition. However, even in mathematics everything that is important is difficult to define. If you don't believe me, try to define probability theory. Or note that "general topology" is defined as "the study of the notion of neighborhood". For general purposes, I follow this example and call fractal geometry "the study of roughness".

If pressed further, I add that it is a theory of self-similarity and related invariances. A self-similar shape is one that looks the same from nearby and afar. Every formal definition I know excludes something very important and I no longer look for a single all-encompassing definition. My first book gave no definition. Then I relented and gave a tentative definition based upon fractal dimension. But a prominent French mathematician, a formalist and a snob, criticized me sharply. He understood fractality and viewed it as a very fundamental notion. To the contrary, fractal dimension is a specialized notion that has many variants. Therefore he told me I should not define a fundamental notion on the basis of a specialized one. Physicists think differently. Fractal is a concept but fractal dimension is a number that quantifies roughness and can be measured empirically. Physicists detest concepts and worship numbers that can be measured.

An important question is whether or not fractal geometry will survive, whether or not it will continue to provide good service to science and economics. But this question has nothing to do with whether we can find an all-embracing definition.

You created a new field, which defies traditional definitions and recognition. What are the highest awards you have received?

The 1993 Wolf Prize for Physics and recently the 2003 Japan Prize.

The Japan Prize is in the field of complexity, a word that fairly describes the search that started in 1952 with my thesis for exciting facts thrown into diverse dumps hidden in dark corners.

The Wolf citation ends by saying that I "changed our view of nature". According to a rumor, the mathematics and physics sections were each looking for candidates that best fulfilled more traditional requirements, therefore kept sending my nomination back and forth. But then one section stopped worrying about definitions and selected me.

Could it be that your receiving the physics prize surprised some colleagues?

Perhaps so. But in my own view physics has always been broad enough to include much of my work. In fact, I had thought of presenting the same dissertation for a Ph.D. in physics. Let me comment on the situation today, first from a practical point of view and then from a fundamental one.

Consider my favorite early topics, finance, turbulence, hydrology, 1/f noises. At the time of my pioneering work (up to forty years ago!), none of these topics was attracting the attention of people calling themselves physicists. In this sense, society viewed them as being outside of physics. But in recent years each of these topics has been, how to say it, invaded by persons transferring from mainstream physics. One might say that the scope of physics has by now extended to include every one of my old favorite topics. This extension resulted from many social forces beyond my example. But the directions it took vindicate what I have been doing all along.

Let me now present a more fundamental second point of view. Historically, "physics" served as a common term to denote several activities, mechanics, optics, acoustics, thermodynamics, each of which arose in response to basic sensations like heaviness, light, sound, and hotness. For example, thermodynamics is the study of hotness. The sensation of taste (sweetness or acidity) gave rise to chemistry.

But what abut the sensation of roughness? It is as important as the other ones but had never been studied properly on its own. The first stage of every scientific approach consists in inventing a way to measure some vague sensation quantitatively and in the case of roughness, there was no quantitative measure until I developed fractal geometry. This would-be theory of roughness is very young and modest but is an unquestionable part of a suitably enlarged part of physics.

In my opinion, science has not reached its end, far from it, but it has reached a stage of sharply increased difficulty. Physics will have to recognize that it has a new frontier in the study of roughness. It seems to move slowly, not because of the shortcomings of the participants, but because the topics are intrinsically difficult. This is, of course, why they were left wide open.

The same is true of a less obvious concept related to roughness that I call "wild randomness". The bulk of known science is devoted to "mild randomness" and I introduced the distinction between mild and wild variability. It is somewhat parallel to the distinction that dynamical systems theory makes between non-chaotic and chaotic solutions.

At the age of 60, you went to Harvard University and soon resigned. What happened?

I felt it was not the place for me. I was uncomfortable, did very little work, and lost sleep. Again, I took chances, followed my own instinct, and walked out. Harvard is the oldest university in America — a great one, no question about that — but to my taste too much of a collection of prima donnas. The faculty spends time fiercely competing for recognition, prizes, and academy memberships. For science, it is also a disadvantage that Boston is an extraordinarily interesting city. Dingy New Haven is much more prone to doing serious work. People can leave their doors open and the level of collegiality is very high especially among mathematicians. I am very happy at Yale.

Do you lecture at Yale?

Yes, of course, but only on a half-time basis. In particular, I am responsible for the introduction of an elementary course on fractals for non-specialists with over a hundred students attending it every year, because they like it, not because they are forced to. With few exceptions, the lectures are given by a colleague who is a superb teacher, which I am not. He spends endless hours with the students, which I could not possibly manage to do. What I am is an effective preacher and without me, the course would not have existed. That course is based on a simple idea. There are many ways of presenting the basic concepts of mathematics and it is the best way to begin with material that students spontaneously find attractive. This happens to be fractal geometry. By now, many colleges have adopted this viewpoint. It is also rapidly spreading to high schools.

Do you use grants?

Of course. This is unavoidable. But I hate a non-stop search for contracts and prefer relative poverty. The impact of funding on the development of science should never be underestimated. Science is defined by its intrinsic value, but it also responds to external effects. Compare the support for atomic physics before and after the first atomic bomb. Statistical physics also became a huge field; then one day, IBM and Bell laid off most of their research physicists and funding became scarce. At Yale, I once had a group of seven people, but I found it too difficult to worry about their individual agendas. One of them was from Hong Kong where a job was waiting for him, whereas in the U.S. there was no job. I have cut down the size of my group, but I have visitors all the time.

Had I written about fractal geometry ten years earlier, I may have been inundated with money, and institutes of fractal geometry may have developed everywhere. So my timing was not very fortunate.

Was it hard to gain recognition for your new field?

Until 1982, recognition was slow, narrow, and very gradual. This made me very impatient but was natural. The risks I took did not only consist in joining IBM but also in interacting with fields in which I was way out of the Establishment. I would enter some existing community, provide my reasons why an approach they took could not possibly work, and propose a very different approach with no attempt to dilute its nature. For instance, economists were unfamiliar with the notions of invariance and symmetry. They found them ridiculous and an affectation. Each time I entered yet another existing field and met its practitioners it seemed that I was voluntarily walking in front of an eager firing squad. One good thing was that my job at IBM was solid.

Actual recognition began in 1962–1963 when I was a visiting professor of economics at Harvard University. Harvard did not mind that I did not have a degree in economics. That visiting position truly shook the world of finance and was an important sign of growing recognition for my work. Before that year had ended, the Harvard Division of Applied Sciences offered me a visiting professorship in 1963–1964. Those two years made me join the best company. But rough episodes at the University of Chicago, then at Harvard, made it clear that nobody wanted me as a regular professor.

I must add that at each stage I enjoyed enormous support from a few people. Take my main paper on turbulence and multifractals, which was

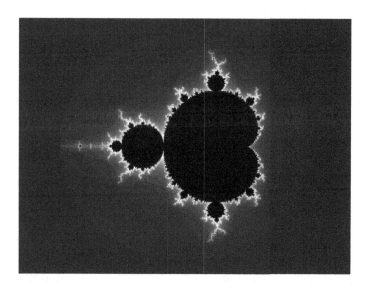

The Mandelbrot set M. According to Mandelbrot, in discovering this set in 1980, he made several visual observations and stated related mathematical conjectures. One took months to be proven, others took five or ten years, but the basic one remains open. It postulates the identity of two definitions of M. One is based on the connectedness of the "Julia set" and the other is physical and involves finite limit sets of quadratic dynamics. Courtesy of B. Mandelbrot.

invited in 1966 but only appeared in 1974. Why so? It took two years to referee the first version, because every possible referee said it was gibberish and contained nothing about turbulence. However, the editor, Professor Keith Moffatt of Cambridge, a brilliant man, advised me not to get discouraged. The second version took another two years and the third version also took two years, and finally the paper appeared eight years from the time of its inception. If not for Moffatt, I would have diluted it out of existence.

This paper is now a classic but it was very difficult and few people could read about multifractals before they knew about fractals. To create an audience for it was truly a long-range enterprise. This is why a significant change in my career came when I decided in 1975 to write the book in French [*Objets fractals*, Flammarion, 1975] which I have already mentioned. For several years it sold poorly, but an English edition that followed did well [*Fractals: Form, Chance and Dimension*. W.H. Freeman, San Francisco, 1977]. Then came *The Fractal Geometry of Nature* [Freeman, San Francisco, 1982] which became an event. Everyone wanted to review that book. Taken together, those books and their many translations sold several hundred thousand copies, to which you may add hundreds of books of all kinds by other authors. My work had become very broadly recognized.

In 1985, I received the Barnard Medal, then other prizes came and many honorary doctorates from all over the world. Recognition grew and for a while something else came along. In addition to being a well-known and respected subject, fractal geometry became a fad. Every computer journal had articles about fractals. Fortunately, fads don't last and fractal geometry survived that period. I found it oppressive. However, some people manifested an absolutely horrendous level of envy.

Was it at IBM or generally?

At IBM nobody felt threatened. Also, it is a collegial community and we made a point to nominate friends for prizes. Outside IBM, it was a different matter. When I received the Barnard Medal, someone warned me at Harvard that I would never be forgiven. Einstein and Fermi had received it in their time, so newspapers called me the second Einstein and second Fermi and what not. Of course, that was pure journalism, but it created enemies.

This is long gone. There was a special semester on fractals in 1999 at the Newton Institute in Cambridge. It was very pleasant, with no fights and most of the talks were about the mathematics of fractals. The mathematicians did not jump on the fractals bandwagon, far from it, but several of my conjectures turned out to be extremely difficult. As a result, many distinct groups are very much involved in different problems raised by fractals.

In physics, the story is different and change is faster. A large group of physicists jumped on the fractals bandwagon. Eventually, the feasible questions were solved. But the diffusion-limited aggregates (DLA) turned out to be very difficult to study. They no longer promise a fast academic career. However, physicists who had been involved with fractals find jobs nowadays in biology and finance and in other areas.

You were the last student of John von Neumann.

In 1953–1954 I was the last person at the Institute for Advanced Study in Princeton sponsored by von Neumann, another Hungarian. He was by then very much involved with defense but I met many other interesting people.

Why did he invite you to the Institute for Advanced Study?

I had sent him my thesis, which he found interesting. He sent word that I could come and see him any time. I visited him on a Saturday. We talked and he asked me if I wanted to come to the Institute for a year. I said that I would love to but it was late May and everything must have been settled for the next academic year. He told me that was not a problem but I should see Warren Weaver at the Rockefeller Foundation. I went there on the next Monday morning. Von Neumann had telephoned with a very strong recommendation and I did not even have to apply.

He was very encouraging. Once I gave a lecture at the Institute for Advanced Study. This lecture was terrible and after I finished someone attacked me bitterly. Oppenheimer jumped up, noted that my lecture did not present my work well, then in his own brilliant way, gave a summary of what I should have said. Then von Neumann stood up and said that he too, was familiar with my work. He agreed that my presentation did not do justice to it then proceeded with warm compliments. On that great day I moved from giving the worst lecture to having both Oppenheimer and von Neumann as lieutenants.

Many years later at IBM, Gomory took off to do research and my situation suddenly became very bad. IBM policy was to never fire anybody, but my temporary boss was someone whom I despised and who handled me so terribly that I was afraid to be forced to accept some project that I would hate. I went back to Warren Weaver for advice and was told that when von Neumann was dying, he asked Weaver to watch out for me because I might get into trouble and need help. As a result, a fellowship was made available on the spot to enable me to visit any university of my choice. In the meantime, the man who had caused my trouble moved to another position, so my problems were solved.

Von Neumann was not exactly a warm person, so I was pleasantly surprised that he could be so completely understanding. Then I realized that despite his fame he was badly treated and unhappy at the Princeton Institute. The mathematicians found computers to be detestable. Some physicists disliked his strong involvement in military affairs. There was also the rumor that von Neumann had expected to be made director, but they elected Oppenheimer instead. This led to resentment. They had similar family and education backgrounds, and could have become either close friends or not friends at all. Von Neumann was very much an insider in Washington but no longer so in academia. In fact, he had accepted a position at U.C.L.A., but died before he could take it.

Did you ever come across Erdős?

I did. Many people admired Paul Erdős and so did I. My uncle thought that Erdős was not on a par with a Henri Poincaré or a David Hilbert because he did not open broad new areas. On the other hand, I am very well known for asking hard new questions. Some people joke that Erdős and I together would have made a truly wonderful combination.

Did you come across Szilard, Wigner, and Teller?

I met Edward Teller once at a meeting on thermodynamics and we had a good conversation. About Leo Szilard, something I wrote in thermodynamics revived and extended something he had done. I wrote him about a paper he wrote in 1925, but he was no longer interested in the topic. As to Eugene Wigner, I was very impressed when I showed him the mountain-like fractal structures. His comment was, "Very nice. Don't tell me how you did it. Let me guess." He thought for a while, mumbling aloud, and he guessed it right.

Who are your heroes?

Most have been mavericks. Among those I have met, von Neumann. Among historical figures, Henri Poincaré. More indirectly but perhaps even more strongly, Louis Pasteur before he became a national hero.

Was Paul Lévy your mentor?

Yes, in many ways. Not my teacher because in truth he never had students. Some of his very best work, not his most difficult but very imaginative, was completed in his late sixties.

You also keep working at a late age.

Yes, but now when I work too much by myself I get bored. I prefer to collaborate with some young fellow who works quickly, and can learn from me. I supply questions and most interpretations, but I am not up to the long and difficult calculations or experiments.

I see no age limitation to creative work of any kind. It was possible for Beethoven to stop composing for ten years and then start again and write greater music than ever. Giuseppe Verdi also stopped, then came back to write his best opera, *Falstaff*, at the age of 80 — and before that his second best opera, *Othello*, at the age of 75. The widely-held thesis is that this could not happen to a scientist but I think that this thesis is deeply flawed. In fact old age *per se* does not exclude creativity. Personal ambition may diminish, but not necessarily creativity, especially for people, like myself, who are prepared to constantly move and start something very different. I think that the importance of youth in science has been absurdly exaggerated.

We have touched upon the question of categorization in science. Which category would be the most appropriate for you?

My long scientific life defied and survived categorization. Several centuries ago, I would have been called a natural philosopher. I love the sound of this term. Today, I am a living fossil, a throwback to ancient ways of being a mathematician which included physics in a broad sense and much else.

Can we talk about "nationalism" among the sciences?

During the 1960s and 1970s, physics and mathematics functioned in a totalitarian mode. Some official academia decided what was proper to do and tolerated nothing else. It was deeper than a question of funding. I am the least nationalistic of all scientists. In my mathematical mode, I speak with a foreign accent, but I am still a mathematician. When writing a physics paper, I am a physicist with a foreign accent. The main reason that I can do work in so many fields is that I am prepared to learn different tongues and do not worry about being unable to get rid of my accent. I do not doubt who I am and see no need to dress into a mathematician's cloth or a physicist's cloth or to put on an economist's hat to think as one.

What is your main activity nowadays?

I continue to write papers in mathematics, physics, and finance and I give many talks on art and music. But I mostly work on books. Together with a brilliant journalist, I am writing a book on finance. On the back burner I continue to compile somewhat idiosyncratic "Selecta" books. Three volumes are out and more are on the way. Half of each book is made of old papers of mine on a defined subject and the other half consists of new papers and discussions. I am also under great pressure to write my memoirs. There are few interesting autobiographies by scientists because most appear to live unsurprising lives.

Not you.

Not me. My story has many elements of high drama. Many people think that I have a large number of co-workers but mostly I have worked by myself and created most of my ideas quite alone.

Symmetry has brought us together. How did you develop your interest in symmetry?

I heard Hermann Weyl lecture on the material of his famous book on symmetry. In physics, symmetry is connected with invariances and groups of transformation. I use this concept both in this sense but also in the ancient Greek sense, mentioned in Weyl's book, of harmony and balance. When I told you that I do not mind the fact that fractals remain ill-defined, I expressed the strong feeling that important concepts should never be defined narrowly. Therefore, I never worry about whether something is or is not symmetric according to a formal definition.

Much of science consists in a search for the proper invariance. Take something as messy as financial prices. There are many well-paid economists but economic knowledge remains very poor. In the early 1960s, I made great progress by injecting the notion that price charts are invariant with respect to certain semi-groups or groups of transformations called dilations or reductions. Those invariances and later generalizations I proposed led to three successive models that I now call mesofractal, unifractal, and multifractal. Clearly, symmetry is not merely a minor wrinkle in this work but a fundamental ingredient.

My position on these things has been controversial but now many physicists have moved into economics and they adopted my invariances quite readily.

The Wenner-Gren Center, where the symmetry symposium is taking place, is very close to our hotel in Stockholm. But we have to cross a big square with a lot of zebra crossings and traffic lights, and a lot of traffic, and the crossing can be done in several different ways. I noticed that although you don't walk too fast, you get from the hotel to the meeting site faster than anybody else. Do you choose your route consciously?

I found the most economical route almost instantly. My wife is continually surprised that when we go to a foreign city and walk a little bit around, we get back to our point of departure faster than would have been estimated from the distance covered. I create spontaneously a mental picture that can be very complicated. For example, Tokyo is not a simple city and many persons have a mental picture of the schematic map of the main subway lines, but not much more. Once my host was driving back to my hotel and got lost. I was able to volunteer as his guide.

In history, the career of several scientists was largely based upon visual acuity and memory. They are my herocs, of course. For example, Santiago Ramón y Cajal [1852–1934] was such a person. He is another of my heroes.

I have a Spanish stamp with his picture. He was a Nobel laureate in medicine in 1906.

An extraordinary story! He was born in Upper Aragon at two days distance (by foot or donkey) from the nearest city, Saragossa. First he became a country nurse, then a doctor. He had very old microscopes and his images were so poor that they could not be photographed, yet he could see structure where others failed, and remember it long enough to draw it. His many important discoveries in neuroanatomy withstand the test of time. Once he travelled on his own expense to a scientific meeting on the nervous system in Germany. He set up his microscope in the meeting corridor and asked all the dignitaries to look into his microscope and then at his drawings. A Herr Professor from Wuerzburg — Albert von Kolliker — stopped and looked into his microscope. On his return to Wuerzburg he confirmed what Cajal was showing and then a miracle happened: he did not appropriate Cajal's results but helped Cajal achieve recognition. This is how a man working in the most backward conditions in a then-backward country was soon awarded the Nobel Prize and became a national hero.

In 1980, when I was in the process of discovering the Mandelbrot set, I felt that I was retracing his steps. As I told you earlier, IBM was not into graphics at all. In any event, I made that discovery at Harvard. There, we had a very primitive machine and we could barely make out what was on the screen. We had to increase the contrast by repeated Xeroxing until we reached a level that could be published.

How do you feel about globalization from the point of view of science?

Today, it does not matter whether I sit at IBM or at Yale, I can communicate instantly with my associates at either place and everywhere else. Unfortunately, globalization also means that fashion in science tends to become the same everywhere.

"Corals" of fractal aggregate or diffusion-limited cluster surrounded by lines of constant Laplacian potential. According to Mandelbrot, this shape catches an aspect of the fundamental but elusive process of random aggregation. It had been half-glimpsed repeatedly but the eye and the tools of fractal geometry were needed for full discovery and later study. The construction algorithm is childishly simple but its outcome remains extremely mystifying to mathematicians and physicists alike. Courtesy of B. Mandelbrot.

Do you think that diversity in a scientist's background is no longer appreciated?

The situation is getting worse and worse. More than ever, I feel like a throwback. In the late nineteenth century, science was not yet professionalized. Chairs were less specialized. If a tenured professor's interests changed, he remained a professor and could be doing something entirely different of what his chair said. Funding was non-existent so it did not impose restrictions.

At some time, many of my papers were being turned down and all my proposals for funding were being refused simply because they did not fulfill certain revoltingly arbitrary and narrow criteria. A community of tenured people form an invisible guild that wants to make sure that incoming people will become very predictable members.

Once I received an article that had many references to much earlier work of mine but claimed to be the first to include realistic mountain reliefs generated on the computer. I called up one of the authors to ask what this meant. He said that I need not worry (not that I worried in the first place) because their claim was narrow: to be the first among their community of computer graphics people. Let me conclude: globalization is one more reason why diversity must be defended.

Do you have many followers?

In a broad sense, they may be counted in millions. No week passes without dozens of letters from all over the world, the majority from young people. I view this as an enormous privilege and burden. For the young, I have become a legend, a part of history, and they are surprised to see me walking and talking.

In a narrow sense, there are no professorships in fractals and no funding is specifically set aside for work on fractals. If you want to work on fractals, you must have a professorship in something else. Even people who are most interested in fractals feel they must be professors of something else. Working at IBM allowed me to produce more by myself but kept me from training others. There was no search for somebody who would follow me. At Harvard and Yale, a group could not be implemented in a mathematics department. Nevertheless, I had several direct students and I am very pleased with them, but each follows up on some specific aspect of what I have done. (There are also charlatans, who are doing very well and have large budgets at their disposal.)

Is there a Mandelbrot School?

Once again, yes and no. A bit like in the case of Gajdusek. The fact that I created no organized institution was unavoidable, given what you heard. But it is extremely unfortunate and a real problem. Many users want to apply my techniques but then proceed very clumsily, then criticize the tool instead of seeking to deepen their own understanding. I regret very much having been deprived of a chance to start a chain reaction by training students, who would train more students and so on. This process could have "infiltrated", if that is the right word, many communities I could not reach.

Considering your career, it is inevitable to ask this question, what do you think will be your legacy?

Nobody doubts that the concept of fractals will survive and diffuse everywhere. If it keeps or expands its foothold in education, my wildest dreams will have been fulfilled many times over.

Within science, fractality is inseparable from roughness and roughness is found everywhere so fractality (and also wild variability) cannot be avoided. Both raise hard problems, therefore their study is bound to remain an increasingly important frontier of science. Unfortunately, this scientific legacy risks being split into seemingly disconnected pieces scattered all around with no trace of their common source. This "worst case" scenario is the normal course of events but I believe that it is not inevitable, is not desirable, and can be slowed or changed.

Today, most people publish the bulk of their work in the form of articles. But I wrote a book, *The Fractal Geometry of Nature* that has been described as a classic of the twentieth century. If it survives in one form or another, it will help preserve a unified fractal geometry.

What will remain of me as an individual who had a strong personality and a well-defined taste? I chose a difficult life quite deliberately but was not a Don Quixote. Schützenberger, whom I have already mentioned, observed that in the study of complexity there exists a Mandelbrot style that is not only adopted by my few former direct students or postdocs but also by a multitude of other writers who do not necessarily quote me — or perhaps even fail to realize that they follow me. This may explain why I invest so much in the memoirs that I mentioned. I know that writing my memoirs is in part a manifestation of plain vanity. But please believe me that my main goal is different. I want to prove by describing actual events that to follow open curiosity can never assure comfort, intellectual or otherwise, but may be feasible.

How would you formulate your present ambitions?

It is too late for new ambitions. But the young need someone to personify science and I am keen on cementing the extraordinary position I attained among them and also among the so-called educated non-professional people. So perhaps my legacy will be to provide some kind of role model. Never before have role models for open curiosity and exploration been fewer and more needed than today. This may be the only grand issue to which I am keenly committed. My own role models, von Neumann, Poincaré, and Pasteur, left no memoirs. What a pity. The official stories have been "sanitized" and often contradict the conclusions I reached from contemporary documents. This sanitizing indicates a disdain of mavericks. But I deeply believe that society needs at least a few, and at this point the conditions for mavericks are far worse than ever. Ramón y Cajal did not equal Poincaré

Benoit Mandelbrot with his wife, Aliette, 1964 (courtesy of B. Mandelbrot).

and Pasteur but his clumsy autobiography might do wonders, if it is well-edited. This is a task I keep pressing publishers to undertake but apparently Cajal is too unclassifiable for them.

Suppose you could be 25 today. How would you chart your career today?

No clue.

Any message?

Not really and only for those who do not want to follow any straight path. Times are bad. You must have clear long-range goals but don't try to pursue them in any logical sequence. Think ten years or more ahead, but on the short run, show flexibility and take advantage of circumstances. Expect to be forced to go to extraordinary lengths to explain and document your claims. Accept being called arrogant. Ask Lady Luck for help and help her.

I would not have amounted to much without a few lucky breaks, IBM, Yale, and, all along, my wife.

Reference

1. Hargittai, I.; Laurent, T. C., eds., *Symmetry 2000*. Parts 1 and 2. Portland Press, London, 2002.

Kenneth G. Wilson, 2000 (photograph by I. Hargittai).

26

KENNETH G. WILSON

Kenneth G. Wilson (b. 1936 in Waltham, Massachusetts) is Hazel C. Youngberg Trustees Distinguished Professor at the Department of Physics, The Ohio State University in Columbus. He received the Nobel Prize in Physics in 1982 "for his theory for critical phenomena in connection with phase transitions".

His father was the distinguished Harvard chemist E. Bright Wilson, Jr. (1908–1992), best known for his molecular structure studies, especially by microwave spectroscopy. Kenneth Wilson obtained his Ph.D. in 1961 at the California Institute of Technology where he was a student of Murray Gell-Mann.

Dr. Wilson was at the Department of Physics of Cornell University between 1963–1988 when he moved to The Ohio State University. He is a member of the National Academy of Sciences of the U.S.A., the American Academy of Arts and Sciences, and the American Philosophical Society. In 1980, he shared the Wolf Prize (Israel) in Physics with Michael Fisher and Leo Kadanoff. Recently Dr. Wilson has been engaged in educational reform. He lives in Maine but continues to work in Columbus, Ohio.

We recorded our conversation at the Physics Department of The Ohio State University in Columbus on May 5, 2000 and the text was finalized by correspondence in July 2002.*

---

*István Hargittai conducted the interview.

You received the Nobel Prize for your work in connection with phase transitions. First I would like to ask you about this work.

What I was concerned with was the critical point. For instance, the boiling of water is a phase transition. When water is boiled and produces steam, there is a finite difference between the density of water and the density of steam. If you put the water and steam both under pressure and increase the temperature so that there is still boiling, you eventually reach the critical point of water. At the critical point, the density of water and the steam become the same. If you increase the pressure and the temperature even further, there is no boiling anymore. As you come close to the critical pressure and temperature, the water and steam become murky, they are no longer transparent. What's happening that underlies the murkiness is that instead of having big bubbles that you can see in the water, you have bubbles of all sizes. The surface tension between water and steam that forces bubbles to be large has gone to zero and there are bubbles as small as the wavelength of visible light. These bubbles scatter light, and the water turns murky. There was, of course, a lot known about the critical point before I came on the scene but there were peculiarities that people found puzzling. There was a simple, standard theory of how the critical point should work. It

Hans Bethe (b. 1906, Nobel Prize in Physics 1967) congratulating Kenneth Wilson on his Nobel Prize, October 1982, at Cornell University (courtesy of K. Wilson).

gave simple rules, one being that the difference in density of water and steam should go to zero at the critical point as the square root of delta T, where delta T is the difference (critical temperature – actual temperature). But the experiments showed more like the one-third power of the delta T than a square root. Before my work, researchers were finding it difficult to make a theory that would enable detailed calculations of this nearly cube root behavior; yet make it natural for the result to be other than a square root. At the same time there was Onsager's remarkable exact solution to the two-dimensional Ising model, which is normally seen as a model of ferromagnetism but can easily be interpreted as a model of a liquid to gas transition, in two dimensions instead of three. In that model, if I remember it correctly, instead of getting a square root of the temperature, you get the eighth root of the temperature. That comes out of a very complicated mathematical procedure by which you solve a model, which gives you no intuitive understanding for why the exponent would be any different from a half. Onsager had a complete solution to the Ising model and the eighth root behavior was one of the things that came out of it.

Was it just a mathematical construct or was it meant to be a true characterization of the critical point?

It was a mathematical construct, but a true one for the case of a simple critical point in two dimensions. The difficulty was that nobody knew how to generalize it to the Ising model in three dimensions, let alone more complex three-dimensional systems. The three-dimensional Ising model has remained unsolved (as an exact mathematical problem) to this day.

Researchers such as Cyril Domb had developed numerical extrapolation procedures working on power series solutions of models such as the three-dimensional Ising model. From these extrapolations, they were able to get numerical results reasonably close to the cube root behavior, and certainly far from a square root behavior. However, that was an extrapolation procedure, and how could anyone be sure that the extrapolation procedure was reliable? In any case, the extrapolations involved made it impossible to explain what was causing the result to be different from a square root.

My role in this area was to help to develop a way of thinking about the problem and then to produce some approximate numerical methods to be applied to the new way of thinking. The new way of thinking, originated by Leo Kadanoff, was to break down the problem in such a way that you started at the atomic scales and worked out to larger and larger scales.

Kadanoff and then I wanted to think of this problem in terms of replacing the initial description of the system at the atomic scale with an effective Hamiltonian that has the atomic scale of length removed, but with a new minimum scale of length that is only a factor of two or so larger. This is accomplished in a statistical mechanics framework. The physics is obtained from the partition function by computing the trace of the exponential of the Hamiltonian; I developed approximate methods for averaging over atomic scale variables as the first step in computing the partition function, leaving an effective Hamiltonian in which instead of having the atomic scale for the basic scale, you would have a new basic scale twice the atomic scale. Kadanoff provided the basic concept for removing the smallest scale but could not find a workable approximation for this process. Then one would carry out the process all over again to remove the new smallest scale and generate yet another Hamiltonian that was valid for a minimum scale four times the atomic scale. So an ordinary mathematical procedure would be the procedure of taking one from one Hamiltonian to the next one. But to understand what happens on the macroscopic scale, which is what you see, you have to determine what happens when one repeats this process many times. I was able to show that unexpected exponents (powers) near to a cube root could result from these computations and are characteristic of the transformation that takes you from one step to the next. Because this is a characteristic of the transformation by which you're changing the scale rather than a characteristic of the original Hamiltonian on the atomic scale, one obtains exponents that will characterize many different systems rather than just one. There are classes of Hamiltonians for different types of systems each of which will have exactly the same exponent. In fact, this was already known to the limits of accuracy with which they could determine these numbers through extrapolation. There were whole classes of systems for which this exponent was one third, for example, except there was no reason to know whether it was precisely one third or some numerical value close to it.

Leo Kadanoff had suggested the idea of thinking this way but had not been able to implement it in any fashion that would give calculable numbers. I did this first with a very crude model. Then Michael Fisher and I set up a formalism with which we could calculate exponents analytically in powers of the difference of the dimensionality from four.

The unfortunate thing about the Nobel Prize was that they gave it to me solo for the crude numerical approximation I had developed first. The development that really caused a revolution enabling newcomers to begin using the new approach to critical phenomena quickly was the expansion in powers of the dimension. It was an analytic procedure that researchers could carry out using known procedures borrowed from quantum field theory. This is the work, published a little later, that I did jointly with Michael Fisher. By all rights he should've shared the Nobel Prize. In fact, Israel's Wolf Prize had been shared by Fisher, Kadanoff, and myself two years earlier.

So it could not be ignorance. What could be the reason?

Only the Nobel Committee knows what the reason is, and it's not for me to substitute my speculation for speculations that have circulated. At some point the archives will be open and we'll know more about the deliberations that took place. But one fact is for sure, namely that there were more than the three of us that deserved to share the award. As Harriet Zuckerman has explained in her excellent book *The Scientific Elite*, it is normally the case that more than three people deserve a specific Nobel Prize, but the rules for the prize prevent awarding it to more than three.

You suspect reasons, you just don't want to speculate about them.

Right.

It's still almost unprecedented that a Nobel laureate would so critically evaluate his own award and would suggest that others should have shared his prize.

Most prizes are shared. The maximum is three and if they had given it to the three of us, many people could have complained that there were more people who deserved the credit. But it is hard to justify picking just one person to win it.

Did you ever talk with the other two, Fisher and Kadanoff about this?

I don't voluntarily bring it up.

Do they know that you feel it would have been more fair to share the Prize?

It's true that I haven't been completely explicit about that. At the beginning I didn't want to second-guess the Nobel Committee. They're responding to the wishes of the community and it's not appropriate for me to insert

Leo Kadanoff (courtesy of Leo Kadanoff, University of Chicago).

Michael Fisher (courtesy of Michael Fisher, University of Maryland at College Park).

my personal beliefs to substitute for that. Now that I'm older and wiser I realize I should've been more explicit about that.

Have the other two ever expressed any bitterness to you?

Not to me directly.

The Nobel Prize is so much higher in prestige than any other recognition. Do you think it's good?

It's not good. It is to the credit of the Nobel people that they have handled the Prize in such a way that it has maintained its incredible reputation. It's good that there is a prize with that reputation. What is bad is that there is no other prize that even comes close. That means that everybody is focused on this one prize whereas there should be just a very slow gradation of prizes so that nobody feels hurt because they had to take one and not another.

The Nobel Prize has a tremendous prestige.

Firstly, it got there first. It was done in such a way that it captured the market for attention for being a really extraordinary prize. It has such a tradition behind it now that when other people set up a prize, they can set up something with a lot of money but they don't have the tradition. What's unfortunate also is the coupling of money with the prize. People who have studied award systems have found that it's not good to couple a lot of money with a prize when the main thing is the prestige of being selected for the award by your most knowledgeable colleagues. A lot of attention gets focused on the Nobel Prize just because there's the money involved. With other awards, like becoming a member of the National Academy of Sciences, there's no money attached but there's still a powerful prestige factor. You shouldn't have as much of a distinction as presently exists between winning a Nobel Prize and becoming a member of an Academy of Sciences.

When did you change your role in research?

I first switched my focus from doing research to supporting other researchers needing supercomputing facilities, and that switch began about six years before the Prize was announced. Actually, when the Prize was announced, I was still heavily into my own research but I was getting ready to make that switch full time. I was working with a community of people who were trying to get the National Science Foundation (NSF) to put a considerable amount of money into supercomputing. Getting the Prize was exceedingly helpful in our efforts. I became one of the de facto spokesmen for the whole supercomputing movement. In this sense my Nobel Prize was an important factor in enabling the NSF to obtain support for supercomputing in the universities. It also coincided with the development of the Internet. Before the supercomputing centers were started, the Internet was entirely a program run by the Defense Department, through what they called the Defense Advanced Research Projects Agency (DARPA, later ARPA). At that point that whole ARPANET program was running out of steam. The NSF-sponsored supercomputer centers (with one directed by me) had to start a network of our own but based on the same principles as the ARPANET because we were unable to negotiate an acceptable arrangement for using the ARPANET directly.

Who provided the support for the independent network?

The NSF paid for it. Before the NSF officially paid for it we bootlegged it with NSF's help. We got together and various people in the centers volunteered their help to get a network started. It took two years for the

NSF to realize it had to do something requiring considerable money to support it. Then there was a stage when the NSF had a solicitation and officially established a new version of the ARPANET, now known as the Internet. It supported that for five or nine years before it was superceded by the commercial growth of the Internet. So the commercial Internet grew out of the network that the NSF had established originally to serve the supercomputer centers.

How has the Internet changed the sociology of science?

One of the most profound sociological changes is that it vastly simplified the opportunities to do collaboration both across disciplines and across different locations in space. It allows people to do instantaneous sharing of papers and conversations. The ability to exchange text (and now figures and videotapes as well) instantaneously makes a big difference to those collaborations.

Has there been any study on this impact?

I'm not aware of it.

The Internet couldn't have been foreseen. What will be the next step?

At the time that I was working on the Internet the pioneers knew that this was important, and I knew that it was important, but they couldn't describe to me anything like the World Wide Web. I don't think anybody expected the World Wide Web and its aftermath to happen so fast. It's very hard to predict timing on these things.

When your Nobel Prize was announced I remember someone saying, "E. Bright Wilson should've received the Nobel Prize and the Nobel Committee may have weighed this also in awarding you the Nobel Prize."

I can't imagine that that had any role for the simple reason that my father was a chemist and this was a physics prize.

People usually do not distinguish between the Nobel Committees and tend to lump them together into one committee. But it was also symptomatic that people held your father in very high regard.

I would sympathize with anybody who claimed that that was true because there is no doubt that in terms of overall influence my father more than deserved the Prize. It's just that the nature of his contribution was of the kind that the rules for the Nobel Prize don't recognize.

Let's play the game that you would have to compile a Nobel citation for him, what would it be?

Chemists, most likely, would have cited his major body of work on molecular spectroscopy. But if the circumstances would've allowed it, I would've given him the Prize for his book, E. Bright Wilson, Jr. *An Introduction to Scientific Research* [McGraw-Hill, New York, 1952; available now from Dover Publications]. It is totally unique, it is unchallenged even after 50 years, it services an extraordinary range of scientific disciplines, but it's not a discovery of the kind that the Nobel rules require. As a citation, I suggest, "for his 1952 characterization of the practices of science, which remains unequalled fifty years later."

At one point he was asked to write a more popular book on the subject and he tried. He wrote a trial chapter over a summer and then abandoned the project. At the time he was investigating Kuhn's book *The Structure of Scientific Revolution*. I find this important to mention because my work on education since 1990 with my current collaborator, Constance Barsky, has involved a very close study of Kuhn and the controversy around his work. I've realized that there really is something new there, which ought to be incorporated into textbooks on scientific methods, not necessarily replacing what my father wrote but maybe as a second volume.

Can you assess the impact of your father's book?

In various places, like in encyclopedias, it is not mentioned when it should be. Unfortunately, very often when people write about the scientific method, they focus on the writings of philosophers and they don't go to knowledgeable scientists such as my father, or Percy Bridgman who wrote *The Logic of Physics* even earlier. But scientists I know who are familiar with my father's book are very grateful for it, and have told me so.

What was the atmosphere in the family of E. Bright Wilson when you were a child?

It was an atmosphere that was larger than just the family. He provided an environment where you were shielded from a lot of nastiness of life at large, where loyalty to the values of science was everything. You didn't get any experience with the backbiting political shenanigans that real life is full of and there were only a few things that he heavily condemned, one being the Republican Party. In his view the evil in these institutions didn't necessarily rub off on every individual member; instead, the evil was in the institutions themselves. It did come as a shock to me later, as I came to understand how bad the whole political system we have actually is, that the Democrats in the end do things for political expediency just as often as the Republicans do.

The book *An Introduction to Scientific Research* has special importance for me now because I've been trying to understand what is the scientific method. The arguments that have been made by some sociologists of science that science is all relative prompted me to look into this matter. There is a book by Pickering called *Constructing Quarks*, which is a very good history of high-energy physics from the late fifties to the early seventies. I was part of that history in that book. But he ends up with presenting a sociological view on what happened. He states that there is no need for anyone to pay attention to what 20th century physics has to say. That has upset the physicists working in elementary particle physics.

It took me a long time to understand why he made that statement. The sociologists are trying to claim that there is an agreement among the physicists about what to claim is true and that the agreement could change tomorrow. Indeed, when you look at the history of physics, physicists have recanted a number of their claims over that history. They used to believe that Newton's laws were exact in the sense that they didn't accept anything that would be a correction to Newton's laws. Then, once there was relativity and physicists accepted the validity of general relativity, they also accepted that Newton's laws were not the final statement, they were only an approximation with computable corrections due to general relativity. The question becomes, how much of what we accept today as exact truth will turn out to be approximate. My father had a very clear attitude for that. He said that anything that we accept as truth today can change tomorrow. An example he offered was that the Sun had risen every day for millions of years but some day the Sun will not rise. Of course, the astronomers now know that. At some point there could be some huge object that wanders into the Solar System and just totally devastates it. It just has to be heavy enough. Or something may come in and knock the Earth out of its orbit. That wouldn't change the basic laws of physics but would prevent the rise of the Sun on some future morning. If we are knocked totally out of our orbit, then as we wander off some place else, the length of the day will change, everything will change, compared to what we take for granted today.

There's an even deeper problem, which the philosophers had understood and that most physicists have simply rejected out of hand, except for a very small sub-set (Percy Bridgman was one of this small subset). The deeper problem is that there is no person and no thing that can guarantee that the laws of physics themselves will not change tomorrow, as opposed to today. Dirac and others have looked for one form of change in the laws of physics, but they have looked for a very restricted version of change. They made the assumption that the laws are not constant but the reason they are not constant is that the parameters in them that we've assumed are constant are actually changing, but they assumed a uniformly slow change that's been continuing ever since the Big Bang. But then they identified enough evidence to rule out such a slow change within present limits of measurement. But what if the fundamental constants of physics don't change uniformly? What if there's some underlying process that is vulnerable to instability, just as a star can go perfectly OK for a billion years and then blows up? Why couldn't there be a similar process in which the fundamental constants stay constant for billions of years and then blow up too?

The reason that this is not quite so absurd as it sounds is that what we call fundamental constants are not fundamental at all. They are phenomenological parameters, which physicists have promoted to call fundamental. The elementary particle physicists all know that the constants that we deal with are phenomenologically defined. The fundamental constants should be computable in a theory that describes what happens at much smaller scales than we are able to measure. For example, we know that at a very small scale of length, gravity would have to be quantized. We do not know yet at what scale size quantum gravity is important but it is certainly smaller than any scale that we can measure in the present time. But very few physicists discuss the possibility that the current laws themselves might break down someday due to some kind of instability in quantum gravity similar to the instabilities of stars. Since the laws of physics have shown no sign of such a breakdown over the past three centuries of the development of our current physics knowledge, it seems an unlikely prospect for the near future, even if it is possible.

Unfortunately, sociologists confuse something that could happen, but is very unlikely, with something that should be commonplace. When Pickering says that you should ignore anything that 20th century physicists say, what he omits to say is that while there is some probability of a breakdown in the next few years in what physicists now claim, that probability is surely very small. Then it comes down to who you are willing to listen to: physicists or sociologists? In an article that Constance Barsky and I contributed to a book called *The One Culture* (edited by Jay Labinger and Harry Collins), we provide our response to the sociologists. The philosophers can say, everything could break down, you have no proofs of anything. That's fine for the philosophers, that's consistent with the way philosophy operates. But the function of the sociologists is to explain why people do certain things. This means that sociologists of science, such as Pickering, have the obligation to explain why the physicists don't accept what the philosophers said. In our paper we say, what the sociologists have to think about is the process by which the accuracy of the scientific predictions keep getting better all the time. They have to explain why things like predicting the future of planetary motions, or estimating the accuracy of the fundamental constants, keeps improving, and given this process of improvement, why it is reasonable for a physicist to think about the possibility that the laws of physics could change tomorrow. It is such a remote possibility that it is reasonable for a physicist to assume that the laws will continue to work for a considerable number of years to come, even though, strictly speaking, there is no proof.

Did the atmosphere in your family prepare you to deal with these questions?

Absolutely. Not only the atmosphere but in the book *An Introduction to Scientific Research*, my father takes seriously the philosophers. He says (p. 27), "Few people worry about the possibility that tomorrow they may be thrown off the Earth because of the failure of the law of gravity. It works today, it worked yesterday, and it has worked for a long time, so that even most philosophers act on the assumption that it will continue to be obeyed." Few physicists would write a sentence like that.

Because for the physicist, it is the law.

Of course.

Did you recognize these things while you were at home or much after you had left?

Much later, as I've been struggling with the social issues.

Did you ever discuss this with your father?

I could've but I never did and that's what I regret most, not having him around to talk about these issues as I wrestle with them. I learn what I can from his book.

There's another important thing that I learned from his book. There is a section on classification. The social sciences are very weak in classifying the phenomena that they study. In one case I had to go to the economics methodology literature. There is a book by Mark Blaug on methodology, called *The Methodology of Economics: Or How Economists Explain* that economists greatly respect. There is not a word about classification in that book. In my father's book there is a whole chapter on classification. I'm working now with a few economists and I'm trying to get the idea across that they need to do classification. I want them to classify technological histories. It's very interesting to find out how technological histories effected economic growth; there are a very large number of technologies that economists could study; and you can't study economic growth in much depth unless you look at historical data on technological change. But I found out that it is premature to propose a full classification of these histories. I found out that in biology, it took a long time to advance to the stage when Linnaeus could propose his full classification scheme, based on his unique scheme for naming plants and animals that is still in use today. I had to look back earlier in the history of biology, when John Ray and his collaborator John Willoughby were cataloging plants and animals, but not yet fully classifying them, and at the same time beginning of the development of the modern technical vocabulary for biology, beginning with a modern definition for the term "species". Based on this historical analogy, an economic historian named Richard Strechel has joined with Contance Barsky and I to propose that histories of technology be cataloged, and Dr. Barsky and I are now developing a sample glossary of technical terms to use in such a catalog. This may sound mundane but in practice it is a quite demanding task.

Did you read the book An Introduction to Scientific Research *while your father was alive?*

I helped him in a very limited way when he was writing it. There's a minor acknowledgment of my role in the book. That was when I was a teenager. So it was not that I could read or digested the whole book. I did not return to it until after his death.

In what way did you help in writing the book?

We built little cards to play games with symbolic logic.

Did your father encourage you to study physics?

There is a very nice anecdote. When my brother David was about to go to college, my father very proudly said, "I want him to have the widest possible choice of career. I've encouraged him to study biology as well as physics and chemistry."

How about you?

Between physics and chemistry he would never have a preference. It was my decision.

What is his legacy in addition to An Introduction to Scientific Research?

There is his body of research in chemistry, such as in molecular spectroscopy. But he would have also pointed to all his students. He had a student who also lived in the same house with us during World War II. During World War II my father went down to Woods Hole to work on underwater explosions and he brought a number of his students with him. Arnold Arons was one of them. Have you heard of him?

No.

Hopefully you will if you live long enough, because of his pioneering work in physics education. Arons switched to physics after World War II after he had got his Ph.D. in chemistry. He also switched to work on the problem of science literacy. He took a job at Stevens Institute of Technology and worked on teaching. When Amherst decided that they wanted to do a very special curriculum for all freshmen and all sophomores based on the ideas of John Dewey, a famous educator, they needed a physicist, and they invited Arons. His course became famous. When the whole Amherst program was cancelled, around 1968, he moved to the University of Washington but he took on a new task, to work with teachers of physics from the schools, to help them improve their teaching. The program he established turned into a real research program on physics education, with a unique focus, for its time, on professional development for teachers. That program continues to be a leading program in physics education today, now led by Lillian McDermott. I so admired what Arnold did that when I started thinking about what I could do to help the bulk of the nation's children with science education — all forty million of them — I decided to imitate and extend what he had been doing. When I started working on education here at OSU, I had a wonderful opportunity because two years after I got here, the State of Ohio was invited by the NSF to submit a proposal whose purpose would be to fix every aspect of math and science education in the State of Ohio. The NSF was offering two million dollars a year for five years for such a proposal. Fortunately for me, I got selected as the representative of The Ohio State University for that proposal. We brought the course of physics for teachers, originally developed by Arnold Arons and greatly improved by Lillian McDermott, to Ohio.

It was very clear when talking to Arnold that he was powerfully influenced by my father in terms of building his own commitment that he would fight on the science literacy front. He retired from the University of Washington in 1980, but continued this commitment afterwards, particularly in publishing a unique book called *A Guide to Introductory Physics Teaching*.

In retrospect his work was much more of an accomplishment than I realized at the time. University physics departments have been reluctant to actually consider teaching as a subject of serious inquiry. I made sure we started a Physics Education research group here and had to use my own reputation to get this concept through the university and department here although the department here was more ready than almost any other physics department elsewhere would have been to support a program in physics education.

What else would you like to add about your father?

There is one more aspect of interactions with my father that's extremely important. This is a gossipy kind of information. I was interested in computers early on. One of the curious things about my Nobel Prize was that it was one of the first ones that was partly based on numerical computations done on computers. The papers they cited for the prize were the papers where I'd set up an equation, which I actually had to solve numerically.

My father was not happy with my interest in computers as a way of doing science and the way he expressed that was he constantly complained to me about what was being done on computers in computational quantum chemistry. He did not think that much of it was good science. He would fulminate about it the way he would fulminate about the Republican Party. This prompted me to look into computational quantum chemistry and see if I could do something about it. From our conversations I got the

E. Bright Wilson and Kenneth Wilson on the occasion of E. Bright Wilson's Robert Welch Award in 1978 (courtesy of K. Wilson).

idea that it would be relatively easy for me to go into it and do something important. I asked him who are the people I should talk to. He instantly gave me four names he respected in that field and I won't tell you who they are because some of them are still alive. I found this a very interesting characteristic of research fields, which are in what Thomas Kuhn called the pre-paradigm phase. It's not simply that outside people don't respect what's going on in the field, in this case computational quantum chemistry; you find the same thing inside, that they don't respect each other. In particular, there were two areas trying to deal with the computation of structures of complexes of atoms. The computational quantum chemists had their way of doing things and the people doing density functional theory in physics were trying to do much larger systems in a much cruder way. These two groups would constantly disparage each other's work. That seems to be a sociological characteristic of pre-paradigm fields, along with the notion that you can always find out who are the good names in such a field even in that disparagement stage. Then comes the first paradigm, just as Kuhn says, and the disparagement disappears, though not as totally as he claims. By the middle of the 1980s, the quantum chemists were publishing reliable computations on very simple molecules, and experimenters were finding these computations to be truly useful.

What was your father's reaction to your Nobel Prize?

For everything I knew it was tremendous pride. The whole family travelled to Stockholm. Once when I was invited to give an after-dinner talk about him, I described my life with him as his 50-year-old graduate student. There was always this question on my mind, "Did I meet his standards?" There were cases when he was seriously disturbed by what I was doing, like playing around with computers. He was upset in the early 1960s that I hadn't published enough yet. One reason that I wasn't publishing things that, in retrospect, I should've published was because of the high standards I inherited from him. He remonstrated with me for that but he didn't make a dent on my publishing behavior.

In terms of just watching him I could imagine that he was just delighted. I never got any sense of him being chewed up the way some people are chewed up because they didn't get this prize or that prize. At the same time I cannot say that I was privy to his innermost thoughts.

Coming back to my being my father's 50-year-old graduate student, this began when I was making trouble for my elementary school teacher. I was in a two-room school-house. The first, second and third grade were in the same room. By the end of second grade I knew much that was taught to the third graders; a year later, after a half year of total boredom I was moved in mid-year into the fourth grade because that put me in the other room. There was one thing that I had to do, I had to catch up in long division. So my father had to work with me at home on long division. I described myself as the only graduate student of my father who had had remedial help with long division.

At the same time as he was writing his book and I was working with him on symbolic logic, he tried to teach me group theory. I was 14. I just couldn't handle it, it was too abstract for me. That was my second failure as his graduate student.

Was he patient?

He was reasonably patient. I don't remember any situation that frustrated him.

Did he pay about the same attention to your brother and sister as to you?

That's not clear. The big difficulty was I didn't pay attention to the other two. They were younger than I; David was four years younger, my sister was six years younger. More recently, he had a second family: two half brothers and a half sister, but that was after I had left home.

About religion. There is some general notion that a higher proportion of physicists is religious than, say, biomedical scientists.

I don't know very many scientists who are religious but, of course, I may have a prejudiced view in that part because a physicist who is religious is not necessarily going to be open about it in discussions with other physicists. I know individual physicists who are religious but we never discuss it.

My father didn't encourage interest in religion of any sort. When I got involved in studying social sciences, I came to realize that I had to view religion as a social phenomenon and it didn't matter what my views on it were, I had to study it as social scientists (such as Max Weber) had already done. I had to be prepared also as a political matter to accept religion if I wanted to make a dent in the educational problem. I couldn't refuse to talk to somebody because his or her business was working in a church. If I wanted to make a difference in education, I had to work with whoever was interested on my side in the education issue, it didn't matter they were religious or not.

You made this interesting comment about the Nobel Prize that it, as an institution, contributes to this sociological phenomenon of attributing everything to individuals, overstressing their roles.

That's right. I'm trying to find a way to counteract that, to help scientists and others recognize this. I don't want to detract from the accomplishment of a Newton, an Einstein, or a Maxwell, but we attribute more to them than they actually did. Let me give you the example of Newton. We all learned Newton's laws, but we also learned methods for solving Newton's laws that are not due to Newton himself. What really astonished me when I learned it was how much more powerful Newton's laws are today because of the incredible amount of research that has been done on the techniques for solving these laws. An early example is the motion of the Moon with which he had trouble in his own time. It's a three-body problem, with the Sun and the Earth both affecting the motion of the Moon. The interaction with the Sun can be treated as a perturbation. In some of the motions of the Moon it is good enough to include the perturbation through second order. But there's one aspect of the motion for which one must include third order terms from a perturbation expansion. Newton didn't know how to do a third order computation; apparently he never realized that a third order computation might be needed. Clairaut, whom I had never previously heard of, did the first successful third order calculation, as did D'Alembert simultaneously. Both were very lucky because that cleared up the problem, with no requirement to go to even higher orders. Clairaut's and D'Alembert's work came around 1740, well after Newton's death.

Until the problem with the Moon was cleared up, a lot of people just paid no attention to Newton. It was difficult and many astronomers of the time said, why bother? The methods that Kepler had developed for computing planetary motions for almanacs were continuing to be used into the late 1700s, until the methodology for using Newton's laws became accurate enough and well enough known that they actually did better than Kepler had done. After that, of course, there have been many further improvements in the way astronomers do the calculations for planetary motions and there are famous people further on in this history (such as Poincaré) but none (except Einstein) as famous as Newton. When I studied mechanics as an undergraduate, I never learned anything nearly as complicated as the three-body calculations that were involved for the Moon. Then in the twentieth century, physicists developed an attitude that studying Newton's laws was not serious research — because for many of them, only quantum mechanics and related topics were at the forefront of research. Then the chaos work came, including the work by Feigenbaum. I knew him because he was a postdoctoral fellow at Cornell but I didn't realize the importance of what Feigenbaum was doing even though he learned some things from me. I didn't know the background of his work and I didn't know about Henri Poincaré's studies of chaos that Feignebaum was building upon.

Whom do you consider to be the greatest scientists in physics?

In their generation, Feynman and Gell-Mann were exceptional. Of more recent times, I'm not giving you names because they are still building their working careers.

It's also interesting to look at the social science situation. It is much harder to accomplish things in the social sciences because they are in the pre-paradigm phase. That's why one cannot get the same kind of recognition that one can get in a science in a post-paradigm phase or if one is the person who creates a paradigm. That's why Kepler is sometimes not considered to be in the same class as Newton even though people who have studied that period say that Kepler was even more extraordinary. I can't make up my mind about that switch, to consider Kepler more extraordinary than Newton. But he is certainly Newton's equal. When people rank Newton versus Kepler, most of them have not studied the work of either. Newton published the *Principia* in 1680 and there followed three hundred years of development in terms of how one actually applies Newton's laws. One can clearly separate all the later developments from what he did himself, which was to write down the three laws along with a small initial set of applications. In most other cases the people who are the pioneers get an area of science started but they don't develop anything nearly as powerful as Newton's laws have been. In these other cases the follow-up work is of even greater importance than even the follow-up work on Newton's laws have been.

I've seen this more than once that the real big names in science are Copernicus, Newton, Einstein, and Watson and Crick. Now you are laughing as many others do. Is this laughter indicative of the lack of perspective for the double helix yet?

No. Obviously, the absolutely central role of DNA and its double helix structure in genetics is beyond question. I laugh in part because you omitted Darwin and the founders of quantum mechanics (such as Schrödinger, Heisenberg, Bohr, and Dirac) from your list. I would place all of them ahead of Watson and Crick, in terms of both the originality and creativity of their work and the breadth of their impact on our understanding of the world that we live in. But I laugh for a more profound reason. There were not all that many scientists working at the time of Copernicus and Newton, and even in the time of Darwin. But in the twentieth century, the dominant story in science is not the contributions of individuals, not even of Einstein, let alone Watson and Crick. Instead the BIG story (at least for me) is the growth of institutions that are dependent on the research of increasing large numbers of scientists, such as the Agricultural Experiment Stations and the Agricultural Extension system in the U.S., that have transformed modern society. (A history of these agricultural institutions is provided in Wallace E. Huffman and Robert E. Evenson's *Science for Agriculture: a Long-Term Perspective.*) The areas of the economy that have been transformed include medicine, transportation, telecommunications and information storage, energy, industrial materials, defense, large-scale construction (including skyscrapers and bridges), and media and entertainment. All of these transformations have been accomplished with the crucial help of a growing base of knowledge: scientific knowledge, engineering knowledge, and professional knowledge in areas such as medicine and law. In fact, knowledge in *toto* has become the dominant source for economic wealth, social progress (if any), and military might. In my view, the person that has told this much bigger story than DNA better than anyone else to date is Peter Drucker, through a long list of books of which *The Age of Discontinuity* (1968), *Innovation and Entrepreneurship* (1985), and *Post-Capitalist Society* (1993) are among the most profound. But Drucker is only a transitional figure. He is exceptional, yet not nearly enough so, to rank with Newton or Einstein. Despite Drucker's work, and that of many other social scientists too, the growth of knowledge from its beginnings long ago to its present dominant role in society is still much farther from being understood than are many of the well-established topics of research in the natural sciences.

Nevertheless, as I learn more about the growth of knowledge as a whole, I become less inclined to take any list of "big names" in science seriously. I admire the accomplishments of Newton and Einstein, and of Darwin and Watson and Crick too. But I give equal weight to seemingly far more mundane matters such as the growth of reference materials (encyclopedias, dictionaries, and the like) in libraries, without which hardly anyone would have a possibility of learning about the accomplishments of the big and not-so-big names in science. My father helped me to understand the importance of reference material for scientists, as part of a whole chapter of *An Introduction to Scientific Research* on searching the literature. His comment about encyclopedias is that they "are surprisingly useful for acquiring a first view of a new field". I could not agree more, based on my own experience with them. More generally, I find that *An Introduction to Scientific Research* provides a far more realistic and balanced view of how science makes progress than one can learn from a list of (or a more detailed study of) the "big names" in science.

Mildred S. Dresselhaus, 2002 (photograph by M. Hargittai).

27

MILDRED S. DRESSELHAUS

M ildred Dresselhaus (b. as Mildred Spiewak in 1930 in Brooklyn) is Institute Professor at the Massachusetts Institute of Technology (MIT). She graduated from the Hunter High School and then from Hunter College. A year at Cambridge University (U.K.) and another year at Harvard University followed, where she received a Master's degree in physics. She did research for her Ph.D. degree in physics at the University of Chicago. Her main results were written up in two sole-authored papers on the surface impedance of a superconductor in a magnetic field. She then joined her husband at Cornell University on an NSF postdoctoral fellowship. During the period 1960–1967, she worked at the MIT Lincoln Laboratory studying magneto-optical effects in semiconductors and semimetals. Then, following one year as Visiting Professor in the Department of Electrical Engineering at MIT, she was appointed full professor and in 1985, Institute Professor at MIT. She has had a joint appointment between the Physics Department and the Department of Electrical Engineering and Computer Science. Mildred Dresselhaus is a member of the National Academy of Sciences of the U.S.A. (1985), the National Academy of Engineering (1974), the American Academy of Arts and Sciences (1974), and the American Philosophical Society. She served as President of the American Physical Society, President of the American Association for the Advancement of Science, as Treasurer of the National Academy of Sciences, and received the National Medal of Science in 1990. She has held many other positions and received numerous awards and distinctions. She has been especially active in women's issues, first at MIT and later nationwide and internationally. We recorded our conversation in Dr. Dresselhaus' office at MIT on February 5, 2002.*

Could we start with the question, what turned you to science originally?

I got into science through music. I grew up in a family that was far away from science. As a child, I had a music scholarship, but somehow I got interested in science and I was mostly self-taught in science all the way. When I was about ten years old, I read Paul de Kruif's book, *Microbe Hunters*, which was a strong influence, yet for quite a while I did not think about science as a career. When I was an undergraduate, I wasn't quite decided yet. I had physics, but I had chemistry and math also, and I could have gone in any of those three directions. Math almost happened because I applied for graduate school in physics and math. I was accepted in math and I accepted a graduate fellowship in math to become a graduate student here, at MIT. I never did it because in the meantime I got a Fulbright scholarship, which was in physics. It seemed like a great occasion to go abroad. That's how I became a physicist and not a mathematician.

You always speak warmly about Rosalyn Yalow.

She was an early mentor. She was having a hard time herself. She'd gotten a Ph.D. and then she couldn't get a suitable research job. She was teaching at Hunter College where the students were mostly going into education. It was not a place of real research interest. Then she got into medical physics, which was not yet a field and it was not clear at all whether that might lead to anything. She had a hard time with it; first of all, she didn't have a medical degree and being a woman in a team of a woman and a man, it was assumed that she was his assistant. The external relationship was like that all the time although it wasn't like that internally. It was a good team and she was the science person and he did not have the physics background, and their joint work depended a lot on physics and physics techniques.

Seeing all that, you nevertheless decided to become a scientist.

---

*Magdolna Hargittai conducted the interview.

Rosalyn Yalow (courtesy of Rosalyn Yalow).

But I didn't see it. I am ten years younger than she is and I didn't see the negatives, and maybe Rosalyn didn't see the negatives so much herself either. Our expectations were not so high. She was happy to have some kind of job and she loved the research. I always had the impression that her teaching experience was also very good and I never heard of her saying anything bad about it. She had a strong personality. She helped me become a scientist. Her academic expectations were very high and she may have frightened away some, whom she did not think should be there. At one time there was a bunch of writers who took the two of us to lunch. They just said a few words and then took notes of everything we said. Rosalyn and I are very different. There are sides of Rosalyn that the public doesn't see but I've seen. She can be very motherly. When I was just starting my career, she and her husband would come to my 10-minute APS talks, and she would bring a shopping bag full of stuff, a little like a housewife. She always dragged her husband along too. I was in a different field, but she wanted to encourage me. Whenever I needed something, she was always there. I was a very independent type person, so I didn't go to her often. She knew that we had very different personalities. I'm a lot less aggressive than she and much more accepting. I've put a lot of time into issues of women and science and she doesn't really like that, but she thinks that I've done the right thing and she supports me. She wouldn't do it herself, but I explained to her why I've done it, and she has approved my efforts.

Have you experienced discrimination?

Mildred Dressselhaus during her postdoc years in her lab at Cornell University in 1960 (courtesy of M. Dresselhaus).

Of course, but you forget about it as quickly as you can. My faculty advisor at the University of Chicago, who was the only advisor in my area of physics, didn't believe women should go to graduate school. He didn't believe that I should be doing what I'm doing. He was very unhappy every time I got a fellowship or any kind of recognition. He said it was a waste of resources. This was the person who was supposed to take care of me. Women have a lot of bad things that happen, but they just have to work around them. I came to MIT as a result of discrimination. I have four children. It was very hard when the children were small to be at the lab at 8 o'clock in the morning. Do you have children?

Yes, two.

There was always a problem in organizing one's family in the morning. My supervisor at Lincoln Lab complained about me so much that I got tired hearing of all the complaints, because I was doing the best that was humanly possible. So I was looking for a year off from all this unpleasantness in my life. It wasn't that I wasn't productive; nobody ever complained about quality, quantity, anything about my work. They didn't like that I came to work at 8:30 instead of 8 o'clock. My oldest child was less than five years. I had a baby essentially every year and it was very hard to make everything work out for an 8 a.m. arrival. The people who were judging me were all bachelors. Then a friend gave me the opportunity to come down to MIT for a year and I came. And I never left.

How about your husband?

Initially Gene was at Cornell University, but there was no opportunity for me at Cornell so he left. They wouldn't even let me work for nothing. This was the 1950s. Ithaca, New York, is not a large place and there was no other opportunity for me to work. I had my own postdoc for two years from the NSF, so I was supported by my own money, but when it was over, that was the end. I had done a good thesis, so people were willing to hire me, but I went where my husband was. The bottom line is that, yes, I had my share of discrimination.

My husband decided that it was more important for me to be able to work, than for him to have a faculty position. So we both went to Lincoln Laboratory, which was a defense lab, a part of MIT, where our only obligation was to help people when they needed help in physics and this was not a heavy obligation. The rest of the time we could do world-class research. It was right after the Sputnik period, which was a wonderful time in the U.S. for science, never to be repeated again.

There was a close connection between the Lincoln lab and the rest of MIT so I had a visiting professorship for one year and very soon after I came, MIT decided to offer me a regular appointment with tenure.

Didn't they mind that you had four children?

They did, but they appointed me anyway, because they needed somebody to teach science to engineering students. At that time the Dean felt that the engineering students should learn more science. He wanted to have a science type person who could work with engineers, which was what I did at Lincoln lab. That's what I have been doing at MIT for all these years.

Was by then the notion gone that a woman shouldn't be teaching engineering students?

That happened to me at Cornell, in 1958, when I taught a course on electromagnetic theory. The reason I taught this course was that the professor who was supposed to teach the course left during the first week of the semester and they couldn't find a person to teach that course. I volunteered to teach the course, with no pay, because I had a fellowship. There was a big uproar. The faculty met every day for a whole week to decide not whether I was qualified to teach this course, but whether the young men would pay attention to me as a young woman. I had a lot of experience in the area of electromagnetic theory. There were no women in the course. It was difficult for these senior faculty to comprehend and deal with having a young woman teach young men. Maybe they decided that it was OK because I was married and already had a baby. I never knew exactly what went on behind the closed doors. However, I did get a chance to teach the course and I did a very good job with that course. I found out more about that many years later when various students who were in the course came to me (they met me in other ways at a later time), and they remembered my teaching, because it was somehow different for them. They told me years later how much that course meant to them.

There is then an anecdote from a later time. I was the Treasurer of the National Academy of the United States, which is quite an important position in Washington. We were having a meeting of the governing board of the National Academy in the lecture hall. A presenter came from the National Aeronautics and Atmospheric Sciences, one of our U.S. agencies, he walked in, he recognized me after all that time, and he said that before he gives his comments, he would like to say something that he meant to say for many years about the influence I had on him. It was a fantastic introduction and terribly embarrassing for me. People do appreciate a teacher who goes out of the way to do something beyond the call of duty.

At MIT, I was very well treated over the years. Within five years after I joined the faculty, I was appointed as Associate Department Head of Electrical Engineering. We had 66 faculty members in my department, all of them men. I had a lot of opportunity for leadership and I never felt that people made a big distinction here about my being a woman. That's one good thing about science, a good thing about physics, a good thing about engineering, that there is some kind of standard; comparisons in performance are more objective than in other fields. I have noticed that physics is much more of a macho thing than chemistry, at least in the U.S., but still once you're doing the work, it is all right to be a woman. Getting through the door is what is hard, and it's still hard. Just look around, we still have very few women Ph.D.s in physics, about 15 percent. We should be much further along, considering what's happened in all the other fields. In the U.S., women in general get Ph.D.s at almost the same rate as men, once they are in physics. So it is not that women would not go on for advanced degrees after college. But they don't choose certain fields. There is still something about certain fields that is not attractive and it's not only the subject matter, I think it's also the sociology of the field. Just across the border from physics, in materials science and in chemistry, there are more than twice the fraction of women.

As you go up the ladder in other fields too, the number of women decreases.

I've done quite a lot of administrative work, but not going further in administrative work was my choice, not MIT's choice. I told them from the beginning that I didn't want a job that doesn't allow me to do my science. When I had administrative work, I always did it in addition to my science. Above a certain level I would've had to give it up. Even when I was working for the government recently, I was here at MIT in the lab on the weekends.

Was it at the Department of Energy? You were the director of the Office of Science.

Yes, I was, and it was a high-level job.

How long did you do it?

When I went into this position I expected it to last for a little over one year. The position was empty and I expected the Senate confirmation to be very quick. I was nominated by various people, not just by one person. However, the process was not fast. It had nothing to do with me, only with politics. So instead of the anticipated 14 months of service, I was actually on the job for 8 months only.

What was the main thrust of the job?

It was really twofold. One was to affect the budget for the physical sciences, to reverse the trend of decreasing support. I think that decreasing support to the physical sciences is a mistake. In this I was successful. The second thing was to try to train the people who worked for me. They were career people whereas I was a political appointee, so to speak. I tried to develop a scientific methodology for making decisions for the funding of science. How to determine the importance of various projects, how do you evaluate projects, what is good quality, how you measure all this, and so forth. This is an important discipline that we have not developed very well. When I was Treasurer of the Academy, we were working on this sort of thing for the U.S. science program. I was four years in that position and I learned a lot. When you are in a position like the one I had at the Department of Energy, people constantly revert back to what they were doing before. I understood that people were not used to thinking about science funding and science management from a fundamental standpoint. There are principles that one can follow and it's also useful to know how the system works. According to me, you don't accept money just because you can get it. There must be some purpose in funding programs.

You tried to make a difference. Do you have any feedback whether after you have left your impact continued?

It probably reverted back to where it had been before I arrived, but that did not surprise me, I expected that.

How did you meet your husband?

I was a student and he was a starting faculty member. We were at the same place at the same time. In the U.S., more than half of women physicists are married to men physicists. When we women physicists are young, we are so busy in our profession doing our work that we don't meet other people.

Did you work together?

Sort of, on and off. There was a ten-year period when I was on the MIT campus and Gene was at Lincoln Lab. Those years we did not collaborate so much. In 1977, when I became Director of this whole building, the Materials Center, I had the largest research program that I ever had. I didn't see how I could manage all of that, the Materials Center and my own research. At that time Gene was very unhappy at Lincoln Lab, because it was no longer a place where he could do the research that he wanted to do. It was a good solution for both of us that he joined me here at MIT and ever since we've been working together. Still, he is free to do anything he wants.

Was there no anti-nepotism rule?

Mildred and Eugene Dresselhaus, 1988 (courtesy of M. Dresselhaus).

MIT never had such rules and I am not formally his boss. Somebody else is his official boss. Our relation is that he can always pick whatever problem he wants to work on. We've worked together for so many years that it's a little different from our interactions with other people. I have a very large number of collaborators. In most cases I work with the collaborators more directly than he does.

Do you talk about science at home a lot?

We do, maybe more than we should. When our children were small, they would use phrases that they had heard from us in talking to each other, having no idea what they meant. At that time I wasn't quite as busy as I was later. For the two younger children, I had less time than for our first two children. When the younger ones became teenagers, I brought them down to the lab. It was a kind of self-defense, to get them busy and to keep an eye on them. This may have contributed to their interest in science. The two younger ones have Ph.D.s, one in physics and the other one in math. When they were about 12, they started participating in research. The older ones did not have that experience. My daughter, whose main job now is taking care of her children, was an MIT student, and she regrets that I didn't have that concept of bringing children to the lab, when she was growing up. The younger ones attended all the colloquia because they were responsible for showing the slides. They started having questions early on, asking us to explain what was that the speaker had said. We thus had a chance to explain what the subject matter of

Mildred Dresselhaus lecturing, 1979 (photograph by and with permission of Georgia Litwack, courtesy of M. Dresselhaus).

the colloquia was. The two younger boys were also working with grad students and they got their names on papers when they had made a sufficient contribution to the work. Now, twenty-some years later, they are still friends with some of the people who were in the lab at that time.

How did you manage to have four children in quick succession and stay in science and in a very successful way at that?

Firstly, I had a husband who did half of the work. He helped me with everything. That made it possible. Also, I had a babysitter, the same woman for 29 years. That was very helpful. To the question of why it was all so compressed, at the time women had the notion that 35 was the cutoff age for having healthy children, and we could not start a family early because we were studying so hard. I had my last child when I was 33. My last child was very sickly; I had a lot of trouble with him. At that time, four children was very common among academics. For men, of course, it did not matter too much to have four children in the family, because mothers generally stayed at home with children.

You have been more famous than your husband. Did it ever create a problem between you?

It didn't start out that way. It started out exactly the opposite. When we married, he was very well known and I was not known. He had a good position and I was just a graduate student and then a postdoc. When we came to Lincoln Lab, it was not that way either. Even though we were only one year apart, he was a senior person and I was much more of a junior person. I was the person who had the breaks because there was discrimination and he did not have so much discrimination. I worked the discrimination to my benefit; it wasn't something I planned that way. I am sort of more of a natural teacher and my talent for teaching was very helpful; students liked me. I was a more outgoing person. I don't know how it happened, it just happened like that. Maybe being a woman was an advantage.

You became a member of the American National Academy of Engineering early on and later also of the National Academy of Sciences of the U.S.A. He is not a member of either.

I don't know how these things happen.

Does it bother him?

You can ask him. It doesn't bother me. I don't consider any of that particularly important. For many of the things we did together, people give me more credit than him; I don't think that this is fair, and I don't go out of my way to get credit for things. I don't think that awards are that important. Our children don't see it that way either. When I get an award, it's very, very surprising to them. When I got the National Medal of Science, they couldn't believe that it was me, the person who washes their clothes. My husband and I do a lot of things together, I never would be doing what I'm doing now without him. In some sense, he has also had an interesting career working with me, because he can participate in all the things we do together, and select what he wants to do. That's not such a bad deal for him. If we had had some opportunity that would have been better for him at other places, I think we would've considered it. I don't know if we ever had such an opportunity. This is not a big issue for us.

I appreciate your response and I like what you said, but I felt I had to ask this question because we are all used to the reverse situation.

Recognition is not all that important for a scientist. We all do the best we can. That society values some particular thing rather than something else, that's a judgment that other people make.

I recently read about the joint work of the Curies and the author concluded that they enhanced each other's potentials.

We do that all the time. It's also a lot more pleasant that somebody close to you understands all your craziness. Devotion to science as we do it is a kind of craziness.

You have written several books together.

In this we are totally complementary. He doesn't like to write, but he likes to comment on what I write.

Who has the ideas?

We talk it over, and have ideas jointly.

You have been very active in women's issues.

I can encapsulate how it all started. When I got my first appointment here at MIT, it was to a chair that Abby Rockefeller Mauzé, the sister of the five Rockefeller brothers, established. Her idea must have been to provide an opportunity for women to be scholars. I felt I had to set aside a little time for the purpose of mentoring women students and I spent about an hour or an hour and a half per week for counseling with women students, trying to help them along. I was doing this from the very outset because part of my appointment, or my interpretation of my appointment, included this kind of activity. However, it has never been a main thing I do here at MIT. Women had very little academic opportunity at that time in the 1960s. Our women students needed role models, so I tried to help.

Rosalyn Yalow was a very good role model for me. Early on, I had one male student, who was very interested in public affairs, science policy, and who was also a very good grad student. He had a girl friend who had social troubles at MIT and her academic troubles came from her social troubles. She needed a role model, she needed somebody to help her. He told me that where I was different from the other MIT professors was that I could do something special for the women students here, especially those who needed help. He convinced me that if I were to spend, say, five percent of my time, which is one or two hours a week, I could have a big impact. I thought about that and I decided that he was right. As a result, I have worked on a lot of projects to help the quality of life for women students.

There has been a big payoff because they tell me, ten years later, that I'm the reason that they stayed on to complete their degree program. I could probably be helpful because of my own experience. Later, the President of MIT, Jerry Wiesner, during his term, asked me to do a number of things, to help women at MIT succeed, and I did that. Then I was asked to do different things for women at the national level too. I didn't go out of my way looking for this kind of work, but I don't say no to it either, just like I do a lot of public service that has nothing to do with women in science.

Do you think that tough conditions in your childhood and early in your career contributed to your becoming successful?

Tough conditions as a child either make you or break you. If you can survive hard times, you have an advantage. I can put it very simply, but I think it's correct. If you overcome those difficulties as a young person, you have a level of maturity that other people don't have, to overcome adversity. It may be too easy for kids nowadays, a lot of things are there already prepared for them, parents look after a lot of things for them, but this doesn't happen for everybody. Young people have their own crises nowadays; today, there are all these different scientific fields, they are very competitive, the world is very competitive. Each generation has their advantages and disadvantages.

Can we apply this to the Jewish experience?

Once I was given an honorary degree in Israel, and they asked me to give a talk. Teddy Kollek, the former mayor of Jerusalem was the other speaker on this occasion and I was glad that I was first. Kollek is a very good speaker and a great man and he had a tremendous impact on Jerusalem and on Israel. I gave my speech about why the Israelis are so good in science. There are two reasons. One is that Jews have a Talmudic tradition of scholarship, which doesn't have to do with science, but it has to do with the Bible. It has to do with trying to find truth in it. You might argue that there is no truth in the Bible, that it's all fiction, but the way it's studied for what it contains, has quite a lot of scholarship that has been around for five thousand years or more and that scholarship has been appreciated. That's an important thing. When I was growing up, I did not see much scholarship in the neighborhood where we lived, but my parents appreciated that I wanted to be a good student. It was a great thing to them and they appreciated my determination in looking for scholarship. It was part of a tradition. This is also present in Israel that they would like to be number one, they would not like to be second best, they would like to be the best. Israel is a small country and yet they want to be ahead in every field of science. This is the second part of the tradition, not accepting anything but excellence. Scholarship and excellence. Other groups have that, the Orientals have that, and you're going to see more and more Nobel Prizes among those groups that value scholarship and excellence. This is my simplistic analysis of why the Jews are so good in science.

What was your family background?

In 1921, my father, Meyer Spiewak, came to the United States from the Polish village where his family had lived for generations. My mother, Ethel Teichteil, was born in Galicia in the Austro-Hungarian Empire. Then her family moved to Holland at the time of World War I, and she came to the United States in 1926. The family members who stayed in Europe perished in the Holocaust. In New York, we lived in the Bronx in a Jewish ghetto. There was hardship, but both my brother and I wanted to study and we did.

Are you religious?

No, but my oldest child is religious. My husband is not Jewish but we go by the Jewish rules as a family.

What was the greatest challenge in your life?

My greatest challenge was surviving as a child. That was the hardest thing I ever did. Once I was mainstreamed, I was like other people.

Women scientists often say that the hardest thing was raising their children and doing science at the same time.

This was not such a hard thing for me. I am a pragmatic person and take things as they come. I don't run my life according to an equation and I don't have any specific goals. I'm not trying to do anything in particular. I have very good relationship with all my children and we talk openly about how they felt to be a child in our family. Because of my positions, wherever my children go, they are recognized, but this vanishes after a few minutes and everything from then on is up to them. You learn how to handle yourself in these situations.

I would like to ask you about your research.

The general carbon science work that I've done has had the biggest impact. I have been involved with it for forty years. It started as a small exercise. I was looking for a material that was different from silicon and germanium, and had simple valence and conducting bands. At that time everybody in the magneto-optics field was working on the same problem — almost everyone was considering the use of magneto-optics to determine effective masses in the valence and conduction bands of semiconductors by studying optical transitions in a magnetic field. The general theory was already developed, and though each material had its own intricacies, the research was not so interesting to me after I had studied one or two materials.

I wanted to do something different. I had little children and I never knew when I would have time for serious work. This is why I wanted to have some problem that was sufficiently hard, but didn't have too many people working on it, a topic that wasn't too attractive to a lot of people. My bosses in Lincoln Lab were skeptical and told me that they did not expect me to make much progress with this project. I thought this was a good challenge. We tried our first experiment in 1961 or 1962 and Gene had had some great ideas that made the whole project a lot more important. We started out with what other people were doing and we also tried to understand the electronic band structure for solids. To understand the graphite system we had to not only take a simple valence and conduction band, but we had to consider what else was going on in the electronic structure, much of which was not known at that time at all. We discovered a lot of things stepwise. My first magneto-optics experiments on graphite were disappointing, since we could not get nice magneto-optical spectra. I soon concluded that my problem with the experiment was materials related. Just at that time I heard that GE had succeeded in producing a new synthetic form of carbon material, called pyrolytic graphite, which had been pioneered at Imperial College in London and was produced at GE under pressure. Once we tried that type of sample, everything came together, and we were able to see beautiful magneto-reflection spectra in graphite. The observed spectra were not quite like other spectra that had been seen before on other materials. So we spent the next few months becoming very familiar with the experimental and theoretical work that had been done up to that time, in the hope that we could interpret the beautiful spectra we had taken. It turned out that the big expert in the theory of the electronic structure of graphite was Joel McClure who was a graduate student at the University of Chicago at the same time I was there. He introduced us to his theoretical model to the special symmetry of graphite.

Symmetry played a large role because the carbon system had a special symmetry that nothing else has. It had some remarkable properties and we started looking at these. Whereas other people had done electronic band structure studies at one or two points in the Brillouin zone, we did electronic structure studies along lines or high symmetry axes. So we sort of generalized the whole symmetry concept and how you do the band structure determination from experiments in a more general way. What we did together in the very early days, was re-discovered by others ten or fifteen years later. At the time we did it, people did not appreciate what we were doing. I was young when I sent in this early paper and the work wasn't appreciated by the referee. I didn't fight to get it published, I said, OK, if it's not good enough, it won't be published. Yes, one way or another many things did get published, but maybe not in the form that was initially intended. We used the generalized symmetry concepts when working first with the electronic energy band structure of graphite, and later for silicon and germanium. Here I was referring to the Shubnikov-de Haas effect. Then when it was re-discovered much later, people thought it was important. This did not discourage me a bit and I continued my research. But I have done other things that were important and were indeed recognized as important. So I've had my share of working on important things.

I started my work in the field of carbon science at that early time and I have continued it for the past forty years. About the buckyball research, we did a paper in the early 1980s, before Smalley and all the other people on some related topics, as did others, I am sure. We figured out that what comes off of a carbon surface when you bombard it with a laser is a big chunk of carbon atoms. It couldn't be just one or two carbon atoms, it had to be a big chunk, a hundred atoms I used to say. Many people laughed at that. They thought it was impossible. We didn't, however, do the key experiment on that emission. The key experiment was the mass spectroscopy measurement. We just didn't think of doing that experiment.

The Exxon group did the mass spectroscopic experiment and yet did not recognize C_{60}.

The Exxon group invited me to give a talk over there on our work on carbon clusters coming off a carbon surface under laser ablation. That was before their crucial experiment. Their involvement on a big scale might have been related to my visit. I showed them that we had found indication that the work they had done on mass spectra involved much too small mass number species. They had to go much higher in mass numbers to explain what we were finding. After I left, they went on to measuring higher mass numbers and they published an important paper. That was a predecessor to the discovery of fullerenes. Whenever there is a big discovery, there are a lot of side discoveries that lead to it. That's the way research seems to work.

Did it ever occur to you that you might have been involved in a Nobel Prize-winning discovery?

I don't think it's very important to be involved in a Nobel Prize-winning discovery. I got the National Medal of Science for my contribution and that's enough. I've gotten a lot of recognition for my work; I'm well known in my field, and that's enough. I've received a lot of awards for work I've done jointly with many others in different countries. I got many of the awards and that's not fair either because others made a lot of contribution. I've gotten more recognition than I deserve for my work. Going after more awards is silly. It's a non-productive activity. This year the Nobel Prize in Physics went for the Bose-Einstein Condensation and those people really deserved it; they did fantastic work. But I've seen people getting the award, and then doing nothing afterward and that's awful. I think that my situation is much better, just to try doing good work and being excited about it.

You published a book on the fullerenes in the year of the fullerene Nobel Prize, in 1996.

You want to know why I wrote that book? I wrote a small book, like a novelette, only 200 pages, on intercalation physics in 1981. My husband and I were co-authors of that book. That book influenced the field and that work is still referred to. Then the fullerenes came along and I started writing this new book in the early 1990s. I had a little bag, like a shopping bag with a lot of reading material and I took that bag along with me for my travels. I would read an article or two a night, and would write a page here and there. What happened was that I was invited to give a talk on fullerenes at Bell Labs. They told me that they had benefited so much from my teaching and papers and in particular from my review article on intercalation compounds, that they thought I should write a book on fullerenes. I thought about this proposition. I had already written a few books and I wrote them for my students, to bring the issues together and to make it easier for them to do their research and to read the literature. When the proposition was made to me at Bell Labs, I thought that I was over 60 years old, that these people are young and they are busy with their research. I felt that now I could afford the time to write the book. I thought it was a good time for me to serve the community. That's how I started to write that book, as a service activity for the people in the field. The book was well received and it was then followed by another book on nanotubes.

You have noted somewhere that the transition between semiconductors and superconductors depends on geometry.

This geometrical aspect has to do with the very special symmetry of graphite. This is something that goes back to the work I did with Gene in the 1960s. Graphite is a two-dimensional zero-gap semiconductor. As you go down to smaller dimensions, there are fewer atoms. The smallest nanotube has only 10 atoms going around the circle of the nanotube. The degeneracy point maintains its importance, and depending on the degeneracy point we may speak about semiconductors or we may speak about semi-metals. We did this work on the symmetry of graphite a long time ago, certainly before the nanotube was discovered.

I understand that you had a sole-author paper in Physical Review *based on your Ph.D. thesis work on superconductivity? Please, tell us something about that.*

When I was a graduate student my papers were all sole-authored because we were expected to submit our sole-authored papers, which had been accepted by journals, as the dossier that was submitted in place of a formal thesis for the thesis defense. My theme was a detailed study of the microwave (30 cm radiation) surface impedance of a superconductor in a magnetic field, and the research project was done at the University of Chicago. The reason for the sole-authorship was to develop independence. The students were expected to develop their own research topic and to carry out the work pretty much on their own. The discovery I made was that even though a large magnetic field could destroy superconductivity altogether, because of quantum confinement effects, a smaller magnetic field could actually lower the surface impedance under selected circumstances. This work attracted quite a bit of attention because it was published soon after the BCS theory for superconductivity, and this theory could not explain my experimental results.

You were one of the first people who used lasers to study magneto-optics effects, already in the 1960s. What were the results of this study?

Yes, we were. We used a laser to deliver right and left circularly polarized radiation, which was conveniently provided by a laser in the infrared range. Ali Javan helped a joint student to build a suitable laser designed to probe the separation between adjacent magnetic energy levels in the valence and conduction bands. This experiment had a large impact on our understanding of the electronic structure of graphite, showing that the previously accepted identification of electrons and holes was wrong and had to be reversed. Once this assignment of the location of electron and hole carrier pockets was interchanged, many unexplained experiments relevant to the electronic structure of graphite fell into place. This laser-based magneto-optics study of transitions between magnetic energy levels led to the model of the electronic structure of graphite that we use today.

When already at MIT, in 1973, you started a new line of research for which you could not find support from funding agencies. What was this project and was it successful?

This project was graphite intercalation compounds. In the mid 1960s Ted Geballe, Bruce Hannay, and others from Bell Labs, who had discovered superconductivity in graphite intercalated with alkali metals, tried to persuade me to carry out magneto-reflection studies on these same systems to try to understand why these materials became superconducting when none of their constituents went superconducting. I did not enter the field right away, because I wanted first to see if such an experiment was feasible. The feasibility was demonstrated in the early 1970s. So when a student interested in the project joined the group in 1973 we started work on this topic. The start date was also related to my appointment as the Abby Rockefeller Mauzé chair, which had a small budget (of perhaps $10K/yr) for scholarly research. This budget provided the seed funds for our entry into graphite intercalation compounds, which turned out to be a very fruitful field, where I was very active for over 15 years, and our group became well known worldwide in this field.

Please, tell us something about your work on intercalation compounds and your studies of carbon fibers.

My early work on intercalation compounds started with magneto-reflection studies of donor and acceptor intercalation compounds, but soon moved into an in-depth study of the graphite host layers, using mostly Raman and infrared spectroscopy and structural characterization techniques. In 1977, I attended a conference in la Napoule, France, on intercalation compounds (actually, the first international conference on this subject) and the conference broadened my vision of this subject and led to the writing of a review article with Gene that both influenced us and influenced the development of the field. After the review article, we did more detailed Fermi surface studies using the Shubnikov-de Haas effect and focused increasingly on the properties of the intercalate layer, and this led to in-depth studies of the magnetic properties of the intercalate layer, which we worked on for more than 5 years.

Studies on carbon fibers actually grew out of our studies on graphite intercalation compounds (GICs). In this connection, I attended the Second International Conference on GICs in 1980, and there I heard a lecture by Morinobu Endo on vapor-grown carbon fibers. This gave me the idea that these fibers would provide an ideal geometry for transport measurements on

Mildred Dresselhaus with her apparatus, 1978 (photograph by and with permission of Georgia Litwack, courtesy of M. Dresselhaus).

intercalated graphite materials. I suggested this idea to Dr. Endo. He also thought this was a good idea, and this started a very active collaboration that continues until the present time. The collaboration started with transport measurements on his fibers in their pristine form and after intercalation, but the research soon moved to the use of many characterization techniques to study the vapor-grown carbon fibers, including Raman scattering characterization, magneto-resistance studies, thermal transport studies, the ion implantation of carbon fibers and many other studies. Our work on carbon fibers became well known, and I started to give many invited talks. On one symposium on carbon fibers, the three speakers were asked to write a tutorial review article, which developed into a text on carbon fibers. This book had considerable impact on the field for over a decade. The book is now out-of-print and needs to be rewritten from a modern standpoint that relates carbon fibers to carbon nanotubes.

What is your present project?

We are doing nanotube spectroscopy at the single nanotube level. I find it exciting and a lot of people are curious about it. About 3 years ago we started on a project to find ways to increase the signal in the Raman spectra from carbon nanotubes. Our first attempt involved the surface enhanced Raman spectroscopy (SERS) effect. Using small colloidal silver particles and the expertise/collaboration of Katrin Kneipp, a visitor to MIT from the Technical University of Berlin, we managed to see Raman spectra from a very few carbon nanotubes, yielding spectra with very much smaller linewidths, indicating very much smaller inhomogeneous broadening effects. However, the small silver particles giving rise to the large enhancement effects, also interact strongly with the nanotubes in an inhomogeneous and complicated way. But if the signal is really so strong, we argued that maybe we could see something without the silver particles. And this led to doing the experiment on isolated nanotubes prepared by Professor Charles Lieber and his group at Harvard Univesity by a vapor-phase method. The reason why we are so excited about making Raman spectroscopy measurements at the single nanotube level is that Raman scattering is a spectroscopic technique that normally gives energy levels. But in the case of a one-dimensional system, such as a carbon nanotube, there is a strong coupling between the vibrational energies that are measured directly and the electronic structure when the photon energy exciting the Raman scattering is equal to the singularities in the joint density of electronic states. Under these resonant conditions, the Raman intensity is enhanced by many orders of magnitude and it is possible to see spectra from individual metallic and semiconducting tubes and to see all the features in the spectra. But the really exciting thing is that every nanotube has a unique set of singularities in its density of electronic states. So working backwards, we use the Raman spectra to determine these singularities in the joint density of states. Once the energies of these singularities are known for a particular tube, we can utilize the unique relation between these singular energies and the nanotube geometrical structure to determine the nanotube diameter and chirality. These structural parameters also determine many other physical properties of the nanotubes.

What is the importance of nanotubes and what are their possible applications?

Nanotubes are important for nanoscience as the simplest known one-dimensional system, and because of their simplicity, they can be used as a model 1-D system, because nanotubes exhibit many remarkable properties. For example, nanotubes can be either semiconducting or metallic, depending on their diameters and chirality. Furthermore, the possibility of doing single nanotube spectroscopy allows for the determination of the geometrical structural parameters of individual nanotubes. The incredibly small size and very high elastic modulus make nanotubes ideal for use in scanning probe tips to facilitate a variety of measurements at the nanoscale. The excellent electron emission properties of nanotubes make them attractive cold electron sources. Much development work has been done with regard to using nanotubes for flat panel displays and efforts at making electronic circuits out of nanotubes have also progressed.

Which of your projects do you remember with the greatest pleasure?

It's hard for me to say. It is like which is your favorite child. We love them all, but maybe in a little different way. And it is the same for research projects at one level. On another level, I am always most excited about the project I am working on. So right now that would be nanotubes and nanowires, or nanoscience more broadly.

What do you do when you are not doing science?

I have a lot of interests. In addition to science I do a lot of service. I'm on many committees and am involved with policy-making. I like that and

Mildred Dresselhaus with her violin, 1978 (photograph by and with permission of Georgia Litwack, courtesy of M. Dresselhaus).

I do it because I feel it's important to do it. At one time in my career, I was a very poor person and society helped me along, and I appreciate that. It's now my turn to help others. My other thing is my music. I've kept that up for many, many years, and I make music almost every night. I have a chamber music group coming to our house most evenings. Tonight, for example, we're doing middle Beethoven quartets. Sorry, I'm wrong. Tonight we're doing Brahms quintets.

Would you have a message?

I have some ideas. I think science is fun. I spend a lot of time with young people whom I'm trying to convince that working in science is a rewarding experience. That's my message and that one should be serving society. I also have a woman's message that you can have a reasonably normal life as a private individual and yet do world-class science.

Catherine Bréchignac, 2000 (photograph by M. Hargittai).

28

CATHERINE BRÉCHIGNAC

Catherine Bréchignac (b. 1946 in Paris) received her Ph.D. at the University of Paris-Sud in 1977. She has worked at the Laboratoire Aime Cotton in Orsay since 1971, between 1989–1995 as its director. She has served as Director General of the CNRS from 1997 till 2000. She is a corresponding member of the French Academy of Sciences (1997), Foreign Member of the American Academy of Arts and Sciences (1999) and Member of the Academie des Technologies of France (2000). She is Chevalier, Legion of Honor (1996) and Officer of the Ordre National du Merite (2000). Presently she is Director of Research of the CNRS. We recorded our conversation in her office in Orsay on October 21, 2000, which was augmented in 2002.*

What made you interested in science?

Originally I was interested in mathematics because I did not have to spend too much time with it — it was logical and easy to do. The other subject I really liked was French literature that was my true love. These were the two things I liked, mathematics and reading.

Later I decided to orientate towards science; I entered the "École Normale" in mathematics and finally decided to do physics. It is difficult to say why I did that — I think it was mostly because among the students I knew, I preferred the physicists, they were more social and provided a

---

*Magdolna Hargittai conducted the interview.

much more pleasant environment. I made this choice purely based on human interactions and not on the scientific disciplines themselves. Then I decided to try research and went for a Ph.D. in physics. I found research exciting and by the time I got my Ph.D. I knew that this is what I wanted to do.

What was your first research project?

For my Ph.D., I worked in the field of atomic physics. As a matter of fact, when I entered the "Laboratoire Aimé Cotton" in Orsay, the only research done in that lab was atomic physics. At that time the director of the laboratory made the decision about the Ph.D. projects, and I did not have any choice, except to work on isotopic shift measurements using Laser spectroscopy. This consists of following the optical transition, of atoms through their different isotopes in order to deduce how the shape of the nucleus affects the atomic optical transition. Conceptually this subject was not of a fundamental interest, and I was in some respect frustrated, but I learned a lot by doing all the experiments just by myself. This was a big advantage for the future. During my Ph.D., I was convinced that I would not stick too long with that subject, and I was tempted to study more complex systems. So I moved to laser induced collisional energy transfer. The idea was to induce selectively a chemical reaction between colliding atoms with a photon at the inter-atomic frequency. The results were nice, but they concerned only a few systems and they cannot be generalized.

As I understand you switched to cluster research around 1980.

That's right. Clusters are the component-precursors of the nano-world. With increasing their size, they can be imagined as bridges between the gas phase and the solid phase. However, because small is different from large even if the properties of solids are usually known, the properties of these clusters are not. So I have decided to work with these clusters by entering a "terra incognita", and it has been a real adventure.

How do you decide what size of clusters you are interested in?

We choose to make the clusters by physical methods. It is also possible to make clusters chemically but we use physical methods, and first focus on metallic clusters. We heat a piece of metal to form a metallic vapor and, like the clouds in the sky, we make "clouds" of atoms, which are cooled to condense into droplets, with sizes, ranging from a few atoms to about 50,000 atoms. By using different temperatures and pressures, we can regulate the mean size of the droplet distribution. Once formed, the clusters are ionized and we choose from them a given cluster size by time-of-flight mass spectrometry. The time-of-flight mass spectrometer is based on an old idea, but it is used in many disciplines, recently even in biology in order to determine the mass of proteins for example. For clusters, it can be used as a mass filter. After being ionized, all the clusters are accelerated with the same energy at a given time, which is the starting point. Then, we leave the clusters to fly in a drift tube, and because they have different masses, the smallest arrive first at the end of the tube, followed in time by the second, the third, and so on. Then we use a gate located at the end of the tube, which can be either closed or open, depending on the mass of interest.

With this flexible experimental set-up I focus my research in two directions. One of them is the study of free clusters to follow their stability, their optical and thermodynamic properties as a function of size, and essentially in a size range where they present unexpected behavior. The other topic is the making of nano-structures from cluster deposition on surfaces. In this case the clusters are considered as elementary building blocks for more complex systems or granular films.

I read that you and your collaborators achieved for the first time to couple cluster beams and synchrotron radiation. What was the goal of this experiment?

The synchrotron radiation allows us to probe locally one atom in a sample. Our idea was to use the synchrotron radiation to investigate the atomic environment in a nano-piece of material, that was an isolated cluster, and we vary the size of the cluster to vary the atomic environment. We focused our research on mercury clusters in order to study, in a new way, the question of the insulator-metal transition; this happened back in 1985. With two electrons per atom mercury should, in principle, be an insulator. Of course, it is well known that mercury is a metal. So we tackled the insulator-metal transition by asking the question: how many mercury atoms are needed to form a metallic droplet? We used the synchrotron radiation to excite a core electron of one atom in the cluster into the valence band to see to what extent this electron becomes delocalized over the whole cluster. The transition is not sharp with respect to the cluster size, but we saw it and these results were a success. Now, an increasing number of groups couple cluster beams and synchrotron radiation.

Are you only interested in clusters of atoms, or also in clusters of molecules?

I am also interested in molecular clusters but not just any kind of molecular clusters. We are concerned with systems, such as oxides, hydroxides, and mixed clusters, generally those with metals or semimetals.

How does the behavior of clusters depend on their size?

There are two types of behavior. One of them evolves regularly with the size — it basically depends on the ratio of the number of surface atoms versus the number of atoms in the whole volume. This is a regular change, which behaves proportionally to the surface to volume ratio. Superimposed to this regular variation, there is another type of behavior, which does not change regularly with the size and occurs at small sizes. This is the case of the quantum size effect in metallic particles. If one assumes that the electrons in metal clusters move in a spherically symmetric potential and are caged by the surface of the cluster, they organize themselves in a shell structure with shell closure at 8, 20, 40, etc. electrons. This quantum behavior manifests itself as a shell structure on the cluster stability and has been experimentally observed since the work of W. Knight in 1984.

What are the cluster sizes at which basic changes in properties occur?

To answer your question I would like to say that it strongly depends on the property you look at. Let us construct the matter atom by atom, and observe its optical properties. One atom has a series of discrete atomic levels, hence, a series of discrete light emissions. Then from two to a few atoms, one constructs a molecule, which can vibrate, rotate, and the levels group to form bands. As the cluster size increases the electrons become delocalized, as we already discussed, and make the glue, which bound the cluster. Its optical response evolves to collective excitation. But this does not mean that you have reached the bulk properties. The frequency of the collective excitation shifts, usually to the blue, as the cluster size increases, from 8 to 10,000 atoms or even more, changing the "color of the metallic clusters", or the tunable frequency of the gap of the semiconductor clusters with size. Such a property has been used a long time ago to make the beautiful colors of the glass-windows of the cathedrals, but the reason why they behave like that was not understood at that time. Now if you consider the mechanical properties, or the stability of the clusters, the change may occur for different sizes. Nevertheless small sizes often present surprising behavior.

Would you care to tell something about your experiment in which you established the Coulombic fission in free metal clusters? Does it have any analogy with nuclear fission?

In the late eighties, I was — and I still am — interested in the stability at the nano-scale, and studying the fragmentation of nano-objects is one of the keys to understand their stability. One possibility to induce the fragmentation of a small droplet is to charge it. Since 1872, when Rayleigh studied the stability of a drop of water, it was commonly accepted that small multiply-charged particles are not stable below a critical size, because the Coulombic repulsive energy between the positive charges exceeds the binding energy of the particle. However, there was yet no clear picture concerning the mechanism of fragmentation and the dissociation channel of small clusters, when we started. We built a special experiment with two time-of-flight mass spectrometers in a row. The first one is used to select a given size of doubly-charged sodium clusters, the second one to study the fragmentation products. We have shown that the critical size, below which multiply-charged clusters are not observable in mass spectrometry, strongly depends on the cluster formation. For hot clusters, we established the competition between evaporation of atoms and Coulombic fission in two singly-charged fragments. For cold clusters, fission always dominates. The fission products depend on the relative ratio between the Coulombic and the surface energies.

It was not obvious that metal clusters should behave like atomic nuclei, but they do, and the model that describes the fission of unstable nuclei can be successively applied to describe the Coulombic fission of metallic clusters. Moreover, clusters are in some respect more flexible since one can independently vary their charge and their mass that is not the case for nuclei.

I noticed that much of your metal cluster research focuses on alkali metals. Any special reason for that?

The reason is that alkali clusters are the simplest metal clusters; each atom gives one electron to the electronic glue. The electrons are arranged in electron shells, with similarities with nuclei. The optical collective plasmon resonance of alkali clusters has their analogue in the giant dipole resonance of nuclei. The fission process presents also analogies with nuclear fission. In many respects these two forms of condensed matter, the alkali metal clusters and the nuclei are amazingly similar. This is because both systems consist of fermions moving, nearly freely in a confined volume. But alkali clusters can grow to macroscopic matter, which is not the case for nuclei, because of the fission process; this is also why I am interested in Coulombic alkali cluster fission.

Please, tell us something about your more recent research, concerning the idea that clusters are precursors of nano-objects.

The production of materials, devices, or systems, by controlling the materials at the nanometer scale, is in full expansion today, and miniaturization is one of the driving forces of the technology. Research on clusters presents a more conceptual approach. When isolated, the cluster can be considered as the prototype of a small finite system, and an ideal object for scaling laws. When interacting, it offers the possibility to use it as an elementary building block for more complex structures. One of the main advantages of using clusters as building blocks is that they can already be themselves composite systems.

Do you do experiments yourself?

I am an experimentalist and I like to do experiments. But now I have students and postdocs working with me. It is very important for me that we work in a team. The work is shared between the seniors and the juniors. However, if you want to be a good and efficient leader in experimental work, you have to know everything about the experimental set-up and how the measurements are done, which means that you have to be there during experiments. I also need theoretical approaches, and I usually do it in cooperation with theoreticians.

What was your most important project during your research career?

I consider the cluster field and its ramification to nanometer scale objects as my most important scientific project. To be more precise the ubiquitous character of fragmentation, fission, and more recently instability at the nano-scale are my favorite ones.

Please, tell us something about your family background.

My father was also a scientist, a nuclear physicist. He was professor at the University of Paris, director of the Institute of Nuclear Physics in France and then "Haut Commissaire" at CEA (Center of Atomic Energy) for 15 years. My mother was a medical doctor and also professor at the University of Paris. Both my parents were academics but both came from working class families. When I was young I spent most of my time with my maternal grandmother. She was very far from the intellectual atmosphere, she was very pragmatic, and wanted to be helpful. She came from a very poor family, in Brittany, where 14 people lived in just two small rooms.

Were you an only child?

No, we were three; I was the oldest and I have two younger brothers.

How could your mother have three children and also be a medical doctor and a professor at the University?

There are three reasons. First of all, my grandmother helped her a lot. Second, we live in France and have nurseries, kindergartens, etc., for children; it is a normal habit that women work. Third, my father also pushed her to work. She was happy.

Did you mind that your mother worked and you probably did not have much time with her?

No, not at all; it was the way it was supposed to be, it was natural. Similarly, it was absolutely obvious for me that I will have to work. It never even occurred to me that I would stay at home and just raise kids.

How did you meet your husband?

We met at the "École Normale". He is also a physicist but closer to chemistry. He has a professor position at the Paris-sud University in Orsay, and recently he became the director of the "Photophysique moleculaire" lab. His research field deals with the chemistry in interstellar medium.

Do you speak about science at home?

We do not see each other often; both of us work a lot. We try to avoid speaking about science at home but of course it happens sometimes.

Catherine Bréchignac with her husband, Philippe, at the 17th Meeting of the Heads of Research Councils of G-7 Countries, Ottawa, Canada, 1998 (courtesy of C. Bréchignac).

Is he also a member of the Academy?

No.

How does he handle the situation that you are apparently more successful than he is?

My husband has always helped me a lot and pushed me to succeed. He knows that I am most happy when I work, he knows that I like the research I am doing, and he is very open minded and understanding. He does not complain if I come home late, if dinner is not ready; on the contrary, he helps to make everything easy. Of course, we have a paid help at home, because both of us are away from home very often. Sometimes he says that too much is too much and then we decide to do something together.

How does he feel about you being a member of the Academy while he is not?

When I was elected, he was very proud, I could read it in his eyes. But our life together is not based on academic success. We try to separate our life from our work and avoid having competition between us. Although, recently, we actually did have some joint work and we have two papers together. But earlier I tried to avoid that. You should ask him what he thought about this but I don't think that it is a real problem for him.

Do you think that your election to the Academy and all your other recognitions and awards were solely on merit or perhaps you being a woman had something to do with it?

Oh, sure, of course. More precisely, about the Academic recognition, I don't know, I don't think so. But the appointment to be Director General of the CNRS definitely happened because I am a woman. Of course, I have a good reputation as a scientist and I also managed well the lab and the Department of Physics at CNRS earlier; but the politicians like to have pride in showing how considerate they are with minorities. Oh, I am sure.

Did you mind that?

No, not at all. I figured that it was fine but after having been chosen, I simply had to prove that I could do the job better than anybody else, men and women alike.

Let us go back to your family life. How many children do you have?

I have three children; they are 27, 25, and 21 years old.

How did you manage when they were small?

The two oldest, both boys, were born before I got my Ph.D. and I have to be thankful to my grandmother who raised me as well, because she also helped me a lot. We lived in the same building in Orsay and she was always there for us. Every morning, I took the boys to the nursery or to the kindergarden, before going to the lab, and picked them up at night. But she was the one who was with them when they were ill, during vacation, and in the evenings. My daughter was born later. Both my husband and myself got a postdoc position in Canada, when I was pregnant with her. We left France when she was one and a half month old. It was in winter, and I remember very well our arrival in Montreal with snow everywhere. Probably that year was the most difficult year in my whole life. We had to spend all of my salary, and even more, on daycare — I realized then that women really have a hard life in the U.S. and Canada if they want to have both a career and children. France is much better in this respect.

How did your children feel about you being a working mother? Have you ever talked about this at home?

Christmas at the Bréchignac home in Orsay, 2001. From left to right: wife of their second son (who is taking the picture), Philippe, Catherine, their son Antoine, and their daughter Charlotte (courtesy of C. Bréchignac).

I suppose that it was more difficult for them than it was for me. They say now that "you never were there for us when we needed you." Often I picked them up in school and took them with me to the lab. I remember a terrible day: my second son, who was about three or four years old that time, was with me while I was doing my experiment. He was terribly bored just sitting there and looking at that machine, so he took a pair of scissors and "bung" he cut the wire and the current. That moment I understood the conflict. But altogether, they are not bitter now; when we talk about those days, they mostly laugh about it.

What are their plans for the future?

None of them wants to be a scientist. The oldest one has a very good position at a water company; he is an engineer. The second one is a professional golfer. The girl, who is the youngest one, studies at "la Sorbonne" — Paris University — in French and comparative literature.

Do you have heroes?

No. I don't like the notion of "heroes". But there is a woman I like very much; it is Madame Germaine de Staël. She was the only daughter of Necker, the Minister of Finance of Louis XVI. She was a woman of letters; independent, liberal but not extreme, and she expressed what she thought. During the terror period of the French revolution she exiled from France to live in Coppet in Switzerland that was a place for gatherings of intellectuals and writers from Europe and there she led deep and free discussions. This is the reason why Coppet was referred to, at that time, as "the salon of Europe". I like her.

Have you experienced any kind of disapproval from people like family doctors or teachers, for being a working mother? Or is it quite natural in France?

No, it is not natural, but it is not rare. I remember that when we were in Canada, the doctor of my daughter tried to make me feel guilty about how I managed my life. He told me that I was not a good mother because my baby needs a full-time mother and not a working mother. He could not convince me; I was so sure that he was wrong, that I did not pay attention to what he said. Now when we talk about this with my children they all say that, at the end, they preferred having a lively working mother than an irritating mother at home.

Looking back at your career, have you experienced any negative discrimination?

Not really with my colleagues in physics. Even in foreign countries, like the U.S. or Japan, I don't feel that. But I don't really care either. It might have happened sometimes but I am so positive that I do not want to see negative discrimination.

I would like to ask you about the CNRS.

It is a huge research institution, the biggest in Europe, I think. In my opinion, it is one of the most interesting institutions even on the world scale, because it provides an environment where research is protected. Research is fragile since it is at the frontier of knowledge. This is why CNRS generates jealousy and heavy criticism.

The CNRS should be open, and now it is — it was not always the case but now it is. Most of the laboratories funded by the CNRS are joint labs with the universities located over the whole country and even from abroad. There is, for example, a joint laboratory between the CNRS and Tokyo University. The CNRS is like an enormous network, a fractal-connecting network like a nervous system of science. Among its 1700 laboratories 80–85% of them are joint laboratories with universities, a few are joint labs with industries, and the rest are CNRS's own laboratories or institutes.

The CNRS has two functions: one of them as an agency to give money for research, and the other to coordinate and to promote interdisciplinary projects, since all the disciplines are gathered in its organization. The main difficulty is the evaluation as it is everywhere, but peer review is not very common in France.

I read on the web that one of the missions of the CNRS is "to enhance the dissemination of scientific information, with an emphasis on the French language". Why? We cannot deny that the international language of science is English.

The reason is the following: about 25% of research at CNRS is in social sciences and the humanities. For these people the language is more important than for a chemist, a biologist, or a physicist, who communicate in English with their colleagues over the world. For researchers in the social sciences and humanities, I can understand that they want to express themselves in French and get a translation into another language if needed. We have talked about this question a lot, especially in connection with the construction of Europe. Europe is a multilingual entity and the communication cannot be exclusively done in English. We have to make it possible for people to express themselves in their own language if they want. The young European generation should learn at least three languages.

You have been Director General of CNRS for three years. Is there anything specific you can single out that you accomplished and are the most proud of?

I am proud of different things. One of them is having CNRS as an open network, with transformations and not destruction. The second point is that we tried to push the units of social sciences to structure themselves. I am in favor of developing the social sciences and the humanities in our world where everything you look at is technology. This is also why I preserved our bookstore "CNRS editions".

The third one is making more connections with industry, having joint laboratories with them. For example, in Marcoussis, 30 kilometers south from Paris, we built a CNRS laboratory with the Alcatel company. That means that a public laboratory has emerged in a private industry to focus its research on nano-technology. In this case it is for optical devices and since many laboratories working in optics are located in the same area, we connected them in a network that is called the "optics-valley".

Another project I really pushed is the VIRGO project. It is an Italian-French project on the gravitational wave detector, which is only funded by two European research organizations: the CNRS, in France, and INFN, in Italy, for a total of 100 millions of euros. This is really a dream. Two such antennas are now under construction in the world: one in the U.S.A., LIGO, and the other is VIRGO.

I also pushed for amplified connection between the different disciplines. However, we always have to keep in mind that to construct an interdisciplinary program we first need disciplines. That interconnection is important in biology, which needs new technology coming from physics and chemistry. Behind biology there is the question of health that makes the society impatient in waiting for new results. But sometimes what we need is more time, and more people to think and not more money.

What is your present ambition?

When you get a position as the Director General of the CNRS, you work a lot, but you are not considered for yourself but for your position. It was clear since the beginning that what I like above everything is freedom. I do not want to be confined in bureaucratic positions because I would not be able to do anything else after leaving the CNRS. So even while I worked there, I kept one day of research per week and it was not a problem to come back to my laboratory. Now I am back in the lab, I am not in charge of anything except doing research. I have a group of about ten people. I am planning to continue cluster research, to make self-assemblies of clusters to build new entities.

What is your advice to young women who want to have a career and a family at the same time?

It is not easy but it is possible. You have to be ambitious. Do not fight against a big mountain which you obviously cannot destroy; it is better to turn and go around the mountain than trying to go through it. It is important to have a goal and then to find the best way to achieve it. My advice is, do what you like, you have only one life.

Many young women think today that it is better to first establish themselves in their career and then, later, to have children.

This is the wrong way, emphatically the wrong way. You should have the children while you are young. First of all, it takes more than ten years to establish a career and then you are almost in your mid or late thirties. It is late. It is much better to have the children as early as you can. You also have more energy when you are very young.

What was the greatest challenge in your life?

To have parallel lives. Private life, working life, having friends, travel, lives that do not always mix easily. I find it important not to focus only on one aspect of life because if something goes wrong then it is very difficult to survive the shock. You have to be happy in many aspects of your life and then when something happens in one, you still have other resources.

Are you religious?

No, I am not; I am an atheist. I believe it is the humans who created the gods. My husband is Catholic but does not go to church. My children — it depends on the particular period of their life.

Do you have any hobbies? Do you have time for them?

I would like to do more than I can do. When I have free time I read, listen to music, or walk in Paris. I like walking in Paris, from north to south or east to west, looking at the architecture, the ancient and the new next to each other, the fountains merging stone and water. I was born there but I have always something to discover.

*****

Since more than 2 years have passed since this interview, I would like to have a follow up. I see that you are very active and publish in top journals. What are your recent results? Could you elaborate on them a little?

Full-time job in Science! Of course I am very active. Continuing what I have done, I am convinced that understanding the behavior of nano-objects or clusters is fundamental in order to understand complex matter, including chemistry and biology. For example, since 1996 a huge effort has been put into forming nano-objects that cannot exist in Nature, but very little is done concerning the questions: are these objects stable? Or how long these objects are stable? This is a very interesting and important issue. We have to keep in mind the fascinating fact that a nano-object has to modify its shape in order to minimize its free energy, and this leads to shape modifications that may be observed on a reasonable time scale. We have recently studied the instability driven fragmentation in fractals at the nanometer scale, and we have discovered how a fractal object breaks down into hundreds of pieces. More generally, studying thermodynamics at the nanometer scale is a real challenge. It covers not only the stability but also measuring the temperature of isolated small systems, phase transitions, phase diagrams — and all this on a scale where the extrapolation from macroscopic systems fails.

Philip W. Anderson, 1999 (photograph by I. Hargittai).

29

PHILIP W. ANDERSON

Philip W. Anderson (b. 1923 in Indianapolis) is Joseph Henry Professor of Physics, Emeritus, at Princeton University. He shared the 1977 Nobel Prize in Physics with Nevill F. Mott (1905–1996) of Cambridge University and John H. Van Vleck (1899–1980) of Harvard University "for their fundamental theoretical investigations of the electronic structure of magnetic and disordered systems". Philip Anderson got both his B.S. (1943) and Ph.D. (1949) degrees from Harvard University. His career has been spent at the Naval Research Laboratory (1943–1945), Bell Laboratories (1949–1984), and Princeton University since 1975. He also held a part-time professorship at Cambridge University between 1967 and 1975. Since 1985 he has been actively involved with the Santa Fe Institute. Dr. Anderson's research interests have included condensed matter physics and more general questions of theory, such as broken symmetry, measurement theory, and the origin of life. His recent interests extend to biophysics, neural nets, computers and complexity, and mixed valence. He has been a member of the National Academy of Sciences of the U.S.A., a foreign member of the Russian Academy of Sciences and a Foreign Fellow of the Royal Society (London), among many others and has received many decorations, including the National Medal of Science (1983). We recorded our conversation in his office at Princeton on March 12, 1999.*

*István Hargittai conducted the interview. This interview was originally published in *The Chemical Intelligencer* 2000, 6(3), 26–32 © 2000, Springer-Verlag, New York, Inc.

You just returned from Santa Fe. I would like to ask you about the Institute there and your involvement with it.

It was originally founded as an institute whose purpose was to work in the spaces between subjects and to foster new areas, interdisciplinary in many cases. Our idea was not to focus on complexity, we just found ourselves doing so. Most areas of our interests have involved complex systems, but not all; actually, we had a long-standing program on quantum theory of measurement, for example. And our work on elementary particles we originally thought of as interdisciplinary between physics and mathematics but it became all mathematics, essentially, and not interdisciplinary, so we dropped it.

I was not one of the founders of the Institute. It was founded by a group of Senior Fellows of the Los Alamos National Laboratory. They talked about an institute, which would not have the departmental structure of a university and would embody some of the broad-ranging thinking that had been characteristic of Los Alamos. One of the good things about the Santa Fe Institute is that many people think of themselves as founders, and in a sense they are, although it was actually founded by a radiation chemist, George Cowan. He is a very eminent scientist, but much of his career was in classified work, and thus he is not as well known as he might be. A great deal of thinking behind the Institute was provided by Murray Gell-Mann. He is permanently in Santa Fe. The Institute is a private organization, funded by a mixture of grants, half of them government. The total budget is between 5 and 6 million dollars.

To my perception, living in Princeton must be an intellectual challenge every day. Can Santa Fe provide such an environment?

I'm not so sure that Princeton is a good place for such things. Here you are busy with your own concerns. There are people who are physically in Princeton, whom I meet and talk to only when I am in Santa Fe. Santa Fe is deliberately organized to bring people from different fields together. My own field is condensed-matter physics, and we never talk about that in Santa Fe. It is a well-organized field with well-defined objectives. In Santa Fe we discuss fields that are not already defined. It's a unique intellectual family. I find myself talking across the table with an archaeologist or a historian, and not just listening to him describe his work, but with the responsibility to assess and contribute if I can. It is totally different in concept from any other institution that existed at the time it was founded.

Wouldn't a British college provide a similar environment?

It's the same experience in communication but not in actually setting up professional collaborations. In 1986 and 1987, I got involved in going into economics; I helped bring 10 scientists and 10 economists together in a room, and we talked until we talked the same language. That meeting was the foundation of an economics program. One result has been that a lot of economics papers are published in non-economics journals, including some physics journals. Even a spin-off company has been formed, called "The Prediction Company", funded by a Swiss bank.

We hold some integrated workshops where we try to figure out what are the common features of various subjects. In one of our first workshops over 10 years ago, we talked a lot about consciousness. I suggested calling it the C-word because until recently it was the death of reputation for psychologists to use the word consciousness. It was a few years before it became respectable for people like Dennett and Crick to write books about it.

Do you think consciousness is the next frontier in science?

Yes, I do.

Any other frontier of comparable importance?

Of course, everyone is trying to guess the frontiers. Ten years ago, you could have said developmental morphology. But they're solving that. There will be more and more penetration of hard science thinking into all kinds of fields.

Eugene P. Wigner was also a physics professor at Princeton. What memories do you have of him?

Eugene and I never really got on. Our politics were totally different but I was a latecomer to Princeton. I only came here in 1975 and Eugene was by then already retired. From 1949 to 1984, I was almost continuously an employee of Bell Labs. My university is Bell Laboratories. In 1967, I went off to the University of Cambridge and had a half-and-half job arrangement between Cambridge and Bell Labs. After eight years, in 1975, I traded my half-and-half with Cambridge for a half-and-half with Princeton. Then finally in 1984 I retired from Bell Labs. So my professional life did not overlap with Wigner's. For the last five of his years, he was manifestly suffering from Alzheimer's disease.

However, I got to know him very early in my career. Wigner was the professor of several of the major figures in condensed-matter physics, Herring was one of his last students in this field, and Herring was my mentor. Bardeen was also Wigner's student. When I first came to Bell Labs, I was working on ferroelectrics with Matthias, who discovered many of the ferroelectric crystals that are now being used. He was a Central European, and he and Wigner met at a party somewhere. Wigner developed some ideas about ferroelectricity, which I thought were pretty nonsensical. I was deeply involved in all the experimental data that were coming out and had my own theory that followed some ideas of Shockley. Bill Shockley was unhappy about Wigner's incursion into the field and suggested to me that I visit Wigner at Princeton and explain to him how ferroelectrics really worked. This was about 1950. It was one year after my Ph.D., and I was naive and brash. Wigner responded to my explanation exactly as you would have expected him to. He was an overly-polite person, but I could see through that very quickly.

Wigner was quite right on the personal situation, that it was rude of me to confront him in this way, but there was no way that he was right on the science, where he was way out of his area of expertise. Somehow I never learned deference to reputation as opposed to expertise. Incidentally, I had great respect for Wigner's earlier work, much of which I studied and used, particularly his work on resonances, which was the mathematical background of one of the "Anderson models".

Then I met him again in Japan, during a meeting in 1954. Wigner was not a Nobel Prize-winner yet but he was Wigner. He was a very honored guest. I was planning to stay in Japan for six months, and my daughter and wife were with me. We got a room with a bathroom, which would have been connected to another room as well but our hosts locked the bathroom to the other room. The occupant of the other room was Wigner. This incident did not help our relationship.

Then there was one more encounter around 1958 or 1959, during one of my talks about the BCS (Bardeen-Cooper-Schrieffer) theory of superconductivity for which I had resolved the problem of gauge invariance. Wigner had written a paper in which he promulgated his superselection rules. One of these said that there can be no phase coherence between two states that have different numbers of particles. A fundamental principle

Anderson at the time of his Nobel Prize, 1977 (courtesy of P. Anderson).

of the BCS theory allows phase coherence between states with different numbers of particles. It explicitly violates Wigner's superselection rule. There is, though, a way you can get around it, but I never believed much in its necessity. I don't think though that I was particularly dismissive of the superselection rules in my talk. In fact, I didn't mention them at all. But Wigner saw through me, and he was very negative in his polite way. He made it clear that he didn't believe the BCS theory, and to his dying day he never accepted it. This was a deeper problem between us whereas the first two I mentioned were merely amusing incidents. Up until that point, Wigner had been very central in theoretical physics, but at about that time, in the 1950s, he put his feet down and refused to go any further. At that point, Wigner, as a person, seemed to become less relevant to physics.

How did your political differences come out?

We never argued about politics, and I doubt if he knew what my politics were but I knew what his were. I remember one party where we had a discussion about the Bohm affair. Wigner was the only member of the department who was not in favor of keeping Bohm. The department voted for Bohm, but the President of Princeton turned down our recommendation and fired him. But Wigner had not voted for him.

Star Wars?

Reagan gave his famous speech about Star Wars in the early spring of 1983, but the scientific community realized only in the summer of 1983 that he was serious about it. The scientists responded with a Pledge Campaign, to make a pledge not to take money from the Star Wars program. By this time, almost all the people had gone away for the summer. I was the only senior faculty member present and found myself leading the Pledge Campaign, giving talks and writing articles about it. We had most of the department to sign.

In hindsight, would you be willing to give Star Wars the benefit of the doubt that it may have contributed to the demise of the Soviet Union?

No, absolutely not. That is an *ex post facto* justification. The Soviet Union was already down in 1983. It was collapsing of itself.

How did you know that? Hadn't the collapse of the Soviet Union been predicted ever since it had been formed?

Anyone who was in the business of studying the strategic situation, and even total outsiders as I was, had to know that the Soviet Union was a paper tiger at that time. For instance, there was a BBC program on the Soviet military, which was taken from interviews with dissidents who had gone to Israel and various other emigres, people who had been in the Soviet armed forces, and they were describing conditions that the general public was beginning to learn about from Afghanistan. They were describing fake equipment and poor morale. When a pilot defected with a MIG, it turned out that some crucial parts were made out of steel rather than titanium alloy, simply because the Russian titanium production facilities weren't working.

Did you bring up these arguments in your discussions?

Yes, occasionally. My finest hour came about in a local Princetonian affair. There was an admiring article about Star Wars in one of the Princeton publications. I wrote a Letter to the Editor in which I was critical of their article, but they didn't publish my letter. Later, they suggested to me to write an article for the *Princeton Alumni Weekly* about the Pledge Campaign. So I did with the — to us — conventional arguments about Star Wars, how completely unworkable it was and how completely it circumvented the fact that in the original planning, some really silly things had been neglected like the curvature of the Earth.

But wasn't the program forcing the Soviet Union to make efforts that were impossible for it?

Yes, but that was not in our best interests either.

Why?

Look at the Soviet Union today. Is it good for us for it to be such a mess?

Isn't this a rather cynical approach?

To have caused a lot of misery in the Soviet Union was a very cynical approach. My argument is that it was not necessary to cause all that misery and, incidentally, to nearly bankrupt ourselves. The Soviet Union was collapsing of its own weight, and it was going to implode whether we did anything or not.

In my first article I gave the conventional arguments. The nice thing is that the then Secretary of State Schultz is a Princetonian, a relatively intelligent man, except that he has a tiger tattooed on his behind. He undertook to answer my article in the *Princeton Alumni Weekly*, which gave me a second chance. He could not possibly have written his article himself because it was straight official propaganda. It made no case at all. In my second article, I made the argument that we had the largest and most effective military alliance that had ever been constructed. The Soviet Union armed forces were a sham and quite ineffective, and the Soviet economy was collapsing, so I asked, "Why are we afraid of them?" The CIA must have known that all of Eastern Europe was completely eager to defect.

The collapse of the Soviet Union had been predicted many times, starting immediately after the formation of the Soviet Union.

There was a time when the Western societies were looking very bad too. That was in the Marshall Plan days. But that was long gone. I felt as early as in 1975 that the Russians could march into Western Europe if they wanted to, but the troops would immediately defect when they saw the contrast in lifestyles.

At this point it would be superfluous for me to ask you about Edward Teller.

No, you need not ask me about him. I have had two opportunities to debate with him, and everything they say is right. It is a hopeless task. We had a debate for National Public Radio (NPR) and it was never aired.

There was this peculiar little man, also Hungarian, I forget his name, who hosted an NPR program of discussions with famous people. One of his segments was with Teller and me, just the two of us. I'm sure I lost. I managed to get him angry though, essentially by saying, "What a wonderful scientific career you could have had if you hadn't wasted all that time on the hydrogen bomb."

You don't consider that contribution positive either?

Has it ever been used?

Isn't that better?

It's wonderful.

At this point I can't help asking you about the atomic bombs of 1945.

I'm not one of those who feel guilt about dropping them on Japan. The one thing that emotionally influences me is that I knew about something which most Americans don't, because, having been there, I knew about the fire bombing from my Japanese friends. The fire bombing of Tokyo

Anderson with Arno Penzias and Robert Wilson at the time of their Nobel Prize, 1978 (courtesy of P. Anderson).

was so close to genocide, killed so many people, that it seemed to me much more of a horror than the atomic bombs. Another thing I was conscious of, and I don't know why so few Americans are conscious of it, is Nanking. Nanking and the Japanese behavior in China and Korea was a horrible thing, unbelievably savage. I don't think I have any complaint whatsoever about the atomic bombs. And I'm not sympathetic to the Germans about Dresden. The old saying is absolutely right, "He that soweth the wind shall reap the whirlwind." That's what both the Germans and the Japanese did. The bombs left them with no illusions about being defeated.

Last year's physics Nobel Prize was awarded for discoveries in a field that is close to yours. There seems to be great variability in the immediately apparent significance among different prize-winning discoveries. Physics prizes are often awarded for very refined and esoteric developments whereas more practical applications seem to be expected of the chemistry discoveries awarded by the Nobel Prize.

It's not always so, take the transistor, the electron microscope, or the scanning tunneling microscope. I do have to defend last year's prize because I was probably the first who nominated the group that was awarded the physics prize last year and kept on nominating them for many years. Their discovery is a very fundamental development and where it eventually leads to we are not going to know for quite a while.[1] The key concept is the fractionalization of quantum numbers. The quantum numbers of a macroscopic system composed of electrons are not necessarily the quantum numbers of the electrons themselves. You can take those quantum numbers of electrons and divide them into pieces, and you can have things which behave like independent excitations but are nonetheless one-third of an electron. That is a very fundamental change, equal in intellectual mystery to how much does it matter whether the world is made of quarks; a quark is one-third of a nucleon. It is fundamentally important too, although one is not sure where the importance is going to end up. I think that high-T_c superconductivity is also related to fractionalization of quantum numbers.

As for the chemistry prizes, it seems to me that they often reward technical virtuosity as opposed to the underlying understanding. For example, some years ago a German team was awarded the chemistry Nobel Prize for the X-ray diffraction determination of the structure of the center of photosynthesis.[2] It was a tour de force, very hard work, but it does not tell you anything about the mechanism that was worked out a number of years before by a very great physicist, George Feher, a Hungarian Jew who fought in the 1948 War of Independence of Israel. He was a graduate of Bell Labs, of course. He had spent a great part of his long career working out the essential structure of those things. He also worked on NMR and EPR. You will find in his papers drawings of the structure of the photosynthesis center, not with each atom in place, but with all the essentials accounted for. The chemists did not believe it until they saw the structure, atom by atom, from X-ray diffraction.

It is of interest that Walter Kohn had often been nominated for the physics Nobel Prize. When I inquired with my colleagues, the usual response was, "Sure, Walter, but is the density functional theory the greatest thing he ever did?" He is very well known among physicists.

You have set up a "central dogma" for high-temperature superconductivity. Would you please summarize it for us?

It was a complex matter, and I was putting myself at considerable risk. It consisted of five or six parts. At least two dozen different substances showed high-temperature superconductivity. They have many different transition temperatures, many different chemical combinations, and even the most fundamental features of their structure may be slightly different. It is more complicated than biology. After all, you only have one DNA and one RNA, and only one code, whereas only of the cuprates we have 30 different ones. You might call though the copper oxide plane our DNA.

I was trying to abstract, from this generality, links that are fundamental to the phenomenon. I was also trying to tell which parts of the structure were important for the phenomenon and which parts were not. I also noticed that the forces that are causing anti-ferromagnetism are also the forces that are causing the superconductivity. In other words, the electrons that are important for superconductivity are only the ones that are in the copper oxide plane. Once you notice that, then you can concentrate on how the process takes place. Another important simplification, still not accepted by a lot of other people, is the idea that there is only one band of those electrons. Others argue that everything is more accurately described by invoking three bands.

Can't this be confronted with experiment?

It has been, but that does not mean that the theorist cannot go on and ignore the experiment. In chemistry, you can also consider all the electrons or be sensible and consider only the valence electrons. A physicist would replace all the core electrons by a pseudopotential, but not all chemists like to do that. The "central dogma" simplifies things.

Quite a while ago, you predicted that less than perfect order would be increasingly important in science in the coming decades. Would you have any comment on quasicrystals?

They are wonderful. Alan Mackay had predicted them and then Dan Shechtman observed them experimentally. Paul Steinhardt had done some modeling before the discovery, but he just put it into his drawer and published it after Shechtman's discovery.

Is this physics or chemistry?

A little bit of each, but mostly physics.

There have been publications, including successful books, linking science and religion. There is also a suggestion that the need for quantum mechanics in describing nature is a manifestation of divine reality. Of course, I do not intend to offend you if you are religious.

I am very much not; in fact, I am an atheist – and I disapprove of the attempt to insert religious meaning into science. In a way, you do offend me a little bit by bringing up religion. I'm an admirer of Weinberg on such questions. On the other hand, I am eager to discuss why quantum mechanics seems so incomprehensible to many people. The problem is in epistemology and semantics. The question is what to understand. In doing that, you also must bring in your intuition. Your brain is constructed to see the three-dimensional world with temporal ordering and full of objects. So that is what our brain can comprehend directly. But that is not necessarily the only possible structure that reality can take. That is the only structure that you can comprehend intuitively. Quantum mechanics does not deal with objects. It deals with fields. It is still three-dimensional, thank God, but it might have also had 33 dimensions. The minute you realize that distinguishable objects may not be the necessary way to describe how the world works, quantum mechanics is a perfectly deterministic theory of the things it deals with.

Why is there not more open discussion of this?

Anderson lecturing with J. Robert Schrieffer (b. 1931, Nobel Prize in Physics 1972) (not) listening (courtesy of P. Anderson).

Partly because even most physicists do not understand quantum mechanics. Broken symmetry is an example where quantum mechanics and everyday reality part company. Quantum mechanics tells us that all eigenstates of systems with a symmetry must be classifiable by that symmetry. But everyday objects, like a pencil, do not represent quantum-mechanical eigenstates. No macroscopic object can be a quantum-mechanical eigenstate, because macroscopic objects have definite orientation and position in space and they are a complicated mixture of eigenstates. But they are not moving, they are not time-dependent, and thus those eigenstates are enormously degenerate, and that is the phenomenon of broken symmetry. What is really complicated is not the behavior of the microscopic object, say, an electron undergoing interference, but the quantum mechanical description of the macroscopic apparatus with which you measure it.

What would be a good example of nonbroken symmetry?

I wonder, what would be a good example?

A dimensionless point?

Yes, that is a good example.

Thus even the symmetry of an ideal crystal is broken symmetry. Is this a point where physics and chemistry diverge?

That is so. In chemistry, you draw structures and think of them as immutable. Of course, that description already takes advantage of the Born–Oppenheimer approximation.

What is your main interest today?

High-temperature superconductivity. I wrote a book about it.[3] Although the title says it is *the* theory, it is only part of it. I had hoped that I could do that and then go on to something else. However, when I performed the crucial test for the theory, it did not always work so I had to go back and continue the work.

Would you care to say something about your family background?

I was brought up in Urbana, Illinois. My father was a plant pathologist. My mother was from Indiana, and she was determined that I would be born in Indiana, and that is what happened, and she took me home after five days. In Urbana, I went to this remarkable high school called University High or Uni High. It has three Nobel Prize-winners. The other two are Jim Tobin, the economist, and Hamilton Smith, the molecular biologist.

My mother died in 1958 when she was 70. I have a sister. She got a Ph.D. in biochemistry, but within a month she had twins and never went back to biochemistry. She married a biochemist, who had a distinguished career at Smith Kline & French. She later got a library degree and became interested in the history of technology: she had a career in both fields.

My wife is "just" a housewife. She was an English major. In our time, it was not unusual that she did not have a job. She was active in political initiatives for the Democratic Party. We have an artist daughter who supports herself by secretarial work.

What did you do with your Nobel Prize money?

It was the smallest amount in history, relative to the cost of living. The Swedish economy was in depression and the finances of the Nobel Foundation had been mishandled, so the Prize was down to well under $100,000, and I got a third of it. We just built this money into our house. My medal is in a safe deposit box.

What is the most important benefit from your close relationship with Bell Labs?

As I said earlier, my "university" was Bell Labs, the place where I learned to do science and did most of my most important work. Now it plays a much less important part in my life.

We have many excellent graduate students, and we have sent some of them to Bell Labs to do their thesis experiments. Since Dan Tsui came to Princeton, the Electrical Engineering Department has also developed a close relationship with Bell Labs. However, during most of my career, Princeton was focused on particle physics, high-energy physics, and Bell Labs was not. Communications have intensified during the past years. Now we have joint appointments and joint seminars. I also have connections with chemists and chemical engineers at the University in the framework of the Princeton Materials Institute.

Any predictions for the next century? Any message?

My wishful thinking is that physics will spread out more toward complexity, geophysics, cosmology, and astrophysics, and most of all, biology.

I recently read a book, Ed Wilson's *Consilience*, which impressed me not so much by teaching me new things but because it resonated with my thinking. The fundamental concept is that emergence is the way the world works. It is reasonable to study by analysis, to study smaller and smaller pieces of matter, to try to analyze how biology works in terms of the molecules. But there is then another way of looking and thinking of how the complex world arises out of the simpler world. The title of an article I wrote a long time ago is "More is Different", and it is about how when you put things together you create more than just their sum. You can follow this through from atoms to molecules and to solids.

Another important thing is that it is all tied together. Ed Wilson calls it consilience, I use the term "seamless web of science", which contains all the different fields fused together by intellectual contacts. The fascinating thing is making the connections. Steve Weinberg said recently that science is not just an evolutionary tree; the branches grow back together, they interconnect. If this is true, then physics is not dead because physics is the one science that mixes in with almost everything.

References and Notes

1. In 1998, the Nobel Prize in Physics was awarded to Robert B. Laughlin, Horst L. Störmer, and Daniel C. Tsui "for their discovery of a new form of quantum fluid with fractionally charged excitation". There is an interview with Daniel Tsui elsewhere in this volume.
2. Johann Deisenhofer, Robert Huber, and Hartmut Michel shared the 1988 Nobel Prize in Chemistry "for the determination of the three-dimensional structure of a photosynthetic reaction centre". Interviews with them have appeared in *Candid Science III: More Conversations with Famous Chemists*, Imperial College Press, London, 2003.
3. Anderson, P. W. *The Theory of Superconductivity in High-T_c Cuprates*. Princeton University Press, Princeton, New Jersey, 1997.

Zhores I. Alferov, 2001 (photograph by M. Hargittai).

30

ZHORES I. ALFEROV

Z hores I. Alferov (b. 1930 in Vitebsk, Byelorussia [Belarus], then the Soviet Union) is Director of the Ioffe Physico-Technical Institute of the Russian Academy of Sciences, St. Petersburg and Vice-President of the Russian Academy of Sciences. He was co-recipient of the Nobel Prize in Physics for 2000, sharing half of the prize with Herbert Kroemer of the University of California at Santa Barbara "for developing semiconductor heterostructures used in high-speed- and opto-electronics". The other half of that Nobel Prize went to Jack S. Kilby of Texas Instruments "for his part in the invention of the integrated circuit".

Zhores Alferov graduated from the V. I. Ulyanov (Lenin) Electrotechnical Institute in Leningrad in 1952 after which he joined the Physico-Technical Institute (today Ioffe Institute) in Leningrad, where he earned his scientific degrees. He has been director of this institute since 1987. He is a Member of the Soviet (later, Russian) Academy of Sciences (1972), Foreign Associate of the National Academy of Sciences of the U.S.A. (1990), and has received many other honors and awards. We recorded our conversation in Stockholm during the Nobel Prize Centennial in December 2001.*

You contributed decisively to the development of communication technologies at the time when East and West were isolated from each other and most of the breakthroughs happened in the West, yet you lived and worked in the East.

---

*Magdolna Hargittai conducted the interview.

It may well be that we overestimate this isolation. The isolation was strong indeed right after the war and through the mid-1950s. However, I did my major work in the 1960s and the isolation at least in scientific work was no longer waterproof.

International exchange and cooperation has existed in our Institute from the very beginning. The Institute was founded by Academician Abram F. Ioffe in 1918, right after the Revolution. In 1921, he made his first trip abroad in the Soviet time. He received money for this trip directly from Lenin, to buy equipment and literature. When the building of the Institute opened in 1923, it was equipped as the best European laboratories at that time. The 1920s and 1930s were a golden age for physics in general and it was also a golden age for physics in our Institute. All the associates were young. All the physicists, who later became famous in the Soviet Union, came from Ioffe's School. Nobel Prize-winners Landau, Kapitsa, and Semenov came from his Institute. Semenov was his deputy for administration. The leaders of our nuclear weapons program came from this school, Kurchatov, Aleksandrov, Khariton, Zeldovich. In our country, the highest award used to be Hero of Socialist Labor. There were only six scientists who got this award three times for the development of nuclear weapons, and five of them were from the Ioffe Institute. In the mid-1930s it was broken, but it was rejuvenated in the 1950s. The traditions for international cooperation were very strong. Even under the Cold War conditions we had good relationships between American and Soviet physicists. The governments on both sides created a lot of barriers, but we frequently succeeded in overcoming them. In the 1960s, the conditions vastly improved. The majority of our journals appeared in English translation, with about a six-month delay. We participated in international conferences and we invited foreign scientists to our conferences in the Soviet Union.

Wasn't much of the so-called sensitive work classified? The Sputnik, for example, came as a surprise outside the Soviet Union.

There was some secrecy; there was definite secrecy in our work on nuclear weapons and on satellites, and on missiles that would carry the nuclear weapons. But in many institutes of the Academy of Sciences, from the end of the 1950s, beginning of the 1960s, this classified work gradually reduced. If you look at the Russian Nobel Prize-winners, the majority were physicists, and all of them came from the Lebedev Institute in Moscow and the Ioffe Institute in Leningrad, and, of course, from the Institute of Physical Problems where Kapitsa used to work. This did not happen by

William Shockley (1910–1989, Nobel Prize in Physics 1956) and Abram F. Ioffe (1880–1960) in Prague, 1960, at a conference (courtesy of Zh. Alferov).

accident, because these institutes were complex physical institutions, which carried out research in all modern branches of physics, from the beginning. Before the war, they operated openly and had excellent international cooperation. Many of their scientists participated in developing the nuclear weapons program. Some of them left the Lebedev Institute and the Ioffe Institute and became the leaders of the new secret, classified center for the nuclear weapons program. In the Ioffe Institute and in the Lebedev Institute, classified research was a minor component of the work from the end of the 1950s. When I came to the Ioffe Institute, and carried out my work on semiconductor technology, it was classified. Then from 1955–1956, I sometimes participated in classified research for some applications of my former results, but I carried out practically all my research in an open way and was allowed to publish my results. I started participating in international conferences in 1960, and I was not alone, there were plenty of young scientists at the semiconductor physics conference in Prague in 1960. The Soviet delegation consisted of 70 people with the majority being young, around 30 years of age.

As I understand, the idea of semiconductor heterostructures came to you and to your co-laureate Herbert Kroemer at about the same time. Was it in the air, so to speak?

It has happened frequently in the history of physics that an important event occurred independently and simultaneously in different places. This happened at the end of 1962. The group of Robert Hall at General Electric in Schenectady, Nick Holonyak at General Electric in Syracuse, Marshall Nathon at IBM, Ben Lax at MIT, and only with a couple of weeks delay, Basov in Moscow, published the first papers on p-n junction homostructure lasers. The p-n junction laser suddenly became the center of attention.

I was young and I did not know how the laser worked but I listened attentively to an important talk by Oleg Krokhin, who, together with Basov and Yurii Popov, had proposed the p-n junction lasers. Just before that time I had finished my research for a high-power silicon and germanium rectifier. I came to the conclusion that there was some misunderstanding in the previous research about p-n-n^+ structures. I understood that the main recombination process is carried out in the high-doped parts of the homostructures. When this paper was published, I and a young theoretician of our Institute, Kazarinov, who had explained to me how the laser worked, were looking for our next research project and started thinking about heterostructures in general. It came to my mind immediately that the main disadvantage of the p-n junction laser could be overcome by using a double heterostructure. We set out to write up an author's certificate patent; at the beginning it was classified, but later I published some papers about it. Kroemer considered the same situation at the same time. It was more or less by accident that we both came to this field but it was not by accident that we both came to the same conclusion. Kroemer had already done much work in the field and I had studied all the available literature about heterostructures.

Although it started as basic research you patented the results.

And so did Kroemer. It was a proposal for a new device. Looking back, I think it was a mistake to patent it because at the time nothing happened with it and it was immediately classified. Ours was really not a patent but something else, called author's certificate. In reality we never patented our discovery.

When I decided to publish some details of the work on heterostructures, I called the device a rectifier rather than a laser because at that time in our country some organization supervising our publishing activities decided that everything in connection with lasers must be classified.

So, you did not earn money from your invention?

Later, we obtained some further author's certificates about some applications of heterostructures that were then implemented in our industry. These certificates have earned some money for us.

This was limited to Russian industry.

That's right. At that time it would have been very complicated to get an international patent. We should have obtained first of all permission from the Russian side, and not only permission, but hard currency as well to pay for patents. Thus, such decisions could not be made at the Institute's level but had to be made at the level of government administration.

Don't you regret in hindsight that you could not patent?

It did not bring money to Kroemer either. Even in the United States, it is not the inventor who gets the income. The right of patent belongs to the company rather than the inventor. In our case all rights belonged to the State and if there was industrial production and you were willing to do a lot of paperwork, you could get some money. What we did usually was that we invited some people from industry as co-authors of our author's certificate, and they did all the paperwork. I have about 60 or 70 author's certificates and some of them later became international patents. They were organized by state and industrial enterprises and not by me. In some cases, in order to publish an article, we had to obtain the right to publish it because there were rules to follow. If an article contained material that could be considered as patent material, then first we had to apply for a patent (that is, author's certificate). Thus, in some cases we did apply for the author's certificate just to make it possible for us to publish our papers. In the Ioffe Institute and in the Russian Academy of Sciences just as most scientists elsewhere, we did not care so much for patents and author's certificates; we cared about publishing our papers and we cared about being the first who did so.

I would like to ask you to summarize the science behind your heterostructures.

It might be best to quote from my Nobel lecture, which was delivered here in Stockholm just one year ago. In fact, that lecture was only slightly different from the lecture I gave, again here, in Sweden, at a Nobel Symposium in 1996. I spent a whole month on writing my Nobel lecture and it is very appropriate both in its details and in general as well. It gives you the real story of heterostructures.

The idea of using heterojunctions in semiconductor electronics was put forward already at the very dawn of electronics. In the first patent concerned with p-n junction transistors, W. Shockley proposed a wide-gap emitter to obtain unidirectional injection. A. I. Gubanov at our Institute first theoretically analyzed volt-current characteristics of isotype and anisotype heterojunctions,

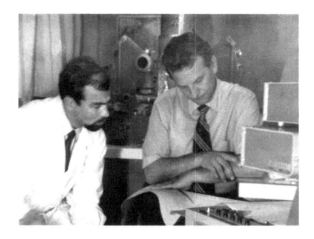

Zhores Alferov (on the right) and colleague in the laboratory, probably in the late 1960s (courtesy of Zh. Alferov).

but the important theoretical considerations at this early stage of heterostructure research have been done by H. Kroemer, who introduced the concept of quasi-electric and quasi-magnetic fields in a graded heterojunction and made an assumption that heterojunctions might exhibit extremely high injection efficiencies in comparison with homojunctions. In the same period there were various suggestions about applying heterostructures in semiconductor solar cells.

The proposal of p-n junction semiconductor lasers, the experimental observation of effective radiative recombination in GaAs p-n structure with a possible stimulated emission and the creation of p-n junction lasers and LEDs (light-emitting diodes) were the seeds from which semiconductor opto-electronics started to grow. However, lasers were not efficient because of high optical and electrical losses. The threshold currents were very high and low temperature was necessary for lasing. The efficiency of LEDs was very low as well due to high internal losses.

An important step was made immediately after the creation of p-n junction lasers when the concept of the double heterostructure (DHS) laser was formulated independently by us and Kroemer. In his article, Kroemer proposed to use the double heterostructures for carriers confinement in the active region. He proposed that "laser action should be obtainable in many of the indirect gap semiconductors and improved in the direct gap ones, if it is possible to supply them with a pair of heterojunctions injectors".

In our patent we also outlined the possibility to achieve high density of injected carriers and inverse population by "double" injection. We specially pointed out that homojunction lasers "do not provide CW (continuous wave) at elevated temperatures" and as an additional advantage of DHS lasers we considered the possibility "to enlarge the emitting surface and to use new materials in various regions of the spectrum".

Initially the theoretical progress was much faster than the experimental realization. In 1966 we predicted that the density of injected carriers could by several orders of magnitude exceed the carrier density in the wide-gap emitter ("superjunction" effect). In the same year I submitted a paper to a new Soviet journal *Fizika i Tekhnika Poluprovodnikov* (*Sov. Phys. Semiconductors*), in which I summarized our understanding of the main advantages of the DHS for different devices, especially for lasers and high power rectifiers: "The recombination, light emitting, and population inversion zones coincide and are concentrated in the middle layer. Due to potential barriers at the boundaries of semiconductors having forbidden bands of different width, the through currents of electrons and holes are completely absent, even under strong forward voltages, and there is no recombination in the emitters (in contrast to p-i-n, p-n-n^+, n-p-p^+ homostructures, in which the recombination plays the dominant role) ... Because of a considerable difference between the permittivities, the light is completely concentrated in the middle layer, which acts as a high-grade wave guide, and thus there are no light losses in the passive regions (emitters)."

The most important peculiarities of semiconductor heterostructures that we already underlined at that time were: superinjection of carriers, optical confinement, and electron confinement.

The fact that we realized the wide-gap window effect was very important for photodetectors, solar cells and LED applications. It permitted to broaden considerably and to control precisely the spectral region for solar cells and photodetectors and to improve drastically the efficiency for LEDs. All we needed then was to find heterostructures where these phenomena could be realized. However, first we had to overcome a certain psychological barrier. AlAs had been synthesized long ago, but many of its properties have not been studied yet, since AlAs was known to be chemically unstable and to decompose in moist air. Thus it did not seem very promising that we could prepare stable and adequate material.

Initially, our attempts to create DHS were related to a lattice-mismatched GaAsP system and we succeeded in fabricating by VPE (vapor phase epitaxy) the first DHS lasers in this system. However, due to lattice mismatch the lasing occurs only at liquid nitrogen temperature.

By the end of 1966, we came to a conclusion that even a small lattice mismatch in heterostructures $GaP_{0.15}As_{0.85}$–$GaAs$ prevents the realization of the potential advantages of DHS. At that time my co-worker D. N. Tret'yakov told me that another colleague of ours, Dr. A. S. Bortshevsky, put in his desk drawer small crystals of $Al_xGa_{1-x}As$ solid solutions of different compositions, which had been prepared two years ago by cooling from a melt, and nothing happened to them. It immediately became clear that $Al_xGa_{1-x}As$ solid solutions turned out to be chemically stable and suitable for preparation of durable heterostructures and devices. Studies of phase diagrams and the growth kinetics in this system and development of the LPE (liquid phase epitaxy) method, especially for heterostructure growth, soon resulted in fabricating the first lattice-matched AlGaAs heterostructures. When we published the first paper on this subject, we were lucky to be the first to find out a unique, practically an ideal lattice-matched system for GaAs. But as it frequently happens, simultaneously and independently the same results were achieved by H. Rupprecht and J. Woodall at the T. Watson IBM Research Center.

After that the progress in the semiconductor heterostructure area was very rapid. First of all, we proved experimentally the unique injection properties of the wide-gap emitters and the superinjection effect, the stimulated emission in AlGaAs DHS, and established the band-diagram of $Al_xGa_{1-x}As-GaAs_x$ heterojunction. We also carefully studied luminescence properties, the diffusion of carriers in a graded heterostructure, and the very interesting peculiarities of the current flow through the heterojunction. These are similar, for instance, to diagonal tunneling-recombination transitions directly between holes of the narrow-band and electrons of the wide-band heterojunction components.

At the same time, we created many important devices in which we realized the main advantages of the heterostructure concepts, which are: low threshold at room temperature DHS lasers; high effective SHS (single heterostructure) and DHS LED; heterostructure solar cells; heterostructure bipolar transistors; heterostructure p-n-p-n switching devices.

One of the first successful applications in industrial scale production in our country was the use of heterostructure solar cells in space research. We transferred our technology to the "Quant" company and since 1974, GaAlAs solar cells have been installed on many of our sputniks. Our space station "Mir" has been using them for 15 years.

Other laboratories achieved many of these results only a year or two later. However, in 1970, the international competition became very strong. Later on one of our main competitors, Izuo Hayashi, who was working together with Morton Panish at the Bell Telephone Lab in Murray Hill, wrote: "In

September 1969, Zhores Alferov of the Ioffe Institute in Leningrad visited our laboratory. We realized he was already getting a $J_{th}^{(300)}$ of 4.3 kA/cm^2 with a DH. We had not realized that the competition was so close and redoubled our efforts ... Room temperature CW operation was reported in May 1970 ..." In our paper, published in 1970, the CW lasing was realized in stripe-geometry lasers formed by photolithography and mounted on copper plates covered by silver. The lowest J_{th} density at 300 K was 940 A/cm^2 for broad area lasers and 2.7 kA/cm^2 for stripe lasers. Independently, CW operation in DHS lasers was reported by Hayashi and Panish (for broad area lasers with diamond heat sinks) in a paper that was submitted only one month after our work. Achievement of CW at room temperature produced an explosion of interest in the physics and technology of semiconductor heterostructures. While in 1969 AlGaAs heterostructures were studied only in a few laboratories, mostly in the U.S.S.R. and the U.S.A. (A.F. Ioffe Institute, "Polyus" and "Quant" — industrial laboratories, where we transferred our technology for applications in the U.S.S.R.; Bell Telephone, D. Sarnoff RCA Research Center, and T. Watson IBM Research Center in the U.S.A.), in the beginning of 1971, many universities, industrial laboratories in the U.S.A., the U.S.S.R., the United Kingdom, Japan, and even in Brazil and Poland started investigations of III — V heterostructures and heterostructure devices.

You must be an extremely busy person. You have many positions.

Too many.

You have held positions such as Dean of the University, Director of the Ioffe Institute, President of the St. Petersburg Scientific Center of the Russian Academy of Sciences, Vice-President of the Russian Academy of Sciences, and I'm sure I'm not aware of all your positions. How can you manage?

It's a very difficult situation. I am considering leaving some of my positions. But you must understand the following. My main position is being the Director of the Ioffe Institute. I do it free of charge, I am not getting any salary for this position. I am Vice-President of the Russian Academy of Sciences and have other positions at the Academy and you can get only one salary from the Academy and you get the salary for the highest position. As Vice-President, I am responsible only for the St. Petersburg Branch of the Academy of Sciences. This is a special responsibility. The

St. Petersburg Branch is responsible for interdisciplinary research, because all institutes belong to specialized divisions of the Academy. Nevertheless, it is too much for me and it takes a lot of my time. Right now, however, our science is in such a difficult position that my authority comes in useful and as long as my authority can help, I feel that I have to carry on.

What can you do?

First of all, when the State Duma [the Russian parliament] considers our budget, and I am a deputy in the State Duma, ...

Yet another position.

Exactly. But I am not involved in any political activities, I only participate in the work of the Committee for Science and Education of the Duma and in its subcommittee on science. I'm most active when the budget is being considered. My aim is to have the share of science in the budget increased. My Nobel Prize last year helped to increase the share of science in the budget of the Russian Federation by 10 percent. Unfortunately, this year even my Nobel Prize did not help. I have to use every possibility to help science. I consider science very important and we have great traditions of science in Russia, especially in basic science, but also in applied science. We have had excellent scientific schools, and Russian science has always distinguished itself by having such scientific schools. The situation is very difficult today. Many of our young scientists are being brain drained, some old ones as well. The budgets of leading research institutions, like the Ioffe Institute in St. Petersburg, the Lebedev Institute in Moscow, and others have decreased manifold since 1992. The international scientific community has helped us a great deal. Our nominal budget at the Ioffe Institute after 1992 has increased three times, but corrected for inflation,

Zhores Alferov after having received a State Medal in 1959 (courtesy of Zh. Alferov).

it is seven or eight times less than it was before 1992. This is why I consider my public involvement very important at the present time. It is even more important than doing research. Many good scientists are doing excellent research and I have my own laboratory where young scientists carry out research. On Fridays I discuss their results with them. However, my job outside the Institute is very important in saving Russian science. The people in power don't always understand that our future development and well-being depends on science.

Do you participate in charting new research in your Institute?

I do, but it is much more complicated than it was before, because at this time the sources of our budget are so diverse. They include international organizations and various foundations. As director, I cannot use financial incentives because I have no money, so I use only my scientific authority in trying to make an impact on future research of the Institute.

You have also been involved in teaching.

This is, again, an old tradition of the Ioffe Institute. Academician Ioffe was not only a great physicist, but he was a genius in understanding the strategic task of physics in science in general. He founded our Institute in 1918. Then, in 1919, he founded the Physico-Mechanical Department of the Polytechnical Institute for educating young people. He was a pioneer of educating physicists with an understanding of applied problems and engineers with a good background in physics and mathematics. I have paid a lot of attention to the question of education especially as I have observed the diminishing prestige of scientific research, beginning in the 1970s. The students of the Physico-Mechanical Department start working in the laboratories of the Ioffe Institute from their third year of education. Many of the graduates of the Physico-Mechanical Department of the Polytechnical Institute become associates of our Institute. All leading scientists of the Ioffe Institute were professors of the Physico-Mechanical Department. The two institutions are geographically also very close, just on the two sides of the same street.

In 1955, Nikita Khrushchev found out that many scientists had two salaries, one from the research institutes and the other from the universities. He declared this incompatible with socialist ideals. He was wrong, of course, because as we learned, the main ideal of socialism was "according to your effort, not your skill or status" (word-for-word translation is as follows:

from everybody by his merits, to everybody by his labor). The result was that it was forbidden for a scientist to be a professor at the same time. In our Institute, the then director, Academician Konstantinov and two professors, Nasledov and Grinberg, decided to remain heads of chairs at the Polytechnic, without salary, just in order to keep the ties. The majority, however, left. These ties, which were very important, were broken.

In 1965, after Brezhnev had come to power, the rules against double appointments were cancelled. Sadly, it is easy to destroy something and it is very difficult to rebuild it. By then all positions were occupied and it was difficult to re-establish positions for the associates of the Ioffe Institute. For instance, I decided to create a new chair for opto-electronics because I developed this area. When I first approached the Polytechnic, they told me that they did not need me although I was already a corresponding member of the Soviet Academy of Sciences, a Lenin-prize-winner, and a recipient of the Franklin Medal. It was the same story at the Leningrad University.

It was only the Electrotechnical Institute, which I graduated from and whose Rector was my friend, that was willing to establish this chair. To me this was very important because I believed that science and education were indivisible. Many of our leading scientists have graduated from my chair. When I became the Director of our Institute, I founded a new faculty at the Polytechnic, the Physico-Technical Faculty. I am the Dean of this Faculty, but it is largely symbolic. Then, in 1987 we founded a physico-technical high school at the Ioffe Institute. This high school does not belong to the Ministry of Education, it belongs to the Academy of Sciences.

It was our idea that pupils of the high school, college students, associates of the Ioffe Institute, and academicians have to live and to work under the same roof. We have joint seminars, various lectures, sporting facilities, and so on. We initiated a new complex of buildings and called it Scientific-Educational Center. I managed to obtain a piece of land near the Institute for the construction. In 1992, the Academy gave us the money for the construction, but the ensuing inflation made this money worthless. We did everything to raise the money and it came from the most diverse sources. We even succeeded getting eight million dollars from Chernomyrdin, the former Prime Minister of Russia, but the money was not government money, it came from his foundation. We ourselves earned four million dollars from our own activities and we spent a lot of time to cut across the red tape to be able to spend this money not for salaries or equipment or other things but for the construction. Primakov gave us additional funds when he was Prime Minister and Putin also gave us some money when his position made it possible for him. This is how we were able to complete the project in 1999. The Center is now in operation.

Just after my Nobel Prize was announced, Putin invited me to the Kremlin. I brought a letter asking him for an additional funding for the construction of a new building with new laboratories. He signed our request because just one day after the announcement of my Nobel Prize it was impossible to refuse our request. Construction has started and I hope that in about two years we will be able to open a new building with modern laboratories for school pupils, university students, and scientific researchers to work in them side by side.

Twice per month we organize public lectures. Leading scientists of the country, and not only of the country, deliver lectures for our students and associates together. For example, this coming Friday, December 14 [2001], Professor Gordeev will deliver a lecture about the centennial of the Nobel Prize and about Roentgen's discovery.

In 1972, you and some of your colleagues received the Lenin Prize. But two of your colleagues were left out.

For me it was a tragedy. There was a strict rule that no more than six persons could be included. Portnoi himself volunteered to be left out. Kazarinov, on the other hand, was excluded because industry pushed very hard its own candidates. Our lasers were just being introduced into production and they had a strong say in the committee that decided on these matters.

You were born in Vitebsk, which is well known for its avant-garde art.

Six months after my birth my father was appointed to a new job in the Archangelsk Region, and we moved away. We returned to Byelorussia after the war, in 1945, after having spent 15 years in many different places of the country, wherever my father was sent as a factory director. In 1945, he became the head of the cellulose-paper industry of Byelorussia and I graduated from a school in Minsk, the capital city. I like Byelorussia, it's my homeland.

What turned you to science originally?

When I was a boy, I was interested in many things. I was interested in literature, in chemistry, I carried out plenty of chemical experiments, including explosions, at home. Then I became interested in building radios. In Minsk, our teacher of physics, Yakov Borisovich Meltserson, delivered excellent lectures, and I became hooked on physics. I graduated from the Electro-technical Institute, from the Faculty of Electronics as he recommended to me to go to Leningrad and to study at this institute. It was his influence. During my studies in Leningrad, I came to the conclusion that I made the wrong choice. Unfortunately, it was too much engineering and too little physics, but it would have been too difficult to change something and therefore I stayed. I started, instead, studying on my own, first of all solid-state physics and others. I also initiated doing research during the third year of my studies. An Associate Professor of our Faculty, Natalya Nikolayevna Sozina had studied semiconductors and photodetectors for her Ph.D. dissertation and she invited me at the beginning of my third year to work in her laboratory. So, since then I started to study semiconductors.

I would like to ask you about your wife.

She is beautiful.

She is. What does she do?

She graduated in Voronezh, a provincial town in Russia, from the Philological Faculty of the Pedagogical Institute. At some time she worked as a sociologist in the Scientific Organization of Labor. We met in 1967, on a beach of the Black Sea, in Sochi. At the time when we met she lived in Moscow and was working in Khimki in an organization headed by Academician Glushko who was responsible for designing missile engines. I started flying every Saturday from Leningrad to Moscow, but after a few months, we decided to get married and she moved to Leningrad. In Leningrad, she continued working in the missile industry, but she stopped to work when our first child was born. By then I was already a Doctor of Science with a sufficiently big salary, four hundred, then five hundred Rubles, which was enough.

You got married relatively late.

It was a second marriage for me and a second marriage for her as well. My first marriage lasted two months but I have a daughter from my first marriage and my wife has a daughter from her first marriage too. We are in very good contact with both. Our son has graduated from the Faculty of Electronics and he is currently in business. He works with a Swedish company — selling and repairing small cranes for our forest industry.

Zhores Alferov, John Bardeen (1908–1991, Nobel Prize in Physics 1956 and 1972), Vladimir Tuchkevich (1904–1997, formerly director of the Ioffe Institute), and Nick Holonyak, Jr., at the University of Illinois, Urbana, in 1974 (courtesy of Zh. Alferov).

Is business in competition with science for gifted young people in Russia? It did not use to be the case.

Now, high-tech industry is not sufficiently developed in Russia. We need a well-developed high-tech industry also for it to represent a demand for science. Science must be requested by society and it must be primarily an economic request. We have developed a lot of high-tech industry; the emphasis used to be on military applications. This high-tech industry has been largely destroyed. A rebirth of high-tech industry should attract a lot of young people and that will include business-oriented people as well. Such a high-tech industry will represent a renewed interest in and demand for scientific results.

How do you compare the Soviet system and the present Russia for science?

Today's circumstances are definitely more difficult. The present system is much more bureaucratic than it was before. Undoubtedly, democracy has its merits. You can speak everything. But it is a strongly bureaucratic system and much of the important things have been captured by the people whom we call oligarchy. The budget of the whole country has also diminished due to the well-known so-called shadow economy and because of the thieves in the national economy. Science used to have a more important position and a more important role than it has today. This importance was, to a great extent, for the military, but not only. I like to note jokingly that in the past the relative importance of a person or an organization could be determined by the positioning of the names signing obituaries. First came always the members of the Politburo of the communist party and the president of the Soviet Academy of Sciences followed immediately. Only after that came the names of deputy prime ministers and the rest of the crowd. This is no longer the case. The relative importance of science has diminished.

Was there any important impact on your life apart from science?

My mother played the most important role in my life. The next ones were my father and my brother who was killed on the front during the Second World War, in a battle. He was killed in 1944, in the Ukraine; before that he participated in the battles of Stalingrad and Kursk. He was a young officer of the infantry and a great person for me. Because of him, I have always been interested in the history of World War II. I have read a lot and was always looking for learning more about it.

Recently, when I received the Nobel Prize, I have donated one third of my award money to establishing a foundation for science and education. Other people and organizations have followed my example. There are other issues too, in which I have involved myself. For example, there are serious faults in our tax system and I am fighting for correcting it. This is also connected with our foundation because we want to make it more advantageous to donate money to it. Our foundation has started awarding stipends to school pupils and university students; we have started to provide financial assistance to widows of deceased members of the Science Academy who meet hardship. This is for people in the St. Petersburg area because we don't have enough funds for doing this nationwide. There is then a school in a small village in the Ukraine, where my brother fought against the Germans and where he was killed. I have visited this place repeatedly and we have established a connection between this school and our Physico-Technical Lyceum. Schoolboys from this small Ukrainian village have visited our school in St. Petersburg and we have established stipends for schoolboys in this village. These stipends are named Marx Alferov Stipends after my brother who helped to bring freedom to this place. After I had received the Nobel Prize, my first visit abroad, because the Ukraine is a different country now, was to this village. I am an honorary citizen of this village.

Do you have other heroes?

In science, my hero was always Academician Ioffe, later also Academician Prokhorov, a Nobel laureate, who was my teacher at some point. He was in Moscow and we carried out a set of joint experiments. In literature, I like Lermontov and Mayakovskii. There was then a novel published around 1938, *Two Captains* by Venyamin Kaverin (in Russian). It was my brother's favorite novel, which my mother bought for him as a birthday present before the war. The main hero of this novel, Grigoryev, is also my hero. I recommend that you read this novel.

May I ask you about your relationship to religion?

I am an atheist.

What was the determining event in your life?

It was that I became a member of the Ioffe Institute in 1953.

Your present ambitions?

It would make me happy if one of my pupils would one day get the Nobel Prize.

Has the Nobel Prize changed your life?

I must give a lot of interviews and I have become much busier than before. I have also received more support for my activities. I have told some members of the Royal Swedish Academy of Sciences that they could make my life easier if they would award further Nobel Prizes to Russian scientists and thereby diverting much of the attention to new laureates.

Are there any candidates?

Definitely.

Would you like to mention some names?

I do this sort of thing in a confidential way.

Daniel C. Tsui, 1999 (photograph by I. Hargittai).

31

Daniel C. Tsui

Daniel Chee Tsui (b. 1939 in Henan, China) left his village and family in 1951 and went to Hong Kong where he graduated from high school in 1957. He continued his studies in 1958 in Augustana College in Rock Island, Illinois, and received his B.A. degree there in 1961. Then he was a Ph.D. student in the Physics Department of the University of Chicago and graduated in 1967. For 13 years he worked for Bell Laboratories in Murray Hill before joining Princeton University in 1982. Today he is Professor in the Department of Electrical Engineering at Princeton University. Dr. Tsui was co-recipient of the Nobel Prize in Physics in 1998 together with Robert B. Laughlin (b. 1950) of Stanford University and Horst L. Störmer (b. 1949) of Columbia University, "for their discovery of a new form of quantum fluid with fractionally charged excitations". Also in 1998 he received the Benjamin Franklin Medal in Physics. He has been a member of the National Academy of Sciences of the U.S.A. and the Academia Sinica. We recorded our conversation on March 8, 1999, in Professor Tsui's office at Princeton University.*

In your experiment you were investigating a two-dimensional system of a large number of electrons using extremely powerful magnetic fields at low temperatures. The press release of the Nobel citation says that you gained a new insight into the structure and dynamics of matter. What is this new insight?

---

*István Hargittai conducted the interview.

Those are not my words.

How would you explain what you could learn from your experiment?

First I would like to comment on the artificiality of science being divided into different fields such as physics, chemistry, biology. Phil Anderson used the word of cross-reference and the analogy of how many different crafts go into building a cathedral. You are asking me about our experiment. At that time it was just a singular experiment. We knew that it was something new and something interesting but for its significance, it takes time to gradually appreciate it. It is related to a lot of beautiful ideas and concepts and you see in it also the clear emergence of physical reality. The most recent example has been the better understanding of composite particles. It has also contributed to various field theories. One can literally imagine now how the electron binds to magnetic fluxes, forming a new particle. By forming a new composite particle the statistics also changes, and instead of a fermion it becomes a boson. It also has relevance to the interaction of electrons, which is so important in chemistry.

What is a fractional charge? The Nobel citation says that you discovered this new form of quantum fluid with fractionally charged excitation.

The electron entity has still the same charge. All the electrons move together in this quantum fluid in a correlated manner and the smallest excitation carries a charge that is a third of the electron charge. It does not mean that the individual electrons carry a third of the charge. The excitation creates a new particle in a sense whose charge is a fraction of the electron charge. You have to consider the whole system. You do something to the system and the result of the excitation is the new particles whose movement you characterize as the particle carrying a third of the electron charge. You make this conclusion from the behavior of the new particle.

In your case, you had the experiment first and the theory followed. In case of the discovery of parity violation, for example, it was the reverse order. What did you have in your mind when you designed the experiment?

I would put it in a different way. We knew that we were looking into a regime where we should expect strong correlation kind of physics. So we were looking for this strong correlation. That part we knew. But we did not know much beyond that and we certainly did not even know what specific questions to ask. In chemistry you always deal with a lot

Robert B. Laughlin (b. 1950) and Horst L. Störmer (b. 1949) in Stockholm, 2001, during the Nobel Prize Centennial. They were co-recipients with Daniel Tsui of the Nobel Prize in Physics 1998 (photographs by I. Hargittai).

of electrons but not in physics. Our system was restricted to a two-dimensional electron plane. If you apply a magnetic field perpendicular to the plane of this electron system, it quenches the kinetic energy. With the kinetic energy disappearing, the interaction energy must show up in some way. Specifically, whether it shows up as a solid, for example, something more familiar, or something less familiar, was not clear at all. Then, once the experiment came out, it became obvious to us as well as to others that the experiment brought out something completely unfamiliar to us.

Your original experiment was in 1982. What do you do now?

The original experiment was at Bell Labs. Then I moved to Princeton and continued to be interested in the same problems and have worked mostly on these so-called two-dimensional electrons. This is also my current interest. Of course a two-dimensional system is part of our three-dimensional world. Although we live in a three-dimensional world, we are most familiar with the two-dimensional part of it. Human activities are in reality mostly two-dimensional because of the gravitational field on us. The same is true for the electrons. In most electronic devices, the device functionality is performed by these two-dimensional electrons.

Do you anticipate extension into the third dimension?

The device people are very keen on this, they work on stacking, and are most interested in third-dimensional integration. The usual devices work in a plane and a common goal is to make them smaller and smaller. Another direction is to stack these devices and accomplish their coupling. This has proved difficult so far.

Do you patent anything?

No. I do not have that frame of mind. But I know that patenting is widespread today. It is so overtly overdone and it does not taste right to me. I belong to the old-fashioned way.

Do you ever go back to China?

The first time I went back was in 1979, then another time in about 1985, and I have not been back since. I no longer have family there. My parents passed away. We have two daughters. One is a graduate student at Harvard University in art history. The other is resident doctor in a hospital in Oregon. My younger daughter is fluent in Chinese. But the older one speaks some too. Our older daughter taught English in China for a year, and our younger daughter taught English in Taiwan for close to a year.

You have been a Nobel laureate for a few months only. Is it changing your life?

It is making me do things in which I am not good at all. I have to handle new situations. Princeton has quite a few Nobel laureates in physics although I may be the first one in the Department of Electrical Engineering.

Leon Lederman recently noted that the brilliant Jewish physicists are being replaced by a new generation of brilliant Oriental physicists.

I think so myself. You can also see in this the continuation of a sociological phenomenon. When you look at third generation Oriental Americans, science is no longer so attractive for them as for earlier generations. It is just too demanding. Of course we cannot generalize too broadly. But the general atmosphere is typical. First came the people from Eastern Europe and the one way to satisfy the kids' curiosity, it was science. They did not have all the toys around them and they did not have the Victorian comforts of life, so they turned to science. Then there was also all the tradition of culture and the parents' missed chances too. That was a very strong force. But that was no longer there after the second or third generation. So then the Asians were coming and the same thing happened and they may also disappear. I have no idea of who may be next. However, in any case you can see many Asian students here and it appears that solid-state physics and solid-state chemistry are popular among them.

You went to a Lutheran college. Was the affiliation important to you?

Yes. I grew up in a Lutheran family and it was important to me and still is. Eugene Wigner went to a Lutheran high school in Budapest. But I went to an American Baptist high school in Hong Kong. I then came to the United States and went to Augustana College in Rock Island, Illinois. Originally it was a Swedish community. When I was there people came to study the Swedish language that was spoken in the late 19th century. It is no longer spoken in Sweden but is still spoken in Rock Island, Illinois.

What is your interest outside physics?

I am very interested in music but I have no talent. I have never had time for anything else but physics and have never done anything worth mentioning outside physics.

Antony Hewish, 2000 (photograph by I. Hargittai).

32

ANTONY HEWISH

Antony Hewish (b. 1924 in England) is Professor of Radioastronomy, Emeritus, at the University of Cambridge. He studied at the University of Cambridge and held various positions before becoming a Professor there. He was elected a Fellow of the Royal Society (London) in 1968. Professor Hewish received the Nobel Prize in Physics for 1974 jointly with the late Sir Martin Ryle (1918–1984) "for their pioneering research in radio astrophysics: Ryle for his observations and inventions, in particular of the aperture synthesis technique, and Hewish for his decisive role in the discovery of pulsars". Antony Hewish is married to Marjorie Richards and they have a computer physicist son and a language teacher daughter who both have their own families. We recorded our conversation in Professor Hewish's office at the Cavendish Laboratory on February 1, 2000. My first question was about his beginnings in science and about the Nobel Prize-winning discovery.*

I've always been a scientist. As a child I was taking things to pieces to see how they worked. I am curious about the physical world. There was nothing in the family that would've pushed me in this direction; my father was a bank manager. I was the youngest of three brothers. One became an engineer during the war and then read English literature at Oxford. We were always making things, models, gunpowder, a little physical chemistry came into it, explosives. In school I got on much better with science than

---

*István Hargittai conducted the interview.

with the arts, in particular history because for history you need a good memory. It's not logical. In science you can reason things out, you can get hold of the principle and then you can apply that principle. Physics appealed to me because you don't have to remember a lot of things.

I came up to Cambridge in 1942 when I left my school and had a year doing natural sciences, physics, chemistry, math. But then there was such a shortage of radar people in this country that I was directed to the Royal Aircraft Establishment at Farnborough. That's where I spent the war years. I was involved with airborne radar equipment for bomber command and it was for jamming German radars. I got back to Cambridge in 1946 and finished my physics degrees in two years. I got a first class degree, which enabled me to get a three-year research studentship to do my Ph.D. I joined the late Martin Ryle whom I had met during the war. He was a brilliant scientist, the father of radio-astronomy in this country. He was picking up radiowaves from the sky and I knew about radiowaves and radio technology. So it was an obvious research choice for me. Although I started research in radio-astronomy, I was on the fringe, and this is where the trail towards the Nobel Prize actually began.

At that time we didn't know the source of the radiowaves but I discovered that the upper atmosphere, the ionosphere, was affecting this radiation on its way through. At our wavelengths the ionosphere is like a transparent layer, like glass of varying thickness. The variations cause random diffraction of the radio waves. You could pick this up on the ground by fluctuations of intensity. The fact that you could actually measure something about the ionosphere, using radio waves from outside, appealed to me because this was the first time you could do that.

The ionosphere itself was discovered at the Cavendish by Edward Appleton before the war. Solar radiation ionizes the upper atmosphere where the radiation is absorbed and generates a layer of ions. That acts like a reflecting screen for radio waves of low enough frequency. A lot of research on the structure and height of the ionosphere had been carried out by Appleton and his successors here at the Cavendish using radio waves transmitted from the ground. The ionization reaches maximum in the middle of the layer and then tails away on either side. As you decrease the radio wavelength, the radiation penetrates further into the layer and, eventually, it goes right through. This is because the refractive index of the plasma depends on the radio wavelength.

The ionosphere is at the height of about 300 kilometers. It's a thick layer of ions of very low density, and it reflects radio waves up to two or three MHz. It's very variable, depending on the solar ultraviolet radiation. There was ionospheric research going on in the Cavendish and I was intrigued by the possibility of using radio-astronomy to study the upper layers of the ionosphere, which cannot be reached from the ground. I'm talking now about the 1950s, long before space research. There was no question of in situ measurements then. Today you can fly satellites, you can get there and measure these things directly. In those days there was none of that.

As I started my research, I worked out the necessary theory for interpreting the data. We observe diffraction patterns, and I call them diffraction rather than refraction because that is a more accurate theory. Refraction is geometrical optics rather than a wave theory. I was measuring these diffraction patterns with radio telescopes and this enabled me to measure the density of these clouds up there, their size, and their speed. This was my Ph.D. project and I worked on this on my own, just as a member of the team. Ryle wasn't interested in this very much. I was using the instruments which were available for a non-astronomical purpose. I was using astronomical techniques to make a study of ionized plasmas.

This was a great project for me because when you are a research student you always want something of your own. I invented this field, it was all mine, and I was happy to go on, developing it further. Next I extended it to the Sun's atmosphere. When you observe any radiation from a distant radio source, it must pass the Sun's atmosphere. We discovered that the atmosphere blowing off the Sun produces similar diffraction of radio waves as does the ionosphere. The Sun is surrounded by plasma, its atmosphere is at a temperature of a million degrees, so it's totally ionized. You can see the Sun's atmosphere at the time of total eclipse when the light is scattered by electrons. I discovered that I could detect the solar atmosphere much further away from the Sun using the same methods as I'd developed for the ionosphere. I could apply the same methodology to the solar atmosphere.

The sources of radiation are radio galaxies and quasars. Inside some galaxies there is a black hole at the center, which is rotating. Black holes occur if you have enough mass in a confined volume. It's a prediction of Einstein's general relativity theory. If the gravitational field is strong enough space-time curvature forms an enclosed region and that's a black hole. In galaxies you have a black hole with a mass of, say, a hundred million solar masses and as it rotates energy is emitted along the rotation axis. It's essentially

Antony Hewish receiving the Nobel Prize on December 12, 1974 from Carl Gustav, the King of Sweden. As is seen, he is actually receiving two prizes, the other one for Martin Ryle who was unable to attend (courtesy of A. Hewish).

a jet, a plasma jet, probably an electron/positron plasma, traveling out at roughly the speed of light.

The material falling into a spinning black hole can release an enormous amount of energy which is beamed in two opposite directions. This energy causes shock waves in space far from the radio galaxy, which release energetic electrons, which radiate. The electrons are moving in random magnetic fields; they are accelerated continuously, so their radiation is a kind of random synchrotron radiation, and this is what we pick up. What you get in a typical radio galaxy is a couple of clouds of radio emitting electrons far from the black hole in the middle. The radiation reaches us unobstructed because space is essentially transparent, once the radiation has left the galaxy source.

I developed the apparatus and the technique to measure the Sun's atmosphere and I also measured the speed with which the Sun's atmosphere is escaping. It's an enormous speed, up to one thousand kilometers per second (up to two million miles per hour). I was highly intrigued by measuring the speed of the outflow all around the Sun.

We're now talking about the mid-1960s when spacecrafts were beginning to venture outside the Earth atmosphere and measure the solar wind directly. Because of the energy required you can't launch them in directions that would measure the solar wind coming off the north and south polar areas of the Sun. That means launching satellites out of the plane of the ecliptic, the plane in which the Earth orbits the Sun. This is because you need the speed of the Earth to launch distant space probes.

We discovered that the solar wind was coming off the poles much faster than from the equator, which was a nice result. It was only confirmed directly by in situ measurements in 1992–1993 when they were able to launch a spacecraft with advanced technology so that it could pass over the Sun's poles. They launched a spacecraft out to Jupiter where the gravitational field turned its orbit around so it could come back over the Sun and it's now orbiting the Sun every eleven years and going north and south of the Sun. It has measured the solar wind directly but in those days my indirect method was the only way and it showed up this high-speed wind from the poles.

Eventually I got a little bored with this work because the wind was usually blowing away from the Sun at about the same speed, and you can't measure the same thing forever. The Americans took up my method and so did the Japanese and they have been vigorously pursuing it ever since. However, I wanted to do something else.

I designed an antenna to use the diffraction effect to learn about the radio galaxies themselves. One of the problems in radio-astronomy is getting sharp enough images. It's very hard to get high angular resolution with a radio telescope. In optical astronomy you take a photograph or a CCD-image and you get a nice picture. If you go back to the 1960s, we knew there were hundreds of radio galaxies in existence but we didn't know what they looked like because we couldn't get a picture. Radio telescopes with enough angular resolving power needed to be far too large; you couldn't actually build them and you had to invent other techniques. This is why Martin Ryle got his Nobel Prize for developing interferometric methods. Instead of actually building a huge radio telescope, Ryle showed how you could get the same results by using small dishes connected in pairs on different baselines.

Because we had no images, I decided that I could use what I called interplanetary scintillation. The flickering of intensity caused by the irregular diffraction through the solar wind gave me a method of measuring how big radio galaxies were with an angular resolution, which far exceeded anything that you could achieve by building an instrument.

I designed a radio telescope to make a survey of the sky to detect radio galaxies of smaller angular size because they would be the ones that showed this diffraction effect most. It needed to be very sensitive because I wanted to see a large number of galaxies. It also needed to measure this scintillation effect, which shows up best — because it's related to the plasma in the solar wind — at longer wavelengths, which radio telescopes were not then using. I had to design a special purpose instrument to study this. It had to work at meter wavelength, it had to be a huge area, because I needed high sensitivity, and for my method it was necessary to make repeated measurements on radio galaxies to get their angular size.

The survey was set up to observe several hundred radio galaxies every week. Now, by a very lucky accident the instrument I designed was ideal for detecting a completely unknown phenomenon called the pulsar. It was totally unexpected, unpredicted, and one of those shattering things that science brings up, like the discovery of X-rays. We couldn't have avoided detecting the pulsars. These are sources, which produce regular flashes of radiation about once a second or sometimes much faster. This phenomenon was totally unanticipated. When we started to see these flashes in November 1967 it took about a month of extremely hard work before I decided it was probably neutron stars, but I wasn't quite sure when we published the discovery in *Nature*. It was that discovery, which led me to the Nobel Prize.

The Prize was awarded for what they said was my decisive role in this discovery. What that means was I conceived the project, designed the apparatus and decided what observations were to be made. Other people were actually involved in carrying it out for their Ph.D.s and so on. It turned out to be the most amazing, fruitful discovery.

Did you accomplish you original goal too?

Oh, yes, but that took much longer. I did that as well. We mapped all these galaxies and that involved several research students over the next decade. It was quite a long program but the pulsar just came out immediately; it was beautiful. I couldn't have been a more lucky astrophysicist.

Do you think you would've got the Nobel Prize without the pulsars?

No, I don't think I would've got the Nobel Prize without the pulsars. You've got to do something very special to get the Nobel Prize.

Was Martin Ryle's discovery primarily methodology?

No, not just that. They mentioned "pioneering work in radio astrophysics" in both our cases. Martin Ryle's invention of the aperture synthesis revolutionized the design of radio telescopes. Everywhere today you find his technique employed. But his observations gave the first evidence for the "big bang" Universe, as opposed to the steady state theory, which was fashionable then.

Didn't you have a particular graduate student who made the first observation of pulsars?

Oh, yes, I did. She was my student doing observations, which I had designed. She was a good student and worked very hard, helping to build the radio telescope and then carefully analyzing hundreds of feet of chart recordings.

What is her name and what happened to her?

She is Jocelyn Bell Burnell; Bell is her family name and Burnell is her married name. She has stayed in the field and after a variety of posts, as she had to follow her husband's jobs at the beginning, she has now a chair in physics (astrophysics), at the Open University in Milton Keynes. She has done very well for herself. There are not many women in this country who have a chair in physics.

The 1993 Nobel Prize in Physics was for the discovery of a new type of pulsar that had ramifications for the study of gravitation. What is its relationship to your work?

That was Joe Taylor and Russell Hulse at Princeton. The really important thing was that they discovered one particular pulsar, which was a superb tool for testing some predictions of Einstein's theory of general relativity. Some aspects of the theory involve conditions that are very rarely met. You can't make them in the lab; you have to use astrophysical environments to get gravitational fields strong enough. The background to Joe Taylor's Nobel Prize was the work he did to verify the existence of gravitational radiation. Crudely speaking, a massive object distorts space and time around it. Einstein called this space-time curvature. To explain it in anything but these crude terms means a lot of mathematics. If you take two massive objects and rotate one around the other you're changing the curvature of space-time periodically. According to Einstein's theory, that's going to radiate waves in space-time, ripples in space-time, or call them space-time curvature waves, which convey energy. They propagate outwards and this is a new form of radiation, which has been discussed theoretically for years but no one had had actually demonstrated that it is real.

634   Hargittai & Hargittai, Candid Science IV

Taylor and Hulse discovered a pulsar that was in orbit about another neutron star. It's a very tight orbit going around every 7 hours, which is fast. A pulsar is a very nice time-keeping device. If your source is moving in a circular or elliptical orbit you can immediately tell because of the huge Doppler effect. It was Hulse's work on the Doppler effect, which actually got him the Nobel Prize. He was Taylor's student and it was Taylor's systematic observations of this source over many years, which showed that the orbit was shrinking. The system is losing energy at exactly the rate corresponding to the radiation of gravitational energy. It confirms that gravitational waves are real. Although the waves are not detected directly, the energy loss fits Einstein's theory exactly. That was an accurate test of general relativity, for which Joe Taylor got the Nobel Prize in Physics.

The original pulsar discovery was more than 30 years ago. Was there anything comparable in your career?

No, of course, not. I have had several projects since then. Pulsar radiation is sufficiently coherent that scintillation due to clouds of interstellar plasma is detectable. This can be used to measure the range of sizes of the interstellar clouds and I developed some relevant theory, and then collaborated with an observer at the Effelsberg radio telescope in Germany who had obtained some nice observational data. More recently, with my group here, we developed a method for mapping space weather patterns in the solar wind. For this we used an improved version of my original radio telescope and measured scintillation on about 1000 radio galaxies each day for several years. This showed space "storms" on their way towards the Earth and I think our method could predict the occurrence of conditions, which damage spacecraft and are hazardous to astronauts. I am still working on this now. Of course, following Martin Ryle's retirement I became head of radio-astronomy at the Cavendish and had plenty to do besides my own research and teaching.

I noted the large number of honors you have received over the years, prizes and memberships in various academies but not titles like the knighthood. Did you turn down anything along those lines?

I never turned down anything, I just never had an offer.

May I ask you about religion?

Antony and Mrs. Hewish in Paris in 1958, during a meeting of the International Astronomical Society (courtesy of A. Hewish).

I'm not a particularly religious person but I support the Church because I believe science is not the answer to all our questions. I was brought up as a Christian. I'm rather similar to Jocelyn on this one. We never discussed religion but I've heard her talk on the radio about her faith as a Quaker. She believes that you may need the assumption of God to explain being alive. I was brought up as an Anglican, the Church of England, and I believe this is important. When I ask the big question, I can't answer it without bringing in religion. It's very hard to put into words but we're trying to find out about things all the time, aren't we; we're forming a picture of the world, which, hopefully, will provide some answer to the big question, what's it all about? I just don't believe that our whole experience, as human beings, is explicable as just some cosmic accident and that we're here because of some fluke arrangement of the fundamental physical constants. That seems to me a wild assumption, which doesn't explain at all the deeper feelings one gets. Let me give you an example. Einstein in America once listened to Yehudi Menuhin, the famous violinist, and when Menuhin finished the concert, Einstein stumbled across the stage, he was an old man by then, and said, "Now I believe that God exists." That's the sort of reaction I'm talking about.

Was it a statement about religion?

Surely it was. When you're suddenly struck by something, shattered, when suddenly you feel very tiny in connection with something that's huge, when you're in awe, I regard that as a religious experience. You're sensing the presence of some reality, which has nothing to do with Newton's laws of motion or the laws of biology, it hits you. Many people feel like this when they are looking at the night sky and say, "What a marvelous thing!" That's not a scientific attitude, is it?

Experiences can make you feel that there are deep realities beyond the reach of science, which are mysterious, something more than the material surface. Science doesn't explain how I feel when I'm listening to a Beethoven quartet. I get uplifted. There's no physics in there and I don't think there's any biology either. Richard Dawkins would say it's all evolution and the apes that liked quartets were more successful in their genes than those that didn't, but that doesn't make any sense for me.

Did you ever engage him in a debate?

No, I didn't. He's a nice chap and very fluent but I don't want to debate with people. He's got a rather closed mind on this one. People who think like I do tend to be physicists, not biologists. We face up to bigger mysteries in physics than anybody else. Biologists have difficulties of complexity. How the proteins fold, all that stuff. That's physical chemistry, which one day will have a solution. But there are the mysteries of quantum physics, actions at a distance, non-localization, the fact that a particle here can instantly affect a particle at a large distance according to quantum laws. Why can there be an instantaneous change in the quantum state of one particle when you change the quantum state of another? Einstein worried about this and it caused him to disbelieve some quantum mechanics. But experiments now are confirming all these amazing things. Coherence of the wave functions is how you talk about this in quantum physics. Two particles made at the same quantum interaction, say, an electron and a positron, and you shoot one of them across the universe. You measure the spin of one here and you can say instantly what is the spin over there. This is the so-called non-localization paradox, which Einstein didn't believe, but it has been demonstrated. This is letting you into a world, which is much deeper than what we can comprehend.

It's mysterious, it doesn't make sense, and, to my mind, if you believe it you can surely believe in the existence of God? God is a concept, which I need to cohere my total experience. Christianity comes nearest to the formal expression of this for me. You've got to have something other than just scientific laws. More science is not going to answer all the questions that we ask.

Jocelyn Bell Burnell, 2000 (photograph by M. Hargittai).

33

JOCELYN
BELL BURNELL

J ocelyn Bell Burnell (b. 1943 in Belfast, Northern Ireland) is a Professor and Dean of Science at the University of Bath. She received
her B.Sc. degree at the University of Glasgow and her Ph.D. at the
University of Cambridge (1968). During her Ph.D. studies she discovered
the first pulsars, for which Antony Hewish, her supervisor, received
half of the Nobel Prize in Physics in 1974 but she was not included
in the Prize (the other half went to Sir Martin Ryle for the aperture
synthesis technique). She has held different part-time positions in
astronomy and physics till 1991, when she became professor at The Open
University and Chair of the Physics Department. She has received
numerous awards: the Michelson Medal (1973), the Robert Oppenheimer
Memorial Prize (1978), The Beatrice M. Tinsley Prize (1987), the Hershel
Medal (1989), the Edinburgh Medal (1999) and the Giuseppi Piazzi
Prize (1999). She is honorary doctor of numerous universities. We
recorded our conversation in April 25, 2000 in her office at the Physics
Department of Princeton University where she was a Distinguished Visiting
Professor for a year. That conversation was augmented in 2002.[*]

---

[*]Magdolna Hargittai conducted the interview.

Can we start with your family background?

I am the eldest of four children. My father was an architect, and this was in Northern Ireland, sometimes called as Ulster, which is part of the United Kingdom.

What made you interested in science?

It became quite clear as soon as we started to have science at school that I was good at the physical sciences, less so at the biological sciences. I was in school in the late 1950s, and in 1957 the Soviets launched the Sputnik satellite. Then the Western world suddenly realized that it was falling behind the Soviets in scientific and technical matters and there was a tremendous push for people to do science. I was caught up in this excitement and was very much encouraged to do science. Then it was the question of what sort of science. My father had a tremendous general knowledge and was very widely read. He brought many different books home from the public library and I would scan through these. One day he brought some astronomy books back and I did not just scan them; I picked them up and took them to my bedroom to read. I was hooked by the scale and grandeur and the excitement that was there in astronomy even in the late fifties, early sixties. I realized that the physics that I was learning in school was a tool that could be applied to help us to understand the cosmos.

Were your parents supportive in your pursuing a career in science?

Yes, they were. My mother was particularly keen that her daughter should get an education because she came of a generation where the girls' education was put second to the boys'. During the depression her family was short of money and her brother got given an education but she did not. So she was very keen that the three girls she had should get a good education.

I did my senior years of high school education in a boarding school in England and then went to university in Glasgow, in Scotland for my first degree. Subsequently I went to Cambridge to get my Ph.D.

Please, tell us something about your famous work there, which eventually led to the discovery of pulsars.

This discovery was an accident because such objects had never been dreamt of. They were unimaginable, literally. I was studying quasars, which are very, very distant objects. The analogy I sometimes use is that you are making a video of a sunset from some vantage point; you have a splendid view of the setting sun. Then along comes a car and parks in the foreground and has its hazard warning lights going and thus spoils the picture that you are making. It was a bit like that with us. We were focusing on some of the most distant things in the Universe and this peculiar signal popped up in the foreground. These turned out to be the pulsars, but it was quite a long trail. First of all we suspected that there was something wrong with the equipment. Then we suspected that we were picking up artificial interference, we suspected all sorts of things. It was really only when I found the second one that we began to believe that these things might be stars, that they might be natural.

I read somewhere that you called them LGM, for Little Green Men. Did you really believe that these might be indications of another intelligent life form?

The naming was in jest, it was a joke, it was tongue in cheek. But radio-astronomers are aware that if there are intelligencies out there, Little Green Men, then it is they, the radio-astronomers who will probably first pick up the signals. That is the basis behind a lot of the SETI project. But the chances of detecting Little Green Men are very-very-very small, so it was not a serious belief. Of course, when I found the second one and then number 3 and 4, the chances totally vanished; it was quite a relief.

Why? Would not that have been an even more astonishing event?

Yes, that's exactly the problem. It would've been. I had a thesis to finish and a limited amount of funding. I was within six months of the end of my funding and needed to get finished.

Do you believe in extraterrestrial civilizations?

The Universe is such a big place, I think there must be but I am not convinced that we'll make contact with them. But I suspect that we are not alone.

What are the pulsars?

Jocelyn Bell with the radio telescope that was used to detect the signals from pulsars in Cambridge, during the mid-1960s (courtesy of J. Bell Burnell).

Pulsars have another name, which is neutron stars, because they are made very largely of neutrons. They are very compact and therefore very dense. There is something like a thousand million million million million tons of material, the mass of the Sun basically, all packed into a ball that has a radius of ten kilometers. So their density is comparable to the density of the nucleus of the atom. They are quite extreme and for that reason have some very unusual physics. The ones that we see as pulsars seem to have some very strong magnetic fields. We believe that a beam of radio waves, a bit like a lighthouse beam, is formed at the magnetic poles of the neutron star. The magnetic poles just as on the Earth are not coincident with the rotational or geographic poles. On the Earth the north magnetic pole is in Northern Canada. There could be something similar on a neutron star but maybe even more so with the magnetic poles, say, in Texas. As the star spins, the beam that comes out from the magnetic pole sweeps around the sky like a lighthouse beam sweeps across the ocean. Every time the beam passes over planet Earth we can pick up a pulse, so we get a series of regular pulses. Pulsar is an abbreviation for Pulsating Radio Star because of that behavior.

Does a neutron star consist only of neutrons?

No; not only neutrons but they are much more neutron rich than material here on Earth. A neutron star is unusual in that we believe it has a solid crust; other stars are balls of plasma, burning gas. But a neutron star has a crust and we believe that the crust is made of iron and the iron atoms are very non-spherical because of a large magnetic field. They stack together first of all to make long fibers, polymers, and the polymers link together to make an extremely strong substance. So the outside of a neutron star is recognizable material but in a rather unusual form. Then as you go inside the neutron star you move to nuclei that are more and more neutron-rich, including many that would be radioactive here, on Earth but are stable in a neutron star. Then as you go in towards the core, the nuclei are touching each other and as they merge, they dissolve and you have a superfluid of neutrons. Finally, right in the core we are not quite sure what there is there because the density is much higher than in the atomic nucleus and higher than we've been able to reach with accelerators. It could be that the neutrons have broken down into their constituent quarks and then we have a sea of quarks in the core.

These are so distant objects; how can you get this information about what's in the core or close to the core of a neutron star.

By theoretical modeling. There are a limited number of observations, which confirm the picture; particularly the superfluid component; we are very clear that that exists in a neutron star. But the rest of it is theoretical modeling. This is the best estimate of the people who work in condensed matter physics; *very* condensed matter physics. They have developed these models, which are self-consistent and fit with what limited data there is. So it's the best that we have but of course, there is no way of being absolutely sure that that is what they are.

Physics and astronomy seem to be totally inseparable nowadays.

Yes. Astronomy is a high-tech branch of modern physics.

Your work over the years was involved with different parts of the electromagnetic spectrum. Would you mind telling us what the difference is between them, say the infrared, gamma ray or X-ray astronomy?

The techniques are very different. To do gamma ray or X-ray astronomy, you have to get your equipment above the Earth's atmosphere. That means a high-altitude balloon, a satellite, or, if it's X-rays, where the fluxes are more copious, you can use a rocket flight. These days it is mostly satellites. We can do infrared astronomy either from the ground or some from space. Radio-astronomy is done from the ground.

What are the differences in their objects of study?

You can study stars and galaxies at all those wavelengths. Some things in it more copiously in the X-rays and gamma-rays than in the optical or the infrared; some things you can study more copiously in the infrared than anywhere else. For instance, if you are interested in studying the dust in the galaxy, you would probably do infrared or millimeter astronomy. If you are interested in studying the nuclei of active galaxies, you would be doing X-rays, radio, optical or ultraviolet astronomy. So you address different questions by using the different wavelength regimes.

What is your idea about the Big Bang?

I teach students the standard Big Bang model, but scientists can only pick up the story shortly after the Big Bang; as a scientist you cannot go back earlier than 10^{-43} of a second after the Big Bang, which is pretty close to it but is not actually at it. The picture seems to be that there was, indeed, a Big Bang.

After the pulsar work, what have been your areas of interest?

I have worked in gamma ray, X-ray, infrared and millimeter wave astronomy during my life time; in some of those fields I had managerial positions, in a few I had research positions.

What are you currently working on?

Here in Princeton I have been teaching this academic year, so I have been mostly involved with writing course material. Last semester I taught the sophomore mechanics course and this semester I am teaching a freshman seminar on where the atoms in our bodies come from. We've looked at the nature of atoms, the nuclei, nuclear synthesis (that is the creation of various nuclei), we've looked at the origin of the Earth, the origin of life, the origin of the Sun, and the origin of the Universe. We'll have our last class tomorrow and I hope that the students learned some physics, some astronomy, some chemistry, some earth sciences and learned that we are dependent on stars for most of the chemical diversity, since most of the atoms in our bodies came from the explosive termination of stars; although the hydrogen in our bodies may date back to the Big Bang. But everything else came from stars and we are dependent on the death of stars for our life. In the early universe after the Big Bang, life would not have been possible because there were no other elements, just hydrogen and helium. All the chemical elements heavier than helium have been created in the course of stars, through nuclear fusion reactions and made available when that star exploded catastrophically at the end of its life. We think that the material in the Sun and in the planets and therefore in us as well had been, on average, through two generations of stars previously. The first generation of stars produced some carbon, nitrogen, oxygen, and so on, then that material was incorporated into another generation of stars, which added more of these elements. That second generation of stars exploded and the material of which the Sun and the planets are made came from them. The Sun is a third generation star.

This was your teaching. Do you do research as well?

Yes, I do but it is a small occupation. This is because when I am back home in Britain, I am chair of the Department of Physics and Astronomy at the Open University and that takes most of my time. But I've managed to have a graduate student and my research time has gone on supervising that grad student. With several recent grad students we have been studying some very exotic, energetic binary systems of stars in our own galaxy. These produce fantastic outbursts and a lot of bizarre behavior. There is a small group of these systems, at the moment we only know about 5 or 6 of them but we are sure we'll find more. These are objects like Cygnus X-3; they are X-ray binaries, they may be gamma ray objects, or they may be radio objects. We try to understand them because they are much more bizarre than any previously known binary systems. They have a much greater diversity of behavior and unpredictable behavior.

Do you think that our time is an especially exciting time for astronomy or is any time exciting?

I think that there is a lot of excitement at the moment. Most of the excitement nowadays is in the area of cosmology, where you are looking at the galaxies, the early Universe, the formation of galaxies, the microwave background and all that kind of thing. There is less interest in stellar astronomy at the moment but I think it'll come back again.

Who are your heroes?

Arthur Stanley Eddington, who was a British astrophysicist and some of the women who have pioneered, such as Cecilia Payne-Gaposhkin, Caroline Hershel, the ones in my area.

Who had the greatest influence on your life?

I think that academically, probably the physics teacher I had for the last three years at school, who was an extremely clear teacher. Generally speaking, I think I learned a lot from some of the Quakers I met as a young woman as I belong to the Quaker church; I was brought up as a Quaker. There were some women there who taught me a lot.

What does it mean to be a Quaker?

Quakerism as we practise it in Britain, is a very good faith for a scientist, because it is a denomination that puts less emphasis on holy writings and less emphasis on tradition, and more emphasis on what you've learnt yourself about the nature of God and the nature of the world. It is strong on ecology, the respect for the Earth, the Planet, animals, and life. One of the testimonies of Quakerism is to live simply, not to consume excessively, which, you can see, is very close to the idea of taking care of the Earth. It is a very exploratory type of religion, a very open one, where you can revise your picture as you go along. You are not required to believe or say a creed or something like that. I think that is why there are so many scientists in Quakerism because it's open and experiential in nature and it suits a research scientist. Interestingly, it formed in Britain in the 1640s, which was just about the same time when science became a recognizable activity and breaking away from theology. That may be a coincidence, I don't know but it is interesting that both got their identities about the same time.

Jocelyn Bell in the mid-1970s
(courtesy of J. Bell Burnell).

Are you active?

Yes. The official title is clerk of the yearly meeting but what that actually means is that you are president of the annual national assembly. I had that role for several years. I gave that up when I came here to Princeton last year.

Are you married?

I was. I was married for about twenty years and I have a son who is now 27 years old. He is a postdoc at the Materials Science Department in Cambridge, England. He is working on squids. My husband was a personnel officer at the local government in England with various county councils. He kept moving around and therefore I had to change my jobs all the time as he moved.

Do you think that having been a mother influenced your scientific career in any way?

Yes. The most obvious way is the problems of combining family responsibilities and career. There were virtually no childcare facilities. In Britain women older than me did not have careers, women younger than me expect to have careers, but my generation is very much at the turning point. I remember that when I became pregnant I went to the chair of my department and asked what maternity leave am I entitled to. He said: "Maternity leave!?! I never heard of it!" The university did not have maternity leave.

So what happened?

I resigned. It was difficult to find childcare, so I worked part time for eighteen years till my son went off to college. That, of course, has dented my career along with the frequent changes as my husband moved jobs. So being married and being a mother has made a very big difference. I think that it has also given me a greater range of skills than I might have had if I had stayed working simply in academia doing research and teaching all my life. Some of my jobs were outside academia and I have held management posts. I have had management training, so I think I have much better interpersonal skills than I would have if I had remained simply in academia. I also have a greater diversity of experience. If I stayed in academia all my life, I would have had more depth of experience but a much narrower range.

You said that your generation was at the changing point with respect to women's work. Is the situation today easier for young women?

Certainly. Today there is more childcare, for example, the Open University has a creche for children and that is becoming normal for universities. It is easier today but I think that it still is very hard work to combine raising children and an academic career. You need a lot of stamina.

What was the greatest challenge in your life?

Combining an academic career with family, I think. I still don't know how you do it. At least not in my time, my generation.

What is your advice to young women concerning starting a family if they want to have an academic career?

I get asked this quite a lot. First of all, there is no right time to have children; there is no such time. It would be too easy, I guess, if there were. You need to be very well organized, you need to have a lot of stamina, you need to have the support of your family, your spouse particularly. I recognize that this is very tough. What I actually find disturbing now is a number of young people saying to me: how do you have a life and be an academic? They see the long, long hours worked by the academics around them and they don't want to have that style of life. They are beginning to question and perhaps rightly so.

This is something not related to being a woman.

Of course, I am also asked this by young men. This is sad. If it provided some kind of corrective to the academic life, that would be good because in Britain, certainly, we have something called research assessment exercises. Your research output is examined and your success and the success of your department depend on that. And it is a very blatant encouragement to be obsessive to work extremely long hours. I do not think that is healthy, so I don't like it.

There have been famous women astronomers over the centuries and even today there are more women in high academic positions among astronomers than in other areas of physics. What could be the reason for this?

You are right about this but I don't know the reason, to be honest. I just would like to mention that the physics situation and probably also the astronomy situation varies from country to country in a very curious way. You will find that in countries like France, or Italy, there is quite a high percentage of women among physicists. But in Japan, Germany, Britain, or the U.S., a much smaller proportion. So partly it has to do with culture. There is nothing absolute about it. But astronomy has now for quite a few decades been perceived as a subject that is more friendly to women or more suitable for women and I don't really know why. It's just that its culture is different from physics. It's quite striking. Also the entry to astronomy is often through physics, so many of the women astronomers have done physics bachelor's degrees and then decided to move to astronomy. It probably has to do with the climate in the astronomy departments compared with the departments of physics. Maybe astronomy is a slightly more friendly field than physics.

If you look at statistics on women in science, obviously as you go from the undergraduate level up the academic ladder, the proportion of women decreases steadily.

Yes, this is the leaky pipeline. At each stage more and more women opt out. I am sure, some of it is the complications of family life. It probably still is that the man's career is put ahead of the woman's career and the woman finds it difficult to keep going. I think it may also be that the women find that the atmosphere is not what they enjoy so they go elsewhere. Maybe there is some direct discrimination and discouragement.

There are many places today that put much emphasis on trying to hire women.

Yes, but not everywhere. There is still discrimination, which is subtle. A group of people like to hire somebody who is like them and anybody who is different, be they colored or female, is at a slight disadvantage. That is the kind of thing that is going on nowadays, it's just that people are more comfortable with people like them. It is perceptions.

You said that you teach at the Open University in England. Is that a continuing education university?

I would not describe it as a continuing education. It is students, who missed out on the chance to do a degree at the normal time in the normal way. Maybe they had to leave school earlier because the family needed the money and they had to take a job. Maybe some of them were told by their schoolteachers that they were stupid and they believed them for the next ten or twenty years, and then began to realize that they were not stupid. Some of them are women at home with young children. There are people who have retired early; there is a tremendous diversity of people.

This must be a very special challenge to a teacher.

There are several special challenges. One is the type of student. The student is mature and anxious; it is a long time since they sat for a test and took exams, so they are very sensitive to criticism, and you have to be very tactful and encouraging when dealing with them. It is also different in that we are teaching at a distance, we are teaching with correspondence, TV and the web and the students are studying at home after work and on their own. So the material has to be very clear.

What kind of interaction do you have with the students?

The central faculty does not much interact with the students, but we have tutors all around the country and they are the main interaction for the students, they will grade their assignments, for example. The central faculty could not handle the number of students. For example, we have 4700 people having physics courses and we have a faculty of 16.

What's the role of the faculty?

They are creating the new courses, maintaining the courses that are now running, they write new test questions, etc.

How long have you been at Open University?

I went there in 1991, so it's 9 years ago. I have also been the chair of the department for 7 or 8 years and there is certainly time to change that. It is not a fixed term but I was made to believe that it would not last this long.

Coming back to the women's question for a moment; it is often seen that women who are successful in their professions are often overwhelmed with invitations to committees, memberships, etc., so it may make life even more difficult for them. Have you experienced this?

Of course. It is certainly a case in Britain that nowadays there is an anxiety to have a woman, not parity just a woman, on a panel or a committee or among the speakers. There are relatively few senior women to fill these roles so I find I am getting many invitations to do that kind of thing. However, it is not necessarily bad. First of all, from my personal point of view it means that I am getting a lot more experience than my male contemporaries. I think it is useful that there is a woman's voice in these areas, so I think it is actually important that there are women. Of course, I would like to see more than one. I have seen it as part of my job to try and make sure that the life of younger women is made easier by creating awareness among my male colleagues. This does take time from everything else but I feel that this is something that I need to do.

The image of science is not very good among laypeople although I believe that astronomy fares much better in this respect than, for example, chemistry. What can be done to improve that image?

Astronomy, indeed, is much more favorably perceived than the other sciences. One of the things I noticed coming to the U.S., is that science in America is much more respected than science in Britain, for example. It is quite striking. I think it's important that those of us who can put science across to the public do so and we get recognition for doing so. There are many academics who would say that, yes, of course, explaining science to the public is extremely important but then will not recognize that it takes time and effort and energy on the part of those of us who are good at explaining things.

Do you do it?

Yes, a lot. I do broadcast, television as well as radio. I do a lot of public speaking and I also speak to amateur astronomical societies.

I think it is inevitable to bring up the question of the Nobel Prize for the pulsars. One often finds the opinion that you should have received it together with Dr. Hewish.

The Nobel Prize was awarded 30 years ago. At that time there was still around the picture that science was done by great men (and they were men). These great men had under them a group of assistants, who were much more lowly and much less intelligent, and were not expected to think, they just carried out the great man's instructions. Maybe that was the way science was done a hundred years ago or maybe even more recently. What has happened in the last 30 years is that we've come to understand that science is much more a team effort, with lots of people contributing ideas and suggestions. But at the time of the Nobel Prize, there was still around the idea that science was done by great men and the awarding of prizes, any prizes, was consistent with that picture. We did not recognize the team nature of science in those days.

Even those days Nobel Prizes were often shared by two or three persons, some of these cases may have been for work done together.

Yes, Antony Hewish shared the prize with Martin Ryle. These were two people who have made significant developments in radio-astronomy. It's actually the Nobel Committee's job to argue this. I think that they would argue that they were awarding a pair of people who'd significantly advanced the subject.

What was your reaction that time?

I was very pleased. Mainly for political reasons. I am a strategist, a politician. That was the first time that a Nobel Prize in Physics was awarded to anything astronomical. There is, of course, no Nobel Prize specifically to astronomy and physics is the nearest. I think it is perhaps debatable whether astronomy is or isn't included in the physics prize. This was the first time that it was clearly signaled that it was included and that was extremely important. It was the opening of a door to a whole new range of things. So I saw that instantly and I was very, very pleased for that reason.

Did it not occur to you that you should have been included?

I was content. And I have discovered subsequently that you can actually do extremely well not getting the Nobel Prize and have a lot of fun, too. And I have received lots of other prizes.

What is your relationship with Dr. Hewish?

He is now semi-retired, he is a professor emeritus in Cambridge. He is in the lab about half of the time and spends the other half, I think, with his grandchildren. I see him from time to time when I visit Cambridge. He stayed in Cambridge and stayed in radio-astronomy all his life, whereas I have gone walkabout so our paths have not crossed as much as if I had stayed in radio-astronomy or if I had stayed in Cambridge. I see him about once every six months or so.

Do you think that he feels somewhat uncomfortable because of this problem?

I think that he has taken upon himself to defend the decision of the Nobel Committee, which is not his job.

Have you ever openly discussed this with him?

Yes, we had several discussions around the time of the Nobel Prize. I don't really remember what he said then; it was, after all, in 1974. With Tony, my relationship has always been kind of casual; it has never been formal. In fact I found it very difficult when I went to do my Ph.D. to call my supervisor "Tony". I thought I would call him Dr. Hewish but that wasn't allowed. So we were on a first-name basis for many, many years and it is very nice. It creates an ambience of friendliness. That is the relationship we pick up whenever we meet.

Did he ever tell you that he was sorry because they did not give you the Prize as well?

I don't remember that.

I find this situation especially ironic in view of the 1993 Nobel Prize in Physics to Dr. Taylor and Dr. Hulse for the double pulsars, where there was exactly the same kind of relationship, a professor and his graduate student.

Joe Taylor invited me to Stockholm as one of his guests for the Nobel Prize ceremony. It was huge fun; I think it was actually much more fun being a guest of a prize winner than being the prize winner.

Come to think of it, why did not Hewish invite you to Stockholm?

Oh, I was either pregnant or with a small child, it was at that stage of life.

There were certain differences between the two cases in spite of their stunning similarities. The Taylor/Hulse prize came in the early 1990s, and during these twenty years our picture of how science is done has changed. It was also different in the sense that there was a much less age difference between Joe and Russel; Joe was a very young faculty member and Russel was the graduate student. They were much closer in age and thus there was much less distinction on those grounds as well. I think you see the change in society, the change in the way science is perceived maybe between the two events.

So you are not unhappy?

No, I am not unhappy at all, thank you.

*****

Since we talked last time you became the Dean of Science at the University of Bath. What is the reason for the move?

I wanted to become involved in strategic management in academia.

Do you still have time to engage in research even if only through grad students or postdocs? If so, what are you currently working on?

No — this is a senior management job and I have neither time for teaching nor research.

Joseph H. Taylor, 2000 (photograph by M. Hargittai).

34

JOSEPH H. TAYLOR

Joseph H. Taylor (b. 1941 in Philadelphia) is James S. McDonnell
Distinguished University Professor of Physics at Princeton University.
He received his B.Sc. degree in physics at Haverford College in 1963 and
his Ph.D. in astronomy at Harvard University in 1968. He was professor
of physics at the University of Massachusetts till 1981. Since then he has
been at Princeton University, where, between 1997 and 2003 he was
the Dean of Faculty. He received the Nobel Prize in Physics, together
with Russell Hulse, in 1993 "for the discovery of a new type of pulsar,
a discovery that has opened up new possibilities for the study of
gravitation". He is a member of the National Academy of Sciences of
the U.S.A. (1981), Fellow of the American Academy of Science and Letters
(1982), and Member of the American Philosophical Society (1992). His
many honors include the Henry Draper Medal of the National Academy
of Sciences (1985), the John J. Carty Award for the Advancement of
Science (1991) and the Wolf Prize (1992). We recorded our conversation
in his office at Nassau Hall at Princeton University on October 9, 2000.*

*By the time you started your research on pulsars they had been discovered
for about eight years. What was the challenge in doing research in this
area?*

You don't quite have it right because I started observing pulsars within
a few weeks after the first one was discovered, or, rather, after their publication

---

*Magdolna Hargittai conducted the interview.

appeared. The Cambridge, England, group discovered the first pulsar in the late summer of 1967 and it took them some time to persuade themselves that these particularly unusual signals were really natural astrophysical phenomena. They published their paper in February of 1968 in *Nature*. I read the article the day it came out and was immediately taken by the interesting characteristics of the signals. My group was one of the first in the United States to confirm the Cambridge observations and then to begin observations of my own. A group of us from the Harvard College Observatory applied for time at the Green Bank, West Virginia, National Radio Astronomy Observatory right in February of 1968. We were observing by April of 1968. One of our goals was to re-detect the signal of the first pulsar that the Cambridge group had published. The other goal was to try to find more of them. We designed a computer-based system that would be capable of detecting pulsars. We started using that system in May or June of 1968. Our first observation was in June of 1968. At that point I was interested already in finding additional pulsars. I had moved from Harvard to the University of Massachusetts in late 1968, early 1969. There, at Massachusetts, a small group of us began to design systems that would detect and measure the characteristics of pulsars.

Why did you move from Harvard to Massachusetts?

I was a postdoctoral fellow and lecturer at Harvard after I had received my Ph.D. in radio-astronomy from Harvard.

Was it a long way from pulsar to the binary pulsar?

The first signals of the binary pulsar were recorded in the early summer of 1974. Our normal mode of operation was to record signals from an area of the sky until we had candidates for possible pulsar-like signals from half a dozen or so regions in that area. Then we would go back and spend one complete observing day re-observing those suspect pulsar detections. The re-observation of the binary pulsar was done in the month of July 1974. We did rediscover the signal, which had essentially the same strength and the same frequency as had been detected in the original discovery except that the pulse frequency was different by about one part in a million. This is not a very large difference, but it's bigger than the normal variations that one sees from one day to the next. That required us to go back and check it again and each time we measured it, the frequency was a little bit different.

It took a total of about a month before we were convinced that the signal was coming from an orbital motion.

Did you immediately realize the significance of it?

Yes, it was very quick to see that there was a system that would be relativistically significant. The total variation in the pulse frequency was about one part in a thousand. That suggests an orbital motion with a velocity of 10^{-3} times the speed of light, a very large velocity for a macroscopic object. The orbital period was quite short, only about 8 hours. We did the calculation of the relativistic effect, and we could see that it was an orbit of an object with a mass close to that of the Sun or a little bigger, around another object with similar mass moving at these relativistic velocities. It was very clear that relativistic effects would be important.

As I understand it, your discovery provided an indirect proof of gravitational field. Has there been any direct proof yet?

No, not yet. There are gravitational wave detectors being built now that may well succeed and there are other ones being planned that would have even higher sensitivity, including some that would fly in space. They are space interferometers. Those are likely to eventually be successful.

What would it be, like two stars collapsing onto each other?

Yes, you need a very big event and the farther away it is the bigger it needs to be.

What was your reaction when the Nobel Prize was announced to be shared between you, who was an authority of the field and your former graduate student, who did not even stay in the field?

Obviously, I was delighted and I was delighted that there was the two of us. We had great fun working together, we were not very far apart in age, I was only a few years older than Dr. Hulse, and we'd had a wonderful time designing this experiment for detecting pulsars. He, in the meantime, moved to a different area of physics, which, in fact, was back toward the area he was planning to work in before I persuaded him to work on this project in radio-astronomy as a graduate student. At the time, in the mid-1970s, when he was finishing his thesis the prospects for him looked better in atomic physics and other areas that he moved back into than they did in astrophysics.

He made a career choice and moved his direction away from following up the discovery that we made together. I think that partly because the nature of the gravitational wave experiment with the binary pulsar required many, many years of observations, neither of us was thinking about Nobel Prizes for this work. For many years afterward, I was certainly enjoying the work I was doing, observing that particular binary pulsar, but many, many others have been discovered in the meantime, by my group and others. We were certainly very interested in what we were doing and very engaged in that particular piece of scientific work, but it was not the kind of research where the results unfolded very quickly. Therefore, after close to twenty years after the original discovery one was no longer thinking about it in a way that it might be recognized with such an honor as the Nobel Prize.

Was it really that unexpected?

Once it became clear that the discovery had the significance of making a very important contribution to experimental relativity, it had been suggested over the years, perhaps just lightly in the nature of conversations over a beer or something, that at some time this discovery will be honored in some important way. My own nature is not to think much about things like that. I think people who do dwell on or hope for or expect that kind of recognition, are often disappointed. There are many important discoveries that are made, so I did not spend much time about such aspects.

Why do you think that it took so many years for the recognition to arrive?

Primarily because the significance of the experiment increased with time. After about five years the radiation effects on the orbit were just barely measurable at the level of about three sigma. After 10 years it was more like 15 sigma; there was a very significant answer, and after 20 years, it was a 1 percent measurement. Over time the measurement got remarkably more precise. When the agreement was only approximate and the uncertainty was relatively large, it was an important result, potentially very significant, but not quite the same as the precision measurement that exists today.

How do you compare your contribution and Dr. Hulse's?

At the time of the discovery, the actual observations on the day-to-day basis were all being made by Russell Hulse. We designed the computer algorithms and the computer hardware for doing the experiment together in the laboratory in Massachusetts. We took the computer and the associated equipment, the receiver, to the Arecibo Observatory in 1972 and I got started on the experiment shortly after that. I was there during many of the early observations, but Hulse stayed there for months at a time. So the actual observations taking place during the discovery were all his own. As soon as he recognized that he was seeing a signal for which the pulse frequency was changing slightly, he communicated immediately, and I returned to the Observatory to help him with the follow-up observations. But I had teaching duties to do in Massachusetts, so I needed to go back and forth. We were closely working together, but he was the man on the spot at the telescope at the time. Soon after that he wanted to finish his thesis, finish his doctor degree, and leave the University of Massachusetts for his first position. The job of following up the experiment and doing additional measurements were left to me and other collaborators, who worked with me later.

Is there any parallelism between your story and the discovery of the pulsar by Jocelyn Bell Burnell and Antony Hewish?

One can only guess at what the Nobel Prize Committee may have been thinking, but it certainly is true that some had thought that the Nobel Prize Committee in 1974 overlooked the significance of the contributions of Jocelyn Bell Burnell. The times have changed over the last several decades in that people are much more aware of the contributions of younger collaborators in large-group efforts. In the future the committees will be more likely to take that into account and to recognize those contributions. It's a difficult decision to make in many instances about the contributions of different individuals in a combined experiment.

She told me that you invited her to the Nobel Prize ceremonies in 1993.

Because she has been a long-time friend and she is a person who I openly admire and I enjoy her company. I simply felt that she would enjoy the experience of the trip and we would enjoy having her with us, and that it might make up a little bit for something that she came very close to once before but did not quite achieved.

What has happened to your field of research?

The field of pulsars has been an active one for a quarter of a century. The total number of pulsars known is now well over a thousand and the total number of binary pulsars known is getting close to a hundred. Because now I'm sitting in this dean's office, I'm not doing very much of this research myself, but I'm enjoying very much seeing many former collaborators and many students and friends working in the field. Every ten years or so a new aspect of the field has been discovered, which in the first instance was the binary pulsar, on another occasion, in the 1980s, it was the very fast spinning millisecond pulsars. There are now a very large number of pulsars known to be harbored in the so-called globular clusters of stars. That is a new field of work in this area. The way in which the formation and evolution of pulsars relates and fills in some of the gaps in our understanding of the evolution of stars in general, particularly the close binary systems of stars, has become an important field. It's difficult to predict what the future may hold. One thing we don't have yet is a good model or understanding of the details of the mechanism, which actually generates the radio waves that we observe. Almost certainly they are generated in the strong magnetic regions above the polar caps of spinning neutron stars, but we don't have a very good picture of the detailed plasma physics that is taking place in those regions. There may still be considerable progress in that area.

Have you worked in other areas too?

We have been involved in radio-astronomy in general, but concentrated on very compact objects and the pulsars are the end points of compact star evolution, and they have been the target of the primary effort in my group. We have lived in a time when the development of the technology for observing electromagnetic radiation from great distances has been improving very rapidly. We have very much improved mechanisms for the storage and quantization of data for measuring high-resolution spectra and wide areas of different portions of the electromagnetic spectrum, and all those things have opened new windows of opportunity for determining what's happening in distant parts of the Universe. We have found ways, in which the observation of pulsars can, for example, give data that are significant cosmologically, because they give us information about the amount of very low frequency gravitational radiation that is kind of rattling around in the Universe as a result of the red-shifting effects happening long after the Big Bang. Those are a subset of results from pulsar observations.

Do you think we will ever fully understand the beginning of the Universe?

We are certainly providing a much better-formed picture of early stages of the evolution of the Universe, but whether we can go back to the beginning point with great confidence, it's too soon to tell.

Do you enjoy the intellectual atmosphere that a place like Princeton provides?

Yes, absolutely. That atmosphere is very much the part of what feeds one's creative instincts and abilities. Very little science today is done by single individuals coming up with uniquely clever ideas. Much more commonly it happens in an environment where at the very least there is frequent discussion going on and by talking through problems with one another we often gain insights. Whether or not there is actual active collaboration, one takes advantage of the ability to talk through a problem with a colleague. That's been very helpful to me. I have not ever myself worked in large groups. I typically have had at least one other faculty colleague and a small cadre of graduates and undergraduates. For me, that size of group has been intellectually very stimulating. Others, because of the nature of the work involved, require larger groups.

There is quite a number of physics Nobel laureates in Princeton. Do you ever get together?

Not as a group of Nobel laureates. I don't think it would be useful, for example, to form a group response to the invitation from the Nobel Committee to submit nominations for new laureates. In all the physics departments there are discussions at this time of the year about who will be the winners this year and who are the ones whose work should be recognized over the next few years.

The announcement is coming up tomorrow.

This year I did not play the game with anyone and I don't know what the betting will be.

Where do you expect great progress?

It's hard for me to single out any one area; there is fascinating work going on in different parts of physics. Conditions are ripe for significant new steps in condense matter physics, in elementary particle physics with the advent of new accelerators and detectors, and astrophysics is also quite ripe for advances, such as direct detection of gravitational waves. But I could not point, right now, to a particular experiment or a particular theoretical advancement, that could be chosen this year for a Nobel recognition.

Being a successful scientist and a Nobel laureate, what motivated you to become dean?

I have a number of reasons. For me fresh challenges have always been important and I was feeling that I had worked on a particular problem for a long enough time and I wasn't sure that new challenges would come up. I was looking for other things to turn my attention to. I have great admiration for Harold Shapiro, who has been the president of Princeton for the last 12 years. When I was approached by him, I gave it a very serious thought. I also knew that I would be doing work of a different kind than I did as a scientist where I could define my own schedule. Most academics enjoy that kind of freedom. As dean of the Faculty, I don't have that at all. I have a rigidly scheduled day with appointments. I knew that there would be a big shift in my life, and I agreed for a five-year term. After that I fully expect to be back in science although I don't know when that will happen. I try to spend one day each week in my physics department office and I go to the physics seminars when I can, and keep in touch with the people who are maintaining an active effort in pulsar research here in Princeton.

I would like to ask you about your family background.

I was one of six children in our family; in effect, eight because my parents raised two young cousins whose parents were not able to take care of them. This group of eight was blessed with parents who cared about us and gave us extremely good education for all of us although they had very modest means. They gave us a lot of freedom to experience life in different ways as we grew up. They deserve the bulk of the credit in forming me. My father died in 1992, one year before my Nobel Prize, and my mother is still alive, doing very well. She enjoyed the trip to Stockholm.

I was raised on a farm here in New Jersey. The principal crop was peaches and we also grew tomatoes and corn and others that do well in this climate. Because we were not living in a town but out on the farm, my closest

Joseph and Marietta Taylor in Stockholm, 2001, during the Nobel Prize Centennial (photograph by I. Hargittai).

companion as a youngster was an older brother. We were slightly less than two years apart. The two of us became interested in electronics and involved in the hobby of hand radio and we taught each other quite a bit of what we didn't quite recognize at the time was physics. We both became physicists. Our early endeavor gave us a lot of independence and ability to work on our own and without much outside direction. That served both of us very well. Then, of course, I had a number of wonderful teachers over the years. If I go back to my school years, I would pick out a particular teacher of English, who was instrumental in teaching me to write well, which has been extremely useful. I wish we had not gone in the direction in which we have gone that high school students enter the university with much less training in writing than many of us had in the past. Of course, there is much more information today than just the written page, information that was not available for my generation. But it is still important for people of high intellect to learn to communicate very well with the written word. When I went to college, I might have become a math major, but I discovered soon that I enjoyed more working in the laboratory than proving theorems. There was a college teacher at Haverford College, who was very influential on me and persuaded me that doing physics was a good thing for me;

it was Faye Eisenberg Seloff, who is now at the University of Pennsylvania. As a graduate student, I was fortunate to have a thesis supervisor, Alan Maxwell, who was extremely good at teaching me good scientific writing techniques. He also let me do things on my own, without too much interaction with others. Alan was the kind of advisor, who opened doors and made things possible, but then got out of the way and let us students proceed on our own. For me this was a very useful technique of guidance.

Do you have heroes?

This is a tough one to answer off the cuff. I certainly have heroes in different aspects of life. In science I can't single out one person or even a small group of individuals.

Your present family?

My wife and I have a combined family with four children, two of each from our former marriages; we have one grandchild and a second one is due in about a month. The children are spread out over the country. Only one is involved in science, in molecular biology. My wife worked in public health as a health care statistician; she is now retired.

Would you care to tell me about your relationship to religion?

We are active in the Religious Society of Friends, that is, the Quakers and it's been an important part of our lives, more so for my wife and me than for our children. My wife and I spend time with our faith group; it's a way for us to make connections with our philosophical views on life, why we are on the Earth, and what we can do for others. The Quakers are a group of Christians who believe that there can be direct communication between an individual and the Spirit, which we may call God. By contemplation and deep inward looking one can effectively commune with this Spirit and to learn things about oneself and about the way one should conduct oneself on Earth. The group believes that war is not the way to settle differences and that peaceful ways are more likely to be lasting. Quakers have refused fighting wars but have been willing to serve their nations in other capacities. We believe that there is something of God in every person and therefore human life is sacrosanct and one needs to look for the depth of spiritual presence in others, even in others with whom you disagree. That is an idea that has a lot of truth in and permits one to face difficult problems.

How do you feel about the special responsibility of Nobel laureates? Are there special responsibilities in the first place?

I believe there are and different individuals may discharge them in different ways. I felt strongly particularly the first few years after my own notoriety in receiving the prize. I made it a special point to make myself available in response to various invitations to other universities, even to high schools. That was a significant distraction for me for three or four years, but I felt it important to communicate the excitement of science to youngsters. Science is a public enterprise enjoying a lot of financial and moral support and I was happy to pay back some of this support to society. This also played a part in my moving into administration. I felt it was important to make use of the credibility and trust that the Nobel Prize had generated. Now I have much fewer public appearances, I choose them carefully but I still do them if I feel they are important. I put a lot of effort in committee work, and I have been a member, recently co-chair, of a national committee, helping the government to identify the major areas of expected development in astronomy and astrophysics for the next ten years.

What are these major areas?

We have a 400-page report on that. Our number one priority for the next decade is the building and launching of a new generation space telescope, something that will follow up on the capabilities of the Hubble Space Telescope, with much improved technology and larger sensitivity.

Women are still under-represented in science, especially in the higher echelons, in the membership of the National Academy of Sciences, for example. What could be done to change this?

We should encourage young women who have the talent and interest in scientific fields to not leave the field too soon. We spend a lot of time at Princeton trying to build up the confidence of the young women who are taking the scientific and engineering courses in departments that are not well represented by women. They are doing just as well as the young men are. There have been improvements. Astronomy has been one of the fields traditionally that has more women than other branches in the physical sciences.

What may be the reason?

It's hard for me to put a finger on except to say that astronomy is something that all segments of society can identify with more easily than some other parts of physics. Anyone can look at the sky and appreciate its beauty and can picture that stars are the same kind of things as the Sun, only farther away. It doesn't need a lot of extrapolation to understand that groups of stars can form into clusters and galaxies. Those things one can visualize and immediately appreciate the beauty as well as the scientific connection with evolution and the planet that we live on. Subatomic physics is not something one can have close connection with so easily. In chemistry, again, one cannot picture what's happening when molecules form. It's one step farther removed and therefore not so close to ordinary human activity. Why that would have a connection with one gender relative to the other is not so easy to identify.

Do you believe in the existence of extraterrestrial intelligence?

It's not like a religious conviction where you either believe it or not; you look for evidence. At the moment we don't have evidence that extraterrestrial biological intelligence exists, but I would be very surprised if life had evolved in only one place in the Universe. It's much more likely that there are other areas where living things exist and quite possibly with developed intelligence at least as far along as ours, and it's quite conceivable that evidence for that will be found in some time. I've never been myself particularly interested in the attempts to build systems that would use radio-telescopes or some other techniques to scan the sky and look for little green men out there, because I think it more likely that we will detect such signals accidentally rather than on purpose. There are too many variables. I may be wrong and maybe there will be somebody with a clever idea. Certainly, such a discovery would be of tremendous importance. My guess is that it's quite plausible that life could've evolved somewhere else with the same chemistry basically that we have. The particular choices made by the evolutionary process with random mutations over a very large number of years could've led to very different beings than what we are accustomed to living on Earth, but I would not be surprised if life elsewhere would have the same basic chemistry that we have on the Earth.

What was the greatest challenge you have met?

It was a most enjoyable puzzle, the solutions based on the few randomly spaced observations of the orbital paths of a few orbital pulsars. Those were problems like brain teasers, when you have a few numbers and you know that in principle there is an underlying regularity that will obey Kepler's laws and the laws are quite simple, and you should be able to fit the data. As far as personal challenges, it was family difficulties, one kind or another, the kind of difficulties that arise with ordinary human events and medical crises.

Is there anything else I should have asked you about?

No, you had very good questions, covered a wide range of territory, and I enjoyed talking with you.

Russell A. Hulse, 2000 (photograph by M. Hargittai).

35

RUSSELL A. HULSE

Russell A. Hulse (b. 1950 in New York City) is a Principal Research Physicist at the Plasma Physics Laboratory at Princeton University (PPPL). He received the Nobel Prize in Physics in 1993, together with Joseph H. Taylor of Princeton University "for the discovery of a new type of pulsar, a discovery that has opened up new possibilities for the study of gravitation". He studied at Cooper Union in New York and did his Ph.D. work at the University of Massachusetts at Amherst, where Joseph Taylor was his thesis advisor. It was during his doctoral studies that they discovered the first binary pulsar, the discovery of which eventually led to the Nobel Prize. A few years after receiving his Ph.D. in physics, he left astronomy and went to work at the Princeton Plasma Physics Laboratory, where he is still today. We recorded our conversation in Dr. Hulse's office at PPPL in April, 2000.*

Although your Nobel Prize-winning discovery took place many years ago, would you care to tell us something about it?

It is a story of scientific serendipity; if you look for new things with greatly increased sensitivity, you often find things that you did not expect. It happened when I was a graduate student at the University of Massachusetts at Amherst in the early 1970s. Joe Taylor was a young assistant professor and my thesis advisor. I was a graduate student in physics, but I always knew that I wanted to do my thesis in radio-astronomy. Joe was doing research on

---

*Magdolna Hargittai conducted the interview.

pulsars, which had been discovered only about 7 years earlier by Hewish and Bell Burnell. I had decided to get my Ph.D. degree in physics in order to give myself a broader background than an astronomy course of study would offer, but at the same time I also knew that I wanted to do my thesis in radio-astronomy; this is why I went to Amherst which had a new radio-astronomy group. Joe got an NSF grant to carry out a new, very high sensitivity search for pulsars, and used it to buy a minicomputer, which was quite fast by the standards of that time. I programmed it to do as thorough a signal analysis as was possible and took it down to Arecibo in Puerto Rico where they had a large radio telescope. Our goal was to use the biggest telescope we could find, along with very intensive signal processing so as to get every last bit of information out of the signals that were captured there. It was a very successful operation, we discovered 40 new pulsars. There were about 100 already known when we started this work but our search was about 10 times as sensitive as any previous work, so it was a very successful project, just as we hoped it would be.

If there were already about 100 pulsars known by that time, was it a really interesting and exciting challenge to discover more of the same thing?

Very interesting question. Some people told me that this is not a very exciting project, we know more or less what pulsars are, how they work, so why work so hard to find more? Actually beyond the always-present possibility of scientific serendipity, there were other very good reasons to do this work. For example, the hundred pulsars that had been found by then were found using a wide variety of techniques, so making generalizations based on this sample was difficult. Thus, first of all we wanted to have a sample that could be analyzed more reliably using the observed distribution of the pulsar pulse periods and other properties. Another reason for doing the search was that by using such high sensitivity we would expect to discover pulsars further away in the galaxy, thereby getting a much deeper sample, which would allow us to better determine what their distribution function was in space. We actually did find new pulsars to the far side of the galaxy. Finally, very fast pulsars with short pulse periods were the most difficult to detect, but also amongst the most interesting, because they were expected to be the youngest pulsars. Our search maintained its sensitivity to very short periods. As it turned out, this was very important, as the binary pulsar was the second fastest pulsar known at that time. Thus, our aim was to carry out a more unbiased search for pulsars, to search farther out into space, and to find shorter-period pulsars, which were expected to be younger. It was the intensive computer processing we used that made this possible, with the result that our search was ten times more sensitive than any other done previously, even more sensitive than previous searches which had used the same telescope. In retrospect I realize now that our work was a good example of the coming computer revolution in science, demonstrating how with computer power we could do much better and more powerful research than ever before.

What is so special about the binary pulsars compared to the simple pulsars?

There was one thing that was strange about this pulsar when it was discovered, and that was that its pulse period was changing. Pulsars have very stable pulse periods, they are very accurate clocks, so such a changing period was most unexpected. So first I thought that there was something wrong with the equipment. To make a long story short, it turned out that it was the Doppler effect, resulting from the fact that this one particular pulsar was moving around in an orbit around another, unseen, companion star. The first exciting thing was just to have found a pulsar in a binary system, because all the previous pulsars had been isolated ones. Due to the origin of pulsars, that is that they are born in supernova explosions, it was expected that they were not to be found as binaries but only as isolated objects. This one had a 59 millisecond period which was the second fastest ever found at that time, so that was also pretty neat. In the end, however, the most important thing about the discovery was that this binary pulsar system proved to be a perfect laboratory for testing Einstein's theory of relativity. The reason for this is that the pulsar moves very fast in its orbit, and the gravitational interaction is very large since they orbit very close to each other, their closest approach being about twice the distance between the Earth and the Moon. So we have high velocities and strong gravitational fields, which is exactly the situation where we expect Einstein's general relativity to be very important in modifying the behavior of the orbit. On Earth we do not have comparable gravitational fields so we cannot test the theory here as well as we can in this system. Thus this was the real payoff of this discovery.

It was obvious from the beginning that we could test some of Einstein's predictions very quickly. One effect that was expected to be large in this system was the advance of periastron. This is an effect where an elliptical orbit slowly rotates in space. This effect had been seen in the detailed behavior of the slightly elliptical orbit of Mercury early in the 20th century, and this observation had been a very important early test of Einstein's theory. Most of the turning of Mercury's orbit is caused by the perturbation of the other planets, but if you take all that out, you still have a 43 second of arc rotation per century, which could not be explained. General relativity predicted and explained that precisely. The binary pulsar, instead of showing a 43 second of arc per century advance of periastron as in the case of Mercury, instead shows a rate of 4 degrees per year of the same kind of orbital change. That effect was readily observable relatively soon after the discovery, providing a very dramatic early demonstration of the value of this discovery in testing Einstein's predictions.

There are other things we can see in this system, such as gravitational red shift, and the transverse Doppler shift (time dilation). The interesting thing is that if you put all these effects together, using general relativity as a tool we can get all the parameters of the system, the masses of both stars, the size and shape and orientation of the orbits, everything. General relativity further makes the prediction that the orbit should decay at a certain rate due to the emission of gravitational waves. Eventually, after many years of careful observations, Joe Taylor and his colleagues found that there is indeed a slight but definite decay of the binary pulsar orbit in precise agreement with Einstein's prediction. This observation of the effect of gravitational radiation was the most dramatic result that showed why the discovery of this system was so important.

Sometimes Nature presents one with a beautiful present. When we realized what sort of orbit this binary pulsar had, we knew that it should, in principle, be an ideal opportunity to test Einstein's relativity theory. But, fine, now that we have this system, half way across the galaxy, how exactly are we going to measure it? The beauty of the system is that here we had one of the most precise measuring instruments in the known Universe, exactly where you needed it, in the orbit, it was the pulsar itself. We could precisely measure the orbit and the predicted relativistic effects simply by precisely measuring the changing "ticking" of this pulsar clock as it moved around in its orbit. In courses on relativity they often say, imagine that you have a clock located in just the ideal place to make a measurement; here Nature had provided us with the most precise clock one could ever wish for, just where one would wish it to be. The pulsar period, the "ticking" of the pulsar clock, is known to 15 significant figures.

Why did it take about 19 years for the Nobel Committee to realize the significance of this discovery?

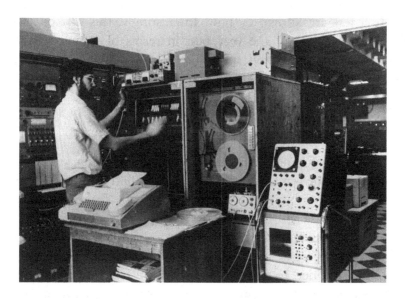

Russell Hulse manning the Modcomp II/25 minicomputer used for pulsar search with the Arecibo radio telescope near the coastal town of Arecibo in Puerto Rico. This photograph was taken shortly after the discovery of the binary pulsar (courtesy of Russell Hulse and the Nobel Foundation).

Well, of course, their deliberations are confidential and nobody outside of the committee knows for sure. But one can make a reasonable guess. Clearly the best way to ascertain the impact of a scientific discovery is to wait. In our case one of the impacts was that it made it possible to test Einstein's theory of general relativity to high precision. It took many years of continued observation to fully verify the effect of gravitational radiation on the orbit. I switched to other research after the discovery of the binary pulsar so that part was done by Joe Taylor and his co-workers. Among other things, they had to make sure that the system was "clean". By clean I mean that other effects, such as tidal interaction between the stars was negligible, so the relativistic effects would not be contaminated. So there needed to be a certain period of time until the world's scientists could think carefully about what might conceivably be wrong with the data, to ask whether there could be other explanations for the observations, etc. Presumably the Nobel Committee was quite properly content to take their time and let time provide the ultimate proof of the value of this discovery.

Have other binary pulsars been found since then?

Yes, quite a few are known by now. Interestingly, this first one we found turned out to be one of the very best in terms of testing relativity. The reason is the nature of its orbit, including its strong ellipticity, which provides a very strong time-variation during the course of the orbit.

You left the field soon after this discovery and started to work on fusion. Why?

It was really only for personal career reasons, not because I did not like astronomy. The issue was that it was very hard to find a position in astronomy at that time. The other issue was my girlfriend, Jeanne Kuhlman, who was working for her master's degree in physics at the University of Pennsylvania. We had the classic "two-body problem", which is how physicists sometimes jokingly refer to a two-career couple's problem trying to stay together. So I decided to take advantage of the fact that my interests were rather broad, in fact I got my degree in physics and not in astronomy in part with the idea to keep some breadth in the event that I ever wanted to switch fields. So in fact that was exactly what I did. I looked around and found out that they were hiring here at the Plasma Physics Laboratory, so I came here. They were expanding the controlled fusion program at that time in the late seventies, and the geography was right, since Princeton was not too far from where Jeanne was in Philadelphia.

In recent years so many things happened in astronomy. Don't you feel sorry for having left it?

I could not agree with you more in that so many things happened there. I tend to remember the state of the field as a reference point at the time when I left, which was in 1977. If I compare what we know now with what we knew back then, it is truly extraordinary, the progress is mind-boggling. One of the examples that comes to my mind is gravitational lensing. When I left the field there were no known examples of that. Later they found these double quasars and thought that this is a good example of gravitational lensing, but one wondered how many examples could there be, it was taken as sort of a fluke. Now we know that it is observable everywhere. So from an unknown to a curiosity, then to an established way of how you think about the Universe, that is a dramatic change. Now you can use it as a tool to map the mass distribution of the Universe. So the development is very impressive.

Would you tell us something about your present work?

Yes. Here we do controlled thermonuclear fusion research. My work is computer modeling; specifically I spent a lot of time developing computer codes to model the behavior of impurity ions in the hot plasmas that we have in these devices. In fusion, impurities are everything except hydrogen, and we get naturally occurring impurities such as carbon, oxygen, etc. that come off the walls of the vacuum vessel. We also have iron and other metals, which are constituents of the vacuum vessel, which erode off the surface and get into the plasma. You can also deliberately inject small amounts of impurities into the plasma, in order to watch how they behave, which gives us information about how particles diffuse through the plasma, which is one aspect of plasma confinement. Obtaining good confinement of the hot plasma is one of the fundamental issues in controlled fusion research. How do you confine the heat in this very hot plasma, and do so efficiently so that you can generate fusion reactions and produce energy at a rate which is greater than the rate at which energy is lost due to imperfect confinement? This work was interdisciplinary, since I needed to understand atomic physics as well as what was happening in the plasma and then create computer models of the interplay between the two.

I also worked on the transport of the electrons in the plasma, particularly following injection of pellets of frozen hydrogen into the plasma. One of the reasons to do so is to fuel the plasma. From my point of view, pellet injection was most interesting as another means of investigating transport within the plasma as an extension of the transport work that I had been doing with impurities. The pellets produce a very nice perturbation in the plasma in the form of a readily observed peak in the electron density profile. Then you can watch how that peak decays away and that gives you information about what transport processes are occurring in the plasma. Literally when the Nobel announcement came I was working on this and was preparing my presentation for the yearly APS meeting. In mid October I was furiously trying to get my poster session together on electron particle transport modeling of the pellet injection experiments. That was what was on my mind that morning when the Nobel Prize announcement was made, and suddenly instead of plasma transport I was thinking once again about pulsars and general relativity — something from my past.

Since the Nobel I have been doing some different things, including trying to figure out what different kinds of useful things I might like to do! I very quickly decided to get into science education. I find it very important and also fun, playing around with some of the experiments and trying to think of ways to get people, especially children, to understand how interesting the natural world around them really is. Especially with younger kids working on science education helps remind me about why I had gotten into science myself. I give talks and I also work with the School of Education at the College of New Jersey, where I work on improving science education skills for education majors, the teachers of tomorrow. For me this is a very satisfying and high-leverage way to make a contribution to education. It is fine to go to the classroom and talk to some of the children but if you can make the teachers better that has much more of a long-term impact. Since the Nobel I am also on a lot more advisory committees.

I would like to stay a little while with plasma research. Would you mind commenting on the importance of plasma research and on the question what justifies a whole institute for such studies?

Sure, that is an interesting question. In recent years we had some serious funding cuts and realignment of programs. We have to find the balance between the fundamental research and the more practical applications of fusion. I think that they do coexist in a well-balanced program. Plasma science clearly is a perfectly valid and useful branch of physics, the behavior of plasmas is important in astrophysics, it is important in various circumstances in the solar system; on the Earth there is a whole range of industrial applications.

One of the things that is interesting about plasma science is that there is a strong coupling between it and the rising interest in nonlinear dynamics in complex systems. Plasmas are very complicated nonlinear systems. When fusion research started back in the fifties, there was some optimism about the potential for rapid development of fusion reactors, but it turned out that it is much more complicated than was first imagined. The reason is that the fundamental behavior of the plasma arises from long-range interactions, and the many degrees of freedom of the system make it hard to understand from first principles. How does a hot plasma of the type that we need to create a fusion reactor really behave? It is not impossible to understand, it just turned out that it is a richer range of behaviors than was imagined at first. Actually it is getting more recognition now that while fusion is an important goal, there is this other facet of plasma science which somehow was not fully recognized because it was hidden in the glare of the fusion mission. But, it also turns out that to reach the fusion energy goal, one in turn really needs to better understand the fundamental behavior of plasmas, which in turn requires the ability to understand complex, nonlinear dynamics. So the more fundamental scientific goals and more applied energy goals are all inter-related, as is the case in most scientific research.

The focus of the research at PPPL has recently been on doughnut-shaped magnetic field configurations called Tokamaks, and related concepts. We have a tokamak here, the Tokamak Fusion Test Reactor (TFTR), which has been very successful, creating the plasma conditions required to produce 10 megawatts of fusion power. It took more megawatts into the machine to make this happen which is why it is not yet a practical energy source in its current form. But a lot of progress has been made in our ability to understand and confine these plasmas.

Although tokamaks such as TFTR have been very successful, it is not clear that the technical details of these tokamak devices scale up to the most desirable practical reactor in the end. So the program has been realigned a little bit to make sure that there would be more emphasis on fundamental understanding, with the hope that this would lead to improved concepts.

A lot of progress has been made not only in gaining experimental results but also in plasma theory. Now we are trying to understand from first principles how the confinement behaves following from basic physics. What is happening in this plasma is that you are trying to confine it at very high temperature, where there is a lot of free energy that wants to get out. You are doing this with a magnetic field in which you try to trap these ionized particles. The problem is that it is like trying to hold together jello with rubber bands, it is very tricky. The plasma has many degrees of freedom, with lots of turbulence inside. Turbulence in fluids is already complicated and we are only starting to get a handle on it computationally. Turbulence in plasmas is much more complicated than that, since in ordinary fluids we are dealing with neutral particles, but in the plasma we have charged particles with long-range interactions. The exciting thing is that during the past few years we have advances at two levels; one of them is theoretical; now we have the ability to do very powerful computer modeling of these turbulences and see what its physics is. The other is that we have wonderful experimental developments where we can actually see the turbulence. Until recently we could not visualize it directly, we knew that it had to be in there but there was no way of measuring this frothing of the plasma. Now we see that the density of the plasma is not constant, it has oscillations, waves going around in there. This is closely related to the fact that the heat does not stay in the plasma the way you would like. So there have been some diagnostic developments in the past several years that allow you to measure the turbulence directly, which is a breakthrough. We have been making real progress at both the experimental and the theoretical level of our understanding.

What temperatures are you using in your plasmas?

We are talking about temperatures in the orders of tens of kiloelectronvolts that is hundreds of millions of degrees, which is, by the way, an order of magnitude hotter than the center of the Sun. So we are not only reproducing the temperature of the Sun but we are capable of doing much better.

Why does it have to be so much hotter than the center of the Sun?

The Sun has an enormous advantage of having a really good confinement, produced by gravity. To make fusion reactions you need to have a couple of different things and the trick is that you need to have them all at the same time. You need high temperatures just to overcome the electrostatic repulsion of the nuclei so that they can get close enough and the nuclear forces can take over and you get the fusion reaction. But you also need confinement, meaning that you have to be able to keep the energy in a reasonably efficient way, otherwise the heat that you put in to create these conditions will leak out too fast and so you do not win. It is indeed the confinement part that is the hard thing.

There are essentially three ways that you can confine a fusion reaction. The Sun and the stars do it gravitationally. You take an awful lot of hydrogen, you let it gravitationally collapse, it heats up enough to start fusion reactions going between the protons, and the plasma is optically thick and that keeps the resulting radiation in. The energy confinement time of the Sun is millions of years so you can afford to do a fusion reaction at a lower temperature because it has such incredibly good confinement. We cannot do this on the Earth. One way we can do it here is by inertial confinement that is how the hydrogen bomb works. Deuterium and tritium are used as fuels because they have the highest cross section and if you compress and heat it very fast, what happens is that before the kinetic energy can disassemble it the reaction is all over. So we have the fusion reaction before the system has the chance to blow apart. This works for the H-bomb. The question is how can we do it for a power plant, and there the physics starts to get even more complicated. There are different places where they do inertial confinement work, at the Lawrence Livermore Laboratory, or at the University of Rochester. Some of this work is classified due to its relationship to weapons physics.

The third possibility is magnetic confinement and that is what we are doing here. This method takes advantage of the fact that at the high temperature of the plasma the particles are not neutral, they are ionized, we have positively charged nuclei and negatively charged electrons. In a magnetic field these charged particles are trapped on magnetic field lines, and that potentially gives you confinement. Our reactors are doughnut-shaped devices in which we have a strong magnetic field running the long way around the doughnut and this lets the particles run along a field-line and still stay in the system. To get gross stability, the field lines have to twist around in a helix-shape and that helical twist is a critical part of the confinement concept. This can be achieved in different ways; one way is by putting in additional magnets. The way the Tokamak works, however, and this is a clever Russian invention, is that you run a very strong current through the plasma, several millions of amperes. This current produces its own magnetic field, which added to the main field produces the desired helical twist to the magnetic field lines. This helical twist in turn provides reasonably good macroscopic stability. While this large-scale stability is still an important area of study, one is now also trying to understand the micro-instabilities in the system; which are the turbulences I mentioned before. Unfortunately, these turbulences and transport of heat and particles out of the plasma are large enough to make the reactors not practical yet. This means that we put more energy in than what we get out; eventually we hope to achieve breakeven, that is, to produce as much energy as the amount we put in. That is obviously an important milestone and we are getting close to break even. Of course, for a practical reactor, we have to get more than just to break even, we have to get a net gain and that is still a challenge for the future. But even more important a goal is that we understand plasmas well enough to be able to come up with a device design, meaning in part a detailed choice of magnetic fields, that can provide a practical fusion reactor; practical meaning attractive from an engineering and economic point of view. There have been remarkable improvements in recent years; people have pointed out that the rate of improvement in fusion devices is similar to or even better than the rate of improvement in computer speed and memory chips. So we cannot make a power plant yet, but there has been an enormous progress made and we understand these plasmas scientifically much, much better than we did before.

How realistic is it to build a power plant based on nuclear fusion?

It is a possibility, but much more needs to be done, particularly in terms of strengthening our base of fundamental understanding of plasma behavior in different magnetic configurations so that we can come up with a really practical type of fusion design. There also needs to be new thinking about how the engineering will work, particularly in terms of making a device which can survive for long periods of time despite the hostile plasma and radiation which will tend to rapidly destroy the inside of the reactor chamber. However, there is a critical need for humankind to develop new energy sources for the future; especially if we think about the enormous future energy needs in the third world. If we think about this realistically, we clearly need more options on the table and fusion is a very exciting, if challenging, option. It has a lot of positive attributes, for example, it is much safer than conventional fission power. Of course, the complication is that we do not really know yet how to build it but that is what research is for, you try to figure it out. Another issue is the public attitude concerning the fact that this is nuclear technology, even if it is much more benign than nuclear fission, but people may not immediately understand that. There is much less intense radioactivity involved with the waste disposal, and one cannot have a runaway reaction with a fusion reactor the way you can in a fission reactor. So we just will have to educate the public and that is the responsibility of the scientists and the politicians. But the public also has a responsibility to make intelligent choices, if they want electricity.

Please, tell me something about your family background and about what made you interested in science.

I was born and raised in New York City. My father worked for the Eastman Kodak Company, and eventually he became the manager of the Kodak exhibit and information center in New York City. He was a practical fellow and always interested in science. My parents were very supportive of my interest in science, which was obvious practically from birth. An important life-experience presented itself when my father started to build a house, originally planned as a summer house on a piece of land that an aunt gave us in upstate New York. Eventually, this house became my grandparent's home and this is where my parents live now that they are retired. I spent many days during weekends and summers as I grew up helping my father with all different stages of building a house and that gave me enormous practice in how to handle tools and how to do things myself. Some of my parents' friends were appalled that this little kid was running around with power tools and other machines, but fortunately I came through the experience with all of my fingers intact. Our spending so much time in this place also made it possible for me, a city kid, to get to know Nature and I learned to love it so much that outdoor activities eventually became some of my favorite recreations.

I went to the New York City public school system; I went to the Bronx High School of Science and then to Cooper Union for college. Cooper is an interesting place. It was founded by Peter Cooper, an industrialist in New York City in the eighteen hundreds. He endowed this college, which is a private, tuition-free college and as such is a rarity. It was obvious that paying my college tuition would have put an enormous burden on my parents, so I wanted to go to a place without tuition, such as Cooper, or City College. Cooper Union at that time had a physics major, which it does not anymore; they discontinued it for financial reasons. Then I went to the University of Massachusetts in Amherst to do my graduate work. My choice of UMass came from the following considerations. I had been interested in radio-astronomy for quite a while; I built a radio telescope during my high school years in my parents' summer house. Despite my strong interest in radio-astronomy, I did not want to get a degree in astronomy per se because I knew that my interests were wider and also that a physics degree would give me a broader, more flexible background which might be useful later in my career. UMass was set up with a joint physics and astronomy department and there I could get a physics Ph.D. but still do my thesis work in radio-astronomy, which is one of the reasons why I applied there.

Do you have heroes?

Just during the past few years I have read a lot about Theodore Roosevelt and have come to admire him enormously. He was an incredible person, an amazingly dynamic and intellectually vibrant person, someone with the wide range of interests and enthusiasms that you don't seem to often find in politics. He was morally strong, he was an expert naturalist, and he was a champion of social reform as well as individual responsibility. He had an incredible influence on the history of the United States. He positioned the country to move into the twentieth century. Although they do not talk about him too much nowadays, much of what he did and circumstances that he dealt with are very relevant today. Just think about conservation, monopolies, the balance of individual morality and government responsibility, these were big issues in his time and the parallels with our times are really uncanny. There is a lot to be learned from history.

Of course, I have scientific heroes as well. I always found Darwin particularly intriguing. I think that there is a part of me, which always wanted to be an explorer, discovering and mapping new lands, observing and marveling at the natural world. You find that in Theodore Roosevelt and also in Darwin. In another direction, Einstein is, of course, a sort of a cliché, but he was a brilliant person; the insight he had in so many different areas particularly in coming up with general relativity and with the whole notion of the geometrical interpretation of space-time is very impressive.

Are you religious? Do you mind if I ask this question?

No, and I am not religious. One of the things that always struck me about religion is that it is such a curious combination of different things, some admirable, some not. It is cultural heritage, it is moral principles, it is also a social unifying force, but it also can be a socially devastating force. I consider myself a very moral person but I do not need a religion for that. I also think of myself as a spiritual person but I find that in doing science and being out in Nature. Religion also has mythology and dogma, which I cannot relate to, which say that you have to believe in certain things just because someone said so despite the evidence or rational analysis. The worst aspect of religion, of course, is when it says that someone is evil and must be destroyed just because they do not have the same cultural origins or because they do not believe in the same things. So it is a very messy package of some things that are admirable, some things that I just can't understand and some things that are just plain destructive.

Do you have a family?

I have my girlfriend, Jeanne, who has been my companion for over 20 years; we do not have any children.

Has the Nobel Prize changed your life?

Oh, yes, it has had a big impact on my life. It has brought me many new opportunities. One example is my involvement with science education. I am also on many more scientific committees than previously. I know that some scientists think that serving on committees is dreadful but I think that some of them can be very useful, depending of course on the committee in question. It is interesting to see, for example, what forces are driving research and development in a certain direction as opposed to some other direction. It is also interesting to be on advisory committees where you can help clarify the forest-versus-tree issue; even the brightest people can be caught up myopically studying the trees and not see the forest. I tend to be a forest person. It is good to feel useful by pointing out the importance of the forest occasionally, or even sometimes an important but different tree.

You were a graduate student of Dr. Taylor when you did your Nobel Prize-winning work. It is not always that a grad student gets such recognition. In fact in a very close area to yours, the discovery of the pulsars, Jocelyn Bell, who was a grad student of Dr. Hewish, and who did the observation, was left out of the prize. What do you think of that case?

One of the wonderful moments of the Nobel ceremony was that I actually got to meet Jocelyn Bell personally for the first time; she was there as the guest of Dr. Taylor. This controversy is something I would not like to comment on. She seemed to me to be going on wonderfully with her life. I respect Jocelyn Bell as a scientist for what she accomplished; on the other hand, I also respect the work of the Academy in Stockholm and it is not my position to second-guess them. Certainly, when I think of pulsars, I always think of Jocelyn Bell, and I respect her and her express desire not to dwell on this subject.

What was your teacher/pupil relationship with Joseph Taylor?

It was very good. I was very fortunate to have him as my thesis advisor; he is a really nice guy. Especially in retrospect I very much appreciate that he let me work the way I wanted, he always gave me the freedom and independence that I needed but also he was always there if I needed to talk with him. He was very easy to talk to and we had a very nice relationship. Even at that time I knew how lucky I was to have him as my advisor.

Do you keep in touch?

On occasion. Although we are only a few miles apart on different campuses of Princeton University, these few miles are sufficient that the staff at the Plasma Lab unfortunately do not get the opportunity for casual, day-to-day interaction with faculty on the main campus.

If you try to evaluate your scientific work, was the pulsar work your most important achievement yet?

Oh, no doubt. At least the Nobel Committee seemed to think that it was a pretty noticeable thing. From among my later work I would mention the whole history of developing this computer code to model impurity transport and all the progress that resulted from that, working with not only the plasma community but also with the atomic physicists who worked to improve our understanding of the atomic processes. This was also a satisfying effort that extended over many years. I do also hope to make some further contributions to science and to society, but exactly how is still to be worked out!

Do you anticipate that the general public will become more science friendly or more alienated from science?

That is a very important question. There are indications both ways. When I was a kid, science was considered to be a very important thing; there were interesting science programs on the TV. Today, at least in the United States, there is a frightening trend towards pseudoscience, which is all over the TV, radio and other mainstream media. All the "snake oils" advertised (this is a U.S. expression meaning phony medicine), that are scientifically ridiculous and people are buying it. There is also an important problem that people are looking for quick, simple answers to complex issues rather than trying to deal with them in a more thoughtful way. This is rather scary because there are some very significant challenges to our social fabric that science is going to be throwing at us during the next not too many years. For example, there will be computers that will think in increasingly human-like ways, and this will force people to deal with the question of what it is that makes them human and different from computers. Then, challenging us from another direction, there is the whole issue of biotechnology. What will it mean to our self-image as human beings when one can manipulate life and our own consciousness at such a detailed level? People may find all this quite disquieting and frightening. I am not sure that our present society has the maturity and the level of discourse necessary to cope with all this. If you look at what passes for news in the media there is no way that they can handle substantive issues. They make a hash out of everything in terms of looking only for ratings and going for pure entertainment and shock value, the emotional response; let's just see what can we get into the news that will get the people most upset or appeal to their basest instincts and they will watch us more. How is a society going to cope with real issues that require some deep thinking and reflection when the news media so utterly fails the society?

If you started your career today what would be the research area of your choice?

Well, that is a dangerous question. When I was a kid I played around with all sorts of things. I had chemistry sets and mechanical engineering things, and built radios, and many other different things. That diversity of changing interests has turned out to be characteristic of me even later in life, even if the realities of having a career tend to make it harder to hop from interest to interest. Among all these interests, however, it was astronomy that I launched into when the time came to focus, partly because there is something special about astronomy. I think many people feel about it the same way. As an aside, I think that astronomy is one of the best ways to interest the general public in science, in part because it is something that people are naturally attracted to and wonder about. Astronomy always had an appeal and I am sure that if I replayed the tape of my life, I would still find it extremely interesting. Of course, if I had to start over, there are now many areas in science that I might find interesting, which were simply not around when I was younger. There is computer science; there is the whole field of self-organization and complex systems just to mention two. It is possible that I might have been drawn into one of these areas. One of the things that I like about self-organization and complex systems research is that they explain a lot about the way the world works. Along those lines, I find the enormous strides made by biology in recent years in yielding a deeper understanding of the structure and nature of life enormously exciting, and there is a good chance that if I were to start again, I would end up in an area of biological science. Of course, physics is still a field that offers its own fundamental insights into how the world works, insights which provide many of the tools and approaches essential to modern biological understanding, for example. I guess that in some sense my instinct always was, and still is, to be a scientist in a general sense rather than a particular kind of scientist. I like to understand how whole systems work, I find the melding of different fields, as physics, chemistry, biology, computer science coming together to create modern biology, very exciting. I am not particularly worried about what the label is; physicist, chemist, or other; one just should do interesting things. That of course, brings up the question of what preparation one requires to participate in such work. It is rather difficult for a student to find the balance between breadth and depth. At present to do interdisciplinary work is difficult from the academic point of view as well; what department would it belong to? The new interdisciplinary centers seem to be a way of trying to solve this problem.

David Shoenberg, 2000 (photograph by I. Hargittai).

36

DAVID SHOENBERG

D avid Shoenberg (b. 1911 in St. Petersburg, Russia) is Emeritus
Professor of Physics at the Cavendish Laboratory of Cambridge
University in England. I visited Professor Shoenberg and his wife in their
home in Cambridge on February 16, 2000. Professor Shoenberg was in
great spirits although he was just recuperating from hip surgery following
an unfortunate fall. The following is a summary of our conversation with
direct quotes about his encounters with Peter Kapitza (1894–1984) and
Lev Landau (1908–1968).*

David Shoenberg's father came from a small mostly Jewish town, Pinsk
in the Pale, Russia [today, Pinsk is in Belarus, southwest from the capital
city Minsk, near the Ukrainian border]. He was trained as an electrical
engineer in Kiev and eventually got a job in the wireless industry, which
was then just beginning. He worked for the Marconi Company, and he
and his family lived in St. Petersburg. At that time, Jews could live in
St. Petersburg only if they had special qualifications. By 1914 he had
saved enough money to fulfill his long-standing desire to go to England
to study for a Ph.D. in mathematics. He had been to England on business
once or twice and was attracted to it because of its liberal atmosphere
and lack of anti-Semitism. The Shoenberg family — the parents and four
children, of whom David was the youngest — moved to England in
1914. Three weeks later World War I began.

The savings of David's father were in Russian investments and he
could no longer receive income from them, so he had to go back to

---

*István Hargittai conducted the interview.

work again at the Marconi Company, at two pounds a week, which was barely enough for the family to live on. However, he rose rapidly in the company. Later, he went into the gramophone business and directed research at the Columbia Company, which was then taken over by HMV (His Master's Voice), whose famous trademark was the dog and the gramophone.

David Shoenberg was an undergraduate at Cambridge with a scholarship at Trinity College. He wanted to study mathematics, but on his father's advice he became a physicist. Switching from mathematics to physics was like "jumping from the frying pan into the fire", from the difficult to something worse, he remembers. However, he did well in physics, stayed on in Cambridge, and became Peter Kapitza's graduate student, although only until 1934, when Kapitza was detained in the Soviet Union and put in charge of the new Institute for Physical Problems in Moscow. Shoenberg then formally became Rutherford's graduate student until he completed his Ph.D. Some time later, in 1937, Shoenberg spent a year at the Moscow Institute at Kapitza's invitation.

In 1994, David Shoenberg attended the celebration of Kapitza's centenary in Moscow, and stayed with his widow, Anna Kapitza. He gave a speech in Russian, which had been translated from his original English. David Shoenberg co-edited a book commemorating his teacher [Boag, J. W.; Rubinin, P. E.; Shoenberg, D., Eds. *Kapitza in Cambridge and Moscow: Life and Letters of a Russian Physicist.* North-Holland, Amsterdam, 1990].

Professor Shoenberg speaks good Russian and so does his wife, Kate. She is also of Russian origin, but her family had left Russia before she was born. She graduated in physiology at University College London and spent her career in biological research in Cambridge. They met at a left-wing political meeting in 1940. The Shoenbergs have three grown-up children, one in Canada and the other two in England. David had thought that things would eventually improve in the Soviet Union and that Stalin was merely a puppet. Eventually, however, he understood that Stalin was in fact the master puppeteer and he became disillusioned with the Soviet system.

David Shoenberg spent his career at the University of Cambridge, eventually rising to a professorship at the Cavendish Laboratory. He was elected Fellow of the Royal Society (London) in 1953. During World War II, he was involved in anti-aircraft research. His peacetime research centered around the magnetic behavior of metals at low temperatures and for a time on superconductivity. I asked Professor Shoenberg about his encounters with Kapitza and with Landau.

My first contact with Peter Kapitza was a very long time ago. When he first came to Cambridge, he was clever at making useful contacts, and my father, who was then working with the Marconi Company in England, was able to help him by supplying a large capacitor, which Kapitza needed for his experiments. When my older brother went to Cambridge, he had an introduction to Kapitza from my father and they played chess together. Kapitza was quite a strong chess player. When he was living in Paris at one time, he used to make a living by going to those small cafés where chess players played for some stake. He pretended that he was just a poor amateur and, in the end, he would usually win.

When I came to Cambridge, I too had an introduction to him. When in 1932 I was completing my undergraduate course, I was intrigued by the lovely new building, the Royal Society Mond Laboratory, which was just then being built for Kapitza, and I very much wanted to work there. Kapitza agreed to take me, but I had a difficult time at first because he was away for some months when I arrived to start work.

At Kapitza's suggestion, I was to study the magnetostriction of bismuth, the slight changes of dimensions when it is magnetized. Kapitza was interested in the magnetostriction of bismuth, which is a peculiar metal. He had earlier made measurements in high impulsive magnetic fields but could measure the change of length only in the direction of the field but not in directions transverse to the field. My task was to develop a method for measuring the transverse magnetostriction in the static field of a conventional

David Shoenberg, 1953 (photograph by Lotte Meitner-Graaf, courtesy of David Shoenberg).

electromagnet. The impulsive fields were of the order of 300,000 gauss whereas the static magnet available could provide only about 15,000 gauss. Since the magnetostriction is proportional to the square of the field, that means about 400 times less effect. So I had to develop a very delicate method for measuring very small changes in length. I nearly gave up because of the difficulties, but in the end, with the help of crucial suggestions from Kapitza, it did work out. Kapitza liked to make rather wild but ingenious suggestions. Often, his suggestions did not prove practical, but one of them proved to be just what was wanted, and I managed to get detailed results on how the magnetostriction of a single crystal varied with crystal orientation.

Kapitza had earlier worked out a scheme by which this variation could be described in forms of a small number of parameters consistent with crystal symmetry, but my measurements could not be fitted to his scheme. I tried to recalculate the scheme by a different method and found that two of Kapitza's parameters, which he had said should be equal, were in fact independent. To my delight, it proved possible to fit all my data by introducing the extra parameter of my version of the theoretical scheme. Rather naively, I told Kapitza that he had made a mistake. To my surprise and dismay, he flew into a rage and said, "How dare you say such a thing to your teacher? Go and talk to Dirac. He checked my calculations and Dirac never makes a mistake." I did subsequently talk to Dirac and it turned out that Kapitza had misinformed him about the crystal symmetry of bismuth. With the correct symmetry, my scheme proved correct.

Peter Kapitza and Paul Dirac playing chess in Cambridge, 1928 (courtesy of David Shoenberg).

Soon afterward, Kapitza went again to Moscow for the summer, but this time he never came back. He was detained there and eventually told to set up a special new institute where he could continue his work. A few years later, I came to Moscow when Kapitza was already building his new institute. He behaved somewhat autocratically in Moscow too and, though his staff respected him, they were a bit afraid of him and he usually got his way. He even stood up to Stalin but in a rather peculiar way. Kapitza had a row with Beria, the infamous head of the secret police, when the atomic bomb was discussed many years later. There was a high-level committee of top physicists, and Kapitza was on this committee. It was not his specialty; he was not a nuclear physicist but he was a very ingenious scientist. Beria was the head of this committee. At some stage, Kapitza made a remark in which he compared Beria to the conductor of an orchestra who had the baton in his hand but had lost the score. Beria was very offended by this, and he got his own back by complaining to Stalin that Kapitza was a dangerous person. Stalin had rather a weakness for people who were able to stand up for themselves, and he told Beria that he could dismiss Kapitza but he was not to be touched otherwise.

Peter Kapitza, his wife, Niels Bohr and Margrethe Bohr and their son in Cambridge (courtesy of David Shoenberg).

Kapitza was then sacked but not arrested and moved to his dacha, where he was relatively free although he was bullied quite a lot. He managed to set up a sort of laboratory there. His institute in Moscow was called Institut Fizicheskikh Problem (Institute for Physical Problems), and his dacha was nicknamed Izba Fizicheskikh Problem (izba means a peasant's hut in Russian). This was a joke but Kapitza and his two sons, who were then teenagers, managed to do quite a lot of good experiments. He corresponded with various high-level officials who were concerned with technology. He was developing a new method to produce a powerful beam of electromagnetic (microwave) radiation. He presented it rather cunningly in his correspondence with government higher-ups, rather in the style of what many years later was called Star Wars technology (an excerpt from one of his letters is quoted below). This may have protected him, and his facilities were greatly improved.

Soon after Stalin died, Kapitza was restored to his former position and returned to normal life. But before this happened, there was a strange incident. Two men came to his dacha and asked to look at his experiments. He soon realized that they belonged to the secret police, but he showed them what he was doing. At some point, they looked at their watches and departed rather abruptly. Later on, Kapitza found out that these secret police people had been sent to protect him because Beria was going to be shot at a certain hour, and the government did not want to risk any last-minute action by Beria against Kapitza. Beria deserved his fate; he was a real beast. He used to travel around Moscow in a car and the chauffeur was instructed to look for pretty girls. He picked them up, took them away, raped them, and sent them off if they were lucky enough not to be put in prison.

When Kapitza was reinstated, he went on working on his intense source of electromagnetic radiation. He was hoping to achieve nuclear fusion by heating a plasma to high enough temperatures to get a nuclear fusion reaction, something nobody has done properly yet. He got generous funds to build up a special new laboratory, but he never achieved his aim. When he got his Nobel Prize in 1978, it was for his low-temperature work. He started his Nobel lecture by saying that his low-temperature work ended 30 years ago so he would rather talk about high-temperature physics — a not-too-subtle hint of his annoyance at having had to wait so long for the award.

I went to work for a whole year in Moscow in 1937 at Kapitza's invitation. By then, all kinds of strange things were happening to people, but it was dangerous to talk to foreigners and nobody talked about such matters to

J. J. Thomson and E. Rutherford outside the Mond Laboratory in Cambridge, 1933 (photograph by and courtesy of David Shoenberg).

me except Landau. I remember his making a remark about "konzlager," the German abbreviation for concentration camp. I thought he was talking about Germany, not Russia, but Landau said, "We invented them." I used to invite Landau for lunch quite often as I had comfortable living quarters and my meals were prepared for me. Over lunch, Landau enjoyed gossiping in a rather indiscreet way. He spoke quite good English with a fairly strong accent but in a quite accurate way.

Landau sounded very much left-wing when he came abroad in 1929 and 1930, but he became much more realistic as things were going wrong. One day, he didn't turn up for lunch anymore and I learned that he had been arrested. He disappeared for a whole year. Later, this brought me into an embarrassing situation. My contact with Landau was very valuable for me because he took up the theory of the problem I was interested in, which was the magnetic behavior of bismuth. I was getting a lot of curious results, which I could not explain. Landau looked at my observations and did some calculations and gave me a formula, but he never showed me the details of his calculations. Everything fitted his formula beautifully.

I finished my work after Landau had been arrested and I wrote up my paper, which I intended to publish in the *Proceedings of the Royal Society*. At that time, it was possible to publish abroad if you published the same thing in Russian at the same time. My manuscript was translated into Russian, and I was suddenly confronted by the Assistant Director of Kapitza's Institute, a woman who was a faithful party member, and she said to me that she was astonished that I had expressed thanks to an "enemy of the people". This was a very ominous expression, "vrag naroda" in Russian. She was referring to Landau, of course, and she said I could not do it. So I went to talk to Kapitza in his huge office. He looked quite sympathetic when I told him my story, but at that moment the Assistant Director marched into his office, and Kapitza immediately turned to me as if he were in mid-sentence and said something like, "So you understand, David, that you have to take that acknowledgment out." He obviously had to show support for his Assistant Director. Then he gave me a hint that, as I was going back to England shortly, there was no reason not to put back the acknowledgment into the English version of my paper. So the Russian version had no mention of Landau at all, only the mysterious formula, which appeared without

Lev Landau in Moscow, 1937 (photograph by and courtesy of David Shoenberg).

his name. When I submitted the English version to the Royal Society, the referee thought I might be plagiarizing since there was no mention of where the basic formula came from. In the meantime, an old friend of Landau, Rudolf Peierls, whom I knew, understood how Landau must have worked out the formula, and he was able to reconstruct the method and to derive Landau's formula. Eventually, it appeared as a one-page appendix to my paper. The paper became quite a classic and, ironically, the Russians had to refer to Landau's theory by quoting Peierls's appendix to my paper. Landau himself was freed from prison after about one year, and Kapitza was very bravely instrumental in getting him out by intervening with Stalin and the KGB. Formally, Landau was released into Kapitza's custody.

*****

KAPITZA'S "STAR WARS"

The first two paragraphs from Peter Kapitza's letter to G. M. Malenkov, on June 25, 1950 (quoted from Kapitza in Cambridge and Moscow: Life and Letters of a Russian Physicist*; Boag, J. W.; Rubinin, P. E.; Shoenberg, D., Eds.; North-Holland, Amsterdam, 1990; p.390).*

I am approaching you not just as one of the leaders of the Party but also because I have always greatly appreciated your interest in my work. I think that the significance of the question I am writing about justifies my giving you a detailed account.

During the war I was already thinking a lot about methods of defense against bombing raids behind the lines more effective than anti-aircraft fire or just crawling into boltholes. Now that atomic bombs, jet aircraft and missiles have got into the arsenals, the question has assumed vastly greater importance. During the last four years I have devoted all my basic skills to the solution of this problem and I think I have now solved that part of the problem to which a scientist can contribute. The idea for the best possible method of protection is not new. It consists in creating a well-directed high-energy beam of such intensity that it would destroy practically instantaneously any object it struck. After two years work I have found a novel solution to this problem and, moreover, I have found that there are no fundamental obstacles in the way of realizing beams of the required intensity.

Name Index

Abel, N. H.   459
Abraham, M.   27
Adams, J.   361
Adams, Q.   361
Ahrens, M.   57
Aleksandrov, A.   604
Alferov, M.   618
Alferov, Zh. I.   **602–619**
Allen, J. S.   224
Allen, J. F.   397
Alon, Y.   36
Alpher, R. A.   276, 277, 289, 290, 297
Alvarez, L.   49, 307, 330, 412, 451, 454
Amaldi, E.   220
Ambler, E.   168, 169, 191
Anaximander   148
Anderson, P. W.   7, 12, 13, 17, **586–601**, 622
Andronikashvili, E. L.   403
Appleton, E.   221, 628
Archimedes   347
Aristotle   66, 148, 430
Arons, A.   538, 539
Ashkin, A.   351
Atkins, P.   492

Attlee, C.   36, 45
Austin, B.   443
Axelrod, J.   144
Bader, A.   9
Bahcall, J. N.   209, **232–259**
Bahcall, N.   255, 258, 259
Bainbridge, K.   330
Bardeen, J.   7, 18, 590, 617
Bargmann, V.   181, 203, 432
Barsky, C.   533, 536, 537
Basov, N. G.   605, 606
Beadle, G.   502
Becker, H.   6, 7
Beethoven, L. van   212, 516, 569, 636
Begin, M.   50
Beier, G.   241
Bell Burnell, J.   477, 633, 635, **638–655**, 661, 672, 685
Ben-Gurion, D.   37, 57–59, 61
Bergman, P.   432
Beria, L.   693, 694
Berlioz, H.   504, 505
Berson, S.   222
Bethe, H.   186, 219, 300, 301, 454, 455
Bevin, E.   45

Page numbers in bold refer to interviews.

Bhaba, H.   431
Bjerge, T.   220
Bjorkolm, J. E.   351
Blaug, M.   537
Blobel, G.   95
Bloch, F.   52, 343, 392
Bloembergen, N.   343
Bogolyubov, N. N.   48
Bohm, D.   12, 591
Bohr, A.   199, 240
Bohr, M.   435, 451, 693
Bohr, N.   11, 14, 19, 166, 198, 313,
    388, 394, 416, 425, 427–429, 431,
    433–436, 450, 451, 455, 456, 544,
    693
Bolton, J.   296
Bondi, H.   38, 236
Boot, H.   326
Born, M.   229, 392
Bortshevsky, A. S.   610
Bowles, E.   329, 331
Boyle, R.   430
Bradbury, N.   413
Brahms, J.   569
Bréchignac, A.   580
Bréchignac, Catherine   570–585
Bréchignac, Charlotte   580
Bréchignac, P.   578, 580
Breit, G.   221, 325, 425, 429
Brezhnev, L.   614
Bridgman, P.   533, 535
Brillouin, L.   362
Brockman   38
Brout, R.   88, 114
Brown, H.   413
Brunauer, S.   421
Bruno, G.   430
Buonaparte, N.   61, 504
Burbidge, E. M.   236
Burbidge, G.   236
Bush, G.   57
Bush, G. W.   98
Bush, V.   326
Butow, R.   451
Calvin, M.   53

Carl Gustav (King of Sweden)   379,
    630
Carnot, L.   60, 61
Carter, B.   456
Chadwick, J.   198, 215, 217–219, 312,
    324
Chain, E.   52, 53
Chamberlain, O.   **298–303**
Chandrasekhar, S.   52, 217, 219, 220,
    292
Cheney, D.   57
Chernomyrdin, V.   614
Choquard, P.   167
Chu, S.   349, 351, 359
Chuang, I.   367
Clairaut, A.   543
Clifford, W.   456
Clinton, W.   249
Cockburn, S.   314
Cockroft, J. D.   323, 324
Cohen, A.   63
Cohen-Tannoudji, C.   351, 359
Collins, H.   536
Compton, A.   335, 431
Conant, J. B.   444
Cooper, P.   683
Copernicus, N.   29, 430, 544
Coquereaux, R.   51
Cornell, E.   354–356, 358, 359, 366,
    367, 369, 377
Cowan, G.   588
Crawford, A.   291
Crick, F.   477, 544, 545, 589
Cronin, A.   348
Cronin, J. W.   193, 203
Curie, M.   188, 189, 558
Curie, P.   152, 188, 189, 504, 505, 558
D'Alembert, J.   543
Darwin, C.   544, 545, 683
Davis, R.   186, 209, 237–241, 244,
    246–248, 251, 252, 254
Dawkins, R.   636
Dayan, M.   36, 38–40
de Broglie, L.   37, 223
de Kruif, P.   144, 548

de Staël, G.   581
de Wit, B.   118
Dehmelt, H. G.   317, 323
Delbruck, M.   502
Democritus   148
Denjoy, A.   499
Dennett, D. C.   589
Descartes, R.   504, 505
Deutsch, M.   74
Dewey, J.   538
Dewhirst, D.   296
DeWitt, B.   456
DeWitt, C.   456
Dicke, R.   290, 456
Dirac, P.   7, 29, 217, 323, 324, 535, 544, 692
Domb, C.   527
Doroshkevich, A. G.   275, 276
Dothan, J.   259
Dresselhaus, E.   551, 554, 555, 561, 564, 566
Dresselhaus, M. S.   **546–569**
Drucker, P.   545
Dunitz, J.   502
Dyson, E.   471, 472
Dyson, F. J.   4, 52, 115, **440–477**
Dyson, G.   471, 472
Dyson, I.   474
Eckart, C.   460
Eddington, A. S.   324, 459, 646
Edlen, B.   236
Ed Salpeter   253
Einstein, A.   15, 26, 27, 38, 46, 61, 67, 68, 71, 97–100, 107, 120, 127–129, 140, 180, 182, 216, 223, 387, 398, 419, 426, 431–433, 444, 447, 449, 450, 514, 542–545, 629, 633–636, 673, 674, 683
Elijah, the Geon of Vilna   57
el-Khilani, R. A.   37
Ellyard, D.   314
Emery, V.   118
Emmett, P. H.   421
Endo, M.   566
Englert, F.   88, 114
Erdös, P.   445, 516

Ernst, M.   118
Eshkol, L.   57
Essen, L.   320, 321
Evenson, R. E.   544
Fairlie, D.   50, 51
Feher, G.   596
Feigenbaum, M.   543
Fermi, E.   34, 48, 70–73, 102, 167, 172, 173, 180, 183, 184, 186, 198, 211, 216, 219, 221, 300–303, 334, 335, 434, 514
Fermi, L.   220
Feyeraband, P.   62
Feynman, R.   46, 47, 52, 86, 87, 90, 106, 107, 115, 177, 183, 188, 237, 251, 252, 342, 425, 426, 454, 455, 457, 458, 481, 543
Fisher, M.   528, 529
Fitch, V. L.   **192–213**
Fleming, A.   53
Flexner, A.   431
Florey, H.   53
Follin, J. W.   289, 297
Foster, J.   413
Fowler, R.   217, 218
Fowler, W.   236–238, 240, 252, 253, 288
Franck, J.   362
Frank, A.   283, 284
French, H. W.   231
Friedman, J. I.   **64–79**, 169, 170, 191, 207
Frisch, O.   46, 433, 434
Fuchs, K.   332
Fukui, K.   9
Fuller, B.   361
Gajdusek, C.   457, 458, 501, 502, 521
Galileo, G.   29, 31, 277, 430
Gallagher, A.   352
Gamow, G.   275–277, 288–290, 334, 429
Gandhi, M.   457
Garson, G.   188
Garwin, R. L.   73, 170, 171, 174, 191, 207
Gavrin, V.   242

Geballe, T.   565
Gell-Mann, M.   41, 42, 46–52, 63, 86, 92, 93, 106, 116, 180, 183, 237, 251, 252, 417, 419, 420, 543, 588
Giacconi, R.   209, 247, 248, 257
Glashow, S. L.   21, 62, 94, 116–118, 135, 339
Glushko, V.   616
Goeppert-Mayer, M.   6, 18, 72, 190
Gold, T.   38, 468
Goldberg, H.   43, 44, 46, 47, 62
Goldhaber, Gerson   48, 49
Goldhaber, Gertrude (Scharff)   215, 221, 222, 225–228
Goldhaber, M.   180, 210, **214–231**, 324, 325
Goldhaber, S.   48, 49
Goldreich, P.   253
Goldring, G.   258
Gomory, R.   506, 507, 515
Gordeev, V. A.   615
Gordon, N.   501
Götze, W.   385
Goudsmit, S.   229
Grass, G.   95
Grenács, L.   181
Greytak, T.   351, 376, 377, 379, 389
Gribbin, J.   47, 48, 62
Gribbin, M.   47, 48, 62
Grinberg, G. A.   614
Groves, L. R.   15, 16, 331, 335
Gubanov, A. I.   607
Guiliani, R.   230
Guth, A.   54
Haber, F.   314
Hadamard, J.   499
Haensch, T.   323, 349
Hahn, O.   14, 216, 433
Haldane, J. B. S.   447, 458
Hall, R.   605
Hamilton, A.   430
Hannay, B.   565
Hanson, H. P.   3, 4
Hardy, G. H.   470
Harkins, W. D.   218
Harteck, P.   306, 314

Hauptman, H.   144
Hawking, S.   130, 132, 456
Hayashi, I.   610, 611
Hayward, R. W.   191
Heisenberg, E.   451, 452
Heisenberg, W.   217, 218, 230, 387–389, 392, 408, 418, 420, 435, 451, 452, 455, 544
Heitler, W.   392
Hellmann, H.   394, 402
Helmerson, K.   352
Herman, R. C.   276, 277, 289, 290, 297
Herring, C.   7, 590
Hershel, C.   646
Herzl, T.   52
Hess, H.   351, 389
Hewish, A.   **626–637**, 639, 652–654, 661, 672, 685
Hewish, M. (Richards)   627, 635
Higgs, P.   40, 88, 99, 114
Hilbert, D.   516
Hitler, A.   14–16, 37, 278, 410, 500
Hoffmann, R.   9
Hofstadter, R.   74
Holonyak, N. Jr.   605, 617
Hooft, B. 't   139
Hooft, E. 't   139
Hooft, G. 't   87–89, 95, 103, 106, **110–141**, 118
Hooft, S. 't   139
Hoppes, D. D.   191
Horowitz, V.   140
Hoyle, F.   38, 236, 237, 275, 276, 469
Hudson, R. P.   191
Huffman, W. E.   544
Hughes, V.   325
Hulet, R.   352, 377
Hulse, R. A.   321, 633, 634, 654, 657, 659–661, **670–687**
Hussein (King of Jordan)   39
Hutchins, R.   65, 66
Huxley, A.   157, 458
Huxley, T.   459
Iliopoulos, J.   117

Ioffe, A. F.   604, 605, 613, 619
Jacob, F.   502
Jaffe, B.   144
Jahn, H. A.   416
Janner, A.   167
Jarlskog, C.   90, 120
Javan, A.   565
Jensen, J. H. D.   6, 18
Johnson, G.   50, 63
Johnson, L.   17
Joliot, F.   218
Joliot-Curie, I.   218
Josephson, B.   490, 491
Jourdan, M.   51
Kadanoff, L.   527–529
Kalckar, F.   429
Kaluza, T.   44
Kantorovich, A.   63
Kapitza, A.   690
Kapitza, P. L.   273, 287, 313, 394,
    397, 604, 689–694, 696, 697
Karle, J.   144
Kármán, T. von   406, 502
Kassem, A. K.   39
Kaverin, V.   619
Kazarinov, R.   606, 615
Kekulé, v. S. F. A.   414
Kellogg, J.   199
Kendall, H. W.   65, 74–76
Kennedy, J. F.   57, 413, 461
Kepler, J.   5, 29, 148, 277, 507, 543,
    544
Ketterle, H.   385
Ketterle, Johanna   385
Ketterle, Jonas   385
Ketterle, W.   345–347, 349, 350, 353,
    355–357, 359, 360, 362, 366, 367,
    368–389, 396
Khariton, Yu. B.   604
Khrushchev, N.   461, 613
Kilby, J. S.   603
King, M. L.   457, 458
Kirsten, T.   242
Kissinger, H.   55
Kistiakowsky, G.   195, 336
Klauder, J.   456

Klein, J.   208
Kleppner, D.   322, 348, 351, 357–360,
    366, 367, 375–377, 379, 389
Kneipp, K.   567
Knight, W.   574
Koestler, A.   61, 409, 410
Kohn, W.   596
Kollek, T.   559
Kolliker, A. von   519
Konopinski, E.   236, 237, 252
Konstantinov   614
Koretz   394
Kornberg, A.   144
Koshiba, M.   186, 241, 247
Kozlovsky, B. Z.   239, 251
Kramers, H. A.   84, 85
Kroemer, H.   603, 605–608
Krokhin, O.   606
Kuhlman, J.   676, 684
Kuhn, T.   62, 533, 540
Kurchatov, I. V.   604
Kurti, N.   396
Labinger, J.   536
Lakatos, I.   62
Lamb, W.   269, 342
Landau, L.   52, 184, 224, 342, 391,
    394–396, 400–402, 408, 409, 415,
    416, 604, 689, 695–697
Larson, C.   14
Laue, M. von   216
Laughlin, R. B.   621, 623
Laurent, T. C.   62
Lauritsen, T.   239
Lawrence, E. O.   218, 300, 306, 309,
    310, 326, 327, 412–414, 454
Lax, B.   605
Layzer, D.   236
Lea, D. E.   218, 219
Lederberg, J.   45
Lederman, L. M.   73, 89, **142–159**,
    170, 171, 176, 191, 201, 207, 337,
    624
Lee, T. D.   73, 86, 106, 145–147,
    168–170, 172, 200, 202, 206, 224,
    340, 342
Lefkowitz, D.   234

Lefkowitz, L.   234
Leibniz, G. W.   430
Leighton, B.   251
Lenin, V.   604
Lermontov, M. Yu.   619
Levich, B.   52
Lévy, P.   503, 516
Lewis, G. N.   217, 306
Liddel, U.   343
Lieber, C.   567
Lieberman, Y.   52
Lifshitz, E. M.   400, 402, 403
Linde, A.   54
Linnaeus, C.   34, 43, 537
Lipkin, H. J.   44, 47, 62
Lofgren, E.   302
London, F.   361, 371, 372, 395–399, 402
London, H.   399
Loomis, W.   327
Lorentz, H.   27, 84
Louis XVI (French King)   581
Mack, J.   365
Mackay, A.   597
Maiani, L.   117
Maier-Leibnitz, H.   264, 270
Maklev, M.   37
Malenkov, G. M.   697
Mandelbrojt, S.   498
Mandelbrot, A.   523
Mandelbrot, B. B.   **496–523**
Mann, A.   241
Marshall, J.   71, 73
Marshall, L.   71
Matthias, B. T.   590
Mauzé, A. R.   558
Maximilian (Emperor)   432
Maxwell, A.   97, 542, 666
Mayakovskii, V. V.   619
Mazarini, L.-J. Mancini   504
McCarthy, J.   165, 307, 395
McClure, J.   561
McDaniel, B.   196, 197
McDermott, L.   538, 539
McMahon, B.   412, 414
McMillan, E.   413

McNamara, R.   413
Meitner, L.   216, 227, 433
Meltserson, Y. B.   616
Mendeleev, D. I.   34, 43, 102, 122
Menuhin, Y.   635
Meselson, M.   470
Metzger   270
Michel, L.   181
Millikan, R.   431
Mills, R.   112, 462, 463
Misener, D.   397
Misner, C.   456
Moffatt, K.   513
Moliere, J.-B.   51
Montel, P.   499
Montgomery, B. L.   37, 45
Moon   270
Mosley, R.   74
Mößbauer, R.   **260–271**
Mott, N.   217, 587
Mottelson, B.   240
Moyer, B.   235
Mundell, R.   95
Myers, E.   348
Myrabo, L. N.   467
Nasledov, D. N.   614
Nasser, G. A.   39
Nathan, M.   234
Nathon, M.   605
Ne'eman, Y.   **32–63**, 258
Necker, J.   581
Neguib, M.   39
Nernst, W.   216
Neumann, J. von   9, 19, 406, 407, 409, 417, 418, 460, 514–516, 522
Neumann, K. von   460
Newton, I.   5, 29, 31, 126, 182, 387, 430, 534, 542–545, 636
Nierenberg, W.   325
Nijboer, B.   118
Noddack, I.   434
Noether, E.   40
Novick, S.   348
Novikov, I.   54, 275, 276
Ohm, G. S.   276
Olendorf, F.   44

Oliphant, M. L. E.   198, **304–315**, 326, 327

Onsager, L.   527

Oppenheimer, R.   186, 198, 220, 300, 313, 330, 331, 333, 334, 412, 429, 453, 515

Ortvay, R.   393

Ostriker, J.   253

Pais, A.   84, 102

Panish, M.   610, 611

Panofsky, E.   428

Panofsky, H.   428

Panofsky, W. K. H.   76, 303, 428

Parsons, W. S.   332, 333

Pasteur, L.   504, 505, 516, 522, 523

Paul, W.   317

Pauli, W.   43, 44, 155, 166, 167, 177, 183, 184

Pauling, L.   263, 414, 415, 502

Payne-Gaposhkin, C.   646

Pearson, R. G.   9

Peebles, P. J. E.   253, 290, 297

Peierls, R.   46, 219, 270, 697

Penrose, R.   464

Penzias, A. A.   51, 52, **272–285**, 287, 288, 291, 294, 594

Penzias, S.   281

Peres, S.   53, 54

Perry, L.   320, 321

Peshkov, V. P.   400

Phillips, W. D.   348, 350, 351, 354, 359, 366, 367, 377

Pickering, A.   534–536

Pinsonneault, M.   241

Piore, E.   506, 507

Placzek, G.   11, 435

Planck, M.   216

Plato   66, 430

Poincaré, H.   27, 516, 522, 543

Polanyi, M.   5–7, 11, 435

Polkinghorne, J. C.   **478–495**

Polkinghorne, R.   485

Pontecorvo, B.   187

Popov, Y.   606

Popper, K.   62

Portnoi   615

Primakov, Y.   614

Pritchard, D. E.   **344–367**, 370, 373, 375–378, 380, 381, 385, 389

Prokhorov, A. M.   619

Pugh, S.   299

Purcell, E. M.   325, 340, 341

Putin, V.   98, 615

Pythagoras   126

Quillen, D.   51

Quixote, D.   522

Rabi, I.   50, 52, 53, 144, 197, 199, 203, 207, 317, 319, 325, 326, 328, 333–335, 366, 377, 429

Rabin, Y.   37, 59

Racah, G.   44

Radnóti, M.   16

Rainwater, J.   199, 200, 211

Ramón y Cajal, S.   519, 522

Ramsey, Elinor   337, 338

Ramsey, Ellie (Welch)   338, 339

Ramsey, N. F.   **316–343**, 360, 366, 377

Randall, J.   326

Ray, J.   537

Rayleigh, Lord (Strutt, J. W.)   575

Reagan, R.   220, 278, 279, 411, 413, 414, 591

Rees, M.   253

Reines, F.   209

Rembrandt, R. van   140

Renner, R.   415, 416

Reynolds, P.   387

Roberts, L.   340–342

Robinson, I.   52

Rockefeller, N.   414

Roentgen, W. C.   615

Roosevelt, T.   15, 44, 45, 220, 683

Rosen, N.   38

Rosenbluth, M.   252

Rosenfeld   11, 435

Rossi, B.   198, 209

Rubbia, C.   95, 96, 119

Ruby, S. L.   224, 225

Ruijgrok, T.   118

Rupprecht, H.   610

Russell, B.   235

Rustad, B. M.   224, 225
Rutherford, E.   215–219, 221, 223, 305–309, 312, 314, 323, 324, 430, 431, 690, 695
Ryle, M.   627–632, 634, 639, 652
Sakata, S.   34
Sakharov, A.   204, 205, 211, 279
Salam, A.   21, 38–42, 49, 50, 62, 94, 95, 116–118, 135, 224, 324, 481
Salpeter, E.   239, 250, 253
Sambursky, S.   44
Schawlow, A. L.   323
Scheck, F.   51
Scherrer, P.   164, 166
Schiff, L.   301
Schmidt, H.   55
Schmidt, M.   251
Schrieffer, J. R.   598
Schrödinger, E.   7, 216, 387, 417, 418, 544
Schubert, F.   212
Schultz, G.   593
Schützenberger, M.-P.   507, 522
Schwartz, M.   143, 155
Schwarzkopf, H. N.   57
Schwarzschild, M.   252, 253
Schwinger, J.   40, 115, 325, 342, 481
Seaborg, G.   413
Sears, D.   241
Segrè, E.   72, 73, 220, 299, 300, 302, 303
Seitz, F.   7
Seloff, F. E.   666
Semenov, N. N.   604
Serber, R.   47
Shakespeare, W.   31
Shamir, Y.   56, 57
Shapiro, H.   664
Sharon, A.   59
Shechtman, D.   597
Sheldon Glashow   62
Shockley, W.   590, 605, 607
Shoenberg, D.   217, **688–697**
Shoenberg, K.   690
Sidles, J.   442

Sijacki, Dj.   63
Silkens, L.   118
Silver, L.   443
Simon, F.   397, 402
Skrinsky, A.   175
Slansky, R.   61
Smalley, R. E.   562
Smith, H.   599
Snow, C. P.   417
Sommerfeld, A.   188
Soneira, R.   249
Sozina, N. N.   616
Spiewak, M.   560
Spinoza, B.   433
Spitzer, L.   252, 253
Stalin, J.   407, 409, 412, 693, 694
Stanley, G.   296
Steinberger, J.   143, 155, 156
Steinhardt, P.   463, 597
Stent, G.   502
Stern, O.   366
Sternberg, S.   51
Stewart, B.   348
Stimson, H. L.   331
Störmer, H. L.   621, 623
Strassmann, F.   14, 433
Strauss, C.   234
Strauss, L.   414
Strechel, R.   537
Streeter, R.   39
Strobel, A.   385
Suzuki, Y.   245
Szenes, H.   36, 44
Szilard, L.   14, 15, 184, 185, 216, 217, 220, 221, 229, 230, 300, 313, 395, 406, 407, 410–412, 460, 469, 516
Szilard, T.   460, 461
Talley, W.   422
Taylor, J. H.   321, 633, 634, 654, **656–669**, 671, 672, 674, 675, 685
Taylor, M.   665
Taylor, R. E.   65, 74–76
Teichteil, E.   560
Telegdi, L.   165
Telegdi, V. L.   71, 73, 74, 107, 145, **160–191**, 207

Teller, E.   9, 10, 19, 54, 71, 185, 186, 190, 198, 220, 229, 302, 314, 333, 334, 391–395, **404–423**, 428, 453, 454, 456, 516, 593, 594
Teller, M.   405, 422, 423
Teller, P.   422
Teller, W.   421
Teresi, D.   156
Thierry-Mieg, J.   51
Thomson, G. P.   226
Thomson, J. J.   223, 226, 312, 323, 695
Ting, S. C. C.   203
Tisza, L.   361, 372, **390–403**, 409, 421
Titterton, E.   195, 197
Tobin, J.   599
Tomonaga, S.-I.   115
Totsuka, Y.   245
Townes, C. H.   322, 325
Tremaine, S.   253
Tret'yakov, D. N.   610
Truman, S.   203
Tsui, D. C.   600, **620–625**
Tuchkevich, V.   617
Tuve, M.   333
Ufford, L.   438
Uhlenbeck, G.   84, 229, 398
Ulam, S.   334
Ulrich, R.   241
Vallarta, M. S.   432
van Beijeren, H.   118
van der Meer, S.   95, 96, 119
van Himbergen, H.   118
van Hove, L.   85
van Kampen, N.   85, 118
Van Vleck, J. H.   14, 587
Veltman, A.   104
Veltman, Helene   104
Veltman, Hugo   104
Veltman, M. J. G.   25, **80–109**, 111, 112, 114
Veltman, Martin   104
Verdi, G.   516
Vernon-Jones, V. S.   217
Voltaire   430

Vörösmarty, M.   16
Vuletic, V.   367
Walker, T.   352
Waller, I.   49
Walther, H.   385
Watson, J.   477, 544, 545
Watson, T. Jr.   506
Weaver, W.   515
Weber, M.   542
Weinberg, S.   5, 11, 13, 17, **20–31**, 62, 94, 116–118, 135, 148, 492, 493, 597, 600
Weinrich, M.   73, 191
Weisskopf, V.   43, 44, 62, 165
Weizsäcker, K. F. von   408
Weyl, H.   10, 19, 86, 149, 518
Wheeler, A.   438
Wheeler, J. A.   11, 19, 199, 200, **424–439**, 455, 456, 477
Wheeler, James   438
Wheeler, Janette   427, 428, 438
Wheeler, Joe   426
Wiegand, C.   301–303
Wieman, C. E.   351, 352, 354, 358, 359, 366, 369, 377
Wiesner, J.   559
Wightman, A. S.   203
Wigner, E. P.   **1–19**, 29, 30, 43, 82, 121, 190, 203, 220, 393, 406, 407, 410, 411, 418, 435, 456, 461, 488, 490, 516, 589–591, 625
Willoughby, J.   537
Wilson, E.   600
Wilson, E. B.   532, 533, 540
Wilson, K. G.   **524–545**
Wilson, R. [former Director of Fermilab]   150, 157, 336
Wilson, R. W.   273, **286–297**, 594
Wineland, D. J.   323
Wingate, O.   36
Witmer, E. E.   9, 18
Witten, E.   436
Wittgenstein, L.   235
Wolf, D.   239
Wolfenson, J.   254
Wolfrum, J.   385

Wollman, E.   502

Woodall, J.   610

Woodward, R. B.   9

Wrubel, M.   236

Wu, C.-S.   73, 145–147, 168, 169, 171, 184, 191, 206, 227, 340, 342

Yadin, Y.   36, 37

Yalow, A.   222

Yalow, R. (Sussman)   222, 227, 548, 549, 558

Yang, C. N.   34, 73, 106, 112, 146, 147, 168–170, 172, 200, 202, 203, 206, 224, 340–342, 462, 463

Ypsilantis, T.   302, 303

Yukawa, H.   34, 431

Zacharias, J. R.   320, 321, 325, 334, 335

Zatsepin, G.   242

Zeldovich, I.   604

Zernike, F.   138

Zewail, A.   95

Zuckerman, H.   529

Zweig, G.   48, 92

Cumulative Index
of Interviewees
Candid Science I–IV

Alferov, Zh. I.   IV/602

Altman, S.   II/338

Anderson, P. W.   IV/586

Bader, A.   III/146

Bahcall, J. N.   IV/232

Balazs, E. A.   III/120

Bartell, L. S.   III/58

Bartlett, N.   III/28

Barton, D. H. R.   I/148

Barton, J. K.   III/158

Bax, A.   III/168

Bell, J. B.   IV/638

Berg, P.   II/154

Bergström, K. S. D.   II/542

Berry, R. S.   I/422

Black, J. W.   II/524

Blobel, G.   II/252

Boyer, P. D.   III/268

Brechignac, C.   IV/570

Brown, H. C.   I/250

Chamberlain, O.   IV/298

Chargaff, E.   I/14

Cohn, M.   III/250

Cornforth, J. W.   I/122

Cotton, F. A.   I/230

Cram, D. J.   III/178

Crutzen, P. J.   III/460

Curl, R. F.   I/374

Deisenhofer, J.   III/342

Dewar, M. J. S.   I/164

Djerassi, C.   I/72

Dresselhaus, M. S.   IV/546

Dunitz, J. D.   III/318

Dyson, F. J.   IV/440

Eaton, P. E.   I/416

Edelman, G. M.   II/196

Eigen, M.   III/368

Elion, G. B.   I/54

Ernst, R. R.   I/294

Ernster, L.   II/376

Eschenmoser, A.   III/96

Finch, J. T.   II/330

Fitch, V. L.   IV/192

Friedman, J. I.   IV/64

Fukui, K.   I/210
Furchgott, R. F.   II/578
Furka, A.   III/220
Gajdusek, D. C.   II/442
Gal'pern, E. G.   I/322
Gilbert, W.   II/98
Gillespie, R. J.   III/48
Gilman, A. G.   II/238
Goldhaber, M.   IV/214
Hassel, O.   I/158
Hauptman, H. A.   III/292
Henderson, R.   II/296
Herschbach, D. R.   III/392
Hewish, A.   IV/626
Hoffmann, R.   I/190
Hooft, G. 't   IV/110
Huber, R.   III/354
Hulse, R. A.   IV/670
Jacob, F.   II/84
Ketterle, W.   IV/368
Klein, G.   II/416
Klug, A.   II/306
Kornberg, A.   II/50
Krätschmer, W.   I/388
Kroto, H. W.   I/332
Kuroda, R.   III/466
Laurent, T. C.   II/396
Lederberg, J.   II/32
Lederman, L. M.   IV/142
Lehn, J.-M.   III/198
Levi-Montalcini, R.   II/364
Lewis, E. B.   II/350
Lipscomb, W. N.   III/18
Mandelbrot, B. B.   IV/496
Marcus, R. A.   III/414
Mason, S.   III/472
McCarty, M.   II/16
Merrifield, B.   III/206
Michel, H.   III/332
Milstein, C.   II/220
Moncada, S.   II/564
Mößbauer, R.   IV/260
Müller-Hill, B.   II/114
Mullis, K. B.   II/182
Nathans, D.   II/142

Ne'eman, Y.   IV/32
Nirenberg, M. W.   II/130
Olah, G. A.   I/270
Oliphant, M. L. E.   IV/304
Orchin, M.   I/222
Osawa, E.   I/308
Ourisson, G.   III/230
Pauling, L.   I/2
Penzias, A. A.   IV/272
Perutz, M. F.   II/280
Pitzer, K. S.   I/438
Polanyi, J. C.   III/378
Polkinghorne, J. C.   IV/478
Pople, J. A.   I/178
Porter, G.   I/476
Prelog, V.   I/138
Prigogine, I.   III/422
Pritchard, D. E.   IV/344
Radda, G. K.   II/266
Ramsey, N. F.   IV/316
Robbins, F. C.   II/498
Roberts, J. D.   I/284
Rowland, F. S.   I/448
Sanger, F.   II/72
Scheuer, P. J.   I/92
Schleyer, P. v. R.   III/80
Seaborg, G. T.   III/2
Semenov, N. N.   I/466
Shoenberg, D.   IV/688
Smalley, R. E.   I/362
Stankevich, I. V.   I/322
Stork, G.   III/108
Taube, H.   III/400
Taylor, J. H.   IV/656
Telegdi, V. L.   IV/160
Teller, E.   IV/404
Tisza, L.   IV/390
Tsui, D. C.   IV/620
Ulubelen, A.   I/114
Vane, J. R.   II/548
Veltman, M. J. G.   IV/80
Walker, J. E.   III/280
Watson, J. D.   II/2
Weinberg, S.   IV/20
Weissmann, C.   II/466

Westheimer, F. H.   I/38
Wheeler, J. A.   IV/424
Whetten, R. L.   I/404
Wigner, E. P.   IV/2
Wilson, K. G.   IV/524

Wilson, R. W.   IV/286
Yalow, R.   II/518
Zare, R. N.   III/448
Zewail, A. H.   I/488
Zhabotinsky, A. M.   III/432